The Mathematical Coloring Book

Alexander Soifer

The Mathematical Coloring Book

Mathematics of Coloring and the Colorful Life of its Creators

Forewords by Branko Grünbaum, Peter D. Johnson Jr., and Cecil Rousseau

 Springer

Alexander Soifer
College of Letters, Arts and Sciences
University of Colorado at Colorado Springs
Colorado Springs
CO 80918, USA
asoifer@uccs.edu

ISBN: 978-0-387-74640-1   e-ISBN: 978-0-387-74642-5
DOI 10.1007/978-0-387-74642-5

Library of Congress Control Number: 2008936132

Mathematics Subject Classification (2000): 01Axx 03-XX 05-XX

Cover illustration: The photographs on the front cover depict, from the upper left clockwise, Paul Erdős, Frank P. Ramsey, Bartel L. van der Waerden, and Issai Schur.

Printed on acid-free paper

springer.com

This coloring book is for my late father Yuri Soifer,
a great painter, who introduced colors into my life.

To Paint a Bird

First paint a cage
With wide open door,
Then paint something
Beautiful and simple,
Something very pleasant
And much needed
For the bird;
Then lean the canvas on a tree
In a garden or an orchard or a forest –
And hide behind the tree,
Do not talk
Do not move. . .
Sometimes the bird comes quickly
But sometimes she needs years to decide
Do not give up,
Wait,
Wait, if need be, for years,
The length of waiting –
Be it short or long –
Does not carry any significance
For the success of your painting
When the bird comes –
If only she ever comes –
Keep deep silence,
Wait,
So that the bird flies in the cage,
And when she is in the cage,
Quietly lock the door with the brush,
And without touching a single feather
Carefully wipe out the cage.
Then paint a tree,
And choose the best branch for the bird
Paint green leaves

Freshness of the wind and dust of the sun,
Paint the noise of animals in the grass
In the heat of summer
And wait for the bird to sing
If the bird does not sing –
This is a bad omen
It means that your picture is of no use,
But if she sings –
This is a good sign,
A symbol that you can be
Proud of and sign,
So you very gently
Pull out one of the feathers of the bird
And you write your name
In a corner of the picture.

by Jacques Prévert[1]

---

[1] [Pre]. Translation by Alexander Soifer and Maurice Stark.

Foreword

This is a unique type of book; at least, I have never encountered a book of this kind. The best description of it I can give is that it is a mystery novel, developing on three levels, and imbued with both educational and philosophical/moral issues. If this summary description does not help understanding the particular character and allure of the book, possibly a more detailed explanation will be found useful.

One of the primary goals of the author is to interest readers—in particular, young mathematicians or possibly pre-mathematicians—in the fascinating world of elegant and easily understandable problems, for which no particular mathematical knowledge is necessary, but which are very far from being easily solved. In fact, the prototype of such problems is the following: If each point of the plane is to be given a color, how many colors do we need if every two points at unit distance are to receive distinct colors? More than half a century ago it was established that the least number of colors needed for such a coloring is either 4, or 5, or 6 or 7. Well, which is it? Despite efforts by a legion of very bright people—many of whom developed whole branches of mathematics and solved problems that seemed much harder—not a single advance towards the answer has been made. This mystery, and scores of other similarly simple questions, form one level of mysteries explored. In doing this, the author presents a whole lot of attractive results in an engaging way, and with increasing level of depth.

The quest for precision in the statement of the problems and the results and their proofs leads the author to challenge much of the prevailing historical "knowledge." Going to the original publications, and drawing in many cases on witnesses and on archival and otherwise unpublished sources, Soifer uncovers many mysteries. In most cases, dogged perseverance enables him to discover the truth. All this is presented as following in a natural development from the mathematics to the history of the problem or result, and from there to the interest in the people who produced the mathematics. For many of the persons involved this results in information not available from any other source; in lots of the cases, the available publications present an inaccurate (or at least incomplete) data. The author is very careful in documenting his claims by specific references, by citing correspondence between the principals involved, and by accounts by witnesses.

One of these developments leads Soifer to examine in great detail the life and actions of one of the great mathematicians of the twentieth century, Bartel Leendert van der Waerden. Although Dutch, van der Waerden spent the years from 1931 to 1945 in the Nazi Germany. This, and some of van der Waerden's activities during that time, became very controversial after Word War II, and led Soifer to examine the moral and ethical questions relevant to the life of a scientist in a criminal dictatorship.

The diligence with which Soifer pursues his quests for information is way beyond exemplary. He reports exchanges with I am sure hundreds of people, via mail, phone, email, visits – all dated and documented. The educational aspects that begin with matters any middle-school student can understand, develop gradually into areas of most recent research, involving not only combinatorics but also algebra, topology, questions of foundations of mathematics, and more.

I found it hard to stop reading before I finished (in two days) the whole text. Soifer engages the reader's attention not only mathematically, but emotionally and esthetically. May you enjoy the book as much as I did!

University of Washington   Branko Grünbaum

Foreword

Alexander Soifer's latest book is a fully fledged adult specimen of a new species, a work of literature in which fascinating elementary problems and developments concerning colorings in arithmetic or geometric settings are fluently presented and interwoven with a detailed and scholarly history of these problems and developments.

This history, mostly from the twentieth century, is part memoir, for Professor Soifer was personally acquainted with some of the principals of the story (the great Paul Erdős, for instance), became acquainted with others over the 18 year interval during which the book was written (Dima Raiskii, for instance, whose story is particularly poignant), and created himself some of the mathematics of which he writes.

Anecdotes, personal communications, and biography make for a good read, and the readability in "Mathematical Coloring Book" is not confined to the accounts of events that transpired during the author's lifetime. The most important and fascinating parts of the book, in my humble opinion, are Parts IV, VI, and VII, in which is illuminated the progress along the intellectual strand that originated with the Four-Color Conjecture and runs through Ramsey's Theorem via Schur, Baudet, and Van der Waerden right to the present day, via Erdős and numbers of others, including Soifer. Not only is this account fascinating, it is indispensable: it can be found nowhere else.

The reportage is skillful and the scholarship is impressive – this is what Seymour Hersh might have written, had he been a very good mathematician curious to the point of obsession with the history of these coloring problems.

The unusual combination of abilities and interests of the author make the species of which this book is the sole member automatically endangered. But in the worlds of literature, mathematics and literature about mathematics, unicorns can have offspring, even if the offspring are not exactly unicorns. I think of earlier books of the same family as "Mathematical Coloring Book" – G. H. Hardy's "A Mathematician's Apology", James R. Newman's "The World of Mathematics", Courant and Robbins' "What Is Mathematics?", Paul Halmos' "I Want to Be a Mathematician: an Automathography", or the books on Erdős that appeared soon after his death – all of them related at least distantly to "Mathematical Coloring Book" by virtue of the attempt to blend (whether successfully or not is open to debate) mathematics with history or personal memoir, and it seems to me that, whatever the merits of those works, they have all affected how mathematics is viewed and written about. And this will be a large part of the legacy of "Mathematical Coloring Book" – besides providing inspiration and plenty of mathematics to work on to young mathematicians, a priceless source to historians, and entertainment to those who are curious about the activities of mathematicians, "Mathematical Coloring Book" will (we can hope) have a great and salutary influence on all writing on mathematics in the future.

Auburn University   Peter D. Johnson

Foreword

What is the minimum number of colors required to color the points of the Euclidean plane in such a way that no two points that are one unit apart receive the same color? *Mathematical Coloring Book* describes the odyssey of Alexander Soifer and fellow mathematicians as they have attempted to answer this question and others involving the idea of partitioning (coloring) sets.

Among other things, the book provides an up-to-date summary of our knowledge of the most significant of these problems. But it does much more than that. It gives a compelling and often highly personal account of discoveries that have shaped that knowledge.

Soifer's writing brings the mathematical players into full view, and he paints their lives and achievements vividly and in detail, often against the backdrop of world events at the time. His treatment of the intellectual history of coloring problems is captivating.

Memphis State University   Cecil Rousseau

Acknowledgments

My first thank you goes to my late father Yuri Soifer, a great painter, who introduced colors into my life, and to whom this book is gratefully dedicated. As the son of a painter and an actress, I may have inherited artistic genes. Yet, it was my parents, Yuri Soifer and Frieda Hoffman, who inspired my development as a connoisseur and student of the arts. I have enjoyed mathematics only because it could be viewed as an art as well. I thank Maya Soifer for restarting my creative engine when at times it worked on low rpm (even though near the end of my work, she almost broke the engine by abandoning the car). I am deeply indebted to my kids Mark, Isabelle, and Leon, and to my cousin and great composer Leonid Hoffman for the support their love has always provided. I thank my old friends Konstantin Kikoin, Yuri Norstein and Leonid Hoffman for years of stimulating conversations on all themes of high culture. I am deeply grateful to Branko Grünbaum, Peter D. Johnson Jr., and Cecil Rousseau, the first readers of the entire manuscript, for their kind forewords and valuable suggestions. My 16-year old daughter Isabelle Soulay Soifer, an aspiring writer, did a fine copy-editing job – thank you, Isabelle!

This is a singular book for me, a result of 18 years of mathematical and historical research, and thinking over the moral and philosophical issues surrounding a mathematician in the society. The long years of writing have produced one immense benefit that a quickly baked book would never fathom to possess. I have had the distinct pleasure to communicate on the mathematics and the history for this book with senior sages Paul Erdős, George Szekeres, Esther (Klein) Szekeres, Martha (Wachsberger) Svéd, Henry Baudet, Nicolaas G. de Bruijn, Bartel L. van der Waerden, Harold W. Kuhn, Dirk Struick, Hilde Brauer (Mrs. Alfred Brauer), Hilde Abelin-Schur (Issai Schur's daughter), Walter Ledermann, Anne Davenport (Mrs. Harold Davenport), Victor Klee, and Branko Grünbaum. Many of these great people are no longer with us; others are near or in their 80s. Their knowledge and their memories have provided blood to the body of my book. I am infinitely indebted to them all, as well as to the younger contributors Ronald L. Graham, Edward Nelson, John Isbell, Adriano Garsia, James W. Fernandez, and Renate Fernandez, who are merely in their 70s.

Harold W. Kuhn wrote a triple essay on the economics of Frank P. Ramsey, John von Neumann and John F. Nash Jr. especially for this book, which can be found in Chapter 30. Steven Townsend wrote and illustrated a new version of his proof especially for this book (Chapter 24). Kenneth J. Falconer wrote a new clear exposition of his proof especially for this book (Chapter 9). Your contributions are so great, and I thank you three so very much!

I am grateful to several colleagues for their self-portraits written for this book: Nicolaas de Bruijn, Hillel Furstenberg, Vadim Vizing, Kenneth Falconer, Paul O'Donnell, Vitaly Bergelson, Alexander Leibman, and Michael Tarsi.

I am thrilled to be able to present results and writings of my colleagues I have known personally: Paul Erdős, Paul O'Donnell, Rob Hochberg, Kiran Chilakamarri, Ronald L. Graham, Joel H. Spencer, Paul Kainen, Peter D. Johnson, Jr., Nicholas Wormald, David Coulson, Paul Seymour, Neil Robertson, Robin J. Wilson, Edward Pegg, Jr., Jaroslav Nešetřil, Branko Grünbaum, Saharon Shelah, Dmitry Raiskii, Dmytro Karabash, Willie Moser, George Szekeres, Esther Klein, Alfred W. Hales, Heiko Harborth, Solomon W. Golomb, Jan Mycielski, Miklós Simonovits, and colleagues I have gotten to know through correspondence: Bruce L. Rothschild, Vadim G. Vizing, Mehdi Behzad, Douglas R. Woodall, and Jan Kyncl.

I am grateful to George Szekeres and MIT Press for their permission to reproduce here the most lyrical George's *Reminiscences* (part of Chapter 29). I thank Bartel L. van der Waerden and Academic Press, London, for their kind permission to reproduce here Van der Waerden's insightful *How the Proof of Baudet's Conjecture Was Found* (Chapter 33).

I thank all those who have provided me with the rare, early photographs of themselves: Paul Erdős (as well as a photograph with Leo Moser); George Szekeres and Esther Klein; Vadim Vizing; Edward Nelson; Paul O'Donnell. Hilde Abelin-Schur has kindly provided photographs of her father Issai Schur. Dorith van der Waerden and Theo van der Waerden have generously shared rare photographs of their uncle Bartel L. van der Waerden and the rest of their distinguished family. Henry Baudet II has generously supplied photographs of his father P. J. H. Baudet and also of his family with the legendary world chess champion Emanuel Lasker. Ronald L. Graham has kindly provided a photograph of him presenting Timothy Gowers with the check for $1000. Alice Bogdán has provided a photograph of her great brother-in-law Tibor Gallai.

I thank numerous archives for documents and photographs and permissions to use them. Dr. Mordecai Paldiel and Yad Vashem, The Holocaust Martyrs' and Heroes' Remembrance Authority, Jerusalem, provided documents related to the granting Senta Govers Baudet the title of a "Righteous Among the Nations." John Webb and Mathematics Department of the University of Cape Town provided a photograph of Francis Guthrie. By kind permission of the Provost and Scholars of King's College, Cambridge, I am able to reproduce here two photographs of Frank Plumpton Ramsey, from the papers of John Maynard Keynes. The Board of Trinity College Dublin has kindly allowed me to reproduce the correspondence between Augustus De Morgan and William R. Hamilton. Humboldt University of Berlin shared documents from the personnel file of Issai Schur.

The following archivists and archives provided invaluable documents and some photographs related to Bartel L. van der Waerden: John Wigmans, *Rijksarchief in Noord-Holland*; Dr. Peter J. Knegtmans, The University Historian,

Universiteit van Amsterdam; Prof. Dr. Gerald Wiemers and Martina Geigenmüller, *Universitätsarchiv Leipzig*; Prof. Dr. Holger P. Petersson and his personal archive; Gertjan Dikken, *Het Parool*; Madelon de Keizer; Dr. Wolfram Neubauer, Angela Gastl and Corina Tresch De Luca, *ETH-Bibliothek, ETH*, Zürich; Dr. Heinzpeter Stucki, *Universitätsarchiv, Universität Zürich*; Drs. A. Marian Th. Schilder, *Universiteitsmuseum de Agnietenkapel*, Amsterdam; Maarten H. Thomp, *Centrale Archiefbewaarplaats, Universiteit Utrecht*; Nancy Cricco, University Archivist, and her graduate students assistants, New York University; James Stimpert, Archivist, Milton S. Eisenhower Library, Special Collections, The Johns Hopkins University; Prof. Mark Walker; Thomas Powers; Nicolaas G. de Bruijn; Henry Baudet II, Dr. Helmut Rechenberg, Director of the Werner Heisenberg Archiv, Munich; G. G. J. (Gijs) Boink, *Het Nationaal Archief*, Den Haag; Library of Congress, Manuscript Division, Washington D.C.; *Centrum voor Wiskunde en Informatica*, Amsterdam; Prof. Nicholas M. Katz, Chair, Department of Mathematics, Princeton University; and Mitchell C. Brown and Sarah M. LaFata, Fine Library, Princeton University.

My research on the life of Van der Waerden could not be based on the archival material alone. I am most grateful to individuals who provided much help in this undertaking: Dorith van der Waerden; Theo van der Waerden; Hans van der Waerden; Helga van der Waerden Habicht; Annemarie van der Waerden; Prof. Dr. Henry Baudet (1919–1998); Prof. Dr. Herman J. A. Duparc (1918–2002); Prof. Dr. Nicolaas G. de Bruijn; Prof. Dr. Benno Eckmann; Prof. Dr. Dirk van Dalen; Prof. Dirk J. Struik (1894–2000); Dr. Paul Erdős (1913–1996); Reinhard Siegmund-Schultze; Thomas Powers; Prof. Mark Walker, Prof. James and Dr. Renate Fernandez; Dr. Maya Soifer; and Princeton University Professors Harold W. Kuhn, Simon Kochen, John H. Conway, Edward Nelson, Robert C. Gunning, Hale Trotter, Arthur S. Wightman, Val L. Fitch, Robert Fagles, Charles Gillispie and Steven Sperber.

I thank my wonderful translators of Dutch, Dr. Stefan van der Elst and Prof. Marijke Augusteijn; and of German: Prof. Simon A. Brendle; Prof. Robert von Dassanowsky; Prof. Robert Sackett; Prof. Dr. Heiko Harborth; and Prof. Dr. Hans-Dietrich Gronau.

I thank those who converted my pencil doodles into computer-aided illustrations: Steven Bamberger, Phil Emerick and Col. Dr. Robert Ewell.

The research quarterly *Geombinatorics* provided a major forum for the essays related to the chromatic number of the plane. Consequently, it is cited numerous times in this book. I wish to thank the editors of *Geombinatorics* Paul Erdős, Branko Grünbaum, Ron Graham, Heiko Harborth, Peter D. Johnson, Jr., Jaroslav Nešetřil, and János Pach.

My University of Colorado bosses provided support on our campus and opportunity to be away for long periods of time at Princeton and Rutgers Universities—I thank Tom Christensen, Tom Wynn, and Pam Shockley-Zalabak.

I am grateful to my Princeton-Math colleagues and friends for maintaining a unique creative atmosphere in the historic Fine Hall, and Fred Roberts for the tranquility of his DIMACS at Rutgers University. Library services at Princeton have provided an invaluable swift service: while working there for 3 years, I must have read thousands of papers and many books.

I thank my Springer Editor Mark Spencer, who initiated our contact, showed trust in me and this project based merely on the table of contents and a single section, and in 2004 proposed to publish this book in Springer. At a critical time in my life, Springer's Executive Editor Ann Kostant made me believe that what I am and what I do really matters—thank you from the bottom of my heart, Ann!

There is no better place to celebrate the completion of the book than the land of Pythagoras, Euclid, and Archimedes. I thank Prof. Takis Vlamos for organizing my visit and lectures on the Island of Corfu, Thessaloniki, and Athens.

Contents

Greetings to the Reader

I bring here all: what have I lived thru,
And that what keeps my soul alive,
My rectitude and aspirations,
And what have seen my own eyes.
                    – Boris Pasternak, *The Waves*, 1931[2]

When the form is realized, it is here to live its own
life.
                    – Pablo Picasso

Pasternak's epigraph describes precisely my work on this book—I gave it all of myself, without reservation. August Renoir believed that just as many people read one book all their lives (the Bible, the Koran, etc.), so can he paint all his life one painting. Likewise I could write one book all my life—in fact, I almost have, for I have been working on this book for 18 years.

It is unfair, however, to keep the book all to myself—many colleagues have been waiting for the birth of this book. In fact, it has been cited and even reviewed many years ago. The first mention of it appears already in 1991 on page 336 of the book by Victor Klee and Stan Wagon [KW], where the authors recommend the book for "survey of later developments of the chromatic number of the plane problem." On page 150 of their 1995 book [JT], Tommy R. Jensen and Bjarne Toft announced that "a comprehensive survey [of the chromatic number of the plane problem]...will be given by Soifer [to appear]." Once in the 1990s my son Mark told me that he saw my *Mathematical Coloring Book* available for $30 for special order at the Borders bookstore. I offered to buy a copy!

I started writing this book when copies of my *How Does One Cut a Triangle?* [Soi1] arrived from the printer, in early 1990. I told my father Yuri Soifer then

---

[2] [Pas], Translated for this book by Ilya Hoffman. The original Russian text is:
Здесь будет все: пережитое,
И то, чем я еще живу,
Мои стремленья и устои,
И виденное наяву

that this book would be dedicated to him, and so it is. This coloring book is for my late father, a great painter and man. Yuri lived with his sketchpad and drawing utensils in his pocket, constantly and intensely looking at people and making sharp momentary sketches. He was a great artist and my lifelong example of searching for and discovering life around him, and creating art that challenged "real" life herself. Yuri never taught me his trade, but during our numerous joint tours of art in museums and exhibitions, he pointed out beauties that only true artists could notice: a dream of harvest in Van Gogh's "Sower," Rodin's distortions in a search of greater expressiveness. These timeless lessons allowed me to become a student of beauty, and discover subtleties in paintings, sculptures, and movies throughout my life.

This book includes not just mathematics, but also the process of investigation, trains of mathematical thought, and where possible, psychology of mathematical invention. The book does not just include history and prehistory of Ramsey Theory and related fields, but also conveys the process of historical investigation—the kitchen of historical research if you will. It has captivated me, and made me feel like a Sherlock Holmes—I hope my reader will enjoy this sense of suspense and discovery as much as I have.

The epigraph for my book is an English translation of Jacques Prévert's genius and concise portrayal of creative process—I know of no better. I translated it with the help of my friend Maurice Starck from *Nouvelle Caledonie*, the island in the Pacific Ocean to which no planes fly from America, but to paraphrase Rudyard Kipling, *I'd like to roll to Nouvelle Caledonie some day before I'm old*!

This book is dedicated to problems involving colored objects, and results about the existence of certain exciting and unexpected properties that occur regardless of how these objects (points in the plane, space, integers, real numbers, subsets, etc.) are colored. In mathematics, these results comprise *Ramsey Theory*, a flourishing area of mathematics, with a motto that can be formulated as follows: any coloring of a large enough system contains a monochromatic subsystem of given in advance structure, or simply put, absolute chaos is absolutely impossible. Ramsey Theory thus touches on many fields of mathematics, such as combinatorics, geometry, number theory, and addresses new problems, often on the frontier of two or more traditional mathematical fields. The book will also include some problems that can be solved by inventing coloring, and results that prove the existence of certain colorings, most famous of the latter being, of course, The Four-Color Theorem.

Most books in the field present mathematics as a flower, dried out between pages of an old dusty volume, so dry that the colors are faded and only theorem–proof narrative survives. Along with my previous books, *Mathematical Coloring Book* will strive to become an account of a live mathematics. I hope the book will present mathematics as a human endeavor: the reader should expect to find in it not only results, but also portraits of their creators; not only mathematical facts, but also open problems; not only new mathematical research, but also new historical investigations; not only mathematical aspirations, but also moral dilemmas of the times between and during the two horrific World Wars of the twentieth century. In my view, mathematics is done by human beings, and knowing their lives and cultures enriches our understanding of mathematics as a product of human activity, rather than an abstraction which exists separately from us and comes to us exclusively as a catalog of theorems and formulas. Indeed, new facts and artifacts will be presented that are related to the history of the Chromatic Number of the Plane problem, the early history of Ramsey Theory, the lives of Issai Schur, Pierre Joseph Henry Baudet, and Bartel Leendert van der Waerden.

I hope you will join me on a journey you will never forget, a journey full of passion, where mathematics and history are researched in the process of solving mysteries more exciting than fiction, precisely because those are mysteries of real affairs of human history. Can mathematics be received by all senses, like a vibrant flower, indeed, like life itself? One way to find out is to experience this book.

While much of the book is dedicated to results of Ramsey Theory, I did not wish to call my book "Introduction to Ramsey Theory," for such a title would immediately lose young talented readers' interest. Somehow, the playfulness of *Mathematical Coloring Book* appealed to me from the start, even though I was asked on occasion whether 5-year olds would be able to color in my book between its lines. To be a bit more serious, and on advice of Vickie Kern of the Princeton University Press, I created a subtitle *Mathematics of Coloring and the Colorful Life of Its Creators*. This book is not a "dullster" of traditional theorem–proof–theorem–proof kind. It explores the birth of ideas and searches for its creators. I discovered very quickly that in conveying "colorful lives of creators," I could not always rely on encyclopedias and biographical articles, but had to conduct historical investigations on my own. It was a hard work to research some of the lives, especially that of B. L. van der Waerden, which alone took 12 years of archival research and thinking over the assembled evidence. Fortunately this produced a satisfying result: we have in this book some definitive biographies, of Bartel L. van der Waerden, Pierre Joseph Henry Baudet, Issai Schur, autobiography of Hillel Furstenberg, and others.

I always attempt to understand who made a discovery and how it was made. Accordingly, this book tries to explore biographies of the discoverers and the psychology of their creative processes. Every stone has been turned: my information comes from numerous archives in Germany, the Netherlands, Switzerland, Ireland, England, South Africa, the United States; invaluable and irreplaceable now interviews conducted with eyewitnesses; discussions held with creators. Cited bibliography alone includes over 800 items—I have read thousands of publications in the process of writing this book. I was inspired by people I have known personally, such as Paul Erdős, James W. Fernandez, Harold W. Kuhn, and many others, as well as people I have not personally met, such as Boris Pasternak, Pablo Picasso, Herbert Read—to name a few of the many influences—or D. A. Smith, who in the discussion after Alfred Brauer's talk [Bra2, p. 36], wrote:

> Mathematical history is a sadly neglected subject. Most of this history belongs to the twentieth century, and a good deal of it in the memories of mathematicians still living. The younger generation of mathematicians has been trained to consider the product, mathematics, as the most important thing, and to think of the people who produced it only as names attached to theorems. This frequently makes for a rather dry subject matter.

Milan Kundera, in his *The Curtain: An Essay in Seven Parts* [Kun], said about a novel what is true about mathematics as well:

> A novelist talking about the art of the novel is not a professor giving a discourse from his podium. Imagine him rather as a painter welcoming you into his studio, where you are surrounded by his canvases staring at you from where they lean against the walls. He will talk about himself, but even more about other people, about novels of theirs that he loves and that have a secret presence in his own work. According to his criteria of values, he will again trace out for you the whole past of the novel's history, and in so doing will give you some sense of his own poetics of the novel.

I was also inspired by the early readers of the book, and their feedback. Stanisław P. Radziszowski, after reviewing Chapter 27, e-mailed me on May 2, 2007:

> I am very anxious to read the whole book! You are doing great service to the community by taking care of the past, so the things are better understood in the future.

In his unpublished letter, Ernest Hemingway in a sense defended my writing of this book for a very long time:[3]

> When I make country, or a city, or a river in a novel it is slow work because you have to always *make* it, then it is alive. But nobody makes anything quickly nor easily if it is any good.

Branko Grünbaum, upon reading the entire manuscript, wrote in the February 28, 2008 e-mail:

> Somehow it seems that 18 years would be too short a time to dig up all this information!

This book will not strike the reader by completeness or most general results. Instead, it would give young active high school and college mathematicians an accessible introduction to the beautiful ideas of mathematics of coloring. Mathematics professionals, who may believe they know everything, would be pleasantly surprised by the unpublished or unnoticed mathematical gems. I hope young and not so young mathematicians alike will welcome an opportunity to try their hand—or mind—on numerous open problems, all easily understood and not at all easy to solve.

If the interest of my colleagues and friends at Princeton-Math is any indication, every intelligent reader would welcome an engagement in solving historical mysteries, especially those from the times of the Third Reich, World War II, and de-Nazification of Europe. Historians of mathematics would find a lot of new information and old errors corrected for the first time. And everyone will experience seeing, for the first time, faces they have not seen before in print: rare photographs of the creators of mathematics presented herein, from Francis Guthrie to Issai Schur as a young man, from young Edward Nelson to Paul O'Donnell, from Pierre Joseph Henry Baudet to Bartel L. van der Waerden and his family, and documents, such as

---

[3] From the unpublished 1937 letter. Quoted from *New York Times*, February 10, 2008, p. AR 8.

the one where Adolph Hitler commits a "micromanagement" of firing the Jew, Issai Schur, from his job of professor at the University of Berlin.

This is a freely flowing book, free from a straight jacket of a typical textbook, yet useable as a text for a host of various courses, two of which I have given to college seniors and graduate students at the University of Colorado: *What is Mathematics?*, and *Mathematical Coloring Course*, both presenting a "laboratory of a mathematician," a place where students learn mathematics and its history by researching them, and in the process realizing what mathematics is and what mathematicians do.

In writing this book, I tried to live up to the high standard, set by one of my heroes, the great Danish film director Carl Theodore Dreyer [Dre]:

> There is a certain resemblance between a work of art and a person. Just as one can talk about a person's soul, one can also talk about the work or art's soul, its personality. The soul is shown through the style, which is the artist's way of giving expression of his perception of the material. The style is important in attaching inspiration to artistic form. Through the style, the artist molds the many details that make it whole. Through style, he gets others to see the material through his eyes. . . Through the style he infuses the work with a soul – and that is what makes it art.

Mathematics is an art. It is a poor man's art: Nothing is needed to conceive it, and only paper and pencil to convey.

This long work has given me so very much, in Aleksandr Pushkin's words, "the heavenly, and inspiration, and life, and tears, and love."[4] I have been raising this book for 18 years, and over the past couple of years, I felt as if the book herself was dictating her composition and content to me, while I merely served as an obedient scribe. At 18, my book is now an adult, and deserves to separate from me to live her own life. As Picasso put it, "When the form is realized, it is here to live its own life." Farewell, my child, let the world love you as I have and always will!

---

[4] In the original Russian it sounds much better:
"И божество, и вдохновенье,
   И жизнь, и слезы, и любовь."

I
Merry-Go-Round

1
A Story of Colored Polygons and Arithmetic Progressions

'Have you guessed the riddle yet?' the Hatter said,
turning to Alice again.
'No, I give it up,' Alice replied. 'What's the answer?'
'I haven't the slightest idea,' said the Hatter.
'Nor I,' said the March Hare.

– Lewis Carroll, A Mad Tea-Party
Alice's Adventures in Wonderland

1.1 The Story of Creation

I recall April of 1970. The thirty judges of the Fourth Soviet Union National Mathematical Olympiad, of whom I was one, stayed at a fabulous white castle, half way between the cities of Simferopol and Alushta, nestled in the sunny hills of Crimea, surrounded by the Black Sea. This castle should be familiar to movie buffs: in 1934 the Russian classic film *Vesyolye Rebyata* (*Jolly Fellows*) was photographed here by Sergei Eisenstein's long-term assistant, director Grigori Aleksandrov. The problems had been selected and sent to printers. The Olympiad was to take place a day later, when something shocking occurred.

A mistake was found in the only solution the judges had of the problem created by Nikolai (Kolya) B. Vasiliev, the Vice-Chair of this Olympiad and a fine problem creator, head of the problems section of the journal *Kvant* from its inception in 1970 to the day of his untimely passing. What should we do? This question virtually monopolized our lives.

We could just cross this problem out on each of the six hundred printed problem sheets. In addition, we could select a replacement problem, but we would have to write it in chalk by hand in every examination room, since there would be no time to print it. Both options were rather embarrassing, desperate resolutions of the incident for the Jury of the National Olympiad, chaired by the great mathematician Andrej N. Kolmogorov, who was to arrive the following day. The best resolution, surely, would have been to solve the problem, especially because its statement was quite beautiful, and we had no counter example to it either.

A. Soifer, *The Mathematical Coloring Book*,
DOI 10.1007/978-0-387-74642-5_1, © Alexander Soifer 2009

Even today, 38 years later, I can close my eyes and see how each of us, thirty judges, all fine problem solvers, worked on the problem. A few sat at the table as if posing for Rodin's *Thinker*. Some walked around as if measuring the room's dimensions. Andrei Suslin, who would later prove the famous Serre's conjecture,[1] went out for a hike. Someone was lying on a sofa with his eyes closed. Silence was so absolute that you could hear a fly. The intense thinking seemed to stop the time inside of the room. However, we were unable, on to stop the time outside. Night fell, and with it our hopes for solving the problem in time.

Suddenly, the silence was interrupted by a victorious outcry: "I got it!" echoed through the halls and the watch tower of the castle. It came from Alexander Livshits, an undergraduate student at Leningrad (St. Petersburg) University, and former winner of the Soviet and the International Mathematical Olympiads (a perfect 42 score at the 1967 IMO in Yugoslavia).[2] His number-theoretic solution used the method of trigonometric sums. However, this, was the least of our troubles: we immediately translated the solution into the language of colored polygons.

Now we had options. A decision was reached to leave the problem in because the problem and its solution were too beautiful to be thrown away. We knew, though, that the chances of receiving a single solution from six hundred bright Olympians were very slim. Indeed nobody solved it.

1.2 The Problem of Colored Polygons

Here is the problem.

Problem 1.1 (N. B. Vasiliev; IV Soviet Union National Olympiad, 1970). Vertices of a regular n-gon are colored in finitely many colors (each vertex in one color) in such a way that for each color all vertices of that color form themselves a regular polygon, which we will call a *monochromatic* polygon. Prove that among the monochromatic polygons there are two polygons that are congruent. Moreover, the two congruent monochromatic polygons can always be found among the monochromatic polygons with the least number of vertices.

I first told the above story and the problem in my 1994 Olympiad book [Soi9]. It appeared in the section *Further Explorations*, and as such I left the pleasure of discovering the proof to the readers. It is time for me to share the solution.

Solution of Problem 1.1 by Alexander Livshits (in "polygonal translation"): Let me divide the problem into three parts: Preliminaries, Tool, and Proof.

Preliminaries: Given a system S of vectors $\vec{v}_1$, $\vec{v}_2, \ldots, \vec{v}_n$ in the plane with a Cartesian coordinate system, all emanating from the origin O. We would call the

---

[1] Daniel Quillen proved it independently, and got Field's Medal primarily for that.

[2] Andrei Suslin informs me that as of 1991 Alexander worked as a computer programmer in Leningrad; I was unable to determine his later whereabouts.

system S *symmetric* if there is an integer k, $1 \leq k < n$, such that rotation of every vector of S about O through the angle $\frac{2\pi k}{n}$ transforms S into itself.

Of course, the sum $\sum \vec{v}_i$ of all vectors of a symmetric system is $\vec{0}$, because $\sum \vec{v}_i$ does not change under rotation through the angle $0 < \frac{2\pi k}{n} < 2\pi$.

Place a regular n-gon P_n in the plane so that its center coincides with the origin O. Then the n vectors drawn from O to all the vertices of P_n form a symmetric system (Fig. 1.1).

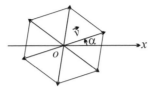

Fig. 1.1

Let $\vec{v}$ be a vector emanating from the origin O and making the angle α with the ray OX (Fig. 1.1). Symbol T^m will denote a transformation that maps $\vec{v}$ into the vector $T^m \vec{v}$ of the same length as $\vec{v}$, but making the angle $m\alpha$ with OX (Fig. 1.2).

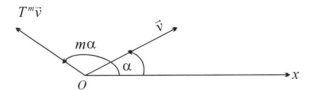

Fig. 1.2

To check your understanding of these concepts, please prove the following tool on your own.

Tool 1.2 Let $\vec{v}_1$, $\vec{v}_2$, ..., $\vec{v}_n$ be a symmetric system S of vectors that transforms into itself under the rotation through the angle $0 < \frac{2\pi k}{n} < 2\pi$, $1 \leq k < n$, (you can think of $\frac{2\pi k}{n}$ as the angle between two neighboring vectors of S). A transformation T^m applied to S produces a system $T^m S$ of vectors $T^m \vec{v}_1$, $T^m \vec{v}_2$, ..., $T^m \vec{v}_n$ that is symmetric if n does not divide km. If n divides km, then $T^m \vec{v}_1 = T^m \vec{v}_2 = \ldots = T^m \vec{v}_n$.

Solution of Problem 1.1: We will argue by contradiction. Assume that the vertices of a regular n-gon P_n are colored in r colors and we got subsequently r monochromatic polygons: n_1-gon P_{n_1}, n_2-gon P_{n_2}, ..., n_r-gon P_{n_r}, such that no pair of congruent monochromatic polygons is created, i.e.,

$$n_1 < n_2 < \ldots < n_r.$$

We create a symmetric system S of n vectors going from the origin to all vertices of the given n-gon P_n. In view of Tool 1.2, a transformation T^{n_1} applied to S produces a symmetric system $T^{n_1} S$. The sum of vectors in a symmetric system $T^{n_1} S$ is zero, of course.

On the other hand, we can first partition S in accordance with its coloring into r symmetric subsystems S_1, $S_2, \ldots,$ S_r, then obtain $T^{n_1} S$ by applying the transformation T^{n_1} to each system S_i separately, and combining all $T^{n_1} S_i$. By Tool 1.2, $T^{n_1} S_i$ is a symmetric system for $i = 2, \ldots,$ r, but $T^{n_1} S$ consists of n_1 identical non-zero vectors. Therefore, the sum of all vectors of $T^{n_1} S$ is not zero. This contradiction proves that the monochromatic polygons cannot be all non-congruent. ∎

Prove the last sentence of Problem 1.1 on your own:

Problem 1.3 Prove that in the setting of Problem 1.1, the two congruent monochromatic polynomials must exist among the monochromatic polynomials with the least number of vertices.

Readers familiar with complex numbers may have noticed that in the proof of Problem 1.1 we can choose the given n-gon P_n to be inscribed in a unit circle, and position P_n with respect to the axes so that the symmetric system S of vectors could be represented by complex numbers, which are precisely all n-th degree roots of 1. Then the transformation T^m would simply constitute raising these roots into the m-th power.

1.3 Translation into the Tongue of APs

You might be wondering what this striking problem of colored polygons has in common with arithmetic progressions (AP), which are part of the chapter's title. Actually, everything! Problem 1.1 can be nicely translated into the language of infinite arithmetic progressions, or APs for short.[3]

Problem 1.4 In any coloring (partition) of the set of integers into finitely many infinite monochromatic APs, there are two APs with the same difference. Moreover, the largest difference necessarily repeats.

Equivalently:

Problem 1.5 Any partition of the set of integers into finitely many APs can be obtained *only* in the following way: N is partitioned into k APs, each of the same difference k (where k is a positive integer greater than 1); then one of these APs is partitioned into finitely many APs of the same difference, then one of these APs (at this stage we have APs of two different differences) is partitioned into finitely many APs of the same difference, etc.

---

[3] An infinite sequence a_1, $a_2, \ldots,$ $a_n, \ldots$ is called an *arithmetic progression* or AP, if there for any integer $m > 1$, we have the equality $a_m = a_{m-1} + k$ for a fixed k, where k is a real number called *the difference* of the arithmetic progression.

It was as delightful that our striking problem allowed two beautiful distinct formulations, as it was valuable: only because of that I was able to discover the prehistory of our problem.

1.4 Prehistory

Indeed, a year after I first published the story of this problem, in 1994 [Soi9], I discovered that this exquisite bagatelle of a problem actually had a prehistory! I became aware of it while watching a video recording of Ronald L. Graham's most elegant lecture *Arithmetic Progressions: From Hilbert to Shelah*. To my surprise, Ron mentioned our bagatelle in the language of integer partitions. Let me present the prehistory through the original e-mails, so that you would discover the story the same way as I have.

April 5, 1995; Soifer to Graham:

In the beginning of your video "Arithmetic Progressions," you present a problem of partitioning integers into APs. You refer to Mirsky–Newman. Can you give me a more specific reference to their paper? You also mention that their paper may not contain the result, but that it is credited to them. How come? When did they allegedly prove it?

April 5, 1995; Graham to Soifer:

Regarding the Mirsky–Newman theorem, you should probably check with Erdős. I don't know that there ever was a paper by them on this result. Paul is in Israel at Tel Aviv University.

April 6, 1995; Soifer to Erdős:

In the beginning of his video "Arithmetic Progressions," Ron Graham presents a problem of partitioning integers into arithmetic progressions (with the conclusion that two progressions have the same difference). Ron refers to Mirsky–Newman. He gives no specific reference to their paper. He also mentions that their paper may not contain the result, but that it is credited to them... Ron suggested that I ask you, which is what I am doing.

I have good reasons to find this out, as in my previous book and in the one I writing now, I credit Vasiliev (from Russia) with creating this problem before early 1970. He certainly did, which surely does not exclude others from discovering it independently, before or after Vasiliev.

April 8, 1995; Erdős to Soifer:

In 1950 I conjectured that there is no exact covering system in which all differences are distinct, and this was proved by Donald J. Newman and [Leon] Mirsky a few months later. They never published anything, but this is mentioned in some papers of mine in the 1950s (maybe in the Summa Brasil. Math. 11(1950), 113–123 [E50.07], but I am not sure).

April 8, 1995; Erdős to Soifer:

Regarding that Newman's proof, look at P. Erdős, on a problem concerning covering systems, Mat. Lapok 3(1952), 122–128 [E52.03].

I am looking at these early Erdős's articles. In the 1950 paper he introduces covering systems of (linear) congruences. Since each linear congruence $x \equiv a \pmod{n}$ defines an AP, we can talk about *covering system of APs* and define it as a set of finitely many infinite APs, all with distinct differences, such that every integer belongs to at least one of the APs of the system. In the 1952 paper [E52.03] Paul introduces the problem for the first time in print (in Hungarian!):[4]

I conjectured that if system [of k APs with differences n_i respectively] is covering then

$$\sum_{i=1}^{k} \frac{1}{n_i} > 1, \tag{1.1}$$

that is, the system does not uniquely cover every integer. However, I could not prove this. For (1.1) Mirsky and Newmann [Newman] gave the following witty proof (the same proof was found later by Davenport and Rado as well).

Wow: Leon Mirsky, Donald Newman, Harold Davenport, and Richard Rado – quite a company of distinguished mathematicians worked on this bagatelle! Erdős then proceeds [E52.03] with presenting this company's proof of his conjecture, which uses infinite series and limits.

In viewing old video recordings of Paul Erdős's lectures at the University of Colorado at Colorado Springs, I found a curious historical detail Paul mentioned in his March 16, 1989 lecture: he created this conjecture in 1950 while traveling by car from Los Angeles to New York!

1.5 Completing the Go-Round

In 1959, Paul Erdős and János Surányi published a book on the Theory of Numbers. In 2003 English translation [ESu2] of its 1996 2nd Hungarian edition, Erdős and Surányi present the result from the Erdős's 1952 paper:

In a covering system of congruences [APs], the sum of the reciprocals of the moduli is larger than 1.

Erdős and Surányi then repeat Mirsky–Newman–Davenport–Rado proof from Erdős's 1952 paper [E52.03]. Then there comes a surprise:

A. Lifsic [sic] gave an elementary solution to a contest problem that turned out to be equivalent to Theorem 3.

---

[4] In English this result was briefly mentioned, without proof, much later, in 1973 [E73.21] and 1980 [EG].

Based again on exercises 9 and 10, it is sufficient to prove that it is not possible to cover the integers by finitely many arithmetic progressions having distinct differences in such a way that no two of them share a common element.

Erdős and Surányi then repeat the trick that was first discovered by us, the judges of the Soviet National Mathematical Olympiad in May 1970, the trick of converting the calculus solution into the Olympiad's original problem about colored polygons! Here is how it goes:

Wind the number line around a circle of circumference d. On this circle, the integers represent the vertices of a regular d-sided polygon... The arithmetic progressions form the vertices of disjoint regular polygons that together cover all vertices of the d-sided polygon.

Erdős and Surányi continue by repeating, with credit, Sasha Livshits's solution of Kolya Vasiliev's Problem of Colored Polygons that we have seen at the start of this chapter.[5] We have thus come to a full circle, a Merry-Go-Round from the Soviet Union Mathematical Olympiad to Erdős and back to the same Olympiad. I hope you have enjoyed the ride!

---

[5] Erdős and Surányi obtained the translation of the problem into the language of polygons and the polygonal proof from the 1988 Russian book [VE] by Vasiliev and Andrei Egorov, which they credit for it. In this book, Vasiliev gives credit for the solution to Sasha Livshits—and in a sign of extreme modesty does not credit himself with creating this remarkable colored polygon problem independently from Erdős and in a different form.

In looking now at the original 1996 Hungarian 2nd edition [ESu1] of Erdős—Surányi book, I realize with sadness that Paul Erdős did not see the beauties of Sasha Livshits's proof—it did not appear in the Hungarian edition of 1996, the year when Paul passed away. Clearly, Surányi alone added Livshits's proof to the 2003 English translation [ESu2] of the book.

II
Colored Plane

2
Chromatic Number of the Plane: The Problem

*A great advantage of geometry lies in the fact that
in it the senses can come to the aid of thought,
and help find the path to follow.*
— Henry Poincaré [Poi]

*[I] can't offer money for nice problems of other
people because then I will really go broke...
It is a very nice problem. If it were mine,
I would offer $250 for it.*
— Paul Erdős Boca Raton, February, 1992

*If Problem 8 [chromatic number of the plane] takes
that long to settle, we should know the answer by the
year 2084.*
— Victor Klee & Stan Wagon [KW]

Our good ole Euclidean plane, don't we know all about it? What else can there be after Pythagoras and Steiner, Euclid and Hilbert? In this chapter we will look at an open problem that exemplifies what is best in mathematics: anyone can understand this problem; yet no one has been able to conquer it for over 58 years.

In August 1987, I attended an inspiring talk by Paul Halmos at Chapman College in Orange, California, entitled "*Some problems you can solve, and some you cannot.*" This problem was an example of a problem "you cannot solve."

"A fascinating problem ... that combines ideas from set theory, combinatorics, measure theory, and distance geometry," write Hallard T. Croft, Kenneth J. Falconer, and Richard K. Guy in their book "*Unsolved Problems in Geometry*" [CFG].

"If Problem 8 takes that long to settle [as the celebrated Four-Color Conjecture], we should know the answer by the year 2084," write Victor Klee and Stan Wagon in their book "*New and Old Unsolved Problems in Plane Geometry*" [KW].

Are you ready? Here it is:

*What is the smallest number of colors sufficient for coloring the plane in such
a way that no two points of the same color are unit distance apart?*

A. Soifer, *The Mathematical Coloring Book*,
DOI 10.1007/978-0-387-74642-5_2, © Alexander Soifer 2009

This number is called *the chromatic number of the plane* and is denoted by χ. *To color the plane* means to assign one color to every point of the plane. Please note that here we color without any restrictions, and are not limited to "nice," tiling-like or map-like coloring. Given a positive integer n, we say that the plane is *n-colored*, if every point of the plane is assigned one of the given n colors.

A *segment* here will stand for just a 2-point set. Similarly, a *polygon* will stand for a finite set of point. *Monochromatic set* is a set whose all elements are assigned the same color. In this terminology, we can formulate the Chromatic Number of the Plane Problem (CNP) as follows: What is the smallest number of colors sufficient for coloring the plane in a way that forbids monochromatic segments of length 1?

I do not know who first noticed the following result. Perhaps, Adam? Or Eve? To be a bit more serious, I do not think that ancient Greek geometers, for example, knew this nice fact, for they simply did not ask this kind of questions!

Problem 2.1 (Adam & Eve?) No matter how the plane is 2-colored, it contains a monochromatic unit distance segment, i.e.,

$$\chi \geq 3.$$

Solution: Toss on the given 2-colored plane an equilateral triangle T of side 1 (Fig. 2.1). We have only 2 colors while T has 3 vertices (I trust you have not forgotten the Pigeonhole Principle). Two of the vertices must lie on the same color. They *are* distance 1 apart. ∎

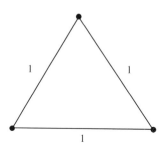

Fig. 2.1

We can do better than Adam:

Problem 2.2 No matter how the plane is 3-colored, it contains a monochromatic unit distance segment, i.e.,

$$\chi \geq 4.$$

Solution by the Canadian geometers, brothers Leo and William Moser, (1961, [MM]) Toss on the given 3-colored plane what we now call *The Mosers Spindle* (Fig. 2.2). Every edge in the spindle has the length 1.

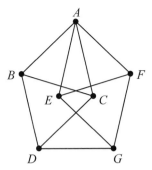

Fig. 2.2 The Mosers Spindle

Assume that the seven vertices of the spindle do not contain a monochromatic unit distance segment. Call the colors used to color the plane red, white, and blue. The solution now will faithfully follow the children's song: "*A B C D E F G...*".

Let the point *A* be red, then *B* and *C* must be one white and one blue, therefore *D* is red. Similarly *E* and *F* must be one white and one blue, therefore *G* is red. We found a monochromatic segment *DG* of length 1 in contradiction to our assumption. ∎

Observe: The Mosers Spindle has worked for us in solving Problem 2.2 precisely because *any* 3 points of the spindle contain two points distance 1 apart. This implies that *in a Mosers spindle that forbids monochromatic distance 1, at most 2 points can be of the same color*. Remember this observation, for we will need it later in Chapters 4 and 40.

When I presented the Mosers' solution to high school mathematicians, everyone agreed that it was beautiful and simple. "But how do you come up with a thing like the spindle?" I was asked. As a reply, I presented a less elegant but a more naturally found solution. In fact, I would call it a second version of the same solution. Here we touch on a curious aspect of mathematics. In mathematical texts we often see "second solution," "third solution," but which two solutions ought to be called distinct? We do not know: it is not defined, and thus is a judgment call. A distinct solution for one person could be the same solution for another. It is interesting to notice that both versions were published in the same year, of 1961, one in Canada and the other in Switzerland.

Second Version of the Solution (Hugo Hadwiger, 1961, [Had4]). Assume that a 3-colored red–white–blue plane does not contain a monochromatic unit distance segment. Then an equilateral triangle *ABC* of side 1 will have one vertex of each color (Fig. 2.3). Let *A* be red, then *B* and *C* must be one white and one blue. The point *A'* symmetric to *A* with respect to the side *BC* must be red as well. If we rotate our rhombus *ABA'C* through *any* angle about *A*, the vertex *A'* will have to remain red due to the same argument as above. Thus, we get a whole red circle of radius

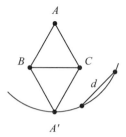

Fig. 2.3

AA' (Fig. 2.3). Surely, it contains a cord d of length 1, both endpoints of which are red, in contradiction to our assumption. ∎

Does an upper bound exist for χ? It is not immediately obvious. Can you find one? Think of tiling the plane with square tiles!

Problem 2.3 There is a 9-coloring of the plane that contains no monochromatic unit distance segments, i.e.,

$$\chi \le 9.$$

Proof We tile the plane with unit squares. Now we color one square in color 1, and its eight neighbors in colors 2, 3, ..., 9 (Fig. 2.4). The union of these 9 squares is a 3 × 3 square S. Translates of S (i.e., images of S under translations) tile the plane and determine how we color it in 9 colors.

You can easily verify (do) that no distance d in the range $\sqrt{2} < d < 2$ is realized monochromatically in the plane. Thus by shrinking all linear sizes by the factor of, say, 1.5, we get a coloring that contains no monochromatic unit distance segments. (Observe: due to the above inequality, we have enough cushion, so that it does not matter in which of the two adjacent colors we color the boundaries of squares). ∎

	6	1	2	6	1	2
	5	4	3	5	4	3
9	7	8	9	7	8	9
2	6	1	2	6	1	2
3	5	4	3	5	4	3
	7	8	9			

Fig. 2.4

Now that a tiling has helped us to solve the above problem, it is natural to ask whether another tiling can help us improve the upper bound. One can indeed!

Problem 2.4 There is a 7-coloring of the plane that contains no monochromatic unit distance segments, i.e.,

$$\chi \leq 7.$$

Solution ([Had3]): We can tile the plane by regular hexagons of side 1. Now we color one hexagon in color 1, and its six neighbors in colors 2, 3, ..., 7 (Fig. 2.5). The union of these seven hexagons forms a "flower" P, a highly symmetric polygon P of 18 sides. Translates of P tile the plane and determine how we color the plane in 7 colors. It is easy to compute (please do) that each color does not have monochromatic segments of any length d, where $2 < d < \sqrt{7}$. Thus, if we shrink all linear sizes by a factor of, say, 2.1, we will get a 7-coloring of the plane that forbids monochromatic unit distance segments. (Observe: due to the above inequality, we have enough cushion so that it does not matter in which of the two adjacent colors we color the boundaries of hexagons). ∎

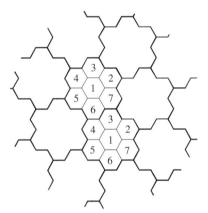

Fig. 2.5 Hexagon based 7-coloring of the plane

This is the way the upper bound is proven in every book I know (e.g., [CFG] and [KW]). Yet in 1982 the Hungarian mathematician László A. Székely found a clever way to prove the upper bound without using hexagonal tiling.

Problem 2.5 (L. A. Szekely, [Sze1]). Prove the upper bound $\chi \leq 7$ by tiling the plane with ... squares again.

Proof This is László Székely's proof from [Sze1]. His original picture needs a small correction in its "Fig. 2.1", and boundary coloring needs to be addressed, which I am doing here. We start with a row of squares of diagonal 1, with cyclically alternating colors of the squares 1, 2, ..., 7 (Fig. 2.6). We then obtain consecutive rows of colored squares by shifting the preceding row to the right through 2.5 square sides.

Upper and right boundaries are included in the color of each square, except for the square's upper left and lower right corners. ∎

3	4	5	6	7	1	2	3

5	6	7	1	2	3	4	5	6

1	2	3	4	5	6	7	1

Fig. 2.6 Square based 7-coloring of the plane

In 1995, my former student and now well-known puzzlist Edward Pegg, Jr. sent to me two distinct 7-colorings of the plane. In the one I am sharing with you (Fig. 2.7), Ed uses 7-gons for 6 colors, and tiny squares for the 7th color. Interestingly, the 7th color occupies only about one third of one percent of the plane.

In Fig. 2.7, all thick black bars have unit length. A unit of the tiling uses a heptagon and half a square.

The area of each square is 0.0041222051899307168162...
The area of each heptagon is 0.62265127164647629646...
Area ratio thus is 302.0962048019455285300783627265828...

If one third of one percent of the plane is removed, the remainder can be 6-colored with this tiling!

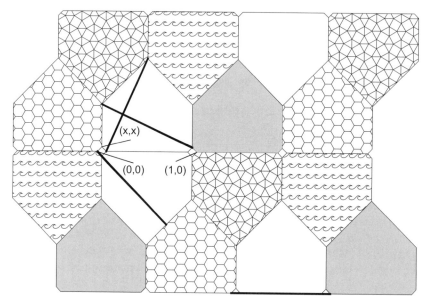

Fig. 2.7 7-coloring of the plane with a minimal presence of the 7th color

The lower bound for the chromatic number of the plane (Problem 2.2) also has proofs fundamentally different from the Mosers Spindle. In the early 1990s, I received from my colleague and friend Klaus Fischer of George Mason University, a finite configuration of the chromatic number 4, different from the Mosers spindle. Klaus had no idea who created it, so I commenced tracing it back in history. Klaus got this configuration from our common friend and colleague Heiko Harborth of Braunschweig Technical University, Germany, who in turn referred me to his source, Solomon W. Golomb of the University of Southern California, the famous inventor of polyomino. He invented this graph as well, and described it in the September 10, 1991 letter to me [Gol1]:

> The example you sketched of a 4-chromatic unit distance graph with ten vertices is original with me. I originally thought of it as a 3-dimensional structure (the regular hexagon below, the equilateral triangle above it in a plane parallel to it), and all connected by unit length toothpicks. The structure is then allowed to collapse down into the plane, to form the final figure [Fig. 2.8]. I have shown it to a number of people, including the late Leo Moser, Martin Gardner, and Paul Erdős, as well as Heiko Harborth. It is possible that Martin Gardner may have used it in one of his columns, but I don't remember. Besides my example and Moser's original example (which I'm reasonably sure I have seen in Gardner's column), I have not seen any other "fundamental" examples. I believe what I had suggested to Dr. Harborth in Calgary was the possibility of finding a 5-chromatic unit distance graph, having a much larger number of edges and vertices.

In the subsequent September 25, 1991 letter [Gol2] Golomb informed me that he likely found this example, which I will naturally call the *Golomb Graph*, in the time period 1960–1965.

Second Solution of Problem 2.2: Just toss the Golomb Graph on a 3-colored (red, white and blue) plane (Fig. 2.8). Assume that in the graph there are no adjacent (i.e., connected by an edge) vertices of the same color. Let the center point be colored red, then since it is connected by unit edges to all vertices of the regular hexagon H, H must be colored white and blue in alternating fashion. All vertices of the equilateral triangle T are connected by unit edges to the three vertices of H of the same color,

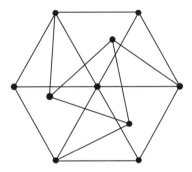

Fig. 2.8 The Golomb Graph

say, white. Thus, white cannot be used in coloring T, and thus T is colored red and blue, which implies that two of the vertices of T are assigned the same color. This contradiction proves that 3 colors are not enough to properly color the ten vertices of the Golomb graph, let alone the whole plane. ∎

It is amazing that the relatively easy solutions of Problems 2.2 and 2.4 give us the best known to today's mathematics. Bounds for the chromatic number of the plane χ. They were published almost half a century ago (in fact, they are older than that: see next chapter for an intriguing historical account). Still, all we know is that

$$\chi = 4, \text{ or } 5, \text{ or } 6, \text{ or } 7.$$

A very broad spread! Which do you think is the exact value of χ? The legendary Paul Erdős thought that $\chi \geq 5$.

The American geometer Victor Klee of the University of Washington shared with me in 1991 an amusing story. In 1980 he lectured in Zürich, Switzerland, where the 77-year-old celebrated algebraist Bartel L. van der Waerden (whom we will meet frequently later in this book— see Part VII) was in attendance. When Vic presented the state of this problem, Van der Waerden became so interested that he stopped listening to the lecture—he started working on the problem. He tried to prove that $\chi = 7$!

For many years I believed that $\chi = 7$, or else 6 (you will find my thoughts on the matter in *Predicting the Future*, Part X of this book). Paul Erdős used to say that "God has a transfinite Book, which contains all theorems and their best proofs, and if He is well intentioned towards those, He shows them the Book for a moment." If I ever deserved the honor and had a choice, I would have asked to peek at the page with the chromatic number of the plane problem. Wouldn't you?

3
Chromatic Number of the Plane: An Historical Essay

> *[This is] a long standing open problem of Erdős.*
> — Hallard T. Croft, 1967

> *[I] cannot trace the origin of this problem.*
> — Paul Erdős, 1961

> *It is often easier to be precise about Ancient Egyptian history than about what happened among our contemporaries.*
> — Nicolaas Govert de Bruijn, 1995[1]

> *It happened long ago and perhaps did not happen at all.*
> — An Old Russian Joke

It is natural for one to inquire into the authorship of one's favorite problem. As a result, in 1991 I turned to countless articles and books. Some of the information I found appears here in Table 3.1 – take a look. Are you confused? I was too!

As you can see in the table, Douglas R. Woodall credits Martin Gardner, who in turn refers to Leo Moser. Hallard T. Croft calls it "a long standing open problem of Erdős," Gustavus J. Simmons credits "Erdős, [Frank] Harary and [William Thomas] Tutte," while Paul Erdős himself "cannot trace the origin of this problem"! Later Erdős credits "Hadwiger and Nelson," while Victor Klee and Stan Wagon state that the problem was "posed in 1960–1961 by M. Gardner and Hadwiger." Croft comes again, this time with Kenneth J. Falconer and Richard K. Guy, to cautiously suggest that the problem is "apparently due to E. Nelson" [CFG]. Yet, Richard Guy did not know who "E. Nelson" was and why he and his coauthors "apparently" attributed the problem to him (our conversation on the back seat of a car in Keszthely, Hungary, when we both attended Paul Erdős's 80th birthday conference in August of 1993).

Thus, at least seven mathematicians—a great group to be sure—were credited with creating the problem: Paul Erdős, Martin Gardner, Hugo Hadwiger, Frank

---

[1] [Bru6].

A. Soifer, *The Mathematical Coloring Book*,
DOI 10.1007/978-0-387-74642-5_3, © Alexander Soifer 2009

Table 3.1 Who created the chromatic number of the plane problem?

Publication	Year	Author(s)	Problem creator(s) or source named
[Gar2]	1960	Gardner	"**Leo Moser**...writes..."
[Had4]	1961	Hadwiger (after Klee)	**Nelson**
[E61.21]	1961	Erdős	"I cannot trace the origin of this problem"
[Cro]	1967	Croft	"A long standing open problem of **Erdős**"
[Woo1]	1973	Woodall	**Gardner**
[Sim]	1976	Simmons	**Erdős, Harary** and **Tutte**
[E80.38] [E81.23] [E81.26]	1980–1981	Erdős	**Hadwiger** and **Nelson**
[CFG]	1991	Croft, Falconer, and Guy	"Apparently due to **E. Nelson**"
[KW]	1991	Klee and Wagon	"Posed in 1960–1961 by **M. Gardner** and **Hadwiger**"

Harary, Leo Moser, Edward Nelson, and William T. Tutte. But it was hard for me to believe that they all created the problem, be it independently or all seven together.

I felt an urge, akin that of a private investigator, a Sherlock Holmes, to untangle the web of conflicting accounts. It took six months to solve this historical puzzle. A good number of mathematicians, through conversations and e-mails, contributed their insight: Branko Grünbaum, Peter D. Johnson, Tony Hilton, and Klaus Fischer first come to mind. I am especially grateful to Paul Erdős, Victor Klee, Martin Gardner, Edward Nelson, and John Isbell for contributing their parts of the puzzle. Only their accounts, recollections, and congeniality made these findings possible.

I commenced my investigation on June 19, 1991 by mailing a letter to Paul Erdős, informing Paul that "I am starting a new 'Mathematical Coloring Book', which will address problems where coloring is a part of a problem and/or a part of solution (a major part),"[2] and then posed the question:

> There is a famous open problem of finding the chromatic number of the plane (minimal number of colors that prevents distance one between points of the same color). Is this your problem?

On August 10, 1991, Paul shared his appreciation of the problem, for which he could not claim the authorship [E91/8/10ltr]:

> The problem about the chromatic number of the plane is unfortunately not mine.

In a series of letters dated July 12, 1991; July 16, 1991; August 10, 1991; and August 14, 1991, Paul also formulated for me a good number of problems related to the chromatic number of the plane that he did create. We will look at Erdős's problems in the following chapters.

Having established that the author was not Paul Erdős, I moved down the list of "candidates," and on August 8, 1991 and again on August 30, 1991, I wrote to

---

[2] This seems to be my first mention of what has become an 18-year long project!

Victor Klee, Edward Nelson, and John Isbell. I shared with them my Table 3.1 and asked what they knew about the creation of the problem. I also interviewed Professor Nelson over the phone on September 18, 1991.

Edward Nelson created what he named "a second 4-color problem" (first being the famous Four-Color Problem of map coloring), which we will discuss in Part IV). In his October 5, 1991, letter [Nel2], he conveyed the story of creation:

Dear Professor Soifer:

In the autumn of 1950, I was a student at the University of Chicago and among other things was interested in the four-color problem, the problem of coloring graphs topologically embedded in the plane. These graphs are visualizable as nodes connected by wires. I asked myself whether a sufficiently rich class of such graphs might possibly be subgraphs of one big graph whose coloring could be established once and for all, for example, the graph of all points in the plane with the relation of being unit distance apart (so that the wires become rigid, straight, of the same length, but may cross). The idea did not hold up, but the other problem was interesting in its own right and I mentioned it to several people.

Eddie Nelson, c. 1950. Courtesy of Edward Nelson

One of the people Ed Nelson mentioned the problem to was John Isbell. Half a century later, Isbell still remembered the story very vividly when on August 26, 1991 he shared it with me [Isb1]:

> ... Ed Nelson told me the problem and $\chi \geq 4$ in November 1950, unless it was October – we met in October. I said what upper bound have you, he said none, and I worked out 7. I was a senior at the time (B.S., 1951). I think Ed had just entered U. Chicago as a nominal sophomore and taken placement exams which placed him a bit ahead of me, say a beginning graduate student with a gap or two in his background. I certainly mentioned the problem to other people between 1950 and 1957; Hugh Spencer Everett III, the author of the many-worlds interpretation of quantum mechanics, would certainly be one, and Elmer Julian Brody who did a doctorate under Fox and has long been at the Chinese University of Hong Kong and is said to be into classical Chinese literature would be another. I mentioned it to Vic Klee in 1958 ± 1 ...

Victor Klee also remembered (our phone conversation, September, 1991) hearing the problem from John Isbell in 1957–1958. In fact, it took place before September 1958, when Professor Klee left for Europe. There he passed the problem to Hugo Hadwiger who was collecting problems for the book *Open Problems in Intuitive Geometry* to be written jointly by Erdős, Fejes–Toth, Hadwiger, and Klee (this great book-to-be has never materialized).

Gustavus J. Simmons [Sim], in giving credit to "Erdős, Harary, and Tutte," no doubt had in mind their joint 1965 paper in which the three authors defined dimension of a graph (Chapter 13). The year of 1965 was too late for our problem's creation, and besides, the three authors have not made or claimed such a discovery.

What were the roles of Paul Erdős, Martin Gardner, and Leo Moser in the story of creation? I am prepared to answer these questions, all except one: I am leaving for others to research Leo Moser's archive (maintained by his brother Willie Moser at McGill University in Montreal) and find out how and when Leo Moser came by the problem. What is important to me is that he did not create it independently from Edward Nelson, as Paul Erdős informed me in his July 16, 1991, letter [E91/7/16ltr]:

> I do not remember whether Moser in 1958 [possibly on June 16, 1958, the date from which we are lucky to have a photo record] told me how he heard the problem on the chromatic number of the plane, I only remember that it was not his problem.

Yet, Leo Moser made a valuable contribution to the survival of the problem: he gave it to both Paul Erdős and Martin Gardner. Gardner, due to his fine taste, recognized the value of this problem and included it in his October 1960 *Mathematical Games* column in *Scientific American* ([Gar2]), with the acknowledgement that he received it from Leo Moser of the University of Alberta. Thus, the credit for the first *publication* of the problem goes to Martin Gardner. It is beyond me why so many authors of articles and books, as far back as 1973 ([Woo1], for example), gave credit for the *creation* of the problem to Martin Gardner, something he himself has never claimed. In our 1991 phone conversation Martin told me for a fact that the problem was not his, and he promptly listed Leo Moser as his source, both in print and in his archive.

Paul Erdős (*left*) and Leo Moser, June 16, 1958. Courtesy of Paul Erdős

Moreover, some authors ([KW], for example) who knew of Edward Nelson still credited Martin Gardner and Hugo Hadwiger because it seems only written, preferably published word, was acceptable to them. Following this logic, the creation of the celebrated Four-Color Map-Coloring Problem must be attributed to Augustus De Morgan, who first *wrote* about it in his October 23, 1852 letter to William Rowan Hamilton, or better yet to Arthur Cayley, whose 1878 abstract included the *first non-anonymous publication* of the problem.[3] Yet we all seem to agree that the 20-year-old Francis Guthrie created this problem, even though he did not publish or even write a word about it! (See Part IV for more on this.)

Of course, a lone self-serving statement would be too weak a foundation for a historical claim. On the other hand, independent disinterested testimonies corroborating each other comprise as solid a foundation for the attribution of the credit as any publication. And this is precisely what my inquiry has produced. Here is just one example of Nelson and Isbell's selflessness. Edward Nelson tells me on August 23, 1991 [Nel1]:

I proved nothing at all about the problem . . .

John Isbell corrects Nelson in his September 3, 1991, letter [Isb2]:

Ed Nelson's statement which you quote, "I proved nothing at all about the problem," can come only from a failure of memory. He proved to me that the number we are talking about is ≥ 4, by precisely the argument in Hadwiger 1961. Hadwiger's attribution (on Klee's authority) of that inequality to me can only be Hadwiger's or Klee's mistake.

This brings us to the issue of the authorship of the bounds for χ

$$4 \leq \chi \leq 7.$$

---

[3] First publication could be attributed to De Morgan, who mentioned the problem in his 1860 book review in *Athenaeum* [DeM4], albeit anonymously – see more on this in Chapter 18.

Once again, the entire literature is off the mark by giving credit for the first proofs to Hadwiger and the Mosers. Yes, in 1961 the famous Swiss geometer Hugo Hadwiger published ([Had4]) the chromatic number of the plane problem together with proofs of both bounds. yet he wrote (and nobody read!):

> We thank Mr. V. L. Klee (Seattle, USA) for the following information. The problem is due to E. Nelson; the inequalities are due to J. Isbell.

Hadwiger did go on to say:

> Some years ago the author [i.e., Hadwiger] discussed with P. Erdős questions of this kind.

Did Hadwiger imply that he created the problem independently from Nelson? We will never know for sure, but I have my doubts about Hadwiger's (co)authorship. Hadwiger jointly with H. Debrunner published an excellent long problem paper in 1955 [HD1] that was extended to their wonderful, famous book in 1959 [HD2]; see also the 1964 English translation [HDK] with Victor Klee, and the 1965 Russian translation [HD3] edited by Isaak M. Yaglom. All of these books (and Hadwiger's other papers) included a number of "questions of this kind," but did not once include the chromatic number of the plane problem. Moreover, it seems to me that the problem in question is somewhat out of Hadwiger's "character": in all problems "of this kind" he preferred to consider closed rather than arbitrary sets, in order to take advantage of topological tools.

I shared with Paul Erdős these two-fold doubts about Hadwiger independently creating the problem. It was especially important because Hadwiger in the quoted above text mentioned Erdős as his witness of sorts. Paul replied in the July 16, 1991 letter [E91/7/16ltr] as follows:

> I met Hadwiger only after 1950, thus I think Nelson has priority (Hadwiger died a few years ago, thus I cannot ask him, but I think the evidence is convincing).

At 9:30–10:30 A.M. on March 10, 1994, during his talk at 25th South Eastern International Conference on Combinatorics, Computing and Graph Theory in Boca Raton, Florida, Paul Erdős summarized the results of my historical research in the characteristically Erdősian style ([E94.60]):[4]

> There is a mathematician called Nelson who in 1950 when he was an epsilon, that is he was 18, discovered the following question. Suppose you join two points in the plane whose distance is 1. It is an infinite graph. What is chromatic number of this graph?
>
> Now, de Bruijn and I showed that if an infinite graph which is chromatic number k, it always has a finite subgraph, which is chromatic number k. So this problem is really [a] finite problem, not an infinite problem. And it was not difficult to prove that the chromatic number of the plane is between 4 and 7. I would bet it is bigger than 4, but I am not sure. And the problem is still open.
>
> If it were my problem, I would certainly offer money for it. You know, I can't offer money for every nice problem because I would go broke immediately. I was asked once

---

[4] Thanks to Prof. Fred Hoffman, the tireless organizer of this annual conference, I have a video tape of this memorable Paul Erdős's talk.

what would happen if all your problems would be solved, could you pay? Perhaps not, but it doesn't matter. What would happen to the strongest bank if all the people who have money there would ask for money back? Or what would happen to the strongest country if they suddenly ask for money? Even Japan or Switzerland would go broke. You see, Hungary would collapse instantly. Even the United States would go broke immediately...

Actually it was often attributed to me, this problem. It is certain that I had nothing to do with the problem. I first learned the problem, the chromatic number of the plane, in 1958, in the winter, when I was visiting [Leo] Moser. He did not tell me from where this nor the other problems came from. It was also attributed to Hadwiger, but Soifer's careful research showed that the problem is really due to Nelson.

The leading researcher of Ramsey Theory, Ronald L. Graham, has also endorsed the results of this historical investigation in his important 2004 problem paper [Gra6] in *Geombinatorics*:

It is certainly not necessary to point out to readers of this journal any facts concerning the history and current status of this problem (which [is] due to Nelson in 1950) since the Editor Alexander Soifer has written a scholarly treatment of this subject in this journal [Soi18], [Soi19], [SS2].

Paul Erdős's and Ron Graham's acceptance of my research on the history of this problem has had a significant effect: most researchers and expositors now give credit to Edward Nelson for the chromatic number of the plane problem. However, there are, unfortunate exceptions. In 2002 László Lovász and K. Vesztergombi, for example, stated [LV] that

in 1944 Hadwiger and Nelson raised the question of finding the chromatic number of the plane.

Of course, the problem did not exist in 1944, in Hadwiger's cited paper or anywhere else. Moreover, Eddie Nelson was just an 11–12-year-old boy at the time! In the same 2002 book, dedicated to the memory of Paul Erdős, one of the leading researchers of the problem and my friend Laszló Székely (who already in 1992 attended my talk on the history of the problem at Boca Raton), goes even further than Lovász and Vesztergombi [Sze3]:

E. Nelson and J. R. Isbell, and independently Erdős and H. Hadwiger, posed the following problem...

The fine Russian researcher of this problem A. M. Raigorodskii repeats from Székely in his 2003 book [Raig6, p. 3], in spite of citing (thus presumably knowing) my historical investigation in his survey [Raig3]:

There were several authors. First of all, already in the early 1940s the problem was posed by the remarkable mathematicians Hugo Hadwiger and Paul Erdős; secondly, E. Nelson and J. P. Isbell worked on the problem independently from Erdős and Hadwiger.[5]

---

[5] My translation from the Russian.

Raigorodskii then "discovers" previously non-existent connection between world affairs and the popularity of the problem:[6]

> In the 1940s there was W.W.II, and this circumstance is responsible for the fact that at first chromatic numbers [sic] did not raise too thunderous an interest.

The two famous Canadian problem people, the brothers Leo and William Moser, also published in 1961 [MM] the proof of the lower bound $4 \leq \chi$ while solving a different problem. Although, in my opinion, their proof is not distinct from those by Nelson and by Hadwiger, the Mosers' emphasis on a finite set and their invention of the seven-point configuration, now called *The Mosers' Spindle*, proved to be very productive (Chapter 2).

Now we can finally give due credit to Edward Nelson for being the first in 1950 to prove the lower bound $4 \leq \chi$. Because of this bound, John Isbell recalls in his letter [Isb1], Nelson "liked calling it a second 4CP!"

From the phone interviews with Edward Nelson on September 18 and 30, 1991, I learned some information about the problem creator. Joseph Edward Nelson was born on May 4, 1932 (an easy number to remember: 5/4/32), in Decatur, Georgia, near Atlanta. The son of the Secretary of the Italian YMCA,[7] Ed Nelson had studied at a *liceo* (Italian prep school) in Rome. In 1949 Eddie returned to the United States and entered the University of Chicago. The visionary Chancellor of the University, Robert Hutchins,[8] allowed students to avoid "doing time" at the University by passing lengthy placement exams instead. Ed Nelson had done so well on so many exams that he was allowed to go straight to the graduate school without working for his bachelor's degree.

Time magazine reported young Nelson's fine achievements in 14 exams on December 26, 1949 [Time], next to the report on the completion of the last war-crimes trials of the World War II (Field Marshal Fritz Erich von Manstein received 18 years in prison), assurances by General Dwight D. Eisenhower that he would *not* be a candidate in the 1952 Presidential election (he certainly was—and won it), and promise to announce *Time*'s "A Man of the Half-Century" in the next issue (the *Time*'s choice was Winston Churchill).

Upon obtaining his doctorate from the University of Chicago in 1955, Edward Nelson became National Science Foundation's Postdoctoral Fellow at the Princeton's Institute for Advanced Study in 1956. Three years later he became—and still is—a professor at Princeton University. His main areas of interest are analysis and logic. In 1975 Edward Nelson was elected to the American Academy of Arts and Sciences, and in 1997 to the National Academy of Sciences. During my 2002–2004 stay at Princeton, I had the pleasure to interact with Professor Nelson almost daily.

---

[6] Ibid.

[7] The Young Mens Christian Association (YMCA) is one of the oldest and largest not-for-profit community service organizations in the world.

[8] Robert Maynard Hutchins (1899–1977) was President (1929–1945) and Chancellor (1945–1951) of the University of Chicago.

My talk on the chromatic number of the plane problem at Princeton's Discrete Mathematics Seminar was dedicated "To Edward Nelson, who created this celebrated problem for us all."

John Isbell was first in 1950 to prove the upper bound $\chi \leq 7$. He used the same hexagonal 7-coloring of the plane that Hadwiger published in 1961 [Had4]. Please note that Hadwiger first used this coloring of the plane in 1945 [Had3], but for a different problem: his goal was to show that there are seven congruent closed sets that cover the plane (he also proved there that no five congruent closed sets cover the plane). Professor John Rolfe Isbell, Ph.D. Princeton University 1954 under Albert Tucker, has been for decades on the faculty of mathematics at the State University of New York at Buffalo, where he is now Professor Emeritus.

Paul Erdős's contribution to the history of this problem is two-fold. First of all, like Augustus De Morgan did for the Four-Color Problem, Erdős kept the flaming torch of the problem lit. He made the chromatic number of the plane problem well-known by posing it in his countless problem talks and many publications, for example, we see it in [E61.21], [E63.21], [E75.24], [E75.25], [E76.49], [E78.50], [E79.04], [ESi], [E80.38], [E80.41], [E81.23], [E81.26], [E85.01], [E91.60], [E92.19], [E92.60] and [E94.60].

Secondly, Paul Erdős created a good number of fabulous related problems. We will discuss one of them in the next chapter.

In February 1992 at the 23rd South Eastern International Conference on Combinatorics, Computing and Graph Theory in Boca Raton, during his traditional Thursday morning talk, I asked Paul Erdős how much he would offer for the first solution of the chromatic number of the plane problem. Paul replied:

> I can't offer money for nice problems of other people because then I will really go broke.

I then transformed my question into the realm of mathematics and asked Paul "*Assume* this is *your* problem; how much would you then offer for its first solution?" Paul answered:

> It is a very nice problem. If it were mine, I would offer $250 for it.

A few years ago the price went up for the improvement of just the lower bound part of the chromatic number of the plane problem. On Saturday, May 4, 2002, which by the way was precisely Edward Nelson's 70th birthday, Ronald L. Graham gave a talk on Ramsey Theory at the Massachusetts Institute of Technology for about 200 participants of the USA Mathematical Olympiad. During the talk he offered $1,000 for the first proof or disproof of what he called, after Nelson, "Another 4-Color Conjecture." The talk commenced at 10:30 AM (I attended the talk and took notes).

Another 4-Color $1000 Problem 3.1 (Graham, May 4, 2002) Is it possible to 4-color the plane to forbid a monochromatic distance 1?

In August 2003, during his talk *What is Ramsey Theory?* at Berkeley [Gra4], Graham asked for more work for $1000:

\$1000 Open Problem 3.2 (Graham, August, 2003). Determine the value of the chromatic number χ of the plane.

It seems that presently Ron Graham believes that the chromatic number of the plane takes on an intermediate value, between its known boundaries, for in his two latest surveys [Gra7], [Gra8], he offers the following open problems:

\$100 Open Problem 3.3 (Graham [Gra7], [Gra8]) Show that $\chi \geq 5$.[9]

\$250 Open Problem 3.4 (Graham [Gra7], [Gra8]) Show that $\chi \leq 6$.

This prompted me to look at all Erdős's published predictions on the chromatic number of the plane. Let me summarize them here for you. First Erdős believes— and communicates it in 1961 [E61.22] and 1975 [E75.24]—that the problem creator Nelson conjectured the chromatic number to be 4; Paul enters no prediction of his own. In 1976 [E76.49] Erdős asks:

Is this graph 4-chromatic?

In 1979 [E79.04] Erdős becomes more assertive:

It seems likely that the chromatic number is greater than four. By a theorem of de Bruijn and myself this would imply that there are n points $x_1, \ldots, x_n$ in the plane so that if we join any two of them whose distance is 1, then the resulting graph $G(x_1, \ldots, x_n)$ has chromatic number > 4. I believe such an n exists but its value may be very large.[10]

A certainty comes in 1980 [E80.38] and [E80.41]:

I am sure that [the chromatic number of the plane] $\alpha_2 > 4$ but cannot prove it.

In 1981 [E81.23] and [E81.26] we read, respectively:

It has been conjectured [by E. Nelson] that $\alpha_2 = 4$, but now it is generally believed that $\alpha_2 > 4$.

It seems likely that $\chi\left(E^2\right) > 4$.

In 1985 [E85.01] Paul Erdős writes:

I am almost sure that $h(2) > 4$.

Once—just once—Erdős expresses mid-value expectations, just as Ron Graham has in his Conjectures 3.3 and 3.4. It happened on Thursday, March 10, 1994 at the 25th South Eastern International Conference on Combinatorics, Computing and Graph Theory in Boca Raton. Following Erdős's plenary talk (9:30–10:30 A.M.), I was giving my talk at 10:50 A.M., when suddenly Paul Erdős said (and I jotted it down):

---

[9] Graham cites Paul O'Donnell's Theorem 45.4 (see it later in this book) as "perhaps, the evidence that χ is at least 5."

[10] If the chromatic number of the plane is 7, then for $G(x_1, \ldots, x_n) = 7$ such an n must be greater than 6197 [Pri].

Excuse me for interrupting, I am almost sure that the chromatic number of the plane is greater than 4. It is not a proof, but any measurable set without distance 1 in a very large circle has measure less than $1/4$. I also do not think that it is 7.

It is time for me to speak on the record and predict the chromatic number of the plane. I am leaning toward predicting 7 or else 4—somewhat disjointly from Graham and Erdős's apparent expectation. Limiting myself to just one value, I conjecture:

Chromatic Number of the Plane Conjecture 3.5 [11]

$$\chi = 7.$$

If you, in fact, prove the chromatic number is 7 or 4, I do not think you would lose Graham's prizes. I am sure Ron will pay his prizes for disproofs as well as for proofs. On January 26, 2007 in a personal e-mail, Graham clarified the terms of awarding his prizes:

> I always assume that we are working in ZFC (for the chromatic number of the plane!). My monetary awards can vary depending on which audience I am talking to. I always give the maximum of whatever I have announced (and not the sum!).

---

[11] See more predictions in Chapter 47.

4
Polychromatic Number of the Plane and Results Near the Lower Bound

When a great problem withstands all assaults, mathematicians create related problems. It gives them something to solve, plus sometimes there is an extra gain in this process, when an insight into a related problem brings new ways to see and conquer the original one. Numerous problems have been posed around the chromatic number of the plane. I would like to share with you my favorite among them.

It is convenient to say that a monochromatic set *S realizes distance d* if *S* contains a monochromatic segment of length *d*; otherwise we say that *S forbids distance d*.

Our knowledge about this problem starts with the celebrated 1959 book by Hugo Hadwiger and Hans Debrunner ([HD2], and subsequently its enhanced translations into Russian by Isaak M. Yaglom [HD3] and into English by Victor Klee [HDK]). Hadwiger reported in the book the contents of the September 9, 1958 letter he received from the Hungarian mathematician A. Heppes:

> Following an initiative by P. Erdős he [i.e., Heppes] considers decompositions of the space into disjoint sets rather than closed sets. For example, we can ask whether proposition 59 remains true in the case where the plane is decomposed into three disjoint subsets. As we know, this is still unresolved.

In other words, Paul Erdős asked whether it was true that if the plane were partitioned (colored) into three disjoint subsets, one of the subsets would have to realize all distances. Soon the problem took on its current "appearance." Here it is.

Erdős's Open Problem 4.1 What is the smallest number of colors needed for coloring the plane in such a way that no color realizes all distances?[12]

This number had to have a name, and so in 1992 [Soi5] I named it the *polychromatic number of the plane* and denoted it by χ_p. The name and the notation seemed so natural that by now it has become standard, and has (without credit) appeared in such encyclopedic books as [JT] and [GO].

Since I viewed this to be a very important open problem, I asked Paul Erdős to verify his authorship, suggested in passing by Hadwiger. As always, Paul was very modest in his July 16, 1991 letter to me [E91/7/16ltr]:

---

[12] The authors of the fine problem book [BMP] incorrectly credit Hadwiger as "first" to study this problem (p. 235). Hadwiger, quite typically for him, limited his study to *closed* sets.

A. Soifer, *The Mathematical Coloring Book,*
DOI 10.1007/978-0-387-74642-5_4, © Alexander Soifer 2009

I am not even quite sure that I created the problem: Find the smallest number of colors for the plane, so that no color realizes all distances, but if there is no evidence contradicting it we can assume it for the moment.

My notes show that during his unusually long 2-week visit in December 1991–January 1992 (we were working together on the book of Paul's open problems, soon to be completed and published by Springer under the title *Problems of pgom Erdős*), Paul confirmed his authorship of this problem. In the chromatic number of the plane problem, we were looking for colorings of the plane such that each color forbids distance 1. In the polychromatic number problem, we are coloring the plane in such a way that each color i forbids a distance d_i. For distinct colors i and j, the corresponding forbidden distances d_i and d_j may (but do not have to) be distinct. Of course,

$$\chi_p \leq \chi.$$

Therefore,

$$\chi_p \leq 7.$$

Nothing else had been discovered during the first 12 years of this problem's life. Then, in 1970, Dmitry E. Raiskii, a student of the Moscow High School for Working Youth[13] 105, published ([Rai]) the lower and upper bounds for χ_p. We will look here at the lower bound, leaving the upper bound to Chapter 6.

Raiskii's Theorem 4.2 (D. E. Raiskii [Rai]) $4 \leq \chi_p$.

Three years after Raiskii's publication, in 1973 the British mathematician Douglas R. Woodall from the University of Robin Hood (I mean Nottingham), published a paper [Woo1] on problems related to the chromatic number of the plane. Among other things, he gave his own proof of the lower bound. As I showed in [Soi17], Woodall's proof stemmed from a triple application of two simple ideas of Hugo Hadwiger ([HDK], Problems 54 and 59).

In 2003, the Russian turned Israeli mathematician Alexei Kanel-Belov communicated to me an incredibly beautiful short proof of this lower bound by the new generation of young Russian mathematicians, all his students. The proof was found by Alexei Merkov, a 10th grader from the Moscow High School 91, and communicated by Alexei Roginsky and Daniil Dimenstein in 1997 at a Moscow Pioneer Palace [Poisk]. The following is the author's proof with my gentle modifications.

Proof of the Lower Bound (A. Merkov): Assume the plane is colored in three colors, red, white and blue, but each color forbids a distance: r, w, and b respectively. Equip the 3-colored plane with the Cartesian coordinates with the origin O, and construct in the plane three seven-point sets S_r, S_w and S_b each being the Mosers Spindle (Fig. 2.2), such that all spindles share O as one of their seven vertices,

---

[13] Students in such high schools hold regular jobs during the day, and attend classes at night.

and have edges all equal to r, w, and b respectively. This construction defines 6 "red" vectors $v_1, \ldots, v_6$ from the origin O to each remaining point of S_r; 6 "white" vectors $v_7, \ldots, v_{12}$ from O to the points of S_w; and 6 "blue" vectors $v_{13}, \ldots, v_{18}$ from O to the points of S_b — 18 vectors in all.

Introduce now the 18-dimensional Euclidean space E^{18} and a function M from R^{18} to the plane R^2 naturally defined as follows: $(a_1, \ldots, a_{18}) \mapsto a_1 v_1 + \ldots + a_{18} v_{18}$. This function induces a 3-coloring of R^{18} by assigning a point of R^{18} the color of the corresponding point of the plane. The first six axes of E^{18} we will call "red", the next six axes "white", and the last six axes "blue."

Define by W the subset in E^{18} of all points whose coordinates include at most one coordinate equal to 1 for each of the three colors of the axes, and the rest (15 or more) coordinates 0. It is easy to verify (do) that W consists of 7^3 points. For any fixed array of allowable in W coordinates on white and blue axes, we get the 7-element set A of points in W having these fixed coordinates on white and blue axes. The image $M(A)$ of the set A under the map M forms in the plane a translate of the original seven-point set S_r. If we fix another array of white and blue coordinates, we get another 7-element set in E^{18}, whose image under M would form in the plane another translate of S_r. Thus, the set W gets partitioned into 7^2 subsets, each of which maps into a translate of S_r.

Now recall the observation we made after the first solution of Problem 2.2 in Chapter 2. It implies here that any translate of the Mosers Spindle S_r contains at most 2 red points out of its seven points. Since the set W has been *partitioned* into the translates of S_r, at most 2/7 of the points of W are red. We can start all over again in a similar way to show that at most 2/7 of the points of W are white, and similarly to show that at most 2/7 of the points of W are blue. But $2/7 + 2/7 + 2/7$ does not add up to 1! This contradiction implies that at least one of the colors realizes all distances, as required. ∎

At the International Congress on Mathematical Education in 1992 in Quebec City, I spent much time with Nikolai N. (Kolya) Konstantinov, whose mathematical circle at the Old Building of Moscow State University I attended as an 8th grader on Saturday afternoons during the 1962–1963 academic year. To my amazement, I learned that the hero of this section Dmitrii Raiskii was Konstantinov's student as well, just 2 years my junior! It took me many years to get "the full story" out of Kolya Konstantinov, but it was worth waiting for his February 23, 2007 e-mail, which I am translating here from the Russian:

> Dima Raiskii entered school Nr. 7 in 1965.[14] He was a part of a very strong group of students, from which several professional mathematicians came out, including Lena Nekhludova, who won gold medal of the International Mathematical Olympiad, Andrej Grjuntal, now chair of a department in the Institute of System Research, Vasilii Kozlov, now professor in the department of statistics of the Mechanics-Mathematical Faculty of the Moscow State University, and several well-known applied mathematicians.

---

[14] This was one of the Soviet Union's best high schools with the emphasis on mathematics, where courses were offered by some of the great Moscow State University professors.

Teachers of main mathematical courses were also very strong, including Joseph Bernstein, Viktor Zhurkin, formerly a graduate of this school and now a well-known biochemist, working in the USA.

The teaching method was based on students proving theorems of a course on their own, and on solving a large number of meaningful problems, which required creative abilities. . .

Dima performed well in mathematics, but was missing classes, and he had difficulties in other disciplines, in which teachers did not want to pass him because of small amount of earned credits. However, the main problem was at home. Dima's father thought his son was inept and insisted that Dima master a profession of a shoemaker, so that he could somehow feed himself. When I got to know Dima's family, I did not see his father, probably because by then he had already left the family, but I did not feel I had the right to ask about it.

Without any help, on his own Dima had read Hadwiger and Debrunner's book on combinatorial geometry.[15] He told me that he solved a problem from that book and wanted to show it to me. His presentation of the proof was in a "hall style" – very careless and informal, and l did not understand it right away – I felt, nevertheless, that the proof seemed plausible.

Dima then wrote down his solution. I made sure that everything was correct. However, Dima did not have an experience of writing articles, and so I undertook the "combing" of the text, and gave it the usual for publication look – I introduced several notations and terms. My work was purely technical; the published text did not contain my single idea. There was, however, an example, inserted by the Editor of *Математические заметки* [*Mathematical Notes*][16] Stechkin.[17] Then a funny episode happened. The inserted paragraph Stechkin ended with the phrase "the author thanks Stechkin for this example." Dima, however, thought that the word "author" refers to Stechkin in this case, and could not understand how Stechkin could thank himself.

Meanwhile clouds were thickening over Dima's head. The school wanted to expel him for absences, and he got into a children section of a psychiatric hospital. I visited him there. I saw lads of a school age behaving themselves quite freely. The counselors looked upon it nonchalantly – what can one ask of the sick ones? One boy, for example, asked, what would happen if to throw Brezhnev[18] into a toilet bowl and flush the toilet? And other silliness of the same kind.

After the release from the hospital, Dima [was expelled from the mathematical school Nr. 7 and] transferred to the school [Nr. 105] for working youth. There his affairs got even worse. He was finishing his senior 11th year, and the teachers' council had to decide whether to graduate the student, who missed countless classes and had almost no grades. At that time, the school received a letter from England. The thing is, at the end of Dima's published article there was the school's number, where he studied at the time of the article's publication. The letter was written by the professor

---

[15] [HD3].

[16] The journal where this article appeared.

[17] Sergei Borisovich Stechkin, a noted Russian mathematician – see his example and more about this story in chapter 6.

[18] Head of the Soviet Union at the time.

who worked on the same problem, but did not succeed. He informed Raiskii that he was sending him all the materials because he would no longer work on this problem, but hoped that Raiskii would be interested in acquainting himself with this unfinished work. This was not just a letter, but a thick packet, and the letter opened with "Dear Professor Raiskii." The lady-principal looked very gloomily during the teachers' council meeting dedicated to the question of Raiskii's graduation. She opened the meeting by acquainting the teachers with the content of this letter. She then said, "Let us graduate him."

In conclusion, let me add that Raiskii's family difficulties continued. Of course, Dima's psyche was not fully normal, but I think that his mother's psyche played a more negative role in his life than his own psyche. Here is one of her tricks. After Dima was released from the hospital, she wrote a letter to the minister of education complaining about me and P. S. Alexandrov.[19] The school [number 7] principal Volkov showed me this letter (which the ministry forwarded to the school). Dima's mother claimed in this letter that Alexandrov and Konstantinov politically corrupted the child, and inoculated the child with the anti-Soviet views. It went on further to claim that Konstantinov established the power over all Moscow psychiatrists and they all dance to his tune. The principal read this letter to me seriously, without any smile, until the last phrase when he finally allowed himself to laugh. I do not think it would be interesting to describe other tricks of Dima's mother.

While a high school student, Dima tried to solve mathematical problems many times. In particular, while participating in the Moscow Mathematical Olympiad, he worked not at all on the problems of the Olympiad, but on his own problems. He then got involved in the Eastern games of the mind – but I am not an expert in them, and do not remember their names. After that, I think, you know more about Dima than I do.

I wish you success [with the book]. Kolya.

On the Christmas day, December 25, 2003, the hero of this section, Dima Raiskii, told me how he came across the polychromatic number of the plane problem:

I learned about our coloring problem while reading the book *Combinatorial Geometry of the Plane* by Hadwiger and Debrunner [HD3]. This book was a part of the 3rd prize that I received at the Moscow Mathematical Olympiad of the 8th graders.

In my phone conversation with Dima Raiskii, I expressed my regret that he left mathematics after such a brilliant first paper. "Mathematicians appeared boring to me," Dima replied, and added: "They were constantly suffering from a feeling of guilt toward each other, or tried to make others repent. I felt much more at ease with *Go* players." And so Dima worked as a programmer and spent his time playing the ancient Chinese game *Go*. Then he gave up the city life, as he informed me on February 6, 2003:

I now settled in a remote village, where there is neither post nor computer. However, when I come to the city, I visit an internet-salon. What is new with your studies of African cultures? Are there meditative practices in Africa?

---

[19] Pavel Sergeevich Alexandrov, one of Russia's great mathematicians.

In his e-mails sent on the go from internet cafés, Dima described his involvement in *Go*, meditation, and writing books to aid others with meditation and spirituality. On March 17, 2003 I read:

> In the latter years I have played *Go*. This is the only game richer than chess; it is popular in China, Japan, Korea, etc. One of my students later became the Russian Champion for players up to the age of 10. According to the tradition, many *Go* players do meditative exercises in the style of Zen because this game equally uses both sides of the brain. In a close circle, I taught Zen meditation. In the East, however, many Buddhist authorities use Christian texts for teaching meditation. I am now preparing a small book of exercises for people raised in the Christian culture...
>
> P.S.: *Go* (brought to Europe by Lasker) is a most interesting object for computer modeling – in this regard, *Go* is richer than chess. One of my acquaintances is the European Champion in *Go* programming. Are people at Princeton involved in it?

Dima asked me several times to publish his results as a joint work of his and Nikolai Nikolaevich (Kolya) Konstantinov, his and my mathematics teacher, who—for better or for worse—influenced my choice of mathematics as profession. Dima insisted on sharing credit with Kolya, and Kolya categorically refused his share, because in his opinion, all of the ideas belonged to Dima.

Dima does not communicate with many people. Even his greetings to his Moscow teacher Konstantinov he sends via me in the USA. His e-mails to me are always inquisitive and warm. In his November 23, 2006 e-mail he expresses an appreciation of our correspondence:

> News from you always improve my mood. Give my regards to Nikolai Nikolaevich [Konstantinov].

In his last e-mail to date, on December 19, 2007, Dima wrote:

> I was always interested in the Eastern culture and studies of the Eastern religions. In the old times, however, I could not have publications [on these subjects], and instead had a lot of troubles. It seems likely that something will be published in the nearest time. This will start my public "biography." Will you be interested in my article? ...
> Yours always, Dima.

Dear Dima Raiskii, through the years of our correspondence, we became not only pen pals, but also friends. The societal pressure altered his life similarly to the changes in the life of Grisha Perelman, who abandoned mathematics at the peak of his creative powers, after conquering the celebrated Poincare and Geometrization conjectures. Their unprotected moral purity and extreme sensitivity made it difficult for them to deal with the ills of the society in general and the mathematical community in particular. Our friendship has provided Dima with an outlet for his thoughts and communication. I hope someone has offered the same to Grisha Perelman.

P.S.: After this book went into production, I informed Dima that his theorem and biography will appear in it, as will Van der Waerden's theorem and biography. On May 3, 2008 Dima replied:

Sasha, thank you very much! My biography and biography of Van der Waerden - not a bad combination. I will be telling my fellow villagers: "Once upon a time I am sitting with Vanya,[20] this Vanya, you know, which is der Waerden, who..." Will definitely read your book. Happy [W.W.II] Victory Day! ∎

Paul Erdős proposed yet another related problem (e.g., see [E85.01]). For a given finite set S of r positive numbers, a set of forbidden distances if you will, we define the graph $G_S(E^2)$, whose vertices are the points of the plane, and a pair of points is adjacent if and only if the distance between them belongs to S. Denote

$$\chi_r = \max_S \chi\left(G_S(E^2)\right).$$

"It is easy to see that $\lim_{r \to \infty} \chi_r / r = \infty$," Erdős writes, and poses a question:

Erdős's Conjecture 4.3 Does χ_r grow polynomially?

It is natural to call the chromatic number $\chi_S\left(E^2\right)$ of the graph $G_S(E^2)$ the *S-chromatic number of the plane*. One can pose a more general and hard problem, and in fact, it is an old problem of Paul Erdős ("I asked long ago," Paul says in [E94.60]):

Erdős's Open Problem 4.4 Given S, find the S-chromatic number $\chi_S\left(E^2\right)$ of the plane.

The difficulty of this problem should be clear to you: for a 1-element set S this is the chromatic number of the plane problem!

---

[20] Vanya is a nickname for Ivan. Dima is playing with the likeness of Van [der Waerden] and Vanya.

5

De Bruijn–Erdős Reduction to Finite Sets and Results Near the Lower Bound

We can expand the notion of the chromatic number to any subset S of the plane. *The chromatic number* $\chi(S)$ *of* S is the smallest number of colors sufficient for coloring the points of S in such a way that forbids monochromatic unit segments.

In 1951 Nicolaas Govert de Bruijn and Paul Erdős published a very powerful tool ([BE2]) that will help us with this and other problems. We will formulate and prove it in Part V. In our setting here, it implies the following.

De Bruijn–Erdős Compactness Theorem 5.1[21] The chromatic number of the plane is equal to the maximum chromatic number of its finite subsets.

Thus, as Paul Erdős used to say, the problem of finding the chromatic number of the plane is a problem about finite sets in the plane.[22]

There are other, easier questions about finite sets in the plane. Solve the following two problems on your own.

Problem 5.2 Find the smallest number δ_3 of points in a plane set whose chromatic number is equal to 3.

Problem 5.3 (L. Moser and W. Moser, [MM]) Find the smallest number δ_4 of points in a plane set whose chromatic number is 4. (Answer: $\delta_4 = 7$).

Victor Klee and Stan Wagon posed the following open problem [KW]:

Open Problem 5.4 When k is 5, 6, or 7, what is the smallest number δ_k of points in a plane set whose chromatic number is equal to k?

Of course, Problem 5.4 makes sense only if $\chi > 4$. In the latter case this problem suggests a way to attack the chromatic number of the plane problem by constructing new "spindles."

When you worked on Problems 5.2 and 5.3, you probably remembered our Problems 2.1 and 2.2. Indeed, those problems provided optimal configurations (Figs. 2.1 and 2.2) for Problems 5.2 and 5.3. Both optimal configurations were built of equilateral triangles of side 1. Can we manage without them?

---

[21] The axiom of choice is assumed in this result.

[22] Or so we all thought until recently. Because of that, I chose to leave this chapter as it was written in the early 1990s. BUT: see Part X of the book for latest developments.

A. Soifer, *The Mathematical Coloring Book*,
DOI 10.1007/978-0-387-74642-5_5, © Alexander Soifer 2009

Problem 5.5 Find the smallest number σ_3 of points in a plane set without unit equilateral triangles whose chromatic number is equal to 3.

Solution: $\sigma_3 = 5$. The regular pentagon of side 1 (Fig. 5.1) delivers a minimal configuration of chromatic number 3.

It is easy to 2-color any four-point set A, B, C, D without equilateral triangles of side 1. Just color A red. All points distance 1 from A color blue; these are the second generation points. All uncolored points distance 1 from any point of the second generation, we color red; these are the third generation points. All uncolored points distance 1 from the points of the third generation, we color blue. If we did not color all four points, we start this process all over again by coloring any uncolored point red. If this algorithm were to define the color of any point not uniquely, we would have an odd-sided n-gon with all sides 1, i.e., an equilateral triangle (since $n \leq 4$), which cannot be present, and thus would provide the desired contradiction.

■

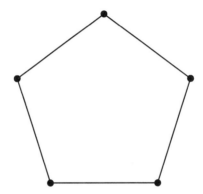

Fig. 5.1 Equilateral pentagon of side 1

For four colors this was for a while an open problem first posed by Paul Erdős in July 1975, (and published in 1976), who, as was usual for him, offered to "buy" the first solution—for $25.

Paul Erdős's $25 Problem 5.6 [E76.49] Let S be a subset of the plane which contains no equilateral triangles of size 1. Join two points of S if their distance is 1. Does this graph have chromatic number 3?

If the answer is no, assume that the graph defined by S contains no C_l [cycles of length l] for $3 \leq l \leq t$ and ask the same question.

It appears that Paul Erdős was not sure of the outcome—which was rare for him. Moreover, from the next publication of the problem in 1979 [E79.04], it is clear that Paul expected that triangle-free unit distance graphs had chromatic number 3, or else chromatic number 3 can be forced by prohibiting all small cycles up to C_k for a sufficiently large k:

Paul Erdős's $25 Problem 5.7 [E79.04] "Let our n points [in the plane] are such that they do not contain an equilateral triangle of side 1. Then their chromatic number is probably at most 3, but I do not see how to prove this. If the conjecture would unexpectedly [sic] turn out to be false, the situation can perhaps be saved by the following new conjecture:

There is a k so that if the girth of $G(x_1, \ldots, x_n)$ is greater than k, then its chromatic number is at most three—in fact, it will probably suffice to assume that $G(x_1, \ldots, x_n)$ has no odd circuit of length $\leq k$."[23]

Erdős's first surprise arrived in 1979 from Australia: Nicholas Wormald, then of the University of Newcastle, Australia, disproved the first, easier, triangle-free conjecture 5.6. Erdős paid \$25 reward for the surprise, and promptly reported it in his next 1978 talk (published 3 years later [E81.23]):

Wormald in a recent paper (which is not yet published) disproved my original conjecture – he found a [set] S for which [the unit distance graph] $G_1(S)$ has girth 5 and chromatic number 4. Wormald's construction uses elaborate computations and is fairly complicated.

In his paper [Wor], Wormald proved the existence of a set S of 6448 (!) points without triangles and quadrilaterals with all sides 1, whose chromatic number was 4, while being aided by a computer. I would like to give you a taste of the initial Wormald construction (or, more precisely, the Blanche Descartes construction that Wormald was able to embed in the plane), but it is a better fit in Chapter 12.

The size of Wormald's example, of course, did not appear to be anywhere near optimal. Surely, it must have been possible to do the job with less than 6448 points! In my March-1992 talk at the Conference on Combinatorics, Graph Theory and Computing at Florida Atlantic University, I shared this Paul Erdős's old question, but I put it in a form of competition:

Open Problem 5.8 Find the smallest number σ_4 of points in a plane set without unit equilateral triangles whose chromatic number is 4. Construct such a set S of σ_4 points.

The result exceeded my wildest dreams: a number of young mathematicians, including graduate students, were inspired by this talk and entered the race I proposed. Coincidentally, during that academic year, with the participation of the celebrated geometer Branko Grünbaum, and of Paul Erdős, whose problem papers set the style, I started a new and unique journal *Geombinatorics*. This journal was dedicated to problem-posing essays on discrete and combinatorial geometry and related areas (it is still alive and well now, 17 years later). The aspirations of the journal were clear from my 1991 editor's page in issue 3 of volume I:

In a regular journal, papers appear 1 to 2 years after research is completed. By then even the author may not be excited any more about his results. In *Geombinatorics* we can exchange open problems, conjectures, aspirations, work-in-progress that is still exciting to the author, and therefore exciting to the reader.

---

[23] The symbol $G(x_1, \ldots, x_n)$ denotes the graph on the listed inside parentheses n vertices, with two vertices adjacent if and only if they are unit distance apart.

A true World Series played out on the pages of *Geombinatorics* around Problem 5.8. The graphs obtained by the record setters were as mathematically significant as they were beautiful. I have to show them to you—see them discussed in detail in Chapters 14 and 15.

Many attempts to increase the lower bound of the chromatic number of the plane were not successful. The Rutgers University's Ph.D. student Rob Hochberg believed (and still does) that the chromatic number of the plane was 4, while his roommate and fellow Ph.D. student Paul O'Donnell was of the opposite opinion. They managed to get alone in spite of this disagreement of the mathematical kind. On January 7, 1994, Rob sent me an e-mail to that effect:

> Alex, hello. Rob Hochbeg here. (The one who's gonna prove $\chi\left(R^2\right) = 4$.). ... It seems that Paul O'Donnell is determined to do his Ph. D. thesis by constructing a 5-chromatic unit distance graph in the plane. He's got several interesting 4-chromatic graphs, and great plans. We still get along.

Two months later, Paul O'Donnell's abstract in the *Abstracts* book of the International Conference on Combinatorics, Graph Theory and Computing in Boca Raton, Florida included the following words:

> The chromatic number of the plane is between four and seven. A five-chromatic subgraph would raise the lower bound. If I discover such a subgraph, I will present it.

We all came to his talk of course (it was easy for me, as I spoke immediately before Paul in the same room). However, at the start of his talk, Paul simply said "not yet," and went on to show his impressive 4-chromatic graph of girth 4. Five years later, on May 25, 1999, Paul O'Donnell defended his doctorate at Rutgers University. I served as the outside member of his Ph.D. defense committee. In fact, it appears that my furniture had something to do with Paul O'Donnell's remarkable dissertation, for in the dissertation's Acknowledgements he wrote:

> Thanks to Alex. It all came to me as I drifted off to sleep on your couch.

The problem of finding a 5-chromatic unit distance graph—or proving that one does not exist—still remains open. However, much was learned about 4-chromatic unit distance graphs. The best of these results, in my opinion, was contained in this doctoral dissertation of Paul O'Donnell. He completely solved Paul Erdős's problem 5.7, and delivered to Paul Erdős an ultimate surprise by negatively answering his general conjecture:

O'Donnell's Theorem 5.9 ([Odo3, Odo4, Odo5]) There exist 4-chromatic unit distance graphs of arbitrary finite girth.

I chose to divide the proof of this result between Parts III and IX. See you there!

6
Polychromatic Number of the Plane and Results Near the Upper Bound

6.1 Stechkin's 6-Coloring

In Chapter 4 we discussed the polychromatic number χ_p of the plane, and looked at the 1970 paper [Rai] by Dmitry E. Raiskii where he was first to prove that 4 is the lower bound of χ_p. The paper also contained the upper bound:

$$\chi_p \leq 6.$$

The example proving this upper bound was found by S. B. Stechkin and published with his permission by D. E. Raiskii in [Rai]. Stechkin has never gotten a credit in the West for his example. Numerous articles and books credited Raiskii (except for Raiskii himself!). How did it happen? As everyone else, I read the English translation of Raiskii's paper [Rai]. It said (italics are mine):

> S. B. *Stechkin noted* that the plane can be decomposed into six sets such that all distances are not realized in any one of them. A corresponding example is presented here with the *author's solution*.

I wondered: the author of what?, The author of the paper (as everyone decided)? But there is very little need for a "solution" once the example is found. I felt as if once again I was a Sherlock Holmes. I ordered a copy of the original Russian text, and I read it in disbelief:

> A corresponding example is presented here with the *author's permission*.

Stechkin *permitted* Raiskii to publish Stechkin's example! The translator mixed up somewhat similarly looking Russian words and "innocently" created a myth (Table 6.1).

Table 6.1 Translator's Folly

Russian word	English translation
решение	solution
*раз*решение	permission

Let us roll back to the mathematics of this example.

A. Soifer, *The Mathematical Coloring Book*,
DOI 10.1007/978-0-387-74642-5_6, © Alexander Soifer 2009

Problem 6.1 (S. B. Stechkin, [Rai]). $\chi_p \leq 6$, i.e., there is a 6-coloring of the plane such that no color realizes all distances.

Solution by S. B. Stechkin [Rai]: The "unit of the construction" is a parallelogram that consists of four regular hexagons and eight equilateral triangles, all of side lengths 1 (Fig. 6.1). We color the hexagons in colors 1, 2, 3, and 4. Triangles of the tiling we partition into two types: we assign color 5 to the triangles with a vertex below its horizontal base; and color 6 to the triangles with a vertex above their horizontal base. While coloring, we consider every hexagon to include its entire boundary except its one rightmost and two lowest vertices; and every triangle does not include any of its boundary points. Now we can tile the entire plane with translates of the "unit of the construction." ∎

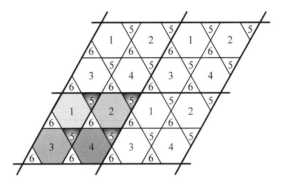

Fig. 6.1 S.B.Stechkin's 6-coloring of the plane

An easy construction solved Problem 6.1—easy to understand after it was found. The trick was to find it, and Sergej B. Stechkin found it first. Christopher Columbus too "just ran into" America! I got hooked.

6.2 Best 6-Coloring of the Plane

I felt that if our ultimate goal was to find the chromatic number χ of the plane or to at least improve the known bounds ($4 \leq \chi \leq 7$), it may be worthwhile to somehow measure how close a given coloring of the plane is to achieving this goal. In 1992, I introduced such a measurement, and named it *coloring type*.

Definition 6.2 (A. Soifer [Soi5], [Soi6]) Given an n-coloring of the plane such that the color i does not realize the distance d_i ($1 \leq i \leq n$). Then we would say that this coloring is of *type* $(d_1, d_2, \ldots, d_n)$.

This new notion of type was so natural and helpful that it received the ultimate compliment of becoming a part of the mathematical folklore: it appeared everywhere without a credit to its inventor (look, for example, p. 14 of the fundamental 991-page long monograph [GO]).

It would have been a great improvement in our search for the chromatic number of the plane if we were to find a 6-coloring of type (1,1,1,1,1,1), or to show that one does not exist. With the appropriate choice of a unit, we can make the 1970 Stechkin coloring to have type $(1, 1, 1, 1, \frac{1}{2}, \frac{1}{2})$. Three years later, in 1973 Douglas R. Woodall [Woo1] found the second 6-coloring of the plane with all distances not realized in any color. Woodall's coloring had a special property that the author desired for his purposes: each of the six monochromatic sets was closed. His example, however, had three "missing distances": it had type $(1, 1, 1, \frac{1}{\sqrt{3}}, \frac{1}{\sqrt{3}}, \frac{1}{2\sqrt{3}})$. Apparently, Woodall unsuccessfully tried to reduce the number of distinct distances, for he wrote "I have not managed to make two of the three 'missing distances' equal in this way" ([Woo1], p. 193).

In 1991, in search for a "good" coloring I looked at a tiling with regular octagons and squares that I saw in many Russian public toilettes (Fig. 6.2).

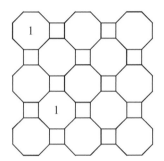

Fig. 6.2 "Russian toilette tiling"

But "The Russian toilette tiling" did not work! See it for yourself:

Problem 6.3 Prove that the set of all squares in the tiling of Fig. 6.2 (even without their boundaries) realizes all distances.

I then decided to shrink the squares until their diagonal became equal to the distance between two closest squares. Simultaneously (!) the diagonal of the now non-regular octagon became equal to the distance between the two octagons marked with 1 in Fig. 6.2. I was in business!

Problem 6.4 (A. Soifer [Soi6]) There is a 6-coloring of the plane of type $(1, 1, 1, 1, 1, \frac{1}{\sqrt{5}})$.

Solution: We start with two squares, one of side 2 and the other of diagonal 1 (Fig. 6.3). We can use them to create the tiling of the plane with squares and (non-regular) octagons (Fig. 6.5). Colors 1, . . . , 5 will consist of octagons; we will color all squares in color 6. With each octagon and each square we include half of its boundary (bold lines in Fig. 6.4) without the endpoints of that half. It is easy to verify (please do) that $\sqrt{5}$ is not realized by any of the colors 1, . . . , 5; and 1 is not realized by the color 6. By shrinking all linear sizes by a factor of $\sqrt{5}$, we get the 6-coloring of type $(1, 1, 1, 1, 1, \frac{1}{\sqrt{5}})$.

To simplify a verification, observe that the unit of my construction is bounded by the bold line in Fig. 6.5; its translates tile the plane. ∎

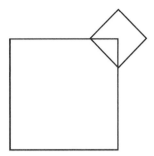

Fig. 6.3 Basis of the construction

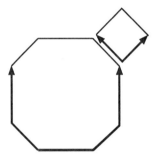

Fig. 6.4 Coloring of the boundaries

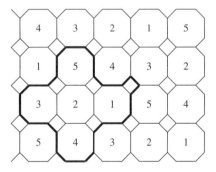

Fig. 6.5 A. Soifer's 6-coloring of the plane

I had mixed feelings when I obtained the result of problem 6.4 in early August 1991. On the one hand, I knew the result was "close but no cigar": after all, a 6-coloring of type (1,1,1,1,1,1) has not been found. On the other hand, I thought that the latter 6-coloring may not exist, and if so, my 6-coloring would be the best possible. There was another consideration as well. While in a Ph.D. program in Moscow, I hoped to produce the longest paper that might still be accepted by a major journal (I had one published in 1973 that in manuscript was 56 pages long :-). This time I was concerned with a "dual record": how short can a paper be and still contain enough "beef" to be refereed in and published? The paper [Soi6] solving problem 6.4 was 1.5 pages long, plus pictures. It was accepted within a day. It also gave birth to a new definition and an open problem.

Definition 6.5 ([HS1]) Almost chromatic number χ_a of the plane is the minimal number of colors that are required for coloring the plane so that almost all (i.e., all but one) colors forbid unit distance, and the remaining color forbids a distance.

We have the following inequalities for χ_a:

$$4 \leq \chi_a \leq 6.$$

The lower bound follows from Dmitry Raiskii's [Rai]. I proved the upper bound in problem 6.4 above [Soi6]. This naturally gave birth to a new problem, which is still open:

Open Problem 6.6 ([HS1]) Find χ_a.

6.3 The Age of Tiling

Hadwiger's, Stechkin's and my ornaments (Figs. 2.4, 6.1, and 6.5 respectively) delivered new mathematical results. They were also aesthetically pleasing. Have

Fig. 6.6 Ancient Chinese Lattice

Fig. 6.7 Ancient Chinese Lattice

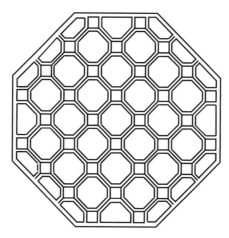

Fig. 6.8 Ancient Chinese Lattice

we contributed something, however little, to the arts? Not really. Nothing is new in the world of art. We can find Henry Moore's aesthetics in pre-Columbian art and Picasso's cubistic geometrization of form in the art of Sub-Saharan Africa. Our ornaments too were known for over 1,000 years to artists of China, India, Persia, Turkey, and Europe. Figures 6.6, 6.7, and 6.8 reproduced with the kind permission of the Harvard-Yenching Institute from the wonderful 1937 book *A Grammar of Chinese Lattice* by Daniel Sheets Dye ([Dye]), show how those ornaments were implemented in old Chinese lattices.

If it is any consolation, I can point out that our Chinese ancestors did not invent the beauty and strength of the honeycomb either: bees were here first!

7

Continuum of 6-Colorings of the Plane

In 1993 another 6-coloring was found by Ilya Hoffman and I ([HS1], [HS2]). Its type was $(1, 1, 1, 1, 1, \sqrt{2}-1)$. The story of this discovery is noteworthy. In the summer of 1993 I was visiting my Moscow cousin Leonid Hoffman, a well-known New Vienna School composer. His 15-year-old son Ilya studied violin at the Gnesin's Music High School. Ilya set out to find out what I was doing in mathematics, and did not accept any general answers. He wanted particulars. I showed him my 6-coloring (Problem 6.4), and Ilya got busy. The very next day he showed me the Stechkin coloring (Fig. 6.1) that he discovered on his own! "Great," I replied, "but you are 23 years too late." A few days later, he came up with a new idea of using a 2-square tiling. Ilya had an intuition of a virtuoso fiddler and no mathematical culture—I calculated the sizes the squares had to have for the 6-coloring to do the job we needed, and the joint work of the unusual musician–mathematician team was born. Today at 30, Ilya has completed the graduate school of Moscow Conservatory in the class of the celebrated violist and conductor Yuri Bashmet, and is now one of Russia's hottest violists and a winner of several international competitions.

Problem 7.1 (I. Hoffman and A. Soifer [HS1], [HS2]) There is a 6-coloring of the plane of type $(1, 1, 1, 1, 1, \sqrt{2} - 1)$.

Solution: We tile the plane with squares of diagonals 1 and $\sqrt{2} - 1$ (Fig. 7.1). We use colors $1, \ldots, 5$ for larger squares, and color 6 for all smaller squares. With each square we include half of its boundary, the left and lower sides, without the endpoints of this half (bold lines in Fig. 7.2).

To easily verify that this coloring does the job, observe the unit of the construction that is bounded by the bold line in Fig. 7.1; its translates tile the plane. ∎

The two examples, found in solutions of Problems 6.4 and 7.1 prompted me in 1993 to introduce a new terminology for this problem, and to translate the results and problems into this new language.

Open Problem 7.2 (A. Soifer [Soi7], [Soi8]) Find the *6-realizable set* X_6 of all positive numbers α such that there exists a 6-coloring of the plane of type $(1, 1, 1, 1, 1, \alpha)$.

A. Soifer, *The Mathematical Coloring Book,*
DOI 10.1007/978-0-387-74642-5_7, © Alexander Soifer 2009

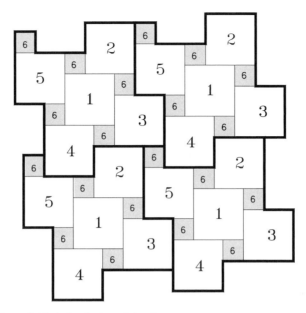

Fig. 7.1 Hoffman-Soifer's 6-coloring of the plane

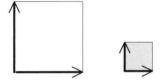

Fig. 7.2

In this new language, the results of Problems 6.4 and 7.1 can be written as follows:

$$\frac{1}{\sqrt{5}}, \sqrt{2} - 1 \in X_6.$$

Now we have two examples of "working" 6-colorings. But what do they have in common? It is not obvious, is it? After a while I realized that they were two extreme examples of the general case, and in fact a much better result was possible, describing a whole continuum of "working" 6-colorings!

Theorem 7.3 (A. Soifer [Soi7], [Soi8])

$$\left[\sqrt{2} - 1, \frac{1}{\sqrt{5}}\right] \subseteq X_6,$$

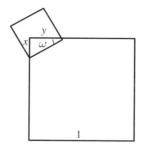

Fig. 7.3

i.e., for every $\alpha \in \left[\sqrt{2} - 1, \frac{1}{\sqrt{5}} \right]$ there is a 6-coloring of type $(1, 1, 1, 1, 1, \alpha)$.[24]

Proof Let a unit square be partly covered by a smaller square, which cuts off the unit square vertical and horizontal segments of lengths x and y respectively, and forms with it an angle ω (Fig. 7.3). These squares induce the tiling of the plane that consists of non-regular octagons congruent to each other and "small" squares (Fig. 7.4).

Now we are ready to color this tiling in 6 colors. Denote by F the unit of our construction, bounded by a bold line (Fig. 7.4) and consisting of 5 octagons and 5 "small" squares. Use colors 1 through 5 for the octagons inside F and color 6 for all "small" squares. Include in the colors of octagons and "small" squares the parts of their boundaries that are shown in bold in Fig. 7.5. Translates of F tile the plane and thus determine the 6-coloring of the plane. We now wish to select parameters to guarantee that each color forbids a distance.

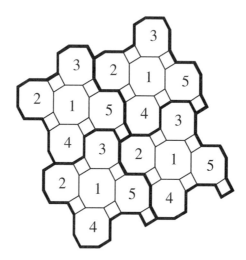

Fig. 7.4

---

[24] Symbol $[a, b]$, $a < b$, as usual, stands for the line segment, including its endpoints a and b.

Fig. 7.5

At first, the complexity of computations appeared unassailable to me. However, a true Math Olympiad approach (i.e., good choices of variables, clever substitutions, and nice optimal properties of the chosen tilings) allowed for successful sailing.

Let $x \le y$ (Fig. 7.3). It is easy to see (Figs. 7.6 and 7.7) that we can split each "small" square into four congruent right triangles with legs x and y and a square of side $y - x$.

The requirement for each color to forbid a distance produces the following system of two inequalities (Fig. 7.6):

$$\begin{cases} d_1 \ge d_2 \\ d_3 \ge d_4 \end{cases} \tag{7.1}$$

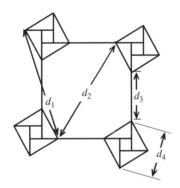

Fig. 7.6

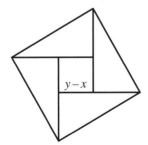

Fig. 7.7

Figures 7.6 and 7.7 allow for an easy representation of all d_i, $(i = 1, 2, 3, 4)$ in terms of x and y. As a result, we get the following system of inequalities:

$$\left.\begin{aligned} \sqrt{(1 + y - x)^2 + (2x)^2} &\geq \sqrt{1 + (1 - 2x)^2} \\ 1 - x - y &\geq \sqrt{2\left(x^2 + y^2\right)} \end{aligned}\right\} \tag{7.2}$$

Solving for x each of the two inequalities (7.2) separately, we unexpectedly get the following system:

$$\left.\begin{aligned} x^2 + 2(1 - y)x + \left(y^2 + 2y - 1\right) &\geq 0 \\ x^2 + 2(1 - y)x + \left(y^2 + 2y - 1\right) &\leq 0 \end{aligned}\right\}$$

Therefore, we get an equation (!) in x and y:

$$x^2 + 2(1 - y)x + \left(y^2 + 2y - 1\right) = 0.$$

Treating this as the equation in x, we obtain a *unique* (!) solution for x as a function of y that satisfies the system (7.2) of inequalities:

$$x = \sqrt{2 - 4y} + y - 1, \quad \text{where } 0 \leq y \leq 0.5. \tag{7.3}$$

Since $0 \leq x \leq y$, we get even narrower bounds for y : $0.25 \leq y \leq \sqrt{2} - 1$. For any value of y within these bounds, x is uniquely determined by (7.3) and is accompanied by the *equalities* (!) $d_1 = d_2$ and $d_3 = d_4$.

Thus, we have showed that for every $y \in [0.25, \sqrt{2} - 1]$ there is a 6-coloring of type $(1, 1, 1, 1, 1, \alpha)$. But what values can α take on? Surely,

$$\alpha = \frac{d_4}{d_2}. \tag{7.4}$$

Let us introduce a new variable $Y = \sqrt{2 - 4y}$, where $Y \in [2 - \sqrt{2}, 1]$, i.e., $4y = -Y^2 + 2$, and figure out x from (7.3) as a function of Y:

$$\left.\begin{aligned} 4y &= -Y^2 + 2 \\ 4x &= -Y^2 + 4Y - 2 \end{aligned}\right\} \tag{7.5}$$

Now substituting from (7.1) and (7.2) the expressions for d_4 and d_2 into (7.4) and using (7.5) to get rid of x and y everywhere, we get a "nice" expression for α^2 as a function of Y (do verify my algebraic manipulations on your own):

$$\alpha^2 = \frac{Y^4 - 4Y^3 + 8Y^2 - 8Y + 4}{Y^4 - 8Y^3 + 24Y^2 - 32Y + 20}.$$

By substituting $Z = Y - 2$, where $Z \in \left[-\sqrt{2}, -1\right]$, we get a simpler function α^2 of Z:

$$\alpha^2 = 1 + \frac{4Z(Z^2 + 2Z + 2)}{Z^4 + 4}.$$

To observe the behavior of the function α^2, we compute its derivative:

$$(\alpha^2)' = -\frac{4}{(Z^4 + 4)^2}(Z^6 + 4Z^5 + 6Z^4 - 12Z^2 - 16Z - 8).$$

Normally there is nothing promising about finding exact roots of an algebraic polynomial of degree greater than 4. But we are positively lucky here, for this sixth degree polynomial can be nicely decomposed into factors:

$$(\alpha^2)' = -\frac{4}{(Z^4 + 4)^2}(Z^2 - 2)\left[(Z + 1)^2 + 1\right]^2.$$

Hence, the derivative has only two zeros. In fact, in the segment of our interest $\left[-\sqrt{2}, -1\right]$, the only extremum of α^2 occurs when $Z = -\sqrt{2}$. Going back from Z to Y to y, we see that on the segment $y \in \left[0.25, \sqrt{2} - 1\right]$ the function $\alpha = \alpha(y)$ decreases from $\alpha = \frac{1}{\sqrt{5}} \approx 0.44721360$ (i.e., 6-coloring of problem 6.4) to $\alpha = \sqrt{2} - 1 \approx 0.41421356$ (i.e., 6-coloring of Problem 7.1). Since the function $\alpha = \alpha(y)$ is continuous and increasing on $\left[0.25, \sqrt{2} - 1\right]$, it takes on *each* intermediate value from the segment $\left[\sqrt{2} - 1, \frac{1}{\sqrt{5}}\right]$, and only *once*.

We have proved the required result, and much more:

For every angle ω between the small and the large squares (Fig. 7.3), there are— and unique—sizes of the two squares (and unique squares intersection parameters x and y), such that the constructed 6-coloring has type $(1, 1, 1, 1, 1, \alpha)$ for a uniquely determined α.

This is a remarkable fact: the "working" solutions barely exist—they comprise something of a curve in a three-dimensional space of the angle ω and two linear variables x and y! We have thus found a continuum of permissible values for α and a continuum of "working" 6-colorings of the plane. ∎

Remark: The problem of finding the 6-realizable set X_6 has a close relationship with the problem of finding the chromatic number χ of the plane. Its solution would shed light—if not solve—the chromatic number of the plane problem:

$$\text{if } 1 \notin X_6, \text{ then } \chi = 7;$$
$$\text{if } 1 \in X_6, \text{ then } \chi \leq 6.$$

I am sure you understand that problem 7.2, formulated in just two words, is extremely difficult.

In 1999, the Russian authorities accused my coauthor and great young musician Ilya Hoffman of computer hacking and imprisoned him before the trial as "danger to the society." I flew to Moscow, met with the presiding judge, met with and received support from members of the Russian Parliament "Duma," human rights leaders, Vice President of the Russian Academy of Sciences and the celebrated jurist Vladimir Nikolaevich Kudriavtsev.[25] When the trial came, Ilya was released home from the courtroom. While in prison, he was not allowed to play viola, so Ilya wrote music and mathematics. This page he sent to me from his cell (page 56):

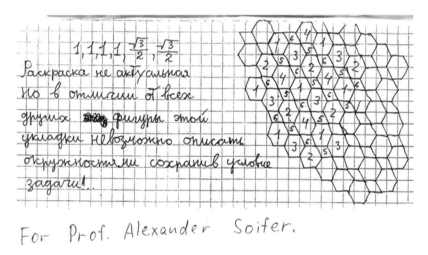

Ilya discovered a new 6-coloring of the plane. Four colors consist of regular hexagons of diameter 1, and two colors occupy rhombuses. By carefully assigning colors to the boundaries, we get the 6-coloring of type

$$(1, 1, 1, 1, \frac{\sqrt{3}}{2}, \frac{\sqrt{3}}{2}).$$

---

[25] In 1951, Stalin's Prosecutor General Vyshinskii announced a new legal doctrine: "one is guilty whom the court finds guilty." The presumption of innocence he called "bourgeois superstition." The young senior lieutenant rose to speak against the new Stalin's doctrine. This amazing hero was V. N. Kudryavtsev. It was unforgettable to meet this heroic man and get his full support.

8
Chromatic Number of the Plane in Special Circumstances

As you know from Chapters 4 and 6, in 1973, 3 years after Dmitry E. Raiskii, Douglas R. Woodall published the paper [Woo1] on problems related to the chromatic number of the plane. In the paper he gave his own proofs of Raiskii's inequalities of Problems 4.1 and 6.1. In the same paper, Woodall also formulated and attempted to prove a lower bound for the chromatic number of the plane for the special case of map-type coloring of the plane. This was the main result of [Woo1]. However, in 1979 the mathematician from the University of Aberdeen Stephen Phillip Townsend found an error in Woodall's proof, and constructed a counterexample demonstrating that one essential component of Woodall's proof was false. Townsend had also found a proof of this statement, which was much more elaborate than Woodall's unsuccessful attempt.

The intriguing history of this discovery and Townsend's wonderful proof are a better fit in Chapter 24, as a part of our discussion of map coloring—do not overlook them! Here I will formulate an important corollary of Townsend's proof.

Chromatic Number of Map-Colored Plane 8.1 The chromatic number of the plane under map-type coloring is 6 or 7.

Woodall showed that this result implies one more meritorious statement:

Closed Chromatic Number of the Plane 8.2 ([Woo1]). The chromatic number of the plane under coloring with closed monochromatic sets is 6 or 7.

I do not like to use the Greek word "lemma" since there is an appropriate English word "tool" :-). And I would like to offer my readers the following tool from topology to prove on their own. We will use this tool in the proof that follows.

Tool 8.3 If a bounded closed set S does not realize a distance d, then there is $\varepsilon > 0$ such that S does not realize any distance from the segment $[d - \varepsilon, \ d + \varepsilon]$.

Proof of Result 8.2 [Woo1]: Assume that the union of closed sets $A_1, \ A_2, \ldots, \ A_n$ covers the plane and for each i the set A_i does not realize a distance d_i. Place onto the plane a unit square lattice L, and choose an arbitrary closed unit square U of L. Choose also i from the set $\{1, \ 2, \ldots, \ n\}$. Denote by $C(U)_i$ the closed set that contains all points of the plane that are at most distance d_i from a point in U. The set

A. Soifer, *The Mathematical Coloring Book*,
DOI 10.1007/978-0-387-74642-5_8, © Alexander Soifer 2009

$A_i \cap C(U)_i$ is closed and bounded, thus by Tool 8.3 there is $\varepsilon_i(U)$ such that no two points of A_i, at least one of which lies in U, realize any distance from the segment

$$[d_i - \varepsilon_i(U), d_i + \varepsilon_i(U)]. \tag{8.1}$$

Denote by $\varepsilon(U)$ the minimum of $\varepsilon_i(U)$ over all $i = 1, 2, \ldots, n$.

Now for the square U we choose a positive integer $m(U)$ such that

$$\frac{1}{2^{m(U)}}\sqrt{2} < \frac{1}{2}\varepsilon(U). \tag{8.2}$$

On the unit square U we place a square lattice L' of little closed squares u of side $\frac{1}{2^{m(U)}}$. The inequality (8.2) guarantees that the diagonal of u is shorter than half of our epsilon $\varepsilon(U)$.

For each little square u contained in each unit square U of the entire plane, we determine $f(u) = \min\{i : u \cap A_i \neq \emptyset\}$, and then for each $i = 1, 2, \ldots, n$ define the monochromatic color set of our new n-coloring of the plane as follows:

$$B_i = \bigcup_{f(u)=i} u. \tag{8.3}$$

As unions of closed squares u, each B_i is closed, and all B_i together cover the plane. The interiors of these n sets B_i are obviously disjoint. All there is left to prove is that the set B_i does not realize the distance d_i. Indeed, assume that the points b, c of B_i are distance d_i apart. The points b, c belong to little squares u_1, u_2 respectively, each little square of side $\frac{1}{2^{m(U)}}$. Due to the definition (8.3) of B_i, the squares u_1, u_2 contain points a_1, a_2 from A_i respectively. With vertical bars denoting the distance between two points, and by utilizing the inequality (8.2) we get:

$$|b, c| - \varepsilon(U) < |a_1, a_2| < |b, c| + \varepsilon(U),$$

i.e.,

$$d_i - \varepsilon(U) < |a_1, a_2| < d_i + \varepsilon(U),$$

which contradicts (8.1).

Thus, the chromatic number under the conditions of result 8.2 is not smaller than the chromatic number under the conditions of result 8.1. ∎

During 1993–1994 a group of three young undergraduate students Nathanial Brown, Nathan Dunfield, and Greg Perry, in a series of three essays, (their first publications,) proved on the pages of *Geombinatorics* [BDP1], [BDP2], [BDP3][26] that a similar result is true for coloring with open monochromatic sets. Now the youngsters are professors of mathematics, Nathan at the University of Illinois at Urbana-Champaign, and Nathanial at Pennsylvania State University.

Open Chromatic Number of the Plane 8.4 (Brown–Dunfield–Perry). The chromatic number of the plane under coloring with open monochromatic sets is 6 or 7.

---

[26] The important problem book [BMP] mistakenly cites only one of these series of three papers. It also incorrectly states that the authors proved only the lower bound 5, whereas they raised the lower bound to 6.

9
Measurable Chromatic Number of the Plane

9.1 Definitions

As you know, the *length* of a segment $[a,b]$, $a < b$, on the line E^1 is defined as $b - a$. *Area* A of a rectangle $R = [a_1, b_1] \times [a_2, b_2]$, $a_i < b_i$ on the plane E^2 is defined as $A = (b_1 - a_1)(b_2 - a_2)$. The French mathematician Henri Léon Lebesgue (1875–1941) generalized the notion of area to a vast class of plane sets. In place of area, he used the term *measure*. For a set S in the plane, we define its *outer measure* $\mu^*(S)$ as follows:

$$\mu^*(S) = \inf \sum_i A(R_i), \qquad (9.1)$$

with the infimum taken over all coverings of S by a countable sequence $\{R_i\}$ of rectangles. When the infimum exists, S is said to be *Lebesgue-measurable* or – since we consider here no other measures—*measurable* set—if for any set B in the plane, $\mu^*(B) = \mu^*(B \cap S) + \mu^*(B \backslash S)$. For a measurable set S, its measure is defined by $\mu(S) = \mu^*(S)$.

Any rectangle is measurable, and its measure coincides with its area. It is shown in every measure theory text that all closed sets and all open sets are measurable. Giuseppe Vitali (1875–1932) was first to show that in the standard system of axioms ZFC for set theory (Zermelo–Fraenkel system plus the Axiom of Choice), there are non-measurable subsets of the set R of real numbers.

We will use the same definition (9.1) for Lebesgue measure on the line E^1, when the infimum is naturally taken over all covering sequences $\{R_i\}$ of segments. For measure of S on the line we will use the symbol $l(S)$. Generalization of the notion of measure to n-dimensional Euclidean space E^n is straight forward; here we will use the symbol $\mu_n(S)$. In particular, for $n = 2$, we will omit the subscript and simply write $\mu(S)$.

9.2 Lower Bound for Measurable Chromatic Number of the Plane

While a graduate student in Great Britain, Kenneth J. Falconer proved the following important result [Fal]:

A. Soifer, *The Mathematical Coloring Book*,
DOI 10.1007/978-0-387-74642-5_9, © Alexander Soifer 2009

Falconer's Theorem 9.1 Let $R^2 = \bigcup\limits_{i=1}^{4} A_i$ be a covering of the plane by four disjoint measurable sets. Then one of the sets A_i realizes distance 1.

In other words, the measurable chromatic number χ_m of the plane is equal to 5, 6, or 7.

I found his 1981 publication [Fal1] to be too concise and not self-contained for the result that I viewed as very important. Accordingly, I asked Kenneth Falconer, currently a professor and dean at the University of St. Andrews in Scotland, for a more detailed and self-contained exposition. In February 2005, I received Kenneth's manuscript, hand-written especially for this book, which I am delighted to share with you.

Before we prove his result, we need to get armed with some basic definitions and tools of the measure theory.

A non-empty collection $\beth$ of subsets of E^2 is called σ-*field*, if $\beth$ is closed under taking complements and countable unions, i.e.,

> *) if $A \in \beth$, then $E^2 \backslash A \in \beth$; and
>
> **) if $A_1, A_2, \ldots, A_n, \ldots \in \beth$, then $\bigcup\limits_{i=1}^{\infty} A_i \in \beth$.

Exercise 9.2 Show that any σ-field $\beth$ is closed under countable intersection and set difference. Also, show that $\beth$ contains the empty set $\emptyset$ and the whole space E^2.

It is shown in all measure theory textbooks that the collection of all measurable sets is a σ-field. The intersection of all σ-fields containing the closed sets is a σ-field containing the closed sets, the minimal such σ-field with respect to inclusion. Its elements are called *Borel sets*. Since closed sets are measurable and the collection of all measurable sets is a σ-field, it follows that all Borel sets are measurable.

(Observe that in place of the plane E^2 we can consider the line E^1 or an n-dimensional Euclidean space E^n, and define their Borel sets.)

The following notations will be helpful:

> $C(x, r)$ – Circle with center at x and radius r;
> $B(x, r)$ – Circular disk (or ball) with center at x and radius r.

For a measurable set S and a point x, we define the *Lebesgue density*, or simply *density*, of S at x as follows:

$$D(S, x) = \lim_{x \to 0} \frac{\mu(S \cap B(x, r))}{\mu(B(x, r))},$$

where $\mu(B(x, r))$ is, of course, equal to πr^2.

Lebesgue Density Theorem (LDT) 9.3 For a measurable set $S \subset E^2$, the density $D(S, x)$ exists and equals 1 if $x \in S$ and 0 if $x \in R^2 \backslash S$, except for a set of points x of measure 0.

For a measurable set A, denote

$$\tilde{A} = \{x \in A : D(A, x) = 1\}.$$

Then due to LDT, we get $\mu(\tilde{A} \triangle A) = 0$, i.e., $\tilde{A}$ is 'almost the same' as A.[27] Observe also that $\mu(S \cap B(x, r))$ is a continuous function of x for $r > 0$; therefore, $\tilde{A}$ is a Borel set.

We will define the *density boundary* of a set A as follows:

$$\partial A = \{x : D(A, x) \neq 0, 1 \text{ or does not exist}\}.$$

By LDT,

$$\mu(\partial A) = 0.$$

You can find on your own or read in [Cro] the proof of the following tool:

Tool 9.4 For a measurable set $A \subset R^2$, such that both $\mu(A) > 0$ and $\mu(R^2 \setminus A) > 0$, we have $\partial A \neq \emptyset$.

Tool 9.5 If $R^2 = \bigcup_{i=1}^{4} A_i$ is a covering of the plane by four disjoint measurable sets, then $\bigcup_{i=1}^{4} \tilde{A}_i$ is a disjoint union with the complement $\mathfrak{M} \equiv \bigcup_{i=1}^{4} \partial A_i$.

Proof follows from Tool 9.4 and the observation that if $x \in \partial A_i$ then also $x \in \partial A_j$ for some $j \neq i$. ∎

The next tool claims the existence of two concentric circles with the common center in $\mathfrak{M}$, which intersect $\mathfrak{M}$ in length 0.

Tool 9.6 Let $\mathfrak{M}$ be as in Tool 9.5; there exists $x \in \mathfrak{M}$ such that

$$l(C(x, 1) \cap \mathfrak{M}) = l\left(C(x, \sqrt{3}) \cap \mathfrak{M}\right) = 0.$$

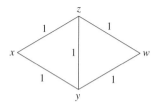

Fig. 9.1

---

[27] Here $A \triangle B$ stands for the symmetric difference of these two sets, i.e., $A \triangle B = (A \setminus B) \cup (B \setminus A)$.

I will omit the proof, but include Falconer's insight: "The point of this lemma is that if we place the "double equilateral triangle" [Fig. 9.1] of side 1 in almost all orientations with a vertex at x, the point x essentially has "2 colors" in any coloring of the plane, and other points just one color. (Note $|xw| = \sqrt{3}$.)"

Tool 9.7 Let $R^2 = \bigcup\limits_{i=1}^{4} A_i$ be a covering of the plane by four disjoint measurable sets, none of which realizes distance 1. Let $x \in \mathfrak{M}$ as in Tool 9.6, say without loss of generality $x \in \partial A_1$ and $x \in \partial A_2$. Then $l\left(C(x, \sqrt{3})\backslash(\tilde{A}_1 \cup \tilde{A}_2)\right) = 0$.

Proof Since $x \in \partial A_1$ and $x \in \partial A_2$, there exists $\varepsilon > 0$ such that

(1) $\varepsilon < \frac{\mu(A_1 \cap B(x,r))}{\pi r^2} < 1 - \varepsilon$ for some arbitrarily small r, and

(2) $\varepsilon < \frac{\mu(A_2 \cap B(x,r))}{\pi r^2} < 1 - \varepsilon$ for some arbitrarily small r.

Consider the diamond (Fig. 9.1) consisting of two unit equilateral triangles xyz and yzw, where x is the point fixed in the statement of this tool, and y, z, $w \notin \mathfrak{M}$ (this happens for almost all orientations of the diamond, by Tool 9.6). Thus suppose $y \in \tilde{A}_{i(y)}$, $z \in \tilde{A}_{i(z)}$, $w \in \tilde{A}_{i(w)}$, where $i(y)$, $i(z)$, $i(w) \in \{1, 2, 3, 4\}$. For sufficiently small r, say $r < r_0$, we get:

(3) $1 - \frac{\varepsilon}{4} < \frac{\mu\left(A_{i(y)} \cap B(y,r)\right)}{\pi r^2} \leq 1$;

(4) $1 - \frac{\varepsilon}{4} < \frac{\mu\left(A_{i(z)} \cap B(z,r)\right)}{\pi r^2} \leq 1$;

(5) $1 - \frac{\varepsilon}{4} < \frac{\mu\left(A_{i(w)} \cap B(w,r)\right)}{\pi r^2} \leq 1$.

We can now choose $r < r_0$ such that (1) holds (as well as (3), (4), (5)). Let v be a vector going from the origin to a point in $B(0, r)$ and consider translation of the diamond x, y, z, w through v, i.e., to the diamond $x + v$, $y + v$, $z + v$, $w + v$. Now (1), (3), (4), (5) imply that

$$\frac{1}{\pi r^2}\mu\left(\{v \in B(0, r) : x + v \in A_1, y + v \in A_{i(y)}, z + v \in A_{i(z)}, w + v \in A_{i(w)}\}\right)$$
$$> \varepsilon - \frac{\varepsilon}{4} - \frac{\varepsilon}{4} - \frac{\varepsilon}{4} > 0.$$

Thus, we can choose $v \in B(0, r)$ such that $x + v \in A_1$, $y + v \in A_{i(y)}$, $z + v \in A_{i(z)}$, $w + v \in A_{i(w)}$. Since by our assumption none of the sets $A_i, i = 1, 2, 3, 4$ realizes distance 1, we conclude (by looking at the translated diamond) that $1 \neq i(y)$, $1 \neq i(z)$, $i(y) \neq i(z)$, $i(z) \neq i(w)$, and $i(w) \neq i(y)$.

The same argument, using (2), (3), (4), (5) produces $2 \neq i(y)$, $2 \neq i(z)$, $i(y) \neq i(z)$, $i(z) \neq i(w)$, and $i(w) \neq i(y)$. Therefore, $i(y)$, $i(z)$ are 3 and 4 in some order, and thus $i(w) = 1$ or 2, i.e., $w \in \tilde{A}_1$ or $w \in \tilde{A}_2$.

By Tool 9.6, this holds for almost every orientation of the diamond. Since $|xw| = \sqrt{3}$, we conclude that for almost all $w \in C(x, \sqrt{3})$, we get $w \in \tilde{A}_1$ or $w \in \tilde{A}_2$. Thus, $l\left(C(x, \sqrt{3})\backslash(\tilde{A}_1 \cup \tilde{A}_2)\right) = 0$, as required. ∎

Tool 9.8 Let C be a circle of radius $r > \frac{1}{2}$ and let E_1, E_2 be disjoint measurable subsets of C such that $l\,(C\backslash(E_1 \cup E_2) = 0$. Then if $\varphi = 2\sin^{-1}\left(\frac{1}{2r}\right)$ is an irrational multiple of π, either E_1 or E_2 contains a pair of points distance 1 apart.

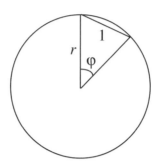

Fig. 9.2

Proof Assume that neither E_1 or E_2 contains a pair of points distance 1 apart. Parameterize C (Fig. 9.2) by angle θ (mod 2π).

Let $l(E_1) > 0$, then by LDT, there is θ and $\varepsilon > 0$ such that

$$l(E_1 \cap (\theta - \varepsilon, \theta + \varepsilon)) > \frac{3}{4}2\varepsilon.$$

Let θ_1 be an angle. Since φ is an irrational multiple of π, there is a positive integer n such that

$$|\theta_1 - (2n\varphi + \theta)| < \frac{1}{4}\varepsilon \text{ (mod } 2\pi).$$

Since neither E_1 or E_2 contain a pair of points distance 1 apart, we get (with angles counted mod 2π):

$$l\,(E_1 \cap (\theta + k\varphi - \varepsilon, \theta + k\varphi + \varepsilon)) = l\,(E_1 \cap (\theta - \varepsilon, \theta + \varepsilon)) \text{ for even } k, \text{ and}$$
$$l\,(E_1 \cap (\theta + k\varphi - \varepsilon, \theta + k\varphi + \varepsilon)) = 2\varepsilon - l\,(E_1 \cap (\theta - \varepsilon, \theta + \varepsilon)) \text{ for odd } k.$$

In particular, $l\,(E_1 \cap (\theta + 2n\varphi - \varepsilon, \theta + 2n\varphi + \varepsilon)) > \frac{3}{4}2\varepsilon$, thus

$$l\,(E_1 \cap (\theta_1 - \varepsilon, \theta + \varepsilon)) > \frac{3}{4}2\varepsilon - \frac{\varepsilon}{4} - \frac{\varepsilon}{4} = \varepsilon.$$

Hence for all θ_1,

$$\frac{l\,(E_1 \cap (\theta_1 - \varepsilon, \theta + \varepsilon))}{2\varepsilon} \geq \frac{1}{2},$$

and by LDT $l\left(C\backslash E_1\right) = 0$. This means that E_1 is almost all of C, and therefore contains a pair of points until distance apart, a contradiction. ∎

Surprisingly, we need a tool from abstract algebra, or number theory.

Tool 9.9 For any positive integer m, $\left(1 - i\sqrt{11}\right)^{2m} \neq (-12)^m$.

Proof It suffices to note that $Q\left(\sqrt{-11}\right)$ is an Euclidean quadratic field, therefore, its integer ring $Z\left(\sqrt{-11}\right)$ (with units $+1/ - 1$) has unique factorization. (See Chapters 7 and 8 in the standard abstract algebra textbook [DF] for a proof).

I believe that an alternative proof is possible: it should be not hard to show that the left side cannot be an integer for any m. ∎

Now we are ready to prove Falconer's Theorem 9.1.

Proof of Falconer's Theorem 9.10 Let $R^2 = \bigcup_{i=1}^{4} A_i$ be a covering of the plane by four disjoint measurable sets, none of which realizes distance 1. Due to Tool 9.6, there is $x \in \mathfrak{M}$ such that $l\left(C(x, \sqrt{3})\backslash(\tilde{A}_1 \cup \tilde{A}_2)\right) = 0$. Taking $E_1 = \tilde{A}_1$, $E_2 = \tilde{A}_2$ and $r = \sqrt{3}$, we get, the desired result by Tool 9.8—if only we can prove that $\varphi = \sin^{-1}\left(\frac{1}{2\sqrt{3}}\right)$ is an irrational multiple of π. We have $\sin\theta = \frac{1}{2\sqrt{3}}$; $\cos\theta = \frac{\sqrt{11}}{2\sqrt{3}}$. Assume $m\theta$ is an integer multiple of 2π for some integral $2m$. Then

$$\left(\frac{\sqrt{11}}{2\sqrt{3}} + i\frac{1}{2\sqrt{3}}\right)^{2m} = 1$$

or

$$\left(1 - i\sqrt{11}\right)^{2m} = (-12)^m.$$

We are done, as the last equality contradicts Tool 9.9. ∎

9.3 Kenneth J. Falconer

I am always interested in learning about the life and personality of the author whose result impressed me, aren't you! Accordingly, I asked Kenneth to tell me about himself and his life. The following account comes from his September 30, 2005, e-mail to me.

> I was born on 25th January 1952 at Hampton Court on the outskirts of London (at a maternity hospital some 100 metres from the gates of the famous Palace). This was two weeks before Queen Elizabeth II came to the throne and when food rationing was still in place. My father had served in India for 6 years during the war while my mother brought up my brother, 12 years my senior, during the London blitz. My parents

were both school teachers, specializing in English, my brother studied history before becoming a Church of England minister, and I was very much the 'black sheep' of the family, having a passionate interest in mathematics and science from an early age...

I gained a scholarship to Corpus Christi College, Cambridge to read mathematics and after doing well in the Mathematical Tripos I continued in Cambridge as a research student, supervised by Hallard Croft. I worked mainly on problems in Euclidean geometry, particularly on convexity and of tomography (the mathematics of the brain scanner) and obtained my PhD in 1977.

I had the good fortune to obtain a Research Fellowship at Corpus Christi College, where I continued to study geometrical problems, including the fascinating problem of the chromatic number of the plane, showing in particular that the chromatic number of a measurable colouring of the plane was at least 5. Also around this time I worked on generalizations of the Kakeya problem (the construction of plane sets of zero area containing a line segment in every direction). Thus I encountered Besicovitch's beautiful idea of thinking of such sets as duals of what are now termed 'fractals', with directional and area properties corresponding to certain projections of the fractals. This led to my 'digital sundial' construction – a subset of R^3 with prescribed projections in (almost) all directions...

In 1980 I moved to Bristol University as a Lecturer, where the presence of theoretical physicist Michael Berry, and analyst John Marstrand were great stimulii. Here I started to work on geometric measure theory, or fractal geometry, in particular looking at properties of Hausdorff measures and dimensions, and projections and intersections of fractals...

It became clear to me that much of the classical work of Besicovitch and his School on the geometry of sets and measures had been forgotten, and in 1985 I published my first book 'The Geometry of Fractal Sets' to provide a more up to date and accessible treatment. This was around the time that fractals were taking the world by storm, following Mandelbrot's conceptually foundational work publicised in his book 'The Fractal Geometry of Nature' which unified the mathematics and the scientific applications of fractals. My book led to requests for another at a level more suited to postgraduate and advanced undergraduate students and in 1990 I published 'Fractal geometry – Mathematical Foundations and Applications' which has been widely used in courses and by researchers, and has been referred to at conferences as 'the book from which we all learnt our fractal mathematics'. A sequel 'Techniques in Fractal Geometry' followed in 1998. In collaboration with Hallard Croft and Richard Guy, I also authored 'Unsolved Problems in Geometry', a collection of easy to state unsolved geometrical problems. Happily (also sadly!) many of the problems in the book are no longer unsolved!...

In 1993 I was appointed Professor at the University of St Andrews in Scotland, where I have been ever since. Although St Andrews is a small town famous largely for its golf, the University has a thriving mathematics department, in particular for analysis and combinatorial algebra, to say nothing of its renowned History of Mathematics web site. I became Head of the School of Mathematics and Statistics in 2001, with the inevitable detrimental effect on research time. I was elected a Fellow of the Royal Society of Edinburgh in 1998, and to the Council of the London Mathematical Society in 2000...

My main leisure activity is long distance walking and hillwalking. I have climbed all 543 mountains in Britain over 2500 feet high. I am a keen member of the Long Distance Walkers Association, having been Editor of their magazine 'Strider' from 1986–91 and Chairman from 2000–03. I have completed the last 21 of the LDWA's annual hundred mile non-stop cross-country walks in times ranging from 26 to 32 hours.

10
Coloring in Space

When in 1958 Paul Erdős learned about the chromatic number of the plane problem, he created a number of related problems, some of which we have discussed in the preceding chapters. Paul also generalized the problem to n-dimensional Euclidean space R^n. On October 2, 1991 I received a letter from him, which contained a historical remark [E91/10/2ltr]:[28]

> I certainly asked for the chromatic number of $E^{(n)}$ long ago (30 years).

He was interested in both asymptotic behavior as n increased, and in the exact values of the chromatic number for small n, first of all $n = 3$.

As we have already discussed in Chapter 4, in 1970 Dmitry E. Raiskii [Rai] proved the lower bound for n-dimensional Euclidean spaces.

Raiskii's Lower Bound 10.1 (Raiskii, 1970).

$$n + 2 \leq \chi\left(E^n\right).$$

For $n = 3$ this, of course, gives $5 \leq \chi\left(E^3\right)$. This lower bound for the three-dimensional space had withstood 30 years, until in 2000 Oren Nechushtan of Tel Aviv University improved it (and published 2 years later [Nec]):

Best Known Lower Bound for R^3 10.2 (Nechushtan, 2000).

$$6 \leq \chi\left(E^3\right).$$

The obvious upper bound of 27 for the chromatic number of 3-dimensional space was reduced to 21 (it is proved in [Cou1], where credit is given to *this* book; but of course, I had nothing to do with it). Then the time had come for David Coulson of Melbourne University, who reduced the upper bound to 18 [Cou1]. Pay attention to the dates, as it seems Coulson's papers are slow to appear in print. The upper bound of 18 was first submitted in 1993 to the *Transaction of the American Mathematical Society* (on September 27, 1993 I received e-mail from Coulson to that effect). Then (I assume due to lack of interest in the *Transactions* for this kind of mathematics)

---

[28] Curiously, Paul wrote an improbable date on the letter: "1977 VII 25".

A. Soifer, *The Mathematical Coloring Book*,
DOI 10.1007/978-0-387-74642-5_10, © Alexander Soifer 2009

Coulson submitted it to *Discrete Mathematics* on April 24, 1995; he revised the paper on August 30, 1996, and finally published it in 1997 [Cou1], 4 years after the initial submission.

Coulson then achieved a truly amazing improvement: he obtained the upper bound of 15 by using face-centered cubic lattice (see Conway and Sloane [CS] for more about 3-dimesional lattices). The upper bound of 15 also took 4 years to appear in print. It was submitted to *Discrete Mathematics* on December 9, 1998. A month later I received this manuscript to referee under number DM 9298. Amazingly, a copy of my February 27, 1999 report survives. I suggested five stylistic improvements, and wrote:

> I found the main result to be of high importance to the field. Indeed, Coulson has dramatically improved his own previous bound of 18 by proving that 15 is an upper bound of the chromatic number of the 3-space. He conjectures that 15 is the best possible upper bound if one uses a lattice based coloring. His argument in favor of this conjecture is good, and we would encourage the author to pursue the proof...
>
> The author hints that his methods may produce similar results in other dimensions. Again, the referee would encourage the pursuit of these results.

I am at a loss to explain why the revised manuscript was received by the editors only about 2 years later, on December 11, 2000. While writing these lines, I am looking at the uncorrected proofs that I received from the author—they are dated to 2001. The paper was published much later yet, in 2002 [Cou2].

In August 2002, David Coulson and I played a very unusual role at the Congress of the World Federation of National Mathematics Competitions in Melbourne: we were co-presenters of an 80-minute plenary talk, entitled *50 Years of Chromatic Number of the Plane* (we did not sing a duet but rather spoke one at a time). I spoke about the problem, its history and results for the plane. In his part, David spoke about his results on upper bounds of the chromatic number of the 3-space. After the talk, I invited David to submit a version of his part of the talk to *Geombinatorics*, where it appeared very quickly, in January issue of 2003 [Cou3].

Best Upper Bound for R^3 10.3 (Coulson, 1998–2002).

$$\chi\left(E^3\right) \leq 15.$$

Curiosity surrounding this result did not end with its publication. It was published again in 2003 by another pair of authors, Rados Radoicic and Géza Tóth [RT]. By the time they received the proofs, the authors saw the Coulson's publication. They added it to the bibliography, and chose to publish their proof based on the same tiling of 3-space. In a copy of this paper downloaded from an author's homepage I read:

> Very recently, Coulson [Cou2] also [sic] proved our [sic] Theorem, moreover, he found essentially the same coloring.

The comment in the published journal version was fairer toward Coulson:

> *Added in proof.* Very recently, Coulson [Cou2] has independently found a very similar 15-coloring of 3-space.

I do not agree with the characterization "very recently," for Coulson first submitted his paper quite a bit earlier, on December 9, 1998. Yet I have no doubts in my mind that Radoicic–Tóth found their proof independently and before reading Coulson's proof. I believe this is a borderline case as far as credit is concerned. I am assigning credit to Coulson alone for the following reasons:

1. Radoicic and Tóth saw Coulson's publication before they received their proofs;
2. Their proof is not essentially different from Coulson's;
3. Coulson first submitted his paper many years prior, in 1998;
4. Coulson circulated his preprint fairly widely ever since 1998 (I was one of the recipients).

As I mentioned in my referee report, Coulson informally conjectured that the upper bound of 15 is best possible for lattice coloring. I dare to conjecture much more: I think it is the exact value for 3-space every bit as likely as 7 is for the plane:

Chromatic Number of 3-Space Conjecture 10.4

$$\chi\left(E^3\right) = 15.$$

Life in 4 and 5 dimensions was studied by Kent C antwell in his 1996 work [Can1]. His lower bounds are still best known today.

Best Lower Bounds for E^4 and E^5 10.5 (Cantwell, 1996).

$$\chi\left(E^4\right) \geq 7;$$
$$\chi\left(E^5\right) \geq 9.$$

On March 31, 2008, a month after this book had been submitted to Springer, I received an impressive submission to *Geombinatorics* from Josef Cibulka of Charles University in Prague. His main result offered the new lower bound for the chromatic number of E^6:

Best Lower Bounds for E^6 10.6 (Cibulka, 2008).

$$\chi\left(E^6\right) \geq 11.$$

In reply to my inquiry, Josef answered on April 1, 2008:

> I am first year graduate student; actually, most results of the submitted paper are from my diploma thesis.

Do not miss Cibulka's paper in *Geombinatorics*: it will appear in issue XVIII(2) in October 2008. Other results of this paper are a better fit in the next chapter.

A long time ago Paul Erdős conjectured, and often mentioned in his problem talks [E75.24], [E75.25], [E79.04], [E80.38], [E81.23], [E81.26] that the chromatic number $\chi\left(E^n\right)$ of the Euclidean n-space E^n grows exponentially in n. In his words:

Erdős's Conjecture on Asymptotic Behavior of the Chromatic Number of E^n
10.7 $\chi\left(E^n\right)$ tends to infinity exponentially.

This conjecture was settled in the positive by a set of two results, the 1972 exponential upper bound, found by D. G. Larman and C. A. Rogers, and the 1981 exponential lower bound established by P. Frankl and R. M. Wilson:

Frankl–Wilson's Asymptotic Lower Bound 10.7 (Frankl and Wilson,1981, [FW])

$$(1 + o\,(1))\,1.2^n \leq \chi\left(E^n\right).$$

Larman–Rogers' Asymptotic Upper Bound 10.8 (1972, [LR])

$$\chi\left(E^n\right) \leq (3 + o\,(1))^n.$$

Asymptotically Larman–Rogers' upper bound remains best possible still today. Frankl–Wilson's Asymptotic Lower Bound has recently been improved:

Raigorodskii's Asymptotic Lower Bound 10.9 (2000, [Raig2])

$$(1.239\ldots + o\,(1))^n \leq \chi\left(E^n\right).$$

Obviously, there is a gap between the lower and upper bounds, and it would be very desirable to narrow it down.

In Chapter 4 you have met the *polychromatic number* χ_p of the plane and in Chapters 4, 6 and 7 seen the results. This notion naturally generalizes to the *polychromatic number* $\chi_p\left(E^n\right)$ of Euclidean n-dimensional space E^n. Dmitry E. Raiskii, whom we met in Chapter 4, was first to publish a relevant result [Rai]:

Raiskii's Lower Bound 10.10

$$n + 2 \leq \chi_p\left(E^n\right).$$

Larman and Rogers [LR] proved the same asymptotic upper bound for the polychromatic number, as they did for the chromatic number:

Larman–Rogers Upper Bound 10.11

$$\chi_p\left(E^n\right) \leq (3 + o\,(1))^n.$$

They also conjectured that $\chi_p\left(E^n\right)$ grows exponentially in n. The positive proof of the conjecture, started by Larman and Rogers, was completed by Frankl and Wilson [FW]:

Frankl–Wilson Lower Bound 10.12

$$(1 + o\,(1))\,1.2^n \leq \chi_p\left(E^n\right).$$

Problem 4.4 can be considered in n-dimensional Euclidean space too. For a given finite set S of r positive numbers, called *a set of forbidden distances*, we define the graph $G_S(R^n)$, whose vertices are points of the Euclidean n-space E^n, and a pair of points is adjacent if and only if the distance between them belongs to S. We will naturally call the chromatic number $\chi_S(E^n)$ of the graph $G_S(R^n)$ the *S-chromatic number of n-space E^n*. The following problem is as general as it is hard:

Erdős's Open Problem 10.13 Given S, find the S-chromatic number $\chi_S(E^n)$ of the space E^n.

By de Bruijn–Erdős compactness theorem that we met in chapter 5, the problem of investigating S-chromatic number of E^n is a finite.[29]

---

[29] De Bruijn–Erdős Theorem assumes the axiom of choice– see Chapters 46–48 for more.

11
Rational Coloring

I would like to mention here one more direction of assault on the chromatic number of the plane. By placing Cartesian coordinates on the plane E^2, we get an algebraic representation of the plane as the set of all ordered pairs (x, y) with coordinates x and y from the set R of real numbers, with the distance between two points defined as usual:

$$E^2 = \{(x, y) : x, y \in R\}. \tag{11.1}$$

Since by De Bruijn–Erdős's Theorem 5.1 it suffices to deal with finite subsets of R^2, we can surely restrict the coordinates in (11.1) to some subset of R. The problem is, which subset should we choose?

A set A is called *countable* if there is a one-to-one correspondence between A and the set of positive integers N.

For any set C, we define C^2 as the set of all ordered pairs (c_1, c_2), where c_1 and c_2 are elements of C:

$$C^2 = \{(c_1, c_2) : c_1, c_2 \in C\}.$$

Open Problem 11.1 Find a countable subset C of the set of real numbers R such that the chromatic number $\chi(C^2)$ is equal to the chromatic number $\chi(E^2)$ of the plane.

The set Q of all rational numbers would not work, as Douglas R. Woodall showed in 1973.

Chromatic Number of Q^2 11.2 (D. R. Woodall, [Woo1])

$$\chi(Q^2) = 2.$$

Proof by D. R. Woodall ([Woo1]): We need to color the points of the rational plane Q^2, i.e., the set of ordered pairs (r_1, r_2), where r_1 and r_2 are rational numbers. We partition Q^2 into disjoint classes as follows: we put two pairs (r_1, r_2), and (q_1, q_2)

A. Soifer, *The Mathematical Coloring Book*,
DOI 10.1007/978-0-387-74642-5_11, © Alexander Soifer 2009

into the same class if and only if both $r_1 - q_1$ and $r_2 - q_2$ have odd denominators when written in their lowest terms (an integer n is written in its lowest terms as $\frac{n}{1}$).

This partition of Q^2 into subsets has an important property: if the distance between two points of Q^2 is 1, then both points belong to the same subset of the partition! Indeed, let the distance between (r_1, r_2), and (q_1, q_2) be equal to 1. This means precisely that

$$\sqrt{(r_1 - q_1)^2 + (r_2 - q_2)^2} = 1,$$

i.e.,

$$(r_1 - q_1)^2 + (r_2 - q_2)^2 = 1$$

Let $r_1 - q_1 = \dfrac{a}{b}$ and $r_2 - q_2 = \dfrac{c}{d}$ be these differences written in their lowest terms. We have

$$\left(\frac{a}{b}\right)^2 + \left(\frac{c}{d}\right)^2 = 1$$

i.e.,

$$a^2 d^2 + b^2 c^2 = b^2 d^2.$$

Therefore, b and d must be both odd (can you see why?), i.e., by our definition above, (r_1, r_2), and (q_1, q_2) must belong to the same subset.

Since any class of our partition can be obtained from any other class of the partition by a translation (can you prove this?), it suffices for us to color just one class, and extend the coloring to the whole Q^2 by translations. Let us color the class that contains the point $(0,0)$. This class consists of the points (r_1, r_2), where in their lowest terms the denominators of both r_1, r_2 are odd (can you see why?). We color red the points of the form $\left(\dfrac{o}{o}, \dfrac{o}{o}\right)$ and $\left(\dfrac{e}{o}, \dfrac{e}{o}\right)$, and color blue the points of the form $\left(\dfrac{o}{o}, \dfrac{e}{o}\right)$ and $\left(\dfrac{e}{o}, \dfrac{o}{o}\right)$, where o stands for an odd number and e for an even number. In this coloring, two points of the same color may not be distance 1 apart (prove this on your own). ∎

Then there came a "legendary unpublished manuscript," as Peter D. Johnson, Jr. referred [Joh8] to the paper by Miro Benda, then with the University of Washington, and Micha Perles, then with the Hebrew University, Jerusalem. The widely circulated and admired manuscript was called *Colorings of Metric Spaces*. Peter Johnson tells its story on the pages of *Geombinatorics* [Joh8]:

> The original manuscript of "Colorings...," from which some copies were made and circulated (and then copies were made of the copies, etc.), was typed in Brazil in 1976. I might have gotten my first or second generation copy in 1977.... The paper was a veritable treasure trove of ideas, approaches, and results, marvelously informative and inspiring.

During the early and mid 1980s "Colorings..." was mentioned at a steady rhythm, in my experience, at conferences and during visits. I don't remember who said what about it, or when (except for a clear memory of Joseph Zaks mentioning it, at the University of Waterloo, probably in 1987), but it must surely win the all-time prize for name recognition in the "unpublished manuscript" category.

Johnson's story served as an introduction and homage to the conversion of the unpublished manuscript into the Benda–Perles publication [BP] in *Geombinatorics'* January 2000 issue.

This paper, dreamed up in the fall of 1975 over a series of lunches the two authors shared in Seattle, created a new, algebraic approach to the chromatic number of the plane problem. Moreover, it formulated a number of open problems, not directly connected to the chromatic number of the plane, problems that gave algebraic chromatic investigations their own identity. Let us take a look at a few of their results and problems. First of all, Benda and Perles prove (independently; apparently they did not know about the Woodall's paper) Woodall's result 11.2 about the chromatic number of the rational plane. They are a few years too late to coauthor the result, but their analysis allows an insight into the algebraic structure that we do not find in Woodall's paper. They then use this insight to establish more sophisticated results and the structure of the rational spaces they study.

Chromatic Number of Q^3 11.3 (Benda & Perles [BP])

$$\chi(Q^3) = 2.$$

Chromatic Number of Q^4 11.4 (Benda & Perles [BP])

$$\chi(Q^4) = 4.$$

Benda and Perles then pose important problems.

Open Problem 11.5 (Benda & Perles [BP]) Find $\chi(Q^5)$ and, in general, $\chi(Q^n)$.

Open Problem 11.6 (Benda & Perles [BP]) Find the chromatic number of $Q^2(\sqrt{2})$ and, in general, of any algebraic extension of Q^2.

This direction was developed by Peter D. Johnson, Jr. from Auburn University [Joh1], [Joh2], [Joh3], [Joh4], [Joh5] and [Joh6]; Joseph Zaks from the University of Haifa, Israel [Zak1], [Zak2], [Zak4], [Zak6], [Zak7]; Klaus Fischer from George Mason University [Fis1], [Fis2]; Kiran B. Chilakamarri [Chi1], [Chi2], [Chi4]; Douglas Jungreis, Michael Reid, and David Witte ([JRW]); and Timothy Chow ([Cho]). In fact, Peter Johnson has published in 2006 in *Geombinatorics* "A Tentative History and Compendium" of this direction of research inquiry [Joh9]. I refer the reader to this survey and works cited there for many exciting results of this algebraic direction.

In the recent years Matthias Mann from Germany entered the scene and discovered partial solutions of Problem 11.5, which he published in *Geombinatorics* [Man1].

Lower Bound for Q^5 11.7 (Mann [Man1])

$$\chi(Q^5) \geq 7.$$

This jump from $\chi(Q^4) = 4$ explains the difficulty in finding $\chi(Q^5)$, exact value of which remains open. Matthias then found a few more important lower bounds [Man2].

Lower Bounds for Q^6, Q^7 and Q^8 11.8 (Mann [Man2])

$$\chi(Q^6) \geq 10;$$
$$\chi(Q^7) \geq 13;$$
$$\chi(Q^8) \geq 16.$$

In reply to my request, Matthias Mann wrote about himself on January 4, 2007:

As I have not spent much time on Unit Distance-Graphs since the last article in Geombinatorics 2003, I do not have any news concerning this topic. To summarize, I found the following chromatic numbers:

$Q^5 >= 7$
$Q^6 >= 10$
$Q^7 >= 13$
$Q^8 >= 16$

The result for Q^8 improved the upper bounds for the dimensions 9–13.

For the Q^7 I think that I found a graph with chromatic number 14, but up to now I cannot prove this result because I do not trust the results of the computer in this case.

Now something about me: I was born on May 12th 1972 and studied mathematics at the University of Bielefeld, Germany from 1995–2000. I wrote my Diploma-thesis (the "Dipl.-Math." is the old German equivalent to the Master) in 2000. It was supervised by Eckhard Steffen, who has worked on edge-colorings. I had the opportunity to choose the topic of my thesis freely, so I red the book "Graph Coloring Problems" by Tommy Jensen and Bjarne Toft (Wiley Interscience 1995) and was very interested in the article about the Hadwiger-Nelson-Problem, and found the restriction to rational spaces even more interesting. After reading articles of Zaks and Chilakamarri (a lot of them in *Geombinatorics*), I started to work on the problem with algorithms.

Unfortunately, I had no opportunity to write a Ph.D.-thesis about unit distance-graphs, so I started work as an information technology consultant in 2000.

In the previous chapter, you have already met Josef Cibulka, a first year graduate student at Charles University in Prague. His essay submitted to *Geombinatorics* on March 31, 2008, a month after this book was sent off to Springer, contained new lower bounds for the chromatic numbers of rational spaces, improving Mann's results:

Newest Lower Bounds for Q^5 and Q^7 11.9 (Cibulka, [Cib])

$$\chi(Q^5) \geq 8;$$
$$\chi(Q^7) \geq 15.$$

Cibulka's paper will be published in the October 2008 issue XVIII(2) of *Geombinatorics*.

We started this chapter with Woodall's 2-coloring of the rational plane (result 11.2). I would like to finish with it, as a musical composition requires. This Woodall's coloring has been used in July 2007 by the Australian undergraduate student Michael Payne to construct a wonderful example of a unit distance graph—do not miss it in Chapter 46!

III
Coloring Graphs

12
Chromatic Number of a Graph

12.1 The Basics

The notion of a graph is so basic, and so unrestrictive, that graphs appear in all fields of mathematics, and indeed in all fields of scientific inquiry.

A *graph* G is just a non-empty set $V(G)$ of *vertices* and a set $E(G)$ of unordered pairs $\{v_1, v_2\}$ of vertices called *edges*. If $e = \{v_1, v_2\}$ is an edge, we say that e and v_1 are *incident*, as are e and v_2; we also say that v_1 and v_2 are *adjacent* or are *neighbors*.

Simple, don't you think?

By all standards of book writing, I am now supposed to give you an example of a graph. Why don't you create your own example instead! As the set V of vertices take the set of all cities you have ever visited. Call two cities a and b from V adjacent if you have ever traveled from one of them to the other. Let us denote Your Travel graph by $T(Y)$.

We can certainly represent $T(Y)$ in the plane. Just take a map of the world, plot the dot for each city you've been in, and draw the lines (edges) of all of your travels (the shape of edges does not matter, but do not connect two adjacent vertices a and b of $T(Y)$ by more than one edge even if you have traveled various routes between a and b).

We often represent graphs in the plane as we have just done for $T(Y)$, where the only things that matter are the set of vertices (but not their positions), and which vertices are adjacent (but not the shape of edges, which we presume have no points in common, except vertices of the graph incident with them).

In fact, you can think of a graph as a set of pins some of which are connected by rubber bands. So we consider the graph unchanged if we reposition the pins and stretch the rubber bands. Thus, we call two graphs *isomorphic* if "pins" of one of them can be repositioned and its "rubber bands" can be stretched so that this graph becomes graphically identical to the other graph.

More formally, two graphs G and G_1 are called *isomorphic* if there is a one-to-one correspondence $f : V \rightarrow V_1$ of their vertex sets that preserves adjacency, i.e., vertices v_1 and v_2 of G are adjacent if and only if the vertices $f(v_1)$ and $f(v_2)$ of G_1 are adjacent.

A. Soifer, *The Mathematical Coloring Book*,
DOI 10.1007/978-0-387-74642-5_12, © Alexander Soifer 2009

For example, the two graphs in Fig. 12.1 are isomorphic while the two graphs in Fig. 12.2 are not (prove both facts on your own, or see the proof, for example, in [BS] pp. 102–105).

Fig. 12.1 Isomorphic graphs

Fig. 12.2 Non-isomorphic graphs

I would like to get to our main interest, coloring, as soon as possible. Thus, I will stop my introduction to graphs here and refer you to [BS] for a little more about graphs; you will find much more in books dedicated exclusively to graphs, such as [BLW], [Har0], [BCL] and a great number of other books. In fact, graph theory is lucky: it has inspired more enjoyable books than most other relatively new fields.

The notion of the chromatic number of the plane (Chapter 2) was clearly motivated by a much older notion of the chromatic number of a graph. As Paul Erdős put it in his 1991 letter to me [E91/10/2ltr]:

Chromatic number of a graph is ancient.

The *chromatic number* $\chi(G)$ of a graph G is the minimum number n of colors with which we can color the vertices of G in such a way that no edge of G is *monochromatic* (i.e., no edge ab has both vertices a and b identically colored). In this case we can also say that G is an *n-chromatic graph*.

A graph G is called *n-colorable* if it can be colored in n colors without monochromatic edges. In this case, of course, $\chi(G) \leq n$.

Let us determine chromatic numbers of some popular (and important) graphs.

A *n-path* P_n from x to y is a graph consisting of n distinct vertices $v_1, v_2, \ldots, v_n$ and edges $v_1 v_2, v_2 v_3, \ldots, v_{n-1} v_n$, where $x = v_1$, $y = v_n$. If $n \geq 3$ and we add the edge $v_n v_1$, we obtain an *n-cycle* C_n.

Problem 12.1 Prove that

$$\chi(C_n) = \begin{cases} 2, & \text{if } n \text{ is even} \\ 3, & \text{if } n \text{ is odd} \end{cases}$$

Problem 12.2 For a graph G, $\chi(G) \leq 2$ if and only if G contains no n-cycles for any odd n.

Such a graph has a special name: *bipartite graph*.

In particular, the *complete bipartite graph* $K_{m,n}$ has m vertices of one color and n vertices of the other, and two vertices are adjacent if and only if they have different colors. In Fig. 12.3 you can find examples of complete bipartite graphs.

Fig. 12.3 Examples of complete bipartite graphs

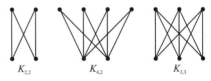

$K_{2,2}$ $K_{4,2}$ $K_{3,3}$

A *complete graph* K_n consists of n vertices every two of which are adjacent. In Fig. 12.4 you will find complete graphs K_n for small values of n.

Fig. 12.4 Examples of complete graphs

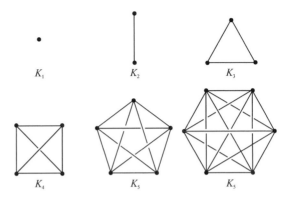

K_1 K_2 K_3

K_4 K_5 K_5

Problem 12.3 Is there an upper limit to chromatic numbers of graphs?

Solution: Since every two vertices of K_n are adjacent, they all must be assigned distinct colors. Thus $\chi(K_n) = n$, and there is no upper limit to chromatic numbers of graphs. ∎

The number of edges incident to a vertex v of the graph G is called the *degree* of v, and is denoted by $deg_G v$. The maximal degree of a vertex in G is denoted by $\Delta(G)$.

If v is a vertex of a graph G, then $G - v$ denotes a new graph obtained from G by deleting v and its incident edges.

Problem 12.4 For any graph G with finitely many vertices

$$\chi(G) \leq \Delta(G) + 1.$$

Proof Let G be a graph of chromatic number $\chi(G) = n$. If there is a vertex v in G, such that $\chi(G - v) = n$, we replace G by $G - v$. We can continue this process of deleting one vertex at a time with its incident edges until we get a graph G_1, such that $\chi(G_1) = n$ but $\chi(G_1 - v) \leq n - 1$ for any vertex v of G_1.

Let v_1 be the vertex of maximum degree in G_1, then

$$\Delta(G) \geq \Delta(G_1) = \deg_{G_1} v_1.$$

If we can prove that $\deg_{G_1} v_1 \geq n - 1$, then coupled with the inequality above, we would get $\Delta(G) \geq n - 1$, which is exactly the desired inequality.

Assume the opposite, i.e., $\deg_{G_1} v_1 \leq n - 2$. Since $\chi(G_1 - v_1) \leq n - 1$, we color the graph $G_1 - v_1$ in $n - 1$ colors. In order to get a $(n - 1)$-coloring of G_1, we have to just color the vertex v_1. We can do it because $\deg_{G_1} v_1 \leq n - 2$, i.e., v_1 is adjacent to at most $n - 2$ other vertices of G_1 (Fig. 12.5), thus at least one of the $n - 1$ colors is unused around v_1. We use it on v_1. Thus, $\chi(G_1) \leq n - 1$, in contradiction to $\chi(G_1) = n$. ∎

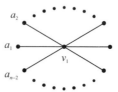

Fig. 12.5

R. Leonard Brooks of Trinity College, Cambridge, in his now classic theorem, reduced the above upper bound by 1 (for most graphs), to the best possible general bound. His result was communicated by William T. Tutte on November 15, 1940 and published the following year [Bro].

Brooks' Theorem 12.5 ([Bro]) If $\Delta(G) = n > 2$ and the graph G has no component K_n then

$$\chi(G) \leq \Delta(G).$$

12.2 Chromatic Number and Girth

W. T. Tutte, R. L. Brooks and Company pulled off the *Blanche Descartes* stint not unlike the better known Nicolas Bourbaki. Arthur M. Hobbs and James G. Oxley convey the story of Blanche Descartes in the memorial article "William T. Tutte 1917–2002" [HO]:

> Not long after he started his undergraduate studies at Cambridge, Tutte was introduced by his chess-playing friend R. Leonard Brooks to two of Brooks's fellow mathematics students, Cedric A. B. Smith and Arthur Stone. The four became fast friends and Tutte came to refer to the group as "the Gang of Four," or "the Four". The Four joined the Trinity Mathematical Society and devoted many hours to studying unsolved mathematics problems together.
>
> They were most interested in the problem of squaring a rectangle or square that is, of finding squares of integer side lengths that exactly cover, without overlaps, a rectangle or square of integer side lengths. If the squares are all of different sizes, the

squaring is called perfect. While still undergraduates at Cambridge, the Four found an ingenious solution involving currents in the wires of an electrical network. . .

The Gang of Four were typical lively undergraduates. They decided to create a very special mathematician, Blanche Descartes, a mathematical poetess. She published at least three papers, a number of problems and solutions, and several poems. Each member of the Four could add to Blanche's works at any time, but it is believed that Tutte was her most prolific contributor.

The Four carefully refused to admit Blanche was their creation. Visiting Tutte's office in 1968, Hobbs had the following conversation with him:

Hobbs: "Sir, I notice you have two copies of that proceedings. I wonder if I could buy your extra copy?"

Tutte: "Oh, no, I couldn't sell that. It belongs to Blanche Descartes."

However, I found at least one occasion when Tutte allowed to use his name in place of Blanche Descartes. Paul Erdős narrates [E87.12]:

Tutte sometimes published his results under the pseudonym Blanche Descartes, and in one of my papers quoting this result I referred to Tutte. Smith wrote me a letter saying that Blanche Descartes will be annoyed that I attributed her results to Tutte (he clearly was joking since he knew that I know the facts), but Richard [Rado] was very precise and when in our paper I wanted to refer to Tutte, Richard only agreed after I got a letter from Smith stating that my interpretation was correct.

You may wonder, what paper by Blanche Descartes does Paul Erdős refer to? Our story commences with the problem [Des1] Blanche Descartes published in April 1947. To simplify the original language used by Descartes, let me introduce here a notion of the *girth of a graph G* as the smallest number of edges in a cycle in G.

Descartes' Problem 12.6 ([Des1], 1947) Find a 4-chromatic graph of girth 6.

Descartes' solution appeared in 1948 [Des2]. This was the start of an exciting train of mathematical thought. In 1949 the first Russian graph theorist Alexander A. Zykov produced the next result [Zyk1]. He limited the restriction to just triangles, but asked in return for arbitrarily large chromatic number:

Zykov's Result 12.7 ([Zyk1], 1949) There exist a triangle-free graph of arbitrarily large chromatic number.

Zykov's 1949 comprehensive publication [Zyk1] contained a construction proving his result. The cold war and the consequent limited exchange of information apparently made Zykov's advance unknown in the West. In 1953, Peter Ungar formulated the same problem in the *American Mathematical Monthly* [Ung], which attracted much of attention and results. *The Monthly* chose not to publish the proposer's solution (which was supposedly similar to that of Zykov). Instead in 1954 *The Monthly* published a solution by Blanche Descartes [Des3], which both generalized Descartes' own 1948 result and solved Zykov–Ungar's problem: Descartes constructed graphs of arbitrarily large chromatic number which contained no cycles of less than six lines. This [Des3] was the Blanche Descartes' paper that Paul Erdős referred to in the quote above, and it was written by William T. Tutte alone.

John B. Kelly and Le Roy M. Kelly obtained a very similar construction in the same year ([KK]). Finally Jan Mycielski, originally from Poland and now Professor Emeritus at the University of Colorado at Boulder, published his original construction [Myc] in 1955.

Let us look at mathematics of this explosion of constructions. We will start with an exercise showing how to increase the chromatic number by attaching 3-cycles.

Problem 12.8 Let T be a 3-cycle with its vertices labeled 1, 2, and 3, and R a set of 7 vertices labeled 1, 2, ..., 7 (Fig. 12.6). For *each* 3-element subset V of the foundation set R we construct a copy T_V of T and attach it to R by joining vertex 1 of T_V with the lowest numbered vertex of V, vertex 2 of T_V with the middle numbered vertex of V, and vertex 3 of T_V with the highest numbered vertex of V. In Fig. 12.7, for example, this connection is drawn for $V = \{2, 3, 6\}$.

Fig. 12.6 A 3-cycle and the foundation set R

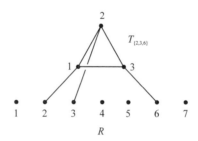

Fig. 12.7 A 3-cycle attached to the foundation set R

Since the number of 3-element subsets of the 7-element set R is equal to $\binom{7}{3} = \frac{7 \cdot 6 \cdot 5}{1 \cdot 2 \cdot 3} = 35$, the resultant graph G has $7 + 3 \cdot 35 = 112$ vertices. Prove that $\chi(G) = 4$.

Proof Four colors suffice to color G, since each T_v in G can be colored with the first three colors and all the vertices of R with the fourth color. Thus $\chi(G) \leq 4$.

Assume now that the graph G is 3-colored. Then by the Pigeonhole Principle, among the seven vertices of R there are three, say vertices 2, 3, and 6 that are colored in the same color, say color A. Then (Fig. 12.7) the color A is not present in the coloring of $T_{\{2,3,6\}}$, thus $T_{\{2,3,6\}}$ is 2-colored. But this is a contradiction since by Problem 12.1 a 3-cycle cannot be 2-colored. Hence, $\chi(G) = 4$. ∎

Problem 12.9 Use the construction of Problem 12.8 with the Mosers Spindle (Fig. 2.2) in place of T and a 25-point foundation set R. What is the chromatic number of the resulting graph G? How many vertices does G have?

The answer should serve as a *hint*: a 5-chromatic graph on $25 + 7 \binom{25}{7} = 3,364,925$ vertices. ∎

In his monograph [Har0], Frank Harary discloses the secret authorship of one result: "this so-called lady [Descartes] is actually a non-empty subset of {Brooks, Smith, Stone, Tutte}; in this [Des3] case {Tutte}." Let us take a look at Blanche Descartes' (Tutte's) construction.

Blanche Descartes's Construction 12.10 [Des3] For any integer $n > 1$ there exists an n-chromatic graph G_n of girth 6.

Proof For the case $n = 2$ we just pick a 6-cycle: $G_2 = C_6$. For $n \geq 3$ we define a sequence of graphs G_3, G_4 ..., G_n, ... by induction. Let G_3 be a 7-cycle: $G_3 = C_7$.

Assume that the graph G_k is defined and has M_k vertices. We need to construct G_{k+1}. The construction is the same as in problem 12.8. Let R be a set of $k(M_k - 1) + 1$ vertices. For *each* M_k-element subset U of R we construct a copy G_k^U of G_k, then pick a one-to-one correspondence f^U between the vertices of U and G_k^U (two M_k-element sets surely have one), and finally connect by edges the corresponding vertices of U and G_k^U. The resulting graph is G_{k+1}.

Thus, we have constructed the graphs G_3, G_4, ..., G_n, ... No graph G_n has a cycle of less than 6 edges (can you prove it?).

By induction we can prove that $\chi(G_n) \geq n$ for every $n \geq 3$. Indeed, G_3 is 3-chromatic as an odd cycle.

Assume that $\chi(G_k) \geq k$ for some $k \geq 3$. We need to prove that

$$\chi(G_{k+1}) \geq k + 1.$$

If to the contrary $\chi(G_{k+1}) \leq k$, then by the Pigeonhole Principle, out of $k(M_k - 1) + 1$ vertices of the set R, there will have to be an M_k-element subset U of vertices all colored in same color, say color A. But then color A is not present in the copy G_k^U of G_k, i.e., the graph G_k can be $(k - 1)$-colored in contradiction to the inductive assumption. The induction is complete.

Please note that we proved the inequality $\chi(G_n) \geq n$. Since we want to have the equality, we may have to delete (one at a time) some vertices of G_n and their incident edges until we end up with G_n' such that $\chi(G_n') = n$. ∎

Mycielski's Construction 12.11 [Myc] For any integer $n > 1$ there exists a triangle-free n-chromatic graph.

Proof You may ask why should we bother to prove the result that is weaker than Descartes' and Kelly & Kelly's result 12.10? It is simply because this is a different construction and it will work best for us in Section 15.1.

Start with a triangle-free $(k - 1)$-chromatic graph G. For each vertex v_i of G add a new vertex w_i adjacent to all neighbors of v_i. Next, add a new vertex z adjacent to all new vertices w_i. The chromatic number of this new graph is k, and it is still triangle-free. ∎

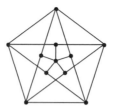

Fig. 12.8 The Mycielski–Grötzsch Graph

Observe that if we were to start with a 5-cycle, then the graph generated by the Mycielski's construction is the unique smallest triangle-free 4-chromatic graph (Fig. 12.8). Three years later, in 1958, this graph was independently found by Herbert Grötzsch [Grö], and thus it makes sense to call it the *Mycielski–Grötzsch Graph*. We will discuss Grötzsch's reasons for discovering this graph in Section 19.3.

The next major advance in our train of thought took place in 1959, when Paul Erdős, using probabilistic methods, dramatically strengthened the result 12.10.

Erdős's Theorem 12.12 (P. Erdős, [E59.06]). For every two integers m, $n \geq 2$, there exists an n-chromatic graph of girth m.

An alternative, non-probabilistic proof of this result was obtained in 1968 by the Hungarian mathematician László Lovász [Lov1].

The greatest result was still to come 30+ years after Lovász. Paul O'Donnell proved the existence of 4-chromatic unit distance graphs of arbitrary girth. We will look at this remarkable piece of work later in the book, in Chapters 14 and 45.

Paul Erdős posed numerous exciting open problems related to the chromatic number of a graph. Let me share with you one such still open problem that I found in Paul's 1994 problem paper [E94.26].

Erdős's Open Problem on 4-Chromatic Graphs 12.13 Let G be a 4-chromatic graph with lengths of the cycles $m_1 < m_2 < \ldots$. Can min $(m_{i+1} - m_i)$ be arbitrarily large? Can this happen if the girth of G is large?

12.3 Wormald's Application

In Chapter 5, I described Paul Erdős's problem conquered by Nicholas C. Wormald [Wor]. Wormald's first step was to construct what I will call the *Wormald Graph*; he then embedded it in the plane. In the construction Wormald used the Descartes construction of Problem 12.10. In Problem 12.8 I showed how one can use this construction; analogously, Wormald uses a 5-cycle in place of T and a 13-point foundation set R. For *each* 5-element subset V of R he constructs a copy T_V of T, fixes a one-to-one correspondence of the vertices of V and T_V and attaches T_V to V by connecting the corresponding vertices.

He ends up with the graph G on $13 + 5 \binom{13}{5} = 6448$ vertices. Wormald uses 5-cycles because his goal is to construct a 4-chromatic graph of girth 5. I leave the pleasure of proving these facts to the reader:

Problem 12.14 (N. C. Wormald, [Wor]) Prove that the Wormald graph G is indeed a 4-chromatic girth 5 graph.

So what is so special about Nicholas Wormald's 1979 paper [Wor]? Even though independently discovered (I think), didn't he use the construction that was published 25 years earlier by Blanche Descartes [Des3]? The real Wormald's trick was to *embed* his huge 6448-vertex graph in the plane, i.e., draw his graph on the plane with all adjacent vertices, and only them, distance 1 apart. In my talk at the conference dedicated to Paul Erdős's 80th birthday in Keszthely, Hungary in July 1993, I presented Wormald's graph as a picture frame without a picture inside it, to indicate that Wormald proved the existence and did not actually draw his graph. Nick Wormald accepted my challenge and on September 8, 1993 mailed to me a drawing of the actual plane embedding of his graph. I am happy to share his drawing with you here. Ladies and Gentlemen, the Wormald Graph! (Fig. 12.9).

Fig. 12.9 The 6448-vertex Wormald Graph embedded in the plane

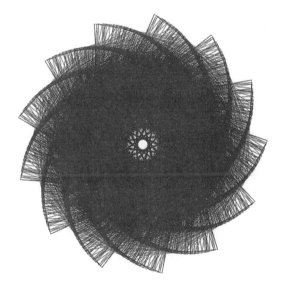

In his doctoral dissertation (May 25, 1999, Rutgers University) Paul O'Donnell offered a much simpler embedding than the Wormald's one—see it in Chapter 14, where I present Paul's machinery for embedding unit distance graphs in the plane.

However, the following problem, is open.

Open Problem 12.15 Find the smallest number λ_4 of vertices in a 4-chromatic unit distance graph without 3- and 4-cycles.

We know, of course, ([Wor]) that $\lambda_4 \leq 6448$. Chapters 14 and 15 will be dedicated to major improvements in this direction.

13
Dimension of a Graph

13.1 Dimension of a Graph

In 1965, a distinguished group of mathematicians consisting of Paul Erdős, Frank Harary, and William Thomas Tutte created a notion of the dimension of a graph ([EHT]).

They defined the *dimension of a graph G*, denoted *dimG*, as the minimum number n such that G can be embedded in the n-dimensional Euclidean space E^n with every edge of G being a segment of length 1. We will call such an embedding here *1-embedding*.

Dimensions of some popular graphs can be easily found.

Problem 13.1 (EHT) Prove the following equalities for complete graphs:

$$\dim K_3 = 2,$$
$$\dim K_4 = 3,$$
$$\dim K_n = n - 1.$$

The symbol $K_n - x$ denotes the graph obtained from the complete graph K_n by deleting one edge x; due to symmetry of all edges, this graph is well defined.

Problem 13.2 ([EHT]) Prove that

$$\dim(K_3 - x) = 1,$$
$$\dim(K_4 - x) = 2,$$

In general,

$$\dim(K_n - x) = n - 2.$$

Now let us take a look at complete bipartite graphs.

A. Soifer, *The Mathematical Coloring Book*,
DOI 10.1007/978-0-387-74642-5_13, © Alexander Soifer 2009

Problem 13.3 ([EHT])[1] Prove that for $m \geq 1$

$$\dim K_{m,1} = 2,$$

except for $m = 1, 2$ when $\dim K_{m,1} = 1$.

Proof Let S be a circle of radius 1 with center in O. By connecting arbitrary m points $A_1, A_2, \ldots, A_m$ of S with O we get a desired embedding of $K_{m,1}$ in the plane (Fig. 13.1).

The graphs $K_{1,1}$ and $K_{2,1}$ can obviously be 1-embedded in the line E^1, thus *dim* $K_{2,1} = 1$. ∎

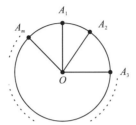

Fig. 13.1

Problem 13.4 ([EHT]) Prove that for $m \geq 2$

$$\dim K_{m,2} = 3,$$

except for $m = 2$ when $\dim K_{2,2} = 2$.

Proof Let ABC be an isosceles triangle with $|AB| = |BC| = 1$. As we rotate ABC about AC, the point B orbits a circle S (Fig. 13.2). By connecting arbitrary m points

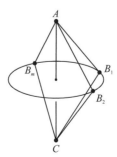

Fig. 13.2

---

[1] The article [EHT] contains a minor oversight: it says "Obviously, for every $n > 1$, $\dim K_{1,n} = 2$."

A_1, A_2, ..., A_m of S with both A and C, we get a desired embedding of $K_{m,2}$ in the 3-space E^3.

Since $m \geq 2$, the graph $K_{m,2}$ can be 1-embedded in the plane E^2 if and only if $m = 2$. Prove the last statement on your own. ∎

Problem 13.5 ([EHT]) Prove that for $m \geq n \geq 3$

$$\dim K_{m,n} = 4.$$

Proof In the solution of Problem 13.4 (Fig. 13.2) we had points of a one-dimensional "circle" (i.e., the two points A and C) distance 1 from the points of a circle S. Similarly in the Euclidean four-dimensional space E^4 we can find two circles S_1 and S_2 such that any point of S_1 is distance 1 from any point of S_2. We pick the circle S_1 in the plane through the coordinate axes X and Y; the circle S_2 in the plane through the coordinate axes Z and W. Both S_1 and S_2 have center at the origin $O = (0, 0, 0, 0)$ and radius $\frac{1}{\sqrt{2}}$. We then just pick m points on S_1 and n points on S_2.

This solution was obtained by Lenz in 1955 according to Paul Erdős. Formally (i.e., algebraically) it goes as follows. Let $\{u_i\}$ be the m vertices of the first color and let $\{v_j\}$ be the n vertices of the second color (remember, we are constructing a complete bypartite graph). We pick coordinates in E^4 for $u_i = (x_i, y_i, 0, 0)$ and for $v_j = (0, 0, z_j, w_j)$ in such a way that they lie on our circles S_1 and S_2 respectively, i.e., $x_i^2 + y_i^2 = \frac{1}{2}$ and $z_j^2 + w_j^2 = \frac{1}{2}$. Then the distance between every pair u_i, v_j will be equal to 1 (verify it using the definition of the distance in E^4).

It is not difficult to show (do) that for $m \geq n \geq 3$ the graph $K_{m,n}$ cannot be 1-embedded in the 3-space E^3. ∎

Problem 13.6 ([EHT]) Find the dimension of the Petersen graph shown in Fig. 13.3.

Solution: I enjoyed the style of the article [EHT]. I quote this solution in its entirety in order to show you what I mean:

"It is easy to see (especially after seeing it) that the answer is 2; Fig. 13.4)."

Paul Erdős told me that Frank Harary wrote this solution for their joint article. ∎

Fig. 13.3 The Petersen Graph

By connecting all vertices of an n-gon ($n \geq 3$) with one other vertex, we get the graph W_n called the *wheel with n spokes*. Figure 13.5 shows some popular wheels.

Fig. 13.4 Embedding of the Petersen Graph in the plane

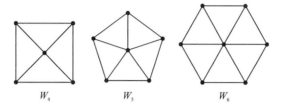

W_4 W_5 W_6

Fig. 13.5 A few wheels

Problem 13.7 (EHT]) The edges and vertices of a cube form a graph Q^3. Find its dimension.

Solution: $\dim Q^3 = 2$. Just think of Fig. 13.6 as drawn in the plane! ∎

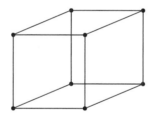

Fig. 13.6 Embedding of the cube's skeleton in the plane

The following two problems are for your own enjoyment.

Problem 13.8 ([EHT]) Prove that

$$\dim W_n = 3,$$

except for the "odd" number $n = 6$ when $\dim W_6 = 2$.

Problem 13.9 ([EHT]) A *cactus* is a graph in which no edge is on more than one cycle. Prove that for any cactus C

$$\dim C \le 2.$$

I hope you have enjoyed finding dimensions of graphs. There is no known systematic method for determining it, but it has its good side. As the authors of [EHT] put it, "the calculation of the dimension of a given graph is at present in the nature of mathematical recreation."

However, there is, one general inequality in [EHT] that connects the dimension and the chromatic number of a graph.

Problem 13.10 ([EHT]) For any graph G,

$$\dim G \leq 2\chi(G).$$

I totally agree with the authors of [EHT] that "the proof of this theorem is a simple generalization of the argument" used in Problem 13.5. However, for the benefit of young readers not too fluent with n-dimensional spaces, I am presenting here both a geometric ideology of the solution as well as a formal algebraic proof.

Geometric Idea: Let $\chi(G) = n$. In the Euclidean $2n$-dimensional space E^{2n} we can find n circles $S_1, S_2, \ldots, S_n$ such that the distance between any two points from distinct circles is equal to 1. We pick the circle S_1 in the plane through the coordinate axes X_1 and X_2; the circle S_2 in the plane through the coordinate axes X_3 and X_4; ...; the circle S_n in the plane through the coordinate aces X_{2n-1} and X_{2n}. All n circles have center at the origin and radius $\frac{1}{\sqrt{2}}$.

Finally, when we color G in n colors, (it can be done since $\chi(G) = n$) we get, say, k_1 points of color 1, k_2 points of color 2, ..., k_n points of color n. Accordingly, we pick arbitrary k_1 points on S_1, k_2 points on S_2, ..., k_n points on S_n for the desired 1-embedding of G.

Algebraic Solution: Let $\{u_i^1\}$ be the k_1 vertices of color 1, $\{u_i^2\}$ the k_2 points of color 2, ..., $\{u_i^n\}$ the k_n vertices of color n. We pick coordinates in E^{2n} for these vertices as follows:

$$u_i^1 = (x_i^1, x_i^2, 0, 0, \ldots, 0)$$
$$u_i^2 = (0, 0, x_i^3, x_i^4, 0, \ldots, 0)$$
$$\ldots\ldots\ldots\ldots$$
$$u_i^n = (0, 0, 0, 0, \ldots, x_i^{2n-1}, x_i^{2n})$$

in such a way that they lie on our circles $S_1, S_2, \ldots, S_n$ respectively (see Geometric Idea above), i.e.,

$$(x_i^1)^2 + (x_i^2)^2 = \frac{1}{2}$$
$$(x_i^3)^2 + (x_i^4)^2 = \frac{1}{2}$$
$$\ldots\ldots\ldots\ldots$$
$$(x_i^{2n-1})^2 + (x_i^{2n})^2 = \frac{1}{2}$$

Then the distance between every pair of points that belong to different circles is equal to 1 (can you see why?). Thus, the distance between any two points of different colors of the graph G in this embedding in E^{2n} is equal to 1. (We do not have to care at all about the distances between two points of G of the same color: they are not adjacent in G.) We got a 1-embedding of G in the space E^{2n}, therefore, $\dim G \leq 2n$. ∎

If it so happened that every vertex of a graph G_1 is also a vertex of a graph G, *and* every edge of G_1 is also an edge of G, then the graph G_1 is called a *subgraph* of the graph G.

Prove on your own the following property of subgraphs.

Problem 13.11 For every subgraph G_1 of a graph G,

$$\dim G_1 \leq \dim G.$$

During his December 1991–January 1992 two-week visit with me in Colorado Springs, Paul Erdős posed the following (quite solvable, I think) problem:

Erdős's Open Problem 13.12 What is the smallest number of edges in a graph G, such that $\dim G = 4$?

13.2 Euclidean Dimension of a Graph

I enjoyed the Erdős–Harary–Tutte paper [EHT] very much. However, there was, one more thing I expected from the notions of 1-embedding and dimension but did not get. I hoped they would unite *the chromatic number of a plane set* (Chapter 5) and *the chromatic number of a graph* (Chapter 12). Here is what I meant. I wanted to consider such embeddings of a graph G in the plane E^2 (and more generally in the n-dimensional space E^n) that *the chromatic number of a plane set V of the vertices of the embedded graph G is equal to the chromatic number of G.*

It was certainly not the case with 1-embeddings discussed above. The chromatic number of the 1-embedded set V of vertices of a graph G may not be uniquely defined. Take, for example, the cycle C_4. We can 1-embed it in the plane so that its vertex set V has the chromatic number 2 (just think of a square), but we can also 1-embed C_4 so that V has the chromatic number 3 (think of a rhombus with a $\pi/3$ angle).

The notions of the chromatic number of the plane and the chromatic number of a plane set have been generalized by Paul Erdős to Euclidean n-spaces nearly half a century ago:

Let S be a subset of the n-dimensional Euclidean space E^n (S may coincide with E^n). The *chromatic number* $\chi(S)$ of S is the smallest number of colors with which we can color the points of S in such a way that no color contains a monochromatic segment of length 1.[2]

Thus, if we adjoin two points a, b of S with an edge if and only if the distance $|ab| = 1$, we will get the graph G, such that *the chromatic number of the graph G is equal to the chromatic number of its vertex set S*:

$$\chi(G) = \chi(S). \quad (*)$$

Two new definitions, as well as most of the problems below, occurred to me on September 9, 1991. I remember this day very well: my daughter Isabelle Soulay Soifer was born on this day at 6 in the evening.

On September 12, 1991, I sent the news to Paul Erdős:

On the Jewish New Year, 9/9/1991 the baby girl Isabelle Soifer was born.

In my September 15, 1991 letter, I shared with Paul the mathematical find of Isabelle's birthday:

I enjoyed Erdős-Harary-Tutte 1965 article where *dimension* of a graph was introduced. (Apparently Harary and Tutte did not particularly like it: dimension of a graph did not appear in their books on graph theory.)

In my book though I am going to introduce a more precise notion. An embedding of a graph G into E^n we call *Euclidean* if two vertices v, w of G are adjacent *if and only if* in E^n the segment vw has length 1. *Euclidean dimension* of a graph G is the minimum n such that there is an Euclidean embedding of G in E^n (notation EdimG). Of course, $\dim G \leq E\dim G$. But a strict inequality is possible: let W_6 be the wheel with 6 spokes, and W_6' [a wheel] without 1 spoke [my drawing in the letter is the Fig. 13.7]. Then $\dim W_6' = 2 < 3 = E\dim W_6'$. Also there is a graph G and its subgraph G_1 such that $E\dim G_1 > E\dim G$. Just take $W_6' \subseteq W_6$.

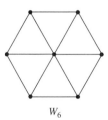

W_6

Fig. 13.7

This Euclidean dimension (rather than dimension) of a graph connects precisely chromatic numbers of a graph and [of] a plane set:

If a graph G is Euclideanly embedded in E^n, then $\chi(G) = \chi(V)$, where $\chi(G)$ is the chromatic number of the graph, and $\chi(V)$ is the chromatic number of the vertex set V of G (i.e., subset of E^n).

---

[2] Victor Klee was first to prove (unpublished) that $\chi(E^n)$ is finite for any positive integer n.

What do you think?

Paul Erdős's reply arrived on October 2, 1991. Following the family affairs, "I am very sorry to hear about your father's death [Yuri Soifer, June 20, 1907–June 17, 1991], but congratulations for the birth of Isabelle," Paul expressed his approval of the new notion of the Euclidean dimension of a graph and commenced posing problems about it [E91/10/2ltr]:

> "Can $E\dim G - \dim G$ be arbitrarily large?"

Little did I know at the time that, in fact, Paul Erdős himself with Miklós Simonovits invented the Euclidean dimension before me—in 1980—they called it *faithful dimension* [ESi], and Paul did not remember his own baby-definition when he discussed it with me! Of course, the credit for the discovery goes to Erdős and Simonovits. However, in my opinion, the term *Euclidean dimension* more faithfully names the essence of the notion, and so I will keep this term here.

Let us summarize the definitions and the early knowledge that we have.

A one-to-one mapping of the vertex set V of a graph G into an Euclidean space E^n we call *Euclidean embedding of G in the E^n* if two vertices v, w of G are adjacent if and only if the distance between $f(v)$ and $f(w)$ is equal to 1.

In other words, to obtain an Euclidean embedding of G in the E^n, we need to draw G in E^n with every edge of G being a segment of length 1 *and the distance between two non-adjacent vertices being not equal to 1.*

We define the *Euclidean dimension* of a graph G, denoted $E\dim G$, as the minimum number n such that G has an Euclidean embedding in the E^n.

Now we do get the desired connection:

Problem 13.13 The chromatic number of a graph G is equal to the chromatic number of its vertex set V when G is "Euclideanly" embedded in E^n for some n.

The two dimensions are connected by the following inequality:

Problem 13.14 Prove that for any graph G

$$\dim G \le E\dim G.$$

For some popular graphs we have the equality:

Problem 13.15 For any complete graph K_n

$$\dim K_n = E\dim K_n$$

i.e., $E\dim K_n = n - 1$.

Problem 13.16 For any complete bipartite graph $K_{m,n}$

$$\dim K_{m,n} = E\dim K_{m,n}.$$

Problem 13.17 For any wheel W_n

$$\dim W_n = E\dim W_n$$

i.e., $E\dim W_n = 3$ except for the "odd" number $n = 6$ when $E\dim W_6 = 2$.

Problem 13.18 For any graph G,

$$\dim G \leq E\dim G \leq 2\chi(G).$$

The new notion makes sense only if there is a graph G for which $\dim G \neq E\dim G$. And it does:

Problem 13.19 Find a graph G such that

$$\dim G < E\dim G.$$

The inequality $\dim G_1 \leq \dim G$ that is trivially true for any subgraph G_1 of a graph G, may not be true at all for the Euclidean dimension:

Problem 13.20 Construct an example of a graph G and its subgraph G_1 such that $E\dim G_1 > E\dim G$.

Solutions to Problems 13.19 *and* 13.20 Take the wheel W_6 with six spokes (Fig. 13.7) and knock out one spoke (Fig. 13.8). Let us prove that the resulting graph W_6' has the Euclidean dimension 3, even though $E\dim W_6 = 2$.

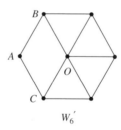

$$W_6'$$

Fig. 13.8

Indeed, when we draw the graph W_6' in the plane so that its every edge is a segment of length 1, the rigid construction of W_6' leaves no options for the distance OA. It is equal to 1 even though the spoke is missing! Thus, there is no Euclidean embedding of W_6' in the plane. It is easy to Euclideanly embed W_6' in 3-space E^3: start with the plane W_6' depicted in Fig. 13.8 and rotate BAC in the space about the axis BC.

We proved that $E\dim W_6' > E\dim W_6$. Thus, problem 13.20 is solved. Problem 13.19 is solved at the same time because $\dim W_6' = 2$, and therefore,

$$E\dim W_6' > \dim W_6'. \quad \blacksquare$$

The question that Paul Erdős posed to me, "Can $E\dim G - \dim G$ be arbitrarily large?"– was answered positively by him and Simonovits 11 years earlier:

Problem 13.21 [ESi]. For any positive n, there is a graph G such that $\dim G \leq 4$ while $n - 2 \leq E\dim G \leq n - 1$.

Hint: In Problem 13.5, we saw that for $n \geq 3$, $\dim K_{n,n} = 4$. Let G be the graph obtained from $K_{n,n}$ by removing a 1-factor, i.e., G is a graph on $2n$ vertices $x_1, \ldots, x_n$ and $y_1, \ldots, y_n$ with edges $x_i y_j$ for all $i \neq j$. Clearly, $\dim G \leq \dim K_{n,n} = 4$. Show that G cannot be Euclideanly embedded in the space R^{n-3}, but can be Euclideanly embedded in R^{n-1}. ∎

Erdős and Simonovits also found an upper bound for the Euclidean dimension of a graph. Their results showed that the Euclidean dimension of a graph G is related to the maximal vertex degree $\Delta(G)$ and not to its chromatic number $\chi(G)$:

Problem 13.22 [ES] For any graph G, $E\dim G \leq 2\Delta(G) + 1$.

Nine years later, this bound was slightly improved by László Lovász, Michael Saks and A. Schrijver:

Problem 13.23 [LSS] For any graph G, $E\dim G \leq 2\Delta(G)$.

Surprisingly, this bound still seems to be the best known.

We are now ready to continue our discussion of Nicholas Wormald's paper [Wor], started in Chapter 5 and continued in Chapter 12. The big deal was not to construct his 4-chromatic graph G without 3- and 4-cycles. The real Wormald's trick was to Euclideanly embed his huge 6448-vertex graph G in the plane, which he accomplished with the use of his ingenuity and a computer. Read more about how he has done it in his paper [Wor]. Here I would like to discuss one approach to the chromatic number of the plane problem.

If you believe that the chromatic number χ of the plane is at least 5, here is what you can do to prove it. You can create a 5-chromatic graph G, and then Euclideanly embed G in the plane. "Easier said than done," you say? Sure, but let us discuss it, then who knows? You just might succeed!

The 3,364,925-vertex graph G that we constructed in Problem 12.9 is surely 5-chromatic. But we constructed it with the use of the Mosers spindle; thus, G has a lot of triangles. It very well may be too rigid to have an Euclidean embedding in the plane.

We can replace the Mosers spindle with, say, the Mycielski–Grötzsch Graph (Fig. 12.8), and use the same construction as we did in Problems 12.8, 12.9, and 12.10. We would get a 5-chromatic graph G with $41 + 11 \binom{41}{11} = 34{,}754{,}081{,}689$ vertices. This graph G has no triangles. But does it have an Euclidean embedding in the plane? To begin with, I do not think (check it out) the Grötzsch graph itself has an Euclidean embedding in the plane.

No, we need to start with a very "flexible" graph having an Euclidean embedding in the plane. Let us start with the Wormald Graph G (Section 12.3) and the foundation set R of $6447 \times 4 + 1 = 25{,}789$ points. For every 6448-element subset V

of R we attach a copy G_V of G. We get a 5-chromatic graph G_1 without triangles, with $25,789 + 6448 \binom{25798}{6448}$ vertices. Does G_1 have an Euclidean embedding? Computers are better today than in 1978 when Nicholas Wormald completed his paper [Wor]. Are computers good enough for this task? Are we, mathematicians of today, good enough to break through these computational walls?

14
Embedding 4-Chromatic Graphs in the Plane

14.1 A Brief Overture

In Chapters 1 and 2 we got acquainted with examples of 4-chomatic unit distance graphs, the Mosers Spindle, and the Golomb Graph. In Chapters 5 and 12 we met Paul Erdős's \$25 Problem 5.6, and its partial solution by Nicholas Wormald, who used Blanche Descartes' construction of a 4-chromatic graph and his own embedding of that graph in the plane. Wormald's result was improved time and time again on the pages of *Geombinatorics* by Paul O'Donnell, Rob Hochberg, and Kiran Chilacamarri. Upon constructing a promising graph G, the authors of the new 4-chromatic unit distance examples used a 2-part approach to complete their task:

1. *Graph-Theoretic Part*. Show that the chromatic number of a graph G is 4 and the graph has no short cycles.
2. *Geometric Part*. Embed G in the plane in such a way that every pair of adjacent vertices is distance 1 apart *and* non-adjacent vertices are not 1 apart (like in the previous chapter dealing with the Euclidean dimension).

In this chapter we will concentrate on the essentials of part 2—tools for embedding in the plane, as developed and presented by Paul O'Donnell [Odo3], [Odo4], [Odo5]. In the next chapter, we will look at the world records in the new sport of embedding. Do use pen and pencil as you read this chapter.

We say that a k-vertex graph G with vertices $V = \{u_1, u_2, \ldots, u_k\}$ is *attached* to a set of vertices $V^* = \{u_1^*, u_2^*, \ldots, u_k^*\}$ if the vertices of G are connected via a *matching* to V (i.e., via a one-to-one correspondence of V to V^* and connection of the corresponding vertices by new edges).

The *shadow of G*, denoted G^*, is the set to which G is attached. We often choose the graph G to be an odd cycle. The odd cycles are attached to subsets of a large *independent* (i.e., no pair of vertices is adjacent) set of size n. The n independent vertices are called *foundation vertices*.

If vertices of G are placed in points of the plane so that adjacent vertices are exactly distance 1 apart, we say this is a *unit distance embedding* of G. Thus in the plane, if the odd cycle $\{u_1, u_2, \ldots, u_k\}$ is attached to $\{u_1^*, u_2^*, \ldots, u_k^*\}$, then the vertices $u_1, u_2, \ldots, u_k, u_1^*, u_2^*, \ldots, u_k^*$ are fixed points in the plane such that for some permutation σ, u_i is distance 1 from $u^*_{\sigma(i)}$ and from u_{i-1} and u_{i+1} (indices are added modulo k) for $1 < i < k$. Since the vertices can be relabeled, we assume that u_i is adjacent to u^*_i in the attachment. Usually we do not want distinct vertices to be placed in the same point in the plane. If vertices of G are placed in distinct points of the plane so that adjacent vertices are exactly distance 1 apart we say this is *a proper unit distance embedding* of G. A graph with a proper unit distance embedding is called a *unit distance graph in the plane*. In this section, higher dimensional analogues are not considered, and so *unit distance graph* will mean unit distance graph in the plane. In our geometric contexts the terms point and vertex will be used interchangeably, while the term edge will mean a unit length edge. The following continuity argument of the attachment procedure is important, and Paul O'Donnell uses it in most of the results of this section:

Continuity Argument 14.1 Given fixed points in the plane:

$$u^*_1 = (x_1, y_1), u^*_2 = (x_2, y_2), \ldots, u^*_k = (x_k, y_k),$$

and a point u_1 on the unit circle centered at u^*_1. Let u_i be a point distance 1 from both u_{i-1} and u^*_1 for $2 \leq i \leq k$. (In the following examples the distance between u_{i-1} and u^*_1 is less than 2, so there are two points satisfying the distance restrictions; let u_i be the one closer to the corresponding point in the attachment.) For an appropriately chosen arc along the unit circle centered at u^*_1, u_i is a continuous function of u_1. If there exist two points u^{short}_1 and u^{long}_1 such that the distance between u^{short}_1 and the corresponding u^{short}_k is less than 1, while the distance between u^{long}_1 and the corresponding u^{long}_k is greater than 1, then due to continuity there must be a point, u^{unit}_1 such that the distance between u^{unit}_1 and the corresponding u^{unit}_k is exactly 1. In other words, the set of points $\{u^*_1, u^*_2, \ldots, u^*_k\}$ has a k-cycle attached, namely, $\{u^{\text{unit}}_1, u^{\text{unit}}_2, \ldots, u^{\text{unit}}_k\}$. ■

The foundation points are distributed among four regions. They are placed in δ-balls centered at the following four points:

$$C_1 = (0, 0)$$
$$C_2 = (0, 0.9)$$
$$C_3 = (0.9, 0.9)$$
$$C_4 = (0.9, 0)$$

Since δ is very close to zero, it is impossible to attach a cycle to k points if they are all inside the same δ-ball. The partitioning of the foundation points is designed to prevent such an occurrence. Can a k-cycle be attached if the points are distributed among at least two of the δ-balls? Yes, they can. First, k-cycles are attached to k foundation points placed exactly at some or all of C_1, C_2, C_3, or C_4. Next, the points are moved slightly so the k-cycles are attached to k distinct points, each placed in the appropriate δ-ball surrounding C_1, C_2, C_3, or C_4. This prevents the foundation vertices from coinciding. Then some of the vertices are moved slightly to eliminate all coincidences.

In *Geombinatorics* [Odo5] (but not in his dissertation [Odo3]), O'Donnell introduces a useful notion of type:

> a set of foundation vertices has *type* $(a_1, a_2, a_3, a_4)_\delta$ if it consists of a_i vertices placed in the δ-ball around C_i, $1 \le i \le 4$.

14.2 Attaching a 3-Cycle to Foundation Points in 3 Balls

Only δ-balls around points C_1, C_2, and C_3 are dealt with for the basic argument. To distinguish between the preliminary and final situations, the foundation vertices coincident with C_1, C_2, C_3 are denoted $\{v_1^*, v_2^*, \ldots, v_k^*\}$, paths or cycles attached to them are denoted $v_1, v_2, \ldots, v_k$, while the foundation vertices inside the δ-balls around C_1, C_2, C_3 are denoted $\{u_1^*, u_2^*, \ldots, u_k^*\}$ and paths or cycles, attached to them, are denoted $u_1, u_2, \ldots, u_k$.

O'Donnell starts by attaching a triangle.

Tool 14.2 A 3-cycle can be attached to the set of foundation points $\{C_1,\ C_2,\ C_3\}$.

Proof Using the points listed in the *Appendix* at the end of this chapter (rounded to five decimal places), two 3-vertex unit distance paths are attached to C_1, C_2 and C_3. In the first path, T_1^{short}, T_2^{short}, T_3^{short}, the distance from T_1^{short} to T_3^{short} is less than 0.99. In the second path T_1^{long}, T_2^{long}, T_3^{long}, the distance from T_1^{long} to T_3^{long} is greater than 1.01 (Fig. 14.1).

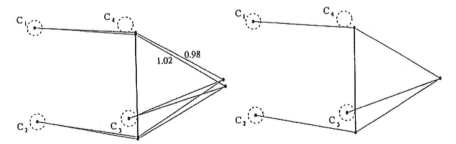

Fig. 14.1 A "too short" attachment and a "too long attachment" are shown together on the left. The "just right" attachment is on the right. (All unlabeled edges have unit length.)

Since one path is obtained from the other by continuously sliding the starting vertex, by the continuity argument there must be a path for which the distance between the first and last vertices is exactly one. This is a required attached 3-cycle. ∎

Now we will relax the condition of Tool 14.2 and allow the 3 foundation vertices to be anywhere inside δ-balls, not just at their centers. Given $\delta > 0$, let u_1^*, u_2^* and u_3^* be foundation vertices placed anywhere inside δ-balls, centered at C_1, C_2 and C_3 respectively. If δ is small enough, we will show that a cycle can be attached to the foundation set $\{u_1^*, u_2^*, u_3^*\}$ which is very close to the cycle attached to the foundation set $\{C_1, C_2, C_3\}$ (which we have already accomplished in Tool 14.2).

Tool 14.3 There exists $\delta > 0$ such that a 3-cycle can be attached to any foundation vertex set of type $(1, 1, 1, 0)_\delta$.

Proof Given $\varepsilon > 0$, choose δ so that we can find "too short" and "tool long" paths whose vertices are less than ε from the corresponding "too short" and "tool long" paths attached to $\{C_1, C_2, C_3\}$. This is possible due to the continuity argument. As in Tool 14.2, we get a "just right" path, which is the required cycle. ∎

14.3 Attaching a k-Cycle to a Foundation Set of Type $(a_1, a_2, a_3, 0)_\delta$

To generalize the above construction to k-cycles, where $k > 3$ is odd, other special points are needed. The three *triangle points*, denoted T_1, T_2, and T_3, are the points of the 3-cycle attached to $C_1 C_2$, and C_3. Three *spoke points*, S_1, S_2, and S_3, are the points such that S_i is unit distance from T_i and C_i, for $1 \le i \le 3$. We define "triangle" points T_i^{short} and T_i^{long} and "spoke points" S_i^{short} and S_i^{long} analogously for $1 \le i \le 3$. At first, cycles or paths are attached which coincide with these triangle and spoke points. The shadows of these cycles coincide with the center points C_1, C_2, and C_3. We then use the continuity argument to show the existence of cycles very close to these.

Tool 14.4 Let $k \ge 3$ be an odd number. For all positive integers a_1, a_2, a_3, such that $a_1 + a_2 + a_3 = k$, a k-cycle consisting only of edges from T_i to T_{i+1} (addition modulo 3), and T_i to S_i, for $1 \le i \le 3$, can be attached to the union of a_i points at C_i, $1 \le i \le 3$.

Proof The work involved in attaching 5-cycles contains all the details of the general case. Suppose, for example, we want to attach a 5-cycle to the set $\{u_1^*, u_{1a}^*, u_{1b}^*, u_2^*, u_3^*\}$ (where the number in the subscript indicates the δ-ball containing the vertex).

By Tool 14.3, for δ small enough, we can attach a 3-cycle u_1, u_2, u_3 to the foundation vertices $\{u_1^*, u_2^*, u_3^*\}$. We just need to insert a "detour" into this cycle. Instead of going from u_1 to u_2, we go from u_1 to u_{1a} to u_{1b}, which is arbitrarily close to u_1. We then continue to u_2 to u_3 and finally back to u_1. Of course, we cannot actually construct the 5-cycle directly from the 3-cycle. Instead, we construct "too short" and "too long" 5-paths with corresponding vertices within ε of the vertices of the "too short" and "too long" 3-paths used to construct the 3-cycle (see Tools 14.2 and 14.3). Given ε, we choose δ such that this is possible. By the continuity argument, we get a "just right" 5-path. This is an attached 5-cycle (Fig. 14.2).

Of course it did not matter that three foundation vertices were in the same δ-ball. Only two were necessary for the argument to work. The basic idea is to take a 3-cycle u_1, u_2, u_3 and construct a 5-cycle u_1, z, u_1, u_2, u_3. It does not matter

Fig. 14.2 A 5-cycle
attached to a set of type δ
$(3, 1, 1, 0)$

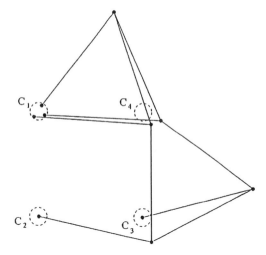

where the foundation vertex z is so long as it's close enough to u_1 so that unit length
edges can be connected to z (i.e., z should be less than 2 units away from u_1).

For example, suppose we want to attach a 5-cycle to $\{u_1^*, u_{1a}^*, u_2^*, u_3^*, u_{3a}^*\}$. We
find "too short" and "too long" 5-paths that are arbitrarily close to the corresponding
"too short" and "too long" 3-paths. By the continuity argument, we get a 5-cycle
(Fig. 14.3).

Similarly (using induction and considering two cases as discussed above),
k-cycles can be attached to k points by first looking at a $(k-2)$-cycle attached to
$k-2$ points, and then performing the insertion procedure described above. The
cycle will look like a triangle with a few spokes coming off some of the vertices. ∎

Fig. 14.3 A 5-cycle
attached to a set of type δ
$(2, 1, 2, 0)$

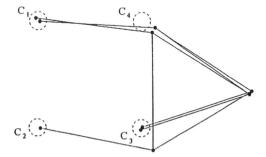

By symmetry, we can now attach k-cycles to sets of types $(a_1, a_2, 0, a_4)_\delta$,
$(a_1, 0, a_3, a_4)_\delta$, and $(0, a_2, a_3, a_4)_\delta$. What if we need to place the foundation vertices
in all 4 of the δ-balls? In fact, for our purposes we need only the case when the
partitioning of the foundation vertices puts just one foundation vertex in the δ-ball
around C^4, so only this case needs to be considered. ∎

14.4 Attaching a k-Cycle to a Foundation Set of Type $(a_1, a_2, a_3, 1)_\delta$

Tool 14.5 Let $k \geq 5$ be an odd number. For all positive integers a_1, a_2, a_3, a_4 such that $a_1 + a_2 + a_3 + a_4 = k$ and $a_4 = 1$; there exists $\delta > 0$ such that a k-cycle can be attached to any foundation set of type $(a_1, a_2, a_3, 1)_\delta$.

Proof The argument of the previous section applies here as well. At least one of a_1, a_2, a_3 is greater than 1, say a_1. We first find a $(k-2)$-cycle attached to a set of type $(a_1 - 1, a_2, a_3, 0)_\delta$ and then replace the vertex u_1 in the cycle by a path u_1, u_4, u_{1a}. This produces an attached k-cycle. Like before, we really do all the work on the "too short" and "too long" paths, and use the continuity argument to prove the existence of the desired "just right" cycle (Fig. 14.4). ∎

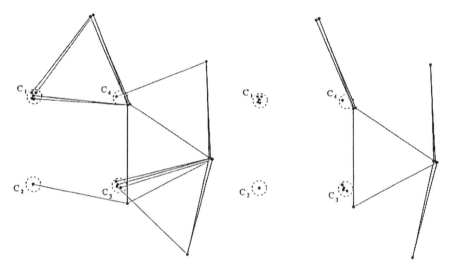

Fig. 14.4 An 11-cycle attached to a set of type δ (5, 1, 4, 1). The cycle and the attaching edges are shown on the left. The cycle alone is shown on the right

14.5 Attaching a k-Cycle to Foundation Sets of Types $(a_1, a_2, 0, 0)_\delta$ and $(a_1, 0, a_3, 0)_\delta$

We have shown that an odd cycle can be attached to k points placed inside δ-balls around any 3 or all 4 of C_1. C_2. C_3. and C_4. But what if the points are distributed between δ-balls around just two of the center points? The crucial step is still attaching a triangle. Once it is shown that a triangle can be attached to the center points, the previous arguments show that a k-cycle can be attached for any odd $k > 3$. We simply think of one of the δ-balls as two overlapping δ-balls (so now we have 3 balls).

Tool 14.6 Let $k \geq 3$ be an odd positive integer. For all positive integers a_1, a_2 such that $a_1 + a_2 = k$ there exists δ such that a k-cycle can be attached to any foundation set of type $(a_1, a_2, 0, 0)_\delta$.

Proof Without loss of generality, assume $a_1 \geq 2$. We attach a 3-cycle to two vertices at C_1 and one vertex at C_2, using the same notation as before for the triangle points, only here triangle points with subscripts 1 or 2 correspond to C_1 while those with subscript 3 correspond to C_2 (see the left drawing of Fig. 14.5).

Using the points listed in the appendix (rounded to five decimal places), two 3-vertex paths are attached to C_1, C_1 and C_2. In the first path, T_1^{short}, T_2^{short}, T_3^{short}, the distance from T_1^{short} to T_3^{short} is less than 1. In the second path, T_1^{long}, T_2^{long}, T_3^{long}, the distance from T_1^{long} to T_3^{long} is greater than 1. Since one path is obtained from the other by continuously sliding the starting vertex, by the continuity argument there must be a path for which the distance between the first and the last vertices is exactly 1. This is the desired attached 3-cycle.

Now we attach the k-cycle. Let a_1' and a_1'' be positive integers such that $a_1' + a_1'' = a_1$. We treat C_1 as if it were two separate vertices C_1' and C_1'' and use the machinery from the previous section to find δ such that any a_1' points in the δ-ball around C_1', a_1'' points in the δ-ball around C_1'', and a_2 points in the δ-ball around C_2, can have a k-cycle attached. In other words, any a_1 points in the δ-ball around C_1 and a_2 points in the δ-ball around C_2 can have a k-cycle attached.

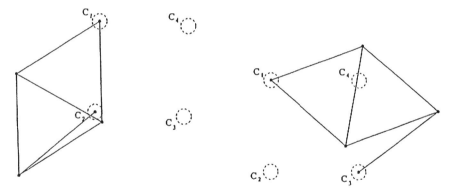

Fig. 14.5 Attaching a 3-cycle to $\{C_1, C_1, C_2\}$ on the left. Attaching a 3-cycle to $\{C_1, C_1, C_3\}$ on the right

This allows the attachment of k-cycles if the center points are distance 0.9 from each other, like C_1 and C_2. The configuration consisting of C_1 and C_3, can be handled similarly. ∎

Tool 14.7 Let $k \geq 3$ be an odd positive integer. For all positive integers a_1, a_3 such that $a_1 + a_3 = k$ there exists δ such that a k-cycle can be attached to any foundation set of type $(a_1, 0, a_3, 0)_\delta$.

Proof We just need to show that we can attach a 3-cycle to 1 vertex at the center of one δ-ball and 2 vertices at the center of the other. As in the proof of the previous tool, we use the "too short," "too long," "just right" continuity argument. (see the right drawing of Fig. 14.5). The *Appendix* at the end of this chapter contains the coordinates of the special points (rounded to five decimal places). ∎

14.6 Removing Coincidences

If two vertices from a graph are placed at the same points in the plane, small cycles may inadvertently be created. We must ensure that no vertices coincide. For δ small enough, the regions containing the foundation vertices are disjoint from the regions containing cycle vertices. Furthermore, the foundation vertices can be placed anywhere in the δ-balls, so we choose distinct locations for all of them. It is possible, however, for cycle vertices to coincide. In small graphs it can be verified computationally that this doesn't occur. For larger graphs Paul O'Donnell develops procedures to remove these coincidences.

If vertices from two different attached cycles coincide, one foundation vertex is moved slightly causing all vertices of the one attached cycle to move slightly while no vertices of the other cycle move. "Slightly" means not enough to introduce any new coincidences. If vertices from the same cycle coincide, a modification of this method is used to remedy it.

Tool 14.8 If there is an embedding of a unit distance graph G with $m \geq 1$ pairs of coincident vertices, then there is an embedding with fewer than m pairs of coincident vertices.

Proof Given an embedding of G with coincident vertices u and w, we shift some of the vertices of G, subject to several restrictions: no foundation vertex can move outside its δ-ball, and no new coincidences may be introduced. Let ε_1 be the minimal distance between any foundation point and the boundary of the containing it δ-ball; let ε_2 be the minimal distance between any two non-coincident vertices; we define

$$\varepsilon = \min\left\{\varepsilon_1, \frac{\varepsilon_2}{2}\right\}.$$

Given $\varepsilon > 0$, we choose δ', $0 < \delta' < \varepsilon$, such that if a foundation vertex is moved a distance less than δ', then no vertex moves a distance ε or greater. Since the foundation vertices are not moved more than ε_1, they remain inside their δ-balls, thus all k-cycles can still be attached. Since all non-coincident pairs of vertices are at least ε_2 apart, the movement by less than $\varepsilon_2/2$ does not create new coincidences. Let us consider two cases.

Case 1 Assume u and w are on different cycles: u is on the cycle $u = u_1, u_2, \dots$ while w is on the cycle $w = w_1, w_2, \dots$.

Let u_j be a vertex such that no w_i is attached to the foundation vertex u_j^*. Moving u_{j+1} along the unit circle centered at u_{j+1}^* causes each vertex in the cycle

$$u_{j+1}, u_{j+2}, \dots, u_k, u_1, \dots, u_{j-1}$$

to move to maintain unit distance from its foundation vertex and from the preceding cycle vertex. We move u_{j+1} so that no vertex has moved more than ε and thus there is a point unit distance from u_{j-1} and u_{j+1} and distance less than δ' from u_j. This point is the new location of u_j. Now we move u_j^* that same distance so it is unit distance from the new u_j. Of course, moving u_j^* may shift vertices of cycles attached to it by distances less than ε, but no new coincidences are introduced. Since u_1 moves and w_1 does not, at least one coincidence is removed.

Case 2 Assume u and w are on the same cycle; to reflect that we call them u_1 and u_i. We choose a cycle vertex u_j different from the coincident vertices and apply the procedure described in case 1. The only foundation vertex that moves is u_j^*. The only point in the ε-ball around the coincident vertices which is distance 1 from u_1^* and u_i^* is the original location of those points (u_1 and u_i). Since u_1 and u_i moved while u_1^* and u_i^* did not, they no longer coincide. As before no new coincidences are introduced. ∎

This has been a display of Paul O'Donnell's embedding machinery and his presentation of it [Odo3], [Odo4], [Odo5]. Can we get an immediate reward from his tool chest? As you know from Chapter 12, Wormald embedded his 6448-vertex graph in the plane. He started with 13 foundation points forming the vertices of a regular 13-gon, attached and embedded whopping $\binom{13}{5}$ 5-cycles, and made sure that no coincidences occurred.

O'Donnell was able to do it much easier—let us take a look.

14.7 O'Donnell's Embeddings

Embedding the Wormald Graph: Place four foundation vertices in each of the δ-balls centered at C_1, C_2, and C_3, plus one foundation vertex in the δ-ball centered at C_4. The embedding tools above allow the attachment of all 5-cycles and elimination of all coincidences that may occur. The unit distance embedding of the Wormald graph is thus accomplished! ∎

Wormald hints that with a considerable effort he probably could embed a larger Blanche Descartes graph, which is constructed by attaching all 7-cycles to the foundation of 19 vertices. No wonder he does not actually deal with it: for one, this is a 352,735-vertext graph, and thus calculations would have grown dramatically; moreover, Wormald admits that he does not see his approach going any further than a graph of girth 6.

The embedding of this 352,735-vertex graph too becomes trivial, compliments of O'Donnell's embedding tools.

Embedding the 352,735-vertex Graph: Indeed, just place six foundation vertices in each of the δ-balls centered at C_1, C_2, and C_3, plus one foundation vertex in the δ-ball centered at C_4. The embedding tools above allow the attachment of all 7-cycles and elimination of all coincidences that may occur. ∎

The next chapter is dedicated to the World Records of Embedding—join me for the exciting World Series!

14.8 Appendix

Vertices used to show a cycle can be attached to vertices at the points C_1, C_2, and C_3.

T_1^{short}	(0.99635, 0.08533)
T_2^{short}	(0.98269, 1.08524)
T_3^{short}	(1.84978, 0.58709)
T_4^{short}	(0.9980, 0.06319)
S_1^{short}	(0.57208, −0.82020)
S_2^{short}	(0.65177, 0.14158)
S_3^{short}	(1.64588, 1.56608)
S_{14}^{short}	(1.60981, −0.70439), distance 1 from C_4 & T_1^{short}
S_{24}^{short}	(1.77788, 0.47888), distance 1 from C_4 & T_2^{short}
S_{34}^{short}	(1.81111, −0.41216), distance 1 from C_4 & T_3^{short}
T_1^{long}	(0.99541, 0.09567)
T_2^{long}	(0.98069, 1.09556)
T_3^{long}	(1.85956, 0.61850)
T_4^{long}	(0.99280, 0.11977)
S_1^{long}	(0.58056, −0.81422)
S_2^{long}	(0.65971, 0.14848)
S_3^{long}	(1.62357, 1.59025)
S_{14}^{long}	(1.65414, −0.65671), distance 1 from C_4 & T_1^{long}
S_{24}^{long}	(1.77374, 0.48640), distance 1 from C_4 & T_2^{long}
S_{34}^{long}	(1.82462, −0.38089), distance 1 from C_4 & T_3^{long}
T_1	(0.99591, 0.09038)
T_2	(0.98173, 1.09028)
T_3	(1.85476, 0.60261)
S_1	(0.57623, −0.81729)
S_2	(0.65565, 0.14494)
S_3	(1.63492, 1.57815)
S_{14}	(1.63230, −0.68098), distance 1 from C_4 & T_1
S_{24}	(1.77587, 0.48255), distance 1 from C_4 & T_2
S_{34}	(1.81794, −0.39671), distance 1 from C_4 & T_3

Vertices used to show a cycle can be attached to vertices at the points C_1, C_2:

T_1^{short}	(−0.06194, 0.99808)
T_2^{short}	(0.83339, 0.55268)
T_3^{short}	(0.75995, 1.54998)

T_4^{short}	$(0.83339, 0.55268)$
S_1^{short}	$(0.83339. 0.55268)$
S_2^{short}	$(-0.06194, 0.99808)$
S_3^{short}	$(0.94288, 0.56685)$
T_1^{long}	$(-0.08916, 0.99602)$
T_2^{long}	$(0.81800, 0.57522)$
T_3^{long}	$(0.74037, 1.57220)$
T_4^{long}	$(0.81800, 0.57522)$
S_1^{long}	$(0.81800, 0.57522)$
S_2^{long}	$(-0.08916, 0.99602)$
S_3^{long}	$(0.95233, 0.59493)$
T_1	$(-0.07551, 0.99715)$
T_2	$(0.82580, 0.56397)$
T_3	$(0.75029, 1.56111)$
S_1	$(0.82580, 0.56397)$
S_2	$(-0.07551, 0.99715)$
S_3	$(0.94768, 0.58079)$

15
Embedding World Records

Around the 1991–1992 New Year, Paul Erdős and I had been writing the book *Problems of p.g.o.m. Erdős* in my home in Colorado Springs,[3] when Ron Graham called, and invited me to come to the Florida Atlantic University in March 1992, so that we can finally meet in person at the South-Eastern International Conference on Combinatorics, Graph Theory and Computing. I had just started publishing the problem-posing quarterly *Geombinatorics*, and at that conference I introduced it to the colleagues for the first time while giving a talk on the chromatic number of the plane problem. As a result, a group of young brilliant Ph. D. students, including Paul O'Donnell and Rob Hochberg, got excited about the problem and the new journal. *Geombinatorics* has become the main home for related problems and results outshining, in regards to these problems, all top journals on combinatorial theory and discrete geometry. One of the most exciting consequences was the competition for the smallest unit distance triangle-free graph, which from now on I will call *Embedding World Series*.

As you recall from Chapter 5, in 1975 Paul Erdős posed a problem to prove or disprove the existence of 4-chromatic unit distance graphs of girth 4, 5, and higher. Nicholas Wormald constructed a girth 5 graph on 6448 vertices (Chapter 12). In my talk I asked for the smallest example, and the World Series began in the earnest on the pages of *Geombinatorics*! New records were set by Paul O'Donnell, Rob Hochberg, and Kiran Chilakamarri; some new record graphs earned names, such as the Moth Graph, the Fish Graph, etc. and appeared on the covers of *Geombinatorics*. Let me, for the first time, present here this competition, current records and record holders. You will also see that, once the mathematical constructions and proofs were out of the way, the record holders went on to find "beautiful," symmetric embeddings of their graphs, the ones to which they—or else I—gave special names.

---

[3] Since this *Coloring Book* is finally finished, I am getting back to finishing *Problems of p.g.o.m. Erdős* book, so stay tuned.

A. Soifer, *The Mathematical Coloring Book*,
DOI 10.1007/978-0-387-74642-5_15, © Alexander Soifer 2009

15.1 A 56-Vertex, Girth 4, 4-Chromatic Unit Distance Graph [Odo1]

As we have touched on in Section 12.2, in 1955 Jan Mycielski [Myc] invented a method of constructing triangle-free graphs of arbitrary chromatic number k: Start with a triangle-free $(k-1)$-chromatic graph G. For each vertex $v_i \in V(G)$ add a vertex w_i adjacent to all vertices in the neighborhood of v_i. Next, add a vertex z adjacent to all of the new vertices. The chromatic number of this new graph is k, and it is still triangle-free. Let us call this graph *Mycielskian of G* and denote $M(G)$. Unfortunately, the resultant graph does not often embed in the plane. Notice that if a vertex of G has degree 3 or more, then the Mycielskian $M(G)$ of G contains a $K_{2,3}$. The plane contains no unit distance $K_{2,3}$ subgraph, so the starting graph G must have maximum degree at most two for the Mycielskian to be a unit distance graph. Thus, the only candidates for the unit distance version of the Mycielski construction are unions of paths and cycles. However, the Mycielskian of an odd cycle does not embed in the plane, so the Mycielski construction does not give a 4-chromatic unit distance graph. The Mycielskian of at least one even cycle does embed.

The 5-cycle u_1, u_2, u_3, u_4, u_5 is said to be *attached* to the set of vertices $\{v_1, v_2, v_3, v_4, v_5\}$ if v_i is adjacent to u_i for $1 \le i \le 5$ (Fig. 15.1). Such an attachment is a useful operation because it can increase the chromatic number of a graph from 3 to 4 without introducing any 3-cycles.

The graph H in Fig. 15.2 is the Mycielskian of the 10-cycle C_{10}. With basic geometry and algebra H can be embedded in the plane, but, O'Donnell reports, Rob Hochberg pointed out a nicer proof which shows *why* this is so. H is a subgraph of the projection of the 5-cube along a diagonal onto the plane. The coordinates of the vertices v_1, v_3, v_5, v_7, v_9 are the fifth roots of unity, while the edges are all unit length since they are translations of these unit vectors. This graph is only 3-chromatic, thus we will attach 5-cycles to make it 4-chromatic.

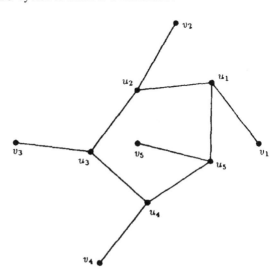

Fig. 15.1 The 5-cycle u_1, u_2, u_3, u_4, u_5 is *attached* to $\{v_1, v_2, v_3, v_4, v_5\}$

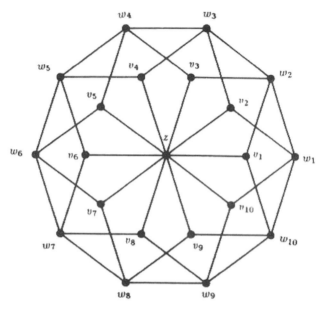

Fig. 15.2 H is the Mycielskian of C_{10}

Construction 15.1 A 5-cycle can be attached to the sub graph $R = \{v_1, \ v_3, \ v_5, \ v_7, \ v_9\}$ of the graph H of Fig. 15.2.

Proof Center a regular pentagon of side length 1 at the origin and rotate it until the distance from one of its vertices to v_1 is 1 (Fig. 15.3). Then the respective distances

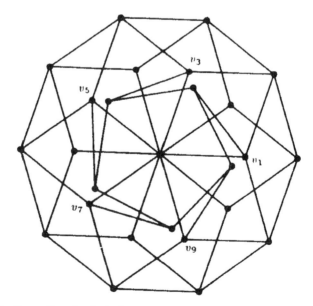

Fig. 15.3 H with one 5-cycle attached

from the other vertices of the pentagon to the other vertices of the graph R will all be 1.

Does this attachment remind you the construction of the Golomb Graph (Fig. 2.8)? It should, for the Golomb Graph was O'Donnell's inspiration for this nice construction. ∎

Construction 15.2 A 5-cycle can be attached to the subgraph $T = \{w_1, w_3, w_6, w_8, z\}$ of the graph H of Fig. 15.2.

Proof The proof relies on the intermediate value theorem and the continuity argument introduced in the beginning of Chapter 14. Described a little less formally, we try to attach a 5-cycle to the five vertices of T so that the cycle edges and the connecting edges are all length one. In fact, we try it twice. The problem with the attachments is that in the first, one of the edges in the cycle is too short, in the second, it, is too long. Since one configuration is obtained from the next by a continuous transformation, there exists an attachment where that same edge has length one. Thus, T can have a 5-cycle attached (Fig. 15.4). ∎

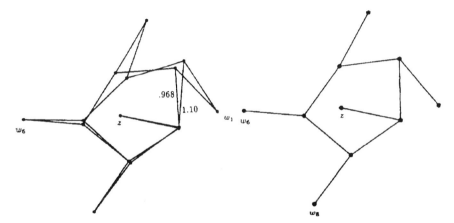

Fig. 15.4 The "Short" Attachment and The "Long" Attachment shown together on the left. The "Just Right" Attachment is on the right. (All unlabeled edges are of unit length.)

The most efficient way to verify these attachments is by computer, although it is necessary to make sure that the error made by approximating the numbers does not affect any of the inequalities. The error in the numbers listed below is $< 10^{-5}$, which does not affect the results.

Let C_i be the unit circle around the ith vertex in H. Let u_1, u_2, u_3, u_4, u_5 be a path with unit length edges and with u_i on C_i. This almost gives an attached 5-cycle. The attaching edges are all unit distance since each u_i is on the unit circle around some vertex in H, and the four-path edges are unit distance. This path can be slid back and forth in a continuous manner with each u_i tracing out an arc on C_i. One such path is approximately (0.95, 0.74413), (−0.04916, 0.70312), (−0.62463, −0.11470), (0.13436, −0.76580), (0.974661, −0.22369). The vertices of this path form a "too short attachment" where all distances are one except from u_5 to u_1 where the distance is about 0.968.

A second path which can be obtained from the first one by continuous sliding is (1.1, 0.85536), (0.13069, 0.60954), (−0.61938, −0.05183), (0.10423, −0.74206), (0.97027, −0.24204). The vertices of this path form a "too long attachment" where the u_5 to u_1 distance is 1.10. By continuity, there is a "just right attachment" where the edge from u_5 to u_1 is exactly one. The exact coordinates of this attachment are unknown, but for claiming their existence, it suffices to show that a 5-cycle can be attached to our set (Fig. 15.4). ∎

Construction 15.1 allows us to attach a 5-cycle to $\{v_1, v_3, v_5, v_7, v_9\}$. Similarly, we can attach another 5-cycle to $\{v_2, v_4, v_6, v_8, v_{10}\}$. We get the new graph, call it H'. In a proper 3-coloring of H' the vertices in $\{v_1, v_3, v_5, v_7, v_9\}$ cannot get the same color since that leaves only 2 colors for the attached 5-cycle. The same holds for $\{v_2, v_4, v_6, v_8, v_{10}\}$. This is enough to rule out most of the 3-colorings of H'. In fact, aside from the vertices of the attached 5-cycles, the coloring of H' is completely determined up to symmetries. This coloring is shown in Fig. 15.5 (the attached 5-cycles are not shown in the figure). Note that there are numerous ways to color the attached 5-cycles, but their attachment forces the rest of the graph to have a unique coloring up to a permutation of the colors and rotation of the graph.

Fig. 15.5 The vertices of H must have this coloring up to symmetries when two 5-cycles are attached. (The attached 5-cycles are not shown.)

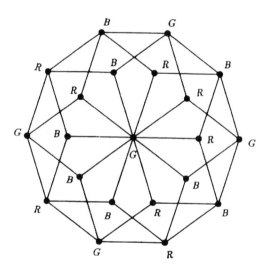

In particular, in every 3-coloring of H', for some j, $1 \le j \le 5$, the set $\{w_j, w_{j+2}, w_{j+5}, w_{j+7}, z\}$ (addition modulo 10) is monochromatic, where z and the w_i are as in Fig. 15.2. By attaching 5-cycles to all five of these sets, we exclude all 3-colorings. The result is a 4-chromatic graph. Moreover, since H is triangle-free, this new graph is also triangle-free. Approximation of the coordinates of the vertices ensures there are no coincident vertices.

Time to count vertices of our construction: H has 21 vertices, then two 5-cycles are added, then five 5-cycles more. The result is a triangle-free, 4-chromatic graph on 56 vertices (Fig. 15.6). ∎

Fig. 15.6 O'Donnell's
56-vertex 4-chromatic
graph in the plane with no
3-cycles

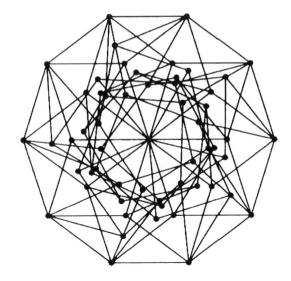

Beating the 6448-vertex Wormald graph with the new world record of a tiny 56-vertex graph was a striking achievement.

In closing, Paul O'Donnell observes: one reason to search for triangle-free graphs is that they seem to be flexible. For example, *H* can be bent into a 4-chromatic graph, containing many Mosers' spindles (Fig. 15.7).

Fig. 15.7 *H* can be bent so
that new edges (unit
distances) are introduced.
The chromatic number of
this new graph is 4

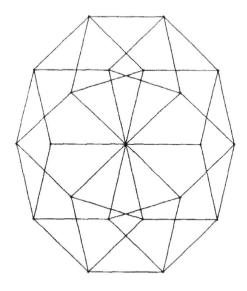

Paul ends [Odo1] with the ultimate goal (or ultimate musing):

Perhaps flexibility will prove useful in a construction of a 5-chromatic graph in the plane!

15.2 A 47-Vertex, Girth 4, 4-Chromatic, Unit Distance Graph [Chi6]

Professor Kiran Chilakamarri, then of the Ohio State University (and presently of Texas Southern University), was one of the early researchers of the chromatic number of the plane. Among other related problems he was very interested in constructing the smallest possible example of a 4-chromatic unit distance graph. I have little doubt that his work was well on the way when Paul O'Donnell published the first, 56-vertex breakthrough in these real (unlike baseball) World Series. In the fall 1995 Kiran responded by beating Paul's world record with the 47-vertex Moth Graph of his own, which I prominently published on the cover of January 1995 issue of *Geombinatorics*.

Chilakamarri constructs his example in stages, at each stage describing the properties, shared by all possible colorings of the graph constructed. He begins with a graph on 12 vertices and 20 edges, which he called the *core graph* shown in Fig. 15.8.

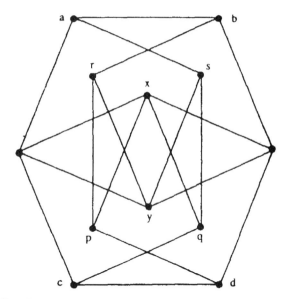

Fig. 15.8 The Core Graph

Chilakamarri then invents the *right wing* graph (Fig. 15.9) on 10 vertices and 12 edges, and symmetrically the left wing.

He then attaches the wings to the core and gets the Butterfly Graph (Fig. 15.10).

Finally, joining two butterflies produces the 47-vertex graph, which is proved to be 4-chromatic (Fig. 15.11).

Kiran then proves "the existence" (i.e., the existence of an embedding in the plane) of the Moth Graph by producing its coordinates. Finally, he proves that the Moth Graph has girth 4 by checking (a) the vertices of the core graph do not form an equilateral triangle (b) the left wing has no equilateral triangle (c) as we add the

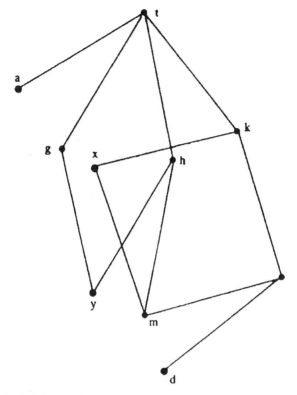

Fig. 15.9 The Right Wing graph

wings to the core no new edges are created and finally (d) as we join two butterfly graphs, no new edges are created other than the edge we have added. ∎

As Kiran Chilakamarri set his new world record of 47, our World Series became so intense in mid 1995 that Chilakamarri in this July-1995 paper mentions in end-notes "Paul O'Donnell tells me he is shrinking the size of the example ($\leq$ 40?)..." Moreover, Robert Hochberg modified O'Donnell's 56-vertex construction to get a 46-vertex unit distance triangle-free 4-chromatic graph, and thus beat Chilaka-marri's World Record of 47, but Rob, to my regret, decided against publishing it because he too learned that O'Donnell was getting ready to roll out yet another new world record, the 40-vertex graph.

15.3 A 40-Vertex, Girth 4, 4-Chromatic, Unit Distance Graph [Odo2]

Similarly to the [Odo1] approach, Paul O'Donnell starts with the Mycielskian of the 5-cycle C_5. This 11-vertex Mycielski—Grötzsch graph (we saw it in Fig. 12.8) is the smallest triangle-free 4-chromatic graph. Since it is not a unit distance graph,

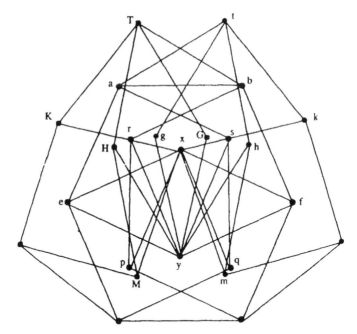

Fig. 15.10 Chilakamarri's Butterfly Graph

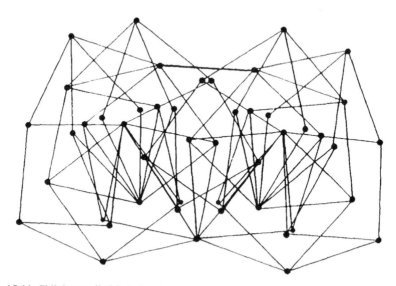

Fig. 15.11 Chilakamarri's Moth Graph

we modify it by taking out the "central" vertex adjacent to the 5 "new" vertices, and replacing it with five vertices each adjacent to a pair of "new" vertices as shown in Fig. 15.12. Let us call this graph H.

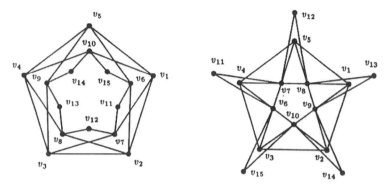

Fig. 15.12 An "instructive" drawing of H is on the left. A unit distance embedding of H is on the right

H is 3-chromatic, but all the 3-colorings share a valuable property. In every 3-coloring, one of the sets $\{v_{1+i}, v_{6+i}, v_{11+(i+1)}, v_{11+(i+2)}, v_{11+(i+3)}\}$, for $0 \leq i \leq 4$ (where parentheses indicate addition modulo 5), is monochromatic. By attaching 5-cycles, one of which is shown in Fig. 15.13, to all such sets, all 3-colorings get excluded. Thus, the resultant graph H' is 4-chromatic and still triangle-free. It remains to show that H' is a unit distance graph.

Fig. 15.13 The 5-cycle u_1, u_2, u_3, u_4, u_5 is attached to $\{w_1, w_2, w_3, w_4, w_5\}$

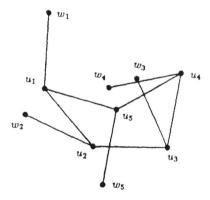

Construction 15.3 A 5-cycle can be attached to $T = \{v_1, v_6, v_{12}, v_{13}, v_{14}\}$, (see the right Fig. 15.12).

Proof We try to attach a 5-cycle w_1, w_2, w_3, w_4, w_5 so that the cycle edges and all the connecting edges are length 1, Fig. 15.14. It is fairly easy to attach a unit distance *path* w_1, w_2, w_3, w_4, w_5 to T. The hardest part is getting w_5 and w_1 to be distance 1 apart to complete the cycle.

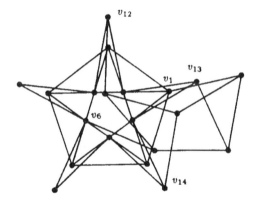

Fig. 15.14 *H* with a 5-cycle attached to *T*

Define a continuous function $f(\theta)$ to be the length of the edge $\{w_1, \; w_5\}$ when vertex w_1 is placed at angle θ and distance 1 from v_1, and each subsequent w_i is placed at distance 1 from both w_{i-1} and its corresponding vertex in *T*. Typically there are two possible positions for w_i, so a precise description of $f(\theta)$ would include how all of the choices are made. It suffices to say, there exists $f(\theta)$ satisfying the description above and continuous on some interval $[a, b]$ on which $f(a) < 1$ and $f(b) > 1$. By the Intermediate Value Theorem, for some $\theta_0 \in [a, \; b]$, $f(\theta_0) = 1$. ∎

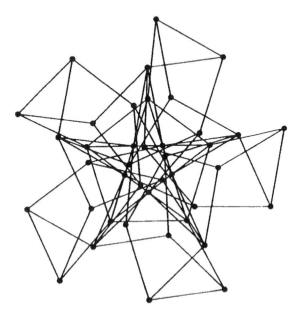

Fig. 15.15 O'Donnell's Pentagonal Graph

By attaching 5-cycles to T and all of its rotations, we obtain a graph with the desired properties. Since H had 15 vertices and we attached five 5-cycles, the result is a 4-chromatic, triangle-free unit distance graph on 40 vertices (Fig. 15.15).

The new world record of 40 was fabulous; I proudly published it on the cover of the July 1995 issue of *Geombinatorics*. However, it was not the end of the World Series of Embedding. Paul ended his essay [Odo2] with the promise of more things to come:

> Some related questions are still wide open. Given k, is there a 4-chromatic unit distance graph with no $< k$-cycles? What is the smallest 4-chromatic triangle-free unit distance graph? And of course, is there a 5-chromatic unit distance graph in the plane? Stay tuned to **Geombinatorics** for further developments.

15.4 A 23-Vertex, Girth 4, 4-Chromatic, Unit Distance Graph [HO]

Indeed, more things have come. It remains a mystery to me why Paul O'Donnell did not include in his doctorate dissertation two world records he has set jointly with Rob Hochberg, records which still stand today. In the dissertation, Paul mentions this great achievement briefly, as if in passing:

> In joint work with R. Hochberg [HO], the upper bounds on the sizes of the smallest 4-chromatic unit distance graphs with girths 4 and 5 were lowered even more. A 23-vertex, girth 4, 4-chromatic unit distance graph was found. The construction involved a generalized version of cycle attachment. A 45 vertex, girth 5, 4-chromatic unit distance graph was found. The construction involved a generalized version of cycle attachment.

And that is all! Fortunately, I published their remarkable paper in *Geombinatorics* in April 1996 [HO], and so we are able to revisit it here.

In [Odo1] and [Odo2] Paul O'Donnell used an idea of attaching odd cycles to specified subsets of vertices of a starting independent set. Here Hochberg and O'Donnell use a more complicated notion of *attaching*: a cycle might not have all of its vertices attached to the independent set, and some vertices in the independent set may have more than one vertex of the cycle attached to them. Figure 15.16 illustrates two applications of this idea.

In Fig. 15.17A the 5-cycle $(u_1, u_2, u_3, u_4, u_5)$ is partially attached (by dashed lines) to $\{w, y\}$. Observe that in any 3-coloring, if w and y get the same color, then u_5 must also receive that color.

To these three vertices $\{w, y, u_5\}$ we then attach the (bold) 5-cycle $(v_1, v_2, v_3, v_4, v_5)$, as shown in Fig. 15.17B.

Now in any 3-coloring of this graph, if w and y (and hence u_5) receive the same color, then there are only 2 colors left for the attached odd cycle making such a 3-coloring impossible. But in any 3-coloring of the square $\{w, x, y, z\}$, one of the pairs $\{w, y\}$ or $\{x, z\}$ must be monochromatic. So we take a copy of the two 5-cycles shown in Fig. 15.17B (flipped about a horizontal axis so that they are now attached to the pair $\{x, y\}$). With the coincidence at the center of the square, this adds only

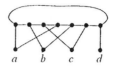

Here, a 5-cycle is partially attached to the independent set $\{v, w\}$. In any 3-coloring of this graph, if v and w get the same color, then x must also get that color.	Here, a 7-cycle is attached to the independent set $\{a, b, c, d\}$. Any coloring of this graph that makes the independent set monochromatic, must use at least 4 colors. Note that this graph has girth 5.

Fig. 15.16 Attaching odd cycles to independent sets

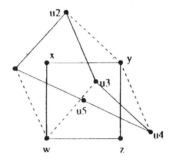

Fig. 15.17A Attaching a 5-cycle

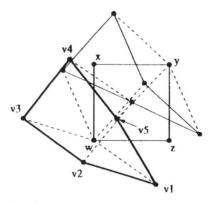

Fig. 15.17B Attaching a 5-cycle

nine new vertices (rather than 10 – every vertex counts when we set world records!), creating a 23 vertex graph with no 3-coloring. This graph is shown in Fig. 15.18. I named it Hochberg–O'Donnell's Fish Graph.

It remains to be shown that the graph is indeed unit distance. Clearly, it suffices to show that the 5-cycles can be attached the way we described. The proof relies

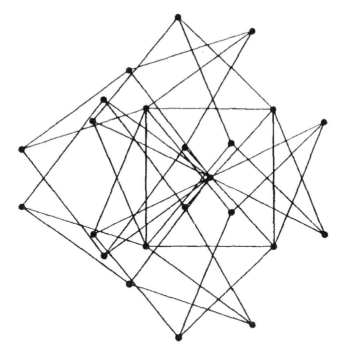

Fig. 15.18 Hochberg–O'Donnell's Fish Graph of order 23

on the Intermediate Value Theorem and the continuity argument. We try to attach a cycle to a specified set of vertices, so that the cycle edges and the connecting edges are all length 1. In fact, we do it twice: in the first, one of the edges in the cycle will be too short, in the second, it will be too long. Since one configuration can he obtained from the other by a continuous transformation (which does not alter the lengths of the unit length edges), there exists an attachment where that same edge has length 1. This works for all the attachments and partial attachments in these constructions. We looked at this argument in greater detail earlier in this chapter where we discussed O'Donnell's 56- and 40-vertex record graphs. ∎

The problem, of course, remains open:

Open Problem 15.4 What is the *smallest* size of a 4-chromatic unit distance graph of girth 4?

As you know, the smallest 4-chromatic triangle-free graph is the Mycielski–Grötzsch Graph of 11 vertices. The Fish satisfies all the Grötzsch conditions plus one extra: it is a unit distance graph. It is remarkable that Rob and Paul managed with merely 23 vertices. Is this the smallest possible number of vertices? I am not sure. I am positive though that 23 is very close to the minimum. And so, in a course of 2 years, on the pages of *Geombinatorics* we traveled from 6448 vertices all the way to 23, an incredible achievement!

15.5 A 45-Vertex, Girth 5, 4-Chromatic, Unit Distance Graph [HO]

Recall the Petersen Graph (Fig. 13.3) and its unit distance embedding in the plane (Fig. 13.4) that was discovered by the distinguished triumvirate of mathematicians Erdős–Harary–Tutte in their famous 1965 article [EHT] (yes that is where they famously remarked "It is easy to see especially after seeing it"). Here Hochberg and O'Donnell pursue their second idea (Fig. 15.16 on the right). Accordingly, in Fig. 15.19, a 7-cycle (shown in bold) is attached to a 4-vertex independent set of the Petersen graph.

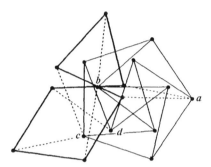

Fig. 15.19 The Petersen graph with a 7-cycle attached (by *dashed lines*)

The authors then simply write: "By the pigeonhole principle, in any 3-coloring of the Petersen Graph, one of the five rotations of the set $\{a, b, c, d\}$ will be monochromatic." Can you figure how the pigeons help here? Upon pondering for a few minutes, I understood it (though not sure whether the authors had the same argument in mind): In a 3-coloring of the Petersen Graph, at least 4 out of its 10 vertices must appear in the same color (that is the Pigeonhole Principle). Now, which 4 vertices could that be (here the Pigeonhole Principle is of no help)? The answer is two vertices on the outer pentagon and two on the inner star. You can now verify (do) that the only pair of the outer monochromatic vertices that allows two inner vertices in the same color, up to a rotation is a, c (Fig. 15.19). It is then clear that a, c must be accompanied in the same color by the vertices b, d of the inside!

When 7-cycles are attached to all five rotations of $\{a, b, c, d\}$, the resulting graph will not be 3-colorable. This gives a 45 vertex 4-chromatic graph with no 3-cycles or 4-cycles. This beautiful graph is shown in Fig. 15.20. I gave it the name honoring its creators, the Hochberg–O'Donnell Star Graph, and published it on the cover of the April 1996 issue of *Geombinatorics*.

Finally, we need to show that the Star Graph is indeed embeddable in the plane. It suffices to show that the 7-cycles can be attached the way we described it. The proof relies again on the Intermediate Value Theorem, and the continuity argument. We need to attach a 7-cycle to a specified set of vertices so that the cycle edges and the connecting edges are all length 1. Instead we do it twice: in the first attachment one

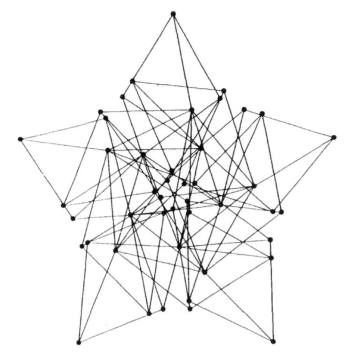

Fig. 15.20 Hochberg–O'Donnell's Star Graph of order 45

of the edges in the cycle will be too short, while in the second one too long. Since one configuration can be obtained from the other by a continuous transformation (which does not alter the lengths of the unit length edges), there exists an attachment where that same edge has length 1. This works for all the attachments and partial attachments in these constructions. We looked at this argument in a greater detail earlier in this chapter when we discussed O'Donnell's 56- and 40-vertex record graphs. ∎

Open Problem 15.5 What is the *smallest* size of a 4-chromatic unit distance graph of girth 5?

I hope you have enjoyed getting acquainted with the beautiful new graphs and the world records they represent. The Tables 15.1 and 15.2 summarize the world

Table 15.1 World records: Smallest unit distance 4-chromatic graph of Girth 4

Num. of vertices	Author	Pub. Date	Journal
6448	N. Wormald	1979	[Wor]
56	P. O'Donnell	July 1994	*Geombinatorics* IV(1), 23–29
47	K. Chilakamarri	January 1995	*Geombinatorics* IV(3), 64–76
46	R. Hochberg	1995	(unpublished)
40	P. O'Donnell	July 1995	*Geombinatorics* V(1), 31–34
23	R. Hochberg & P. O'Donnell	April 1996	*Geombinatorics* V(4),137–141

Table 15.2 World records: Smallest unit distance 4-chromatic graph of Girth 5

Num. of vertices	Author	Pub. Date	Journal
6448	N. Wormald	1979	[Wor]
45	R. Hochberg & P. O'Donnell	April 1996	*Geombinatorics* V(4),137–141

records history, and underscore the role of *Geombinatorics* as the playing field of this World Series.

It is now time to move on: we still have a lot of exciting colored and coloring mathematics to experience. Armed with great results on colored integers in Part VII, we will return to Paul O'Donnell's dissertation: Part IX will be dedicated to his main results.

16
Edge Chromatic Number of a Graph

16.1 Vizing's Edge Chromatic Number Theorem

We can assign a color to each edge of a graph instead of its vertices. This gives birth to the following notion.

A graph G is called *n-edge colorable* if we can assign one of the n colors to each edge of G in such a way that the adjacent edges are colored differently.

The *edge chromatic number* $\chi_1(G)$ also known as *chromatic index* of a graph G is the smallest number n of colors for which G is n-edge colorable.

The following two statements follow straight from the definition.

Problem 16.1[4] For any graph G

$$\chi_1(G) \geq \Delta(G).$$

Problem 16.2 For any subgraph G_1 of a graph G

$$\chi_1(G_1) \leq \chi_1(G).$$

In 1964, the Russian mathematician Vadim G. Vizing published [Viz1] a wonderful result about the edge chromatic number of a graph. His proof is fairly long, but so nice that I am going to present it here completely. *Do read it with pencil and paper!*

Vizing's Theorem 16.3 (V. G. Vizing, [Viz1]) If G is a non-empty graph, then

$$\chi_1(G) \leq \Delta(G) + 1, \ (^*)$$

i.e., the edge chromatic number $\chi_1(G)$ of a graph is always equal to Δ or $\Delta + 1$, where $\Delta = \Delta(G)$.

---

[4] $\Delta(G)$ is defined in Chapter 12.

A. Soifer, *The Mathematical Coloring Book*,
DOI 10.1007/978-0-387-74642-5_16, © Alexander Soifer 2009

Vadim G. Vizing in early 1960s, when he worked on his classic theorem

Proof I enjoyed the version of the Vizing's proof in [BCL]. My presentation is based on theirs—I tried to make it more visual by including many illustrations, splitting one case into two, and adding a number of elucidations.

Part I. Preparation for the Assault: We will argue by contradiction. Assume that the inequality (*) is not true. Then among the graphs for which (*) is not true, let G be a graph with the *smallest* number of edges. In other words, G is not $(\Delta + 1)$-edge colorable; but the graph G' obtained from G by removing one edge e, is $[\Delta(G') + 1]$-edge colorable. Since obviously $\Delta(G') \leq \Delta(G)$, the graph G' is $(\Delta + 1)$-edge colorable.

Let G' be actually edge colored in $\Delta + 1$ colors, i.e., every edge of the graph G except $e = uv$ (this equality simply denotes that the edge e connects vertices u and v) is colored in one of the $\Delta + 1$ colors in such a way that adjacent edges are colored differently. For each edge $e' = uv'$ of G that is incident with u (including e), we define its *dual color* as any one of the $\Delta + 1$ colors that is not used to color edges incident with vertex v'. (Since the degree of any v' does not exceed Δ, we always have at least one color to chose as dual. It may so happen that distinct edges have the same dual color—it is all right).

We are going to construct a sequence of distinct edges $e_0, e_1, \ldots, e_k$ all incident with u as follows (Fig. 16.1). Let $e = e_0$ have dual color α_1 (i.e., α_1 is not the color of any edge of G incident with v). There must be an edge, call it e_1, of color α_1 incident with u (for if not, then the edge e could be colored α_1, thus producing a $(\Delta + 1)$-edge coloring of G). Let α_2 be the dual color of e_1. If there is an edge of color α_2 incident with u and distinct from e_0 and e_1, we denote it by e_2 and its dual color by α_3, etc. We have constructed a *maximal* (i.e., as long as possible) sequence $e_0, e_1, \ldots, e_k,\ k \geq 1$ of *distinct* edges. The last edge e_k by construction is colored α_k and has dual color α_{k+1}.

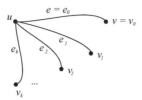

Fig. 16.1

If there were no edge of color α_{k+1} incident with u, then we would recolor each edge of our sequence $e_0, e_1, \ldots, e_k$ into its dual color, and thus achieve a $(\Delta + 1)$-edge coloring of G (do verify that!). This contradicts our initial assumption.

Therefore, there is an edge e_{k+1} of color α_{k+1} incident with u; but since we have constructed the longest sequence of distinct edges $e_0, e_1, \ldots, e_k$, the edge e_{k+1} must coincide with one of them: $e_{k+1} = e_i$ for some $i,\ 1 \leq i \leq k$. Since the edges coincide, so do their colors: $\alpha_{k+1} = \alpha_i$. The color α_k of the edge e_k may not be the same as the dual color α_{k+1} of e_k : $\alpha_{k+1} \neq \alpha_k$. Thus we get $\alpha_{k+1} = \alpha_i$ for some $i,\ 1 \leq i < k$. Denote $t = i - 1$, then the last equality can be written as follows:

$$\alpha_{k+1} = \alpha_{t+1}$$

for some $t,\ 0 \leq t \leq k - 1$. Finally, this means that *the edges e_k and e_t have the same dual color.*

And now the last preparatory remarks.

a. For each color α among the $\Delta + 1$ colors, there is an edge of color α adjacent with the edge $e = uv$ (for if not, e could be colored α, thus producing $(\Delta + 1)$-edge coloring of G). But since there are at most Δ edges incident with the vertex u, there is a color, call it β, assigned to an edge incident with the vertex v that is not assigned to any edge incident with u.

b. The color β must be assigned to at least one edge incident with the vertex v_i for each $i = 1, 2, \ldots, k$ (Fig. 16.1). Indeed, if we assume that there is a vertex $v_m,\ 1 \leq m \leq k$, such that no edge incident with v_m is colored β, then we can change the color of e_m to β and change the color of each $e_i,\ 0 \leq i \leq m$ to its dual color to obtain a $(\Delta + 1)$-edge coloring of G (verify that).

Part II The Assault: A sequence of edges $a_1, a_2, \ldots, a_n$ of a graph is called a *path of length n* if the consecutive edges of the sequence are adjacent (Fig. 16.2). You can trace a path with a pencil without taking it off the paper all the way from the *initial vertex of the path* v_0 to the *terminal vertex of the path* v_n. The edge a_1 is called *initial*, while the edge a_n *terminal edge* of the path.

Define two paths P and R as follows: their initial vertices are v_k and v_t respectively, and each of the paths has *maximum possible length with edges alternately colored β and $\alpha_{k+1} = \alpha_{t+1}$* (we established in Part I that colors α_{k+1} and α_{t+1} coincide). Denote the terminal vertices of the paths P and R by w and w' respectively and consider five possibilities for w and w'.

Fig. 16.2

Case 1: $w = v_m$ for some m, $0 \le m \le k - 1$ (Fig. 16.3).

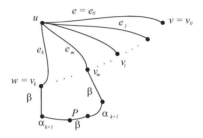

Fig. 16.3

Observe that the color α_{k+1} as the dual color of the edge e_k, may not be adjacent to e_k, therefore the initial edge of the path P must be colored β and $m \neq k$.

The terminal edge of P must be colored β as well. Indeed, if alternatively the terminal edge of P were colored α_{k+1}, then we would be able to make P longer by adding one more edge incident with v_m and colored β (it exists as we noticed in (b) at the end of Part I of this proof).

Note that the vertex v_t is *not* on P unless $v_m = v_t$. Indeed, assume that v_t is on P and $v_t \neq v_m$, then v_t is incident with edges of P (Fig. 16.4). One of them must be colored α_{k+1} (and the other β), but the dual color of e_t is $\alpha_{t+1} = \alpha_{k+1}$, therefore no edge of color α_{k+1} may be adjacent to e_t. This contradiction proves that v_t is not on P unless $v_t = v_m$.

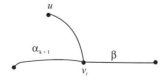

Fig. 16.4

We are ready to finish Case 1. Interchange the colors β and α_{k+1} on the edges of P. Please note (and prove) that as a result of this interchange, we do *not* alter the dual colors of edges e_i for any $i < m$, and end up with no edge of color β incident with v_m. Now to obtain a $(\Delta + 1)$-edge coloring of G, we just change the color of e_m to β and change the color of every e_i for $0 \le i < m$ to its dual. (Do verify that we get a $(\Delta + 1)$-edge coloring of G.) We have reached a contradiction, for G is not $(\Delta + 1)$-edge colorable.

Case 2: $w' = v_m$ for some m, $0 \le m \le k$ (Fig. 16.5).

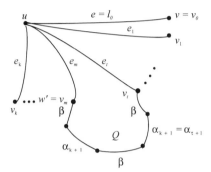

Fig. 16.5

Observe that the color $\alpha_{k+1} = \alpha_{t+1}$ as the dual color of the edge e_t, may not be adjacent to e_t, therefore the initial edge of the path R must be colored β, and $m \ne t$.

The terminal edge of R must be colored β as well. Indeed if alternatively the terminal edge of R is colored α_{k+1}, then we would be able to make R longer by adding one more edge incident with v_m and colored β (it exists as we showed in (b) at the end of Part I of the proof).

The vertex v_k is *not* on R unless $v_m = v_k$ (the proof is identical to a relevant argument in Case 1 above). Now we interchange the colors β and α_{k+1} of the edges of R. As a result of this interchange, we do *not* alter the dual colors of edges e_i for any $i \ne t$, and end up with no edge of color β incident with v_m.

If $m < t$, we finish as in Case 1. If $m > t$, we change the color of e to β and change the color of every e_i, $0 \le i < m$ to its dual. In either case we get a $(\Delta + 1)$-edge coloring of G, which is a contradiction.

Case 3: $w \ne w_m$ for any m, $0 \le m < k$ *and* $w \ne u$. As in Case 1, the initial edge of P must be colored β.

We interchange the colors β and α_{k+1} of the edges of P. As a result (just like in Case 1), we do not alter the dual colors of edges e_i for any $i < k$, and end up with no edge of color β incident with v_k. As in the previous cases, we can now obtain a $(\Delta + 1)$-edge coloring of G, a contradiction.

Case 4: $w' \ne v_m$ for any $m \ne t$, and $w' \ne u$. This case is similar to Case 3— consider it on your own.

Case 5: $w = w' = u$. (Figs. 16.6 and 16.7)

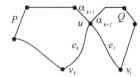

Fig. 16.6

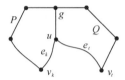

Fig. 16.7

Since by definition of β, u is incident with no edge colored β, the terminal edge of both paths P and R is colored α_{k+1}.

If P and R have no edges in common (Fig. 16.6), then u is incident with two edges colored α_{k+1}, which cannot occur in edge coloring of a graph. But if P and R do have an edge in common, then there is a vertex (g in Fig. 16.7) incident with at least three edges of P and R. Since each of these three edges is colored β or α_{k+1}, two of them must be assigned the same color which cannot occur with two adjacent edges of an edge colored graph. In either case we have obtained a contradiction. ∎

This remarkable theorem partitions graphs into two classes: *class one*, when $\chi_1(G) = (G)$; and *class two*, when $\chi_1(G) = \Delta(G) + 1$.

Each class does contain a graph. The graph in Fig. 16.8 is of class one and the graph in Fig. 16.9 is of class two. Can you prove it?

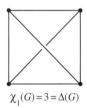

$$\chi_1(G) = 3 = \Delta(G)$$

Fig. 16.8 A class one graph

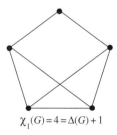

$$\chi_1(G) = 4 = \Delta(G) + 1$$

Fig. 16.9 A class two graph

Problem 16.4 Prove that an n-cycle C_n, $(n \geq 3)$ is of class one if n is even, and of class two if n is odd.

Problem 16.5 Prove that a complete graph K_n is of class one if n is even, and of class two if n is odd.

Proof This problem does not sound exciting, does it? You are in for a nice surprise, true mathematical recreation! In fact, do not read any further just yet, try to solve it on your own. Then read this solution which comes from [BCL].

1. Assume the graph K_n is an edge colored in $\Delta(K_n) = n - 1$ colors. Every vertex is incident with $n - 1$ edges, which must be colored differently. Therefore, *every vertex is incident with an edge of every color.*

   Now take color 1. Every vertex of K_n is incident to an edge of color 1, and edges of color 1 are not adjacent. Therefore, edges of color 1 partition the n vertices of K_n into disjoint pairs. Hence, n must be even.

   We proved that if K_n is a graph of class one, then n is even.

2. Now let us prove that, conversely, the graph K_{2n} is of class one.

   It is true for $n = 1$. Assume $n \geq 2$. Denote the vertices of K_{2n} by $v_0, v_1, \ldots, v_{2n-1}$. We arrange the vertices $v_0, v_1, \ldots, v_{2n-1}$ in a regular $(2n - 1)$-gon, and place v_0 in its center. We join every two vertices by a straight line segment, thereby creating K_{2n}.

We are ready to color the edges of K_{2n} in $2n - 1$ colors. We assign the color i $(i = 1, 2, \ldots, 2_{n-1})$ to the edge $v_0 v_i$ and to all edges that are perpendicular to $v_0 v_i$. We are done! All of the edges are colored: indeed we assigned n edges to each color for a total of $n(2n - 1)$ *edges* which is the number of edges of K_{2n}. No two edges of the same color are adjacent: they clearly do not share a vertex. Figure 16.10 shows all edges of color 1 for K_8. Edge sets of other colors are obtained from this one by rotations about the center v_0 – this fact is true for the general case of K_{2n}. ∎

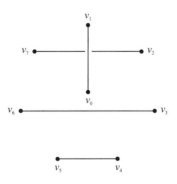

Fig. 16.10

Which class of graphs is "larger?" It does not appear obvious at all! Paul Erdős and Robin J. Wilson showed in 1977 ([EW]) that *almost all graphs are of class one.* "Almost all" is made precise by the authors of [EW]:

Problem 16.6 ([EW]) If U_n is the number of graphs with n vertices of class one, and V_n is the total number of graphs with n vertices, then $\frac{U_n}{V_n}$ approaches 1 as n approaches infinity.

But how do we determine which graph belongs to which class? Nobody knows!

In 1973, Lowell W. Beineke and Robin J. Wilson published [BW] the following simple sufficient condition for a graph to be of the second class.

The *edge independence number* $\beta_1(G)$ of a graph G is the maximum number of mutually non-adjacent edges of G. ("Mutually non-adjacent edges" means every two edges are non-adjacent.)

Problem 16.7 ([BW]) Let G be a graph with q edges. If

$$q > \Delta(G) \cdot \beta_1(G),$$

then G is of class two.

Proof Assume G is of class one, i.e., $\chi_1(G) = \Delta(G)$, hence, we can think of G as being $\Delta(G)$-edge colored. How many edges of the same color can we have in G? At most $\beta_1(G)$ because the edges of the same color must be mutually non-adjacent. Therefore, the number of edges q in G is at most $\Delta(G) \cdot \beta_1(G)$, which contradicts the given inequality. G is of class two. ∎

Problem 16.8 For any graph G with p vertices

$$\beta_1(G) \le \left[\frac{p}{2}\right]$$

where $\left[\frac{p}{2}\right]$ denotes the maximum integer not exceeding $\frac{p}{2}$.

Proof Assume that the graph G has $\beta_1(G)$ mutually non-adjacent edges. The p vertices of G are thereby partitioned into $\beta_1(G)$ two-vertex subsets plus perhaps one more subset (of vertices non-incident with any of the $\beta_1(G)$ edges). Therefore, $\beta_1(G) \le \frac{p}{q}$, but as an integer $\beta_1(G) \le \left[\frac{p}{q}\right]$. ∎

Problems 16.7 and 16.8 join in for an immediate corollary.

Problem 16.9 Let G be a graph with p vertices and q edges. If

$$q > \Delta(G) \cdot \left[\frac{p}{2}\right]$$

then G is of class two.

The last problem shows that graphs with relatively large ratio of their number of edges and the number of vertices are "likely" to be of class two.

Yet, conditions of Problems 16.7 and 16.9 are far from being necessary. Can you think of a counterexample? Here is one for you:

Problem 16.10 Show that the Peterson graph (Fig. 13.3) is of class two even though it does not satisfy the inequalities of Problems 16.7 and 16.9.

This is a mysterious, relatively rare class two: can we use another approach to gain an insight? We can gain an insight if we limit our consideration to planar graphs, i.e., those that can be embedded in the plane without intersection of edges. It is easy to find (do) class two planar graphs G with the maximum degree $\Delta(G)$ equal to 2, 3, 4, and 5. We do not know whether the maximum degree 6 or 7 can be realized in a class two planar graph. In 1965 Vadim G. Vizing [Viz2] proved that higher than 7 maximum degrees are impossible.

Problem 16.11 (Vizing, [Viz2, Theorem 4]) If G is a planar graph with $\Delta(G) \geq 8$, then G belongs to class one.

The following problem is still awaiting its solution:

Open Problem 16.12 Find criteria for a graph G to belong to class two.

16.2 Total Insanity around the Total Chromatic Number Conjecture

In February 1992, I gave my first talk at the International Southwestern Conference on Combinatorics, Graph Theory and Computing at Florida Atlantic University, Boca Raton, Florida. I gave a talk about chromatic number of the plane, and my research into the authorship of the problem. My investigative skills must have looked good, for the British graph theorist Hugh R. Hind shared with me another controversy. In his manuscript on total chromatic number conjecture, Hugh gave credit for the conjecture to Vizing and Behzad. As a condition of publication, the referee demanded that the credit be given to Behzad alone. While Hind thought that both mathematicians authored the conjecture independently and deserved credit, he felt that he had no choice but to comply with the referee's demand. Hugh asked me to investigate the authorship of the total chromatic number conjecture.

I was shocked. The referee's ultimatum, backed by the editor (who sent the referee report to the author), seemed to be nothing short of the cold war on the mathematical front. What were the referee's and the editor's motives? Was it retaliation for the Soviet anti-Semitism and other violations of scientific norms? Was it retaliation for the leading Soviet graph theorist A. A. Zykov's ridiculously giving in his book [Zyk3] credit for the Kuratowski Planarity Theorem to both Pontryagin and Kuratowski? (Of course, Zykov's crediting Pontryagin was outrageous, and Pontryagin deserved no credit whatsoever.) However, life is no math—it does not multiply two negatives to get positive—two wrongs make no right. Surely, the referee and the editor of Hind's manuscript acted every bit as wrongly as the Soviet apparatchiks—unless they had historical factual grounds to deny Vizing credit, grounds they never disclosed. I accepted the call to investigate. What follows is my report.

Total chromatic number $\chi_2(G)$ of a graph G is the minimum number of colors required for coloring vertices *and* edges of G so that incident and adjacent elements are never assigned the same color.

Total Chromatic Number Conjecture 16.13 For any graph G,

$$\chi_2(G) \le \Delta(G) + 2.$$

I started my investigation right away, in Boca Raton during the same conference (February 1992). I asked the well-known graph theorist Mark K. Goldberg, professor of computer science at Rensselaer Polytechnic Institute, whether he knew anything about the authorship of the total chromatic number conjecture. This was a very lucky choice, for Mark was an eye-witness to the story. Goldberg told me that in December 1964 he arrived in Academgorodok (Academy Town), located just outside Novosibirsk, a city in Russian Siberia, to apply for their Ph.D. program in mathematics. During this trip he interacted with the Junior Research Staff member Vadim G. Vizing, who shared with Goldberg his edge-chromatic number theorem and the total chromatic number conjecture.

Three years later I was able to ask Vadim Vizing himself about the total chromatic number conjecture. I learned from Bjarne Tofts, a professor at Odense University in Denmark that Vadim G. Vizing was visiting him, and on March 12, 1995 I asked Toft to pass my e-mail with numerous questions to Vizing. I asked biographical questions and, of course, questions about the conjecture. Two days later, on March 14, 1995, I received the following reply (my translation from Russian):

Dear Alexander!

At the present time I am in Odense on B. Toft's invitation.

I was born March 25, 1937 in Kiev. I commenced my work on Graph Theory in 1962 as a Junior Research Staff if the Institute of Mathematics in Novosibirsk, in the Department of Computing Techniques [Computer Science]. As part of my job I had to write a program for coloring conductors in circuits. I discovered C. E. Channon's work, dedicated to this question, published in 1949 (Russian translation was published in 1960). Having studied Channon's work, I began to think about the precision of his bound. I knew only one type of multigraphs on which his bound was precise [best possible]. This is why I assumed that for ordinary graphs (without multiple edges) Channon's bound could be strengthened. It took a year and a half for me to prove my theorem for ordinary graphs.

In early 1964 the article was sent to "Doklady AN USSR," but was rejected by the editorial board. In the fall of 1964 I obtained the generalization of the result to p-graphs and published an article about it in the antology "Diskretnyi Analiz", issue 3 [Viz1] that was released in December 1964 in Novosibirsk (I am mailing to you a copy of this article).

In early 1964, while presenting the theorem about coloring edges of a graph at A. A. Zykov's' Seminar (present were A. A. Zykov, L. S. Melnikov, K. A. Zaretskij, V. V. Matjushkov, and others), I formulated the conjecture on the total chromatic number, which we called then *conjecture on the simultaneous coloring of vertices and edges*. Many of my colleagues in Novosibirsk attempted to prove the conjecture but without success. By the time of publication [Viz3] of my article on unsolved problems of Graph Theory in "Uspekhi Mat. Nauk" (1968), in which I first published the conjecture, the conjecture already had wide distribution among Soviet mathematicians. In the nearest future I will mail to you the article in Russian, in which the conjecture on the total chromatic number of multigraph appears on p. 131.

Thus, Vizing's recollection of creating the total chromatic number conjecture by early 1964, verified independently by Mark Goldberg, leaves no doubts about his authorship. Vizing's total chromatic number conjecture was also presented by Alexander A. Zykov at the problem session of the Manebach Colloquium in May 1967 (published in 1968 [Zyk2], p. 228). Vizing published this conjecture himself among many other open problems in 1968 [Viz3]. In addition, any impartial expert would agree that this conjecture was a natural continuation of the train of thought emanating from Vizing's famous theorem on chromatic index of the graph (Theorem 16.3 above).

I then looked at articles of specialists working on the total chromatic number of a graph. Hugh R. Hind [Hin1], [Hin2], Anthony J. W. Hilton and Hind [HH], and Amanda G. Chetwynd almost universally credited Behzad with the conjecture. Chetwynd even "explained" what led Behzad to discover the conjecture [Che]:

This [i.e., Brooks' Theorem and Vizing's Theorem] led Behzad to conjecture a similar result for the total chromatic number.

What is wrong with this "explanation?" Everything:

1. In reading Mehdi Behzad's 1965 thesis (Chetwynd obviously did not read it before writing about it), it is obvious that Behzad did not know Vizing's Theorem: Behzad conjectures the statement of Vizing's Theorem, but is able to prove it only for graphs of maximum degree 3.
2. If Vizing's Theorem led even Behzad to the total chromatic conjecture, it would have surely led (and did!) Vizing himself to formulate the conjecture. Then why does Chetwynd give no credit to Vizing?

On December 2, 2007 I contacted Professor Behzad, and asked him to present a case in support of his sole authorship. He kindly submitted his final text of the reply in the December 14, 2007 e-mail to me:

I started to think about my Ph.D. thesis in 1963—1964, at Michigan State University, to be written under the supervision of Professor E. A. Nordhaus. In those days there was only one book in the field of graph theory in English, and no courses were offered on the subject. I was interested in vertex coloring and then line coloring. For several months, naively, I tried to solve the 4-color problem. Then I thought of combining these two types of colorings. I mentioned the notion, which was later called "total chromatic number of a graph,"[5] to Nordhaus. He liked the idea, but for several months he did not allow me to work on the notion. Later he told me this idea was so natural that he thought someone might have worked on the subject. Thanks to Professor Branko

---

[5] According to Behzad, it was Nordhaus who coined the term.

Grünbaum who resolved the problem. In my thesis I introduced this notion and presented the related conjecture. In addition, I introduced the total graph of a graph in such a way that the total chromatic number of G was equal to the vertex chromatic number of its total graph. My Thesis was defended in the Summer of 1965. Prior to 1968, when Professor Vizing's paper entitled "Some Unsolved Problems in Graph Theory" appeared, several papers were published on topics related to total concepts;[6] I informally talked about TCC in two of the conferences that I attended in 1965 and 1966 held at The University of Michigan, and the University of Waterloo. As I mentioned before, aside from my thesis, the Proceedings of the International Symposium in the Theory of Graphs – Rome, 1966 contains the subject and the TCC...

As far as I know, out of several hundred articles, theses, books, and pamphlets containing TCC, none omit my name, and very many authors provide only one reference for TCC and that is my thesis. I am not aware of a single work mentioning TCC and giving reference to Vizing alone. There are authors who have given credit to the two of us, but have decided to stop doing so.

I am reading Mehdi Behzad's Ph. D. thesis [Beh]. Perhaps, due to the field being relatively new in 1965, to me Behzad's thesis appears light on deep proofs. However, the author, demonstrates a fine intuition: he conjectures (already published by Vizing a year earlier) Vizing's Theorem on edge-chromatic number of a graph (conjecture 1, p. 18), and formulates the total chromatic number conjecture (conjecture 1, p. 44).

Behzad and Chartrand submitted their "expository article" on total graphs to the 1966 Rome Symposium, and it was published [BC1] in 1967. It says (I just replaced notations to contemporary ones):

It was conjectured in [Beh] that

i. $\Delta(G) \le \chi_1(G) \le \Delta(G) + 1$, and
ii. $\Delta(G) + 1 \le \chi_2(G) \le \Delta(G) + 2$

The conjecture (i) has been proved by Vizing [Viz1], but (ii) remains an open question.

The good news is that Behzad (with Chartrand) published the total chromatic number conjecture. As to "conjecture (i)", we are told above that Behzad conjectured the chromatic index theorem, and Vizing proved Behzad's conjecture. In reality, Vizing's paper already came out in 1964, a year before Behzad ever conjectured this result. Of course, Vizing worked on *his* conjecture on the chromatic index much earlier, during 1962–1963, for as he says, it took him a year and a half to prove his conjecture. In early 1964 he finally submitted his paper.

---

[6] Indeed, the following papers, authored or coauthored by Behzad, address total graphs, but do not include the total chromatic number conjecture: M. Behzad and G. Chartrand, Total graphs and traversability, *Proc. Edinburgh Math. Soc.* (2) 15 (1966). 117–120. M. Behzad, G. Chartarand, and J.K. Cooper Jr. The colour numbers of complete graphs, *J. Lond. Math. Soc.* 42 (1967) 225–228. M. Behzad, A criterion for the planarity of the total graphs of a graph, *Proc. Cambridge Philos. Soc.* 63 (1967) MR35#2771. M. Behzad and H. Radjavi, The total group of a graph, *Proc. Amer. Math. Soc.* 19 (1968), 158–163.

M. Behzad had no way of knowing about my findings showing that Vizing formulated the total chromatic number conjecture in early 1964, i.e., well before Behzad. However, Behzad knew about Vizing's 1968 paper [Viz3], where the total chromatic number conjecture was published. Surely, it took time, prior to the submission of this paper, for Vizing to assemble such a large survey of unsolved problems of graph theory. Thus, independent authorship of Vizing should not have been questioned. Yet, Behzad's 1971 book [BC2], p. 214] joint with Gary Chartrand, his former fellow Ph.D. student of E. A. Nordhaus, gave the sole credit to Behzad for the total chromatic number conjecture, as did the 1979 fine book written by Behzad, Chartrand, and Linda Lesniak-Foster [BCL, p. 252].

I informed Professor Behzad of my findings in our phone conversation on January 2, 2008. He was pleased that at long last someone had taken the time and effort to investigate the credit for this famous conjecture, and that the credit was rightfully due to the two discoverers.

Summing it all up, the total chromatic number conjecture has been first formulated by Vadim G. Vizing in early 1964, and published in 1968. Mehdi Behzad independently formulated the conjecture in the unpublished thesis in summer of 1965, and published it (jointly with Gary Chartrand) in 1967. In my opinion, this unquestionably merits the joint credit to Vizing and Behzad.

I hope this analysis will end editorial room bias, threats and politicking, and will restore the joint credit for the conjecture. Joint credit and correct publication dates were given by Tommy R. Jensen and Bjarne Toft in their enlightened 1995 problem book [JT] and repeated in Reinhard Diestel's textbook [Die]. In later papers, e.g., [HMR], Hugh Hind, Michael Molloy, and Bruce Reed give credit to both Vizing and Behzad for the concept of the total chromatic number and the conjecture. Yet, even in 2005 the latest, 4th edition of *Graphs & Digraphs* [CL] by Chartrand and Lesniak (Behzad is not listed as a coauthor), still credits M. Behzad, and Behzad alone, for total coloring and the total chromatic number conjecture. I hope that, having read these lines, the authors will correct the credit in their next edition.

Mehdi Behzad is a professor of mathematics at Shahid Beheshti University in Iran. Following his visit of Denmark in 1995, Vadim Vizing, who recently worked on the theory of scheduling, wrote to me that he was going to renew "intensive work on graph theory," and has indeed, as his publications show. He lives in Odessa, Ukraine.

In spite of active work and partial results, the total chromatic number conjecture remains as challenging as it is open. With an ease of formulation and apparent difficulty of proving, this conjecture now belongs to mathematics' classic open problems.

17
Carsten Thomassen's 7-Color Theorem

One day in 1998, I was asked by *The American Mathematical Monthly* to referee a manuscript submitted by one of the world's leading graph theorists, Professor Carsten Thomassen of The Technical University of Denmark. The paper offered a fresh, purely graph-theoretic approach to finding the chromatic number of the plane. I was very impressed, and asked the author to expand his too concise (for *The Monthly*) presentation, and informed him of my research that proved that Edward Nelson, and Nelson alone (without Hadwiger) was the author of the problem. In this chapter I will present Thomassen's attempt to find the chromatic number of the plane. Though he has not found it—no one has—he still obtained a fine result, and in the process showed how graph theory proper can be utilized for an assault on this problem. I will present Thomassen's proof with minor editorial revisions. The use of paper and pencil is a must while reading a proof written in Thomassen's style.

Thomassen offers a vast generalization of the popular hexagonal coloring that we used to prove the upper bound 7 (Chapter 2) to the class of colorings that he calls *nice*. He considers a graph G on a surface S that is a metric space (i.e., curve-wise connected Hausdorff space in which each point has a neighborhood homeomorphic to an open circular disc of the Euclidean plane). The graph G on the surface S creates a map $M(G, S)$, in which a region is an edge-connected component of $S \backslash G$. For his purposes, Thomassen assumes that each region that has diameter of less than 1 is homeomorphic to a Euclidean disc and is bounded by a cycle in G. I choose to avoid a detour into basics of topology, and offer the unfamiliar reader to simply think that S is the plane or a sphere, i.e., the graph G is drawn on the Euclidean plane or a sphere—coloring the plane is, after all, our main goal.

The *area* of a subset A of S is the maximum number of pairwise disjoint open discs of radius $1/2$ that can be packed in A. (If this maximum does not exist we say that A has infinite area.) A simple closed curve C is *contractible* if $S \backslash C$ has precisely two edge-wise connected components such that one of them is homeomorphic to an open disc in the Euclidean plane. That component is called the *interior* of C and is denoted $int(C)$. (If S is a sphere, then $int(C)$ denotes any component of $S \backslash C$ of the smallest area).

Given a graph G on a surface S, Thomassen defines *nice coloring* of S as a coloring in which each color class is the union of regions (and part of their boundaries) such that the distance between any two of these regions is greater than 1.

A. Soifer, *The Mathematical Coloring Book*,
DOI 10.1007/978-0-387-74642-5_17, © Alexander Soifer 2009

Finally, I need to introduce here a *map-graph duality*, which we will use not only in this chapter but in the next part as well. Given a map M, we can define the graph of the map, or *map graph* $\Gamma(M)$ as the graph whose vertices are the regions of M with two vertices adjacent if and only if the corresponding regions share part of their boundary, which is not merely a finite number of points. If the map $M(G, S)$ is induced by the graph G on the surface S, we will simplify the notation for the map graph to $\Gamma(G, S)$.

Thomassen's 7-Color Theorem 17.1 Let G be a connected graph on a surface S satisfying (i), (ii), and (iii) below. Then every nice coloring of S requires at least 7 colors.

   (*) Every non-contractible simple closed curve has diameter at least 2.
   (**) If C is a simple closed curve of diameter less than 2, then the area of $int(C)$
        is at most k.
   (***) The diameter of S is at least $12k + 30$.

Before proving his theorem, Thomassen introduces Tool 17.2, for which he needs a few notations and definitions.

If $V(G)$ is the vertex set of a graph G and $x \in V(G)$, then $D_1(x)$ stands for the set of neighbors of x. For $n > 2$, we define $D_n(x)$ inductively as the set of vertices in $V(G) \setminus \big[\{x\} \cup D_1(x) \cup \ldots \cup D_{n-1}(x) \big]$ that have a neighbor in $D_{n-1}(x)$. A graph G is called *locally finite* if $D_1(x)$ is finite for each vertex x of G; and *locally connected* if the minimal subgraph of G that contains $D_1(x)$ is connected for each vertex x of G. We call G *locally Hamiltonian* if G has a cycle with vertex set $D_1(x)$ for each vertex x of G.

Tool 17.2 Any connected locally finite, locally Hamiltonian graph with at least 13 vertices has a vertex of degree at least 6.

Proof If no vertex of the graph G satisfying all conditions has degree at least 6, pick a vertex x of maximum degree. Clearly $deg(x) \geq 3$.

Assume $deg(x) = 3$. Since G contains a cycle with the vertex set $D_1(x)$, the subgraph of G induced by $\{x\} \cup D_1(x)$ is the graph of the tetrahedron, Since maximum degree in G is 3, $D_2(x)$ is empty. Since G is connected, G is the graph of the tetrahedron, i.e., has just four vertices, in contradiction to the assumption that G has at least 13 vertices.

Assume now $deg(x) = 4$. Since vertices of $D_1(x)$ form a cycle, we can conclude that each vertex $y \in D_1(x)$ has at most one neighbor $z \in D_2(x)$. Since vertices of $D_1(y)$ form a cycle, z has at least three neighbors in $D_1(x)$. Thus, there are at most four edges from $D_1(x)$ to $D_2(x)$, and therefore every vertex in $D_2(x)$ has at least three neighbors in $D_1(x)$. Hence $D_2(x)$ has at most one vertex z. Since vertices of $D_1(z)$ form a cycle, it follows that $D_3(x) = \emptyset$. Thus, G has at most six vertices, a contradiction.

Finally assume $deg(x) = 5$. Each vertex $y \in D_1(x)$ has at most two neighbors in $D_2(x)$ because vertices of $D_1(x)$ form a cycle and this cycle uses up two points out of the maximum degree 5 of y. Since vertices of $D_1(y)$ form a cycle, every neighbor

z of y in $D_2(x)$ has at least two neighbors in $D_1(x)$. Observe that z cannot have two or more neighbors in $D_3(x)$ because then a cycle with vertex set $D_1(z)$ will show that z has at least two neighbors in $D_2(x)$ that is, z has a total of at least six neighbors, a contradiction. So z has at most one neighbor in $D_3(x)$ and that neighbor has at least three neighbors in $D_2(x)$. Since there are at most 10 edges from $D_1(x)$ to $D_2(x)$, and every vertex in $D_2(x)$ has at least two neighbors in $D_1(x)$, it follows that $D_2(x)$ has at most five vertices. Hence there are at most five edges from $D_2(x)$ to $D_3(x)$. Since each vertex in $D_3(x)$ has at least three neighbors in $D_2(x)$, it follows that $D_3(x)$ has at most one vertex, and thus $D_4(x) = \emptyset$. Hence G has at most 12 vertices, a contradiction that completes the proof. ∎

Now we are ready to prove the theorem.

Proof of Thomassen's Theorem Given a graph G on a surface S that satisfies (i), (ii), and (iii). Assume the opposite, i.e., there is a nice coloring utilizing at most 6 colors. Let x be a vertex of the map graph $\Gamma = \Gamma(G, S)$ and let C_x be the cycle in G bounding the corresponding region. Let us choose an orientation of C_x and let $x_1, x_2, \ldots, x_k, x_1$ be the vertices of $D_1(x)$ listed in the order in which they appear as we traverse C_x.

Thomassen first considers a simple case, which illustrates the idea of his proof. Assume that for each vertex x, all vertices $x_1, x_2, \ldots, x_k$ are distinct. In this case, Γ is locally Hamiltonian. Since the surface S is edgewise connected, it follows that Γ is connected. Since S has diameter greater than 13, Γ has more than 12 vertices, and hence, by Tool 17.2, Γ has a vertex of degree at least 6. Now x and its neighbors must have distinct colors because x corresponds to a face of diameter < 1 on S. This contradiction completes the proof in this case.

In the general case, a vertex may appear more than once in the sequence $x_1, x_2, \ldots, x_k, x_1$ above. Omit those appearances (except possibly one) of x_i for which C_{x_i} and C_x have only one vertex in common. In other words, if x_i appears more than once in the new sequence, we list only those appearances for which C_{x_i} and C_x share an edge. Then any two consecutive vertices in the sequence $x_1, x_2, \ldots, x_k, x_1$ are neighbors in Γ, and thus Γ is locally connected. It follows that $\Gamma - x$ is connected. Moreover, if y is any other vertex of Γ, then $\Gamma - x - y$ is connected unless y appears twice in the sequence $x_1, x_2, \ldots, x_k$, that is, C_x and C_y have at least two edges in common.

Let now x and y be vertices such that C_x and C_y have at least two edges e and f in common, i.e., $y = x_i = x_j$ for $1 \leq i < j - 1 < k - 1$). Let R be a simple closed curve in the regions bounded by C_x and C_y such that R crosses each of e and f precisely once and has no other point in common with G. By (i), R is contractible. Hence $\Gamma - x - y$ is disconnected. We say in this case that $\{x, y\}$ is a *2-separator* in Γ. For each vertex z in Γ such that C_z is in $int(R)$ and has color 1, we pick a point P_z in $int(C_z)$. By (ii), there are at most k points P_z and hence there are altogether at most $6k$ vertices z such that $int(C_z) \subseteq int(R)$.

Let $int(\Gamma, x, y)$ stand for the subgraph of $\Gamma - x - y$ induced by all those vertices z in Γ such that C_z is in $int(R)$ for some R. Then each connected component of $int(\Gamma, x, y)$ has at most $6k$ vertices. Since S has diameter at least $12k + 3$, it follows that G has two vertices whose graph distance is at least $12k+2$. Hence $\Gamma - x - y$ has some component that is not in $int(M, x, y)$. We claim that $\Gamma - u - v$ has precisely one such component, which we call $ext(M, x, y)$. To see this, let $e_1, e_2, \ldots, e_m$ be the edges in $C_x \cap C_y$ occurring in this cyclic order on C_x. Then $e_1, e_2, \ldots, e_m$ divide $D_1(x) \backslash \{y\}$ into m classes $A_1, A_2, \ldots, A_m$. By letting $\{e, f\} = \{e_i, e_{i+1}\}$, $1 \le i < m$ in the preceding argument, we conclude that for each $i = 1, 2, \ldots, m$, either $A_i \subseteq int(\Gamma, x, y)$ or $A_i \cap int(\Gamma, x, y) = \emptyset$. Since the former cannot hold for each $i \in \{1, 2, \ldots, m\}$, the latter must hold for some i, and hence the former holds for all other $i \in \{1, 2, \ldots, m\}$. Thus, we proved that for any two vertices x, y in Γ, $\Gamma - x - y$ has precisely one connected component $ext(\Gamma, x, y)$ with more than $6k$ vertices.

If $\{u, v\}$ is a 2-separator in Γ such that either x or y or both are in $int(\Gamma, u, v)$, then clearly $int(\Gamma, x, y) \subset int(\Gamma, u, v)$. (To see this, we use the properties of Γ established previously and forget about S.) If no such 2-separator $\{u, v\}$ exists, then we say that $\{x, y\}$ is a *maximal 2-separator* and that xy is a *crucial edge*. Since each connected component of $int(\Gamma, x, y)$ has at most $6k$ vertices, a maximal 2-separator exists (provided a 2-separator exists). Let H be the subgraph of Γ obtained by deleting $int(\Gamma, x, y)$ for each maximal 2-separator $\{x, y\}$. Then $H \ne \emptyset$. Moreover, since a shortest path in Γ between two vertices in H never uses vertices in $int(\Gamma, x, y)$, H is connected. We can similarly prove that H is locally connected. We now claim that H is locally Hamiltonian. Consider again a vertex x in H and the sequence $x_1, x_2, \ldots, x_k, x_1$ in $D_1(x)$; (taken in Γ). If this sequence forms a Hamiltonian cycle in $D_1(x)$ in H, we are done. By definition of H, $k \ge 3$. So assume that $x_i = x_j$ where $1 \le i < j - 1 < k - 1$. Then $\{x, x_i\}$ is a 2-separator and vertices can be re-indexed so that $int(\Gamma, x, x_i)$ contains all the vertices $x_{i+1}, x_{i+2}, \ldots, x_{j-1}$. We repeat this argument for each pair i, j such that $x_i = x_j$ where $1 \le i < j - 1 < k - 1$. Then the vertices in $x_1, x_2, \ldots, x_k, x_1$ that remain after we delete all vertices in the interiors of the 2-separators form a cyclic sequence with no repetitions. As H is connected and locally connected and has at least three vertices (by (iii)), the preceding reduced cyclic sequence has at least two distinct vertices. It cannot have precisely two vertices u, v because then $H - u - v$ is disconnected, and hence $\Gamma - u - v$ is disconnected (because Γ is obtained from H by "pasting graphs on edges of H"). Since one of the edges xu or xv is crucial (because $D_1(x)$ is smaller in H than in Γ), the maximality property of the 2-separator $\{x, u\}$ or $\{x, v\}$ implies that $ext(\Gamma, u, v)$ is the connected component of $\Gamma - u - v$ containing x. For each vertex z in that component, Γ has a path of length at most $6k$ from z to either x, u, or v. Hence Γ has diameter at most $12k + 1$, a contradiction that proves that H is locally Hamiltonian.

If H has a vertex x of degree at least 6, we are done because x and its neighbors must have different colors in the nice coloring. Assume now that each vertex of H has degree at most 5. By Tool 17.2, H has at most 12 vertices. Hence H has at most 30 edges. Since Γ is obtained from H by "pasting" $int(\Gamma, x, y)$ on the crucial edge xy for each crucial edge of H, we conclude that the diameter of Γ is at most $12k + 29$, a contradiction to (iii). ∎

Observe: all three conditions in the theorem are essential. If *any* of these conditions (i), (ii), (iii) were dropped, then the number of colors needed may decrease:

A thin two-way infinite cylinder has a nice 6-coloring, which shows that (i) cannot be omitted.

A thin one-way infinite cylinder (with a small disc pasted on the boundary of the cylinder to form the bottom) shows that (ii) cannot be omitted.

A sphere of diameter less than 1 has a nice coloring in 2 colors, hence (iii) cannot be omitted.

In Chapter 24 we will look at an analogous Townsend–Woodall's 5-Color Theorem, obtained by different means.

IV
Coloring Maps

G. D. Birkhoff once told one of the authors that every great mathematician had at some time attempted the Four Colour Conjecture, and had for a while believed himself successful.
-- Hassler Whitney and W. T. Tutte [WT]

The word disease is quite appropriate for a puzzle which is easy to comprehend, apparently impossible for anyone to solve, infectious, contagious, recurrent, malignant, painful, scarring, and sometimes even hereditary!
-- Frank Harary[1]

If I may be so bold as to make a conjecture, I would guess that a map requiring five colors may be possible.
-- H. S. M. Coxeter [Cox2]

In this part, we will color regions of maps. The following few definitions will help us formalize our intuitive notion of a map.

By allowing more than one edge to connect two vertices, we slightly generalize a notion of a graph: what we get is called a *multigraph*. A multigraph that *can be* drawn in the plane without intersection of its edges is called *planar*, while a multigraph that *is* drawn in the plane without intersection of its edges is called *plane*. A multigraph is called *connected* if for any two vertices there is a path connecting them. An edge x of a connected multigraph G is called a *bridge* if the multigraph $G - x$ is not connected.

A plane connected multigraph without bridges is called a *map*. A map divides the plane into *regions*. Regions are *adjacent* if they share at least one edge.

Coloring a map is an assignment of colors to each of the regions of the map, such that no adjacent regions get the same color. Let n be a positive integer; a map M is called *n-colorable* if there is a coloring of M in n colors.

---

[1] From the appropriately entitled paper [Har1] "The Four Color Conjecture and other graphical diseases," appropriately "supported in part by a grant from the National Institute of Mental Health."

A natural question then is what is the *minimum* number of colors we must use to color any map? You can easily construct (do) an example showing that four colors are necessary. You have likely heard the puzzle and the conjecture I am going to introduce here as an overture to this map-coloring part.

The following puzzle originated in discussions between August Ferdinand Möbius and his amateur mathematician friend Adolph Weiske, and "was perhaps originated by Weiske" [Tie2]. In his 1840 lecture, Möbius shared the puzzle with the public. It was apparently solved even by the Bishop of London, later Archbishop of Canterbury (see Chapter 19 for details). Wouldn't you like to solve it on your own and try to help the brothers!

Möbius–Weiske's Puzzle IV-1 (Circa 1840) Once upon a time in the Far East there lived a Prince with five sons. These sons were to inherit the kingdom after his death. But in his will, the Prince made the condition that each of the five parts into which the kingdom was to be divided must border on every other. ... After the death of the father, the five sons worked hard to find a division of the land which would conform to his wishes; but all their efforts were in vain.[2]

The following conjecture, together with Fermat's Last Theorem, had been the two most popular open problems of mathematics.

The Four-Color Conjecture (4CC) IV-2 (Francis Guthrie, 1852 or before) Any map in the plane is 4-colorable.

My friend Klaus Fischer of George Mason University once asked me in the early 1990s, why would one want to write about the conjecture so celebrated that everything has been written about it? Well, *everything* is never written, I replied, and every little bit helps.

---

[2] [Tie2].

18
How the Four-Color Conjecture Was Born

18.1 The Problem is Born

It takes time and effort to gain access and read manuscripts. The letter containing the first mention of the 4CC is of high importance, yet to the best of my knowledge, its complete facsimile has never been reproduced. Selected transcriptions and fragment facsimiles served a purpose, but as we will see in Section 18.2, they contained certain shortcomings. In view of this, I am reproducing here, for the first time, the facsimile of De Morgan's letter to Hamilton; the relevant fragment of the latter's reply, analysis of these documents, and the corrected transcription of De Morgan's letter. I am grateful to The Board of the Trinity College Dublin, whose kind permission made reproducing of the letters [DeM1] and [Ham] possible (see the facsimiles in this chapter on pp. 148–151 and 154).

A. Soifer, *The Mathematical Coloring Book*,
DOI 10.1007/978-0-387-74642-5_18, © Alexander Soifer 2009

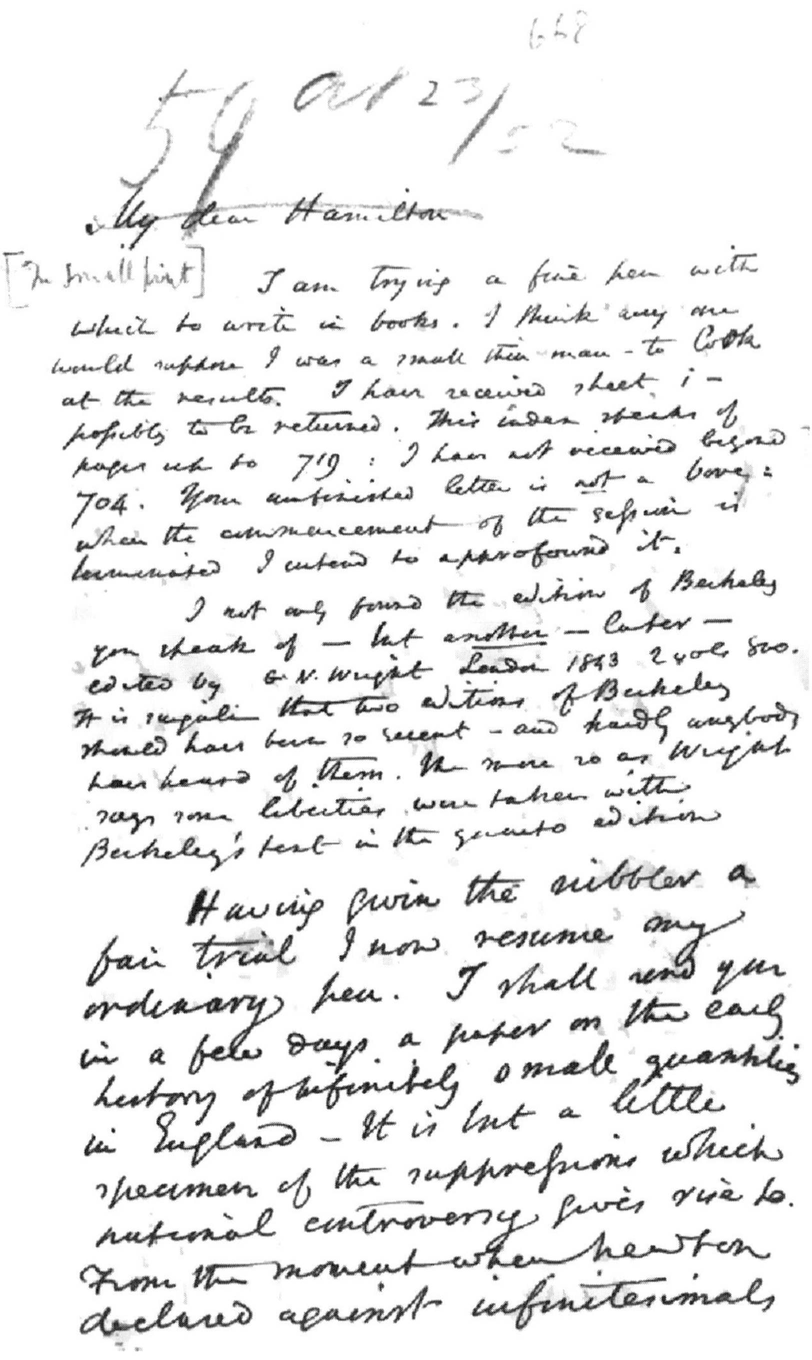

648

5 9   Oct 23/52

My dear Hamilton

[In small print] I am trying a fine pen with
which to write in books. I think any one
would suppose I was a small thin man – to look
at the results. I have received sheet i –
possibly to be returned. This index steals of
paper up to 719 : I have not received beyond
704. Your unfinished letter is not a bore :
when the commencement of the sequin is
terminated I intend to approfound it.

        I not only found the edition of Berkeley
you speak of – but another – later –
edited by G. N. Wright London 1843 2 vol. 8vo.
It is singular that two editions of Berkeley
should have been so recent – and hardly anybody
have heard of them. The more so as Wright
says some liberties were taken with
Berkeley's text in the quarto edition

        Having given the nibbler a
fair trial I now resume my
ordinary pen. I shall send you
in a few days a paper on the early
history of infinitely small quantities
in England – It is but a little
specimen of the suppressions which
national controversy gives rise to.
From the moment when newton
declared against infinitesimals

(1704) which title then he had
exclusively used in fluxions,
the English world agreed to
suppose that they never had
been used here — and to forget
the works in which they had been
used — All these works are now
absent from the Roy: Soc: library,
except Newton's Principia

A student of mine asked
me to Day to give him a reason
for a fact which I did not
know was a fact — and do
not yet. He says that, if
a figure be any how divided
and the compartments differently
Coloured so that figures with
any portion of common boundary
line are differently coloured
— four colours may be wanted
but not more — The following
is his case in which four
are wanted

A B C &c are
names of
colours

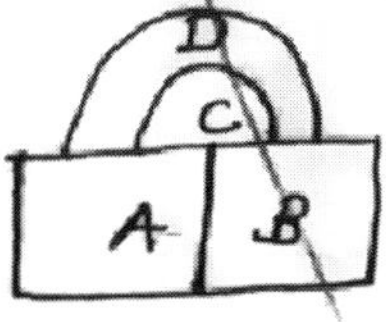

'Query cannot a necessity for
five or more be invented?
As far as I see at this moment,
if four ultimate compartments have each
boundary line in common with
one of the others, three of them
inclose the fourth, and prevent
any fifth from connexion
with it. If this be true, four
colours will colour any possible
map without any necessity
for colour meeting colour
except at a point.

Now it does seem that
drawing three compartments
with common boundary   A B C
two and two – you cannot
make a fourth
take boundary from
all, except by
inclosing one – But
it is tricky work 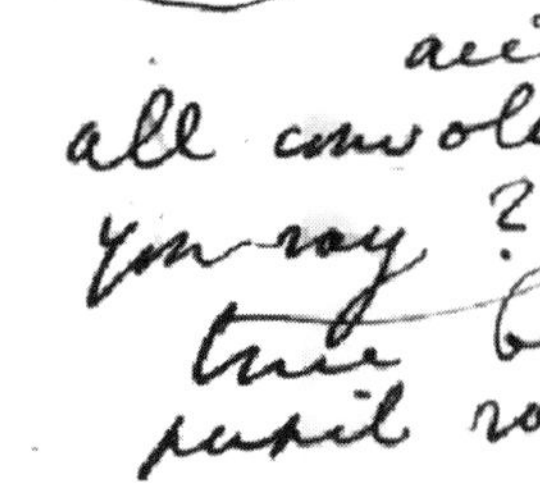
and I am not sure of
all convolutions – What do
you say? And has it, if
true been noticed? My
pupil says he guesses it in

colouring a map of England,

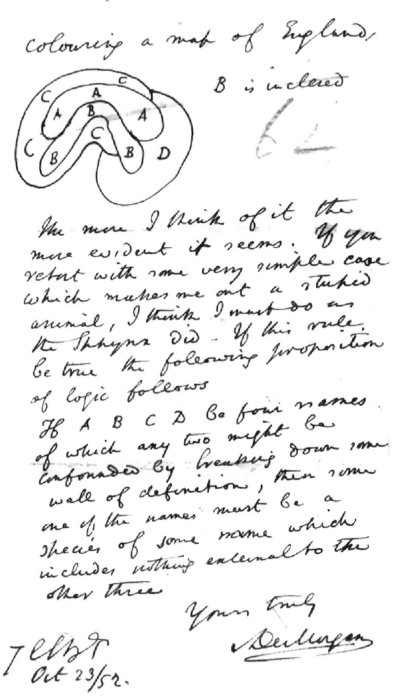

B is inclosed

the more I think of it the
more evident it seems. If you
retort with some very simple case
which makes me out a stupid
animal, I think I must do as
the Sphynx did — If this rule
be true the following proposition
of logic follows

If A B C D be four names
of which any two might be
confounded by breaking down some
wall of definition, then some
one of the names must be a
species of some name which
includes nothing external to the
other three

                    Yours truly
                        DeMorgan

7 ElbT
  Oct 23/52.

Augustus De Morgan, Letter to W. R. Hamilton, October 23, 1852. Courtesy of the Trinity College, Dublin

The written record of the problem begins with the October 23, 1852 letter that Augustus De Morgan, Professor of Mathematics at University College, London, writes [DeM1] to Sir William Rowan Hamilton, Professor of Mathematics at Trinity College, Dublin (underlining in the manuscript):

My dear Hamilton[3].

  ...A student of mine asked me to day to give him a reason for a fact which I did not know was a fact – and do not yet. He says that if a figure be any how divided and the compartments differently coloured so that figures with any portion of common boundary <u>line</u> are differently coloured – four colours may be wanted but not more – the following is *his* case in which four <u>are</u> wanted [.]

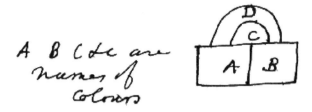

Query [:] cannot a necessity for five or more be invented [?] As far as I see at this moment, if four <u>ultimate</u> compartments have each boundary line in common with one of the others, three of them inclose the fourth, and prevent any fifth from connexion with it. If this be true, four colours will colour any possible map without any necessity for colour meeting colour except at a point.

  Now it does seem that drawing three compartments with common boundary A B C two and two – you cannot make a fourth take boundary from all, except by inclosing one – But it is tricky work and I am not sure of all convolutions – What do you say? And has it, if true [,] been noticed? My pupil says he guessed it in colouring a map of England [.]

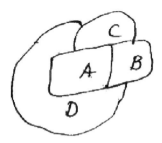

---

[3] De Morgan, A., Letter to W. R. Hamilton, dated Oct. 23, 1852; TCD MS 1493, 668; Trinity College Dublin Library, Manuscripts Department

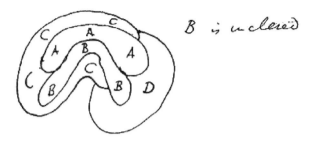

B is unclear

The more I think of it the more evident it seems. If you retort with some very simple case which makes me out a stupid animal, I think I must do as the Sphynx did – If this rule be true the following proposition of logic follows [.]

If A B C D be four names of which any two might be confounded by breaking down some wall of definition, then some one of the names must be a species of some name which includes nothing enternal to the other three [.]

     Yours truly
     ADeMorgan [Signed]

7 CSCT[4]
Oct 23/52

Twenty-eight years later, Frederick Guthrie, the student mentioned by De Morgan in this letter, published his own account [GutFr], which publicly revealed for the first time that the author of the 4CC was his 2-year senior brother, Francis:

Some thirty years ago, when I was attending Professor De Morgan's class, my brother, Francis Guthrie, who had recently ceased to attend them (and who is now professor of mathematics at the South African University, Cape Town), showed me the fact that the greatest necessary number of colors to be used in coloring a map so as to avoid identity of color in lineally contiguous districts is four. I should not be justified, after this lapse of time, in trying to give his proof, but the critical diagram was as in the margin.

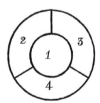

With my brother's permission I submitted the theorem to Professor De Morgan, who expressed himself very pleased with it; accepted it as new; and, as I am informed by those who subsequently attended his classes, was in the habit of acknowledging whence he had got his information.

---

[4] These four letters must stand for De Morgan's address, which was 7 Camden-Street, Camden Town.

If I remember rightly, the proof which my brother gave did not seem altogether satisfactory to himself; but I must refer to him those interested in the subject.

Thus, we learn from the younger brother Frederick Guthrie, by now a Professor of Chemistry and Physics that the 4CC was created by the 20-year-old student Francis Guthrie (of course, he may have been younger when the conjecture first occurred to him), and that Francis Guthrie found a configuration showing that four colors are necessary, shared this simple configuration with his brother Frederick, who passed it to De Morgan. There was likely more to Francis's proof, but it "did not seem altogether satisfactory" to Francis, as Frederick reports above. We will likely never learn what else Francis Guthrie deduced about this incredible for his tender age and the state of mathematics conjecture.

Let us go back to De Morgan. The day he received the 4CC from Frederick Guthrie, October 23, 1852, he immediately wrote about it to William Rowan Hamilton, who was not only one of the leading mathematicians, but also De Morgan's "intimate friend" and lifelong correspondent.[5] Hamilton's October 26, 1852 reply (Royal Post must have worked very well, as there are only 3 days between the dates of De Morgan's letter and Hamilton's reply) is also preserved in the manuscript collection of the Trinity College, Dublin. Hamilton was apparently so obsessed with the *quaternions*[6] he discovered that he could not make himself interested in coloring maps [Ham]:

My dear De Morgan[7]
I am not likely to attempt your "quaternion of colors" very soon...

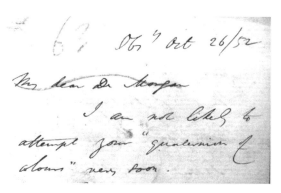

William R. Hamilton, Letter to A. De Morgan, October 26, 1852; a fragment. Reproduced with kind permission by the Board of Trinity College Dublin

---

[5] When W. R. Hamilton died, De Morgan wrote about it in his September 13, 1865 letter to Sir J. F. W. Hershel [DeM5]: "W. R. Hamilton was an intimate friend whom I spoke to once in my life – at Babbage's, about 1830; but for thirty years we have corresponded."

[6] Arguably, this obsession prevented Hamilton from inventing linear algebra.

[7] Hamilton, W. R., Letter to A. De Morgan, October 26, 1852; TCD MS 1493, 669; Trinity College Dublin Library, Manuscripts Department.

And that is all! Just the Victorian, cordial way of saying "I am not interested, lay off my back." De Morgan was left alone to keep the 4CC alive, and he succeeded. He repeatedly mentioned the problem in his lectures at University College ([GutFr]), formulated it in his letters (we know a few such instances: [DeM1], [DeM2][8] and [DeM3][9]). As was discovered in 1976 by John Wilson, a high school teacher from Eugene, Oregon [WilJ], De Morgan was also first to publish the problem in his April 14, 1860 unsigned long review [DeM4] of W. Whewell's *The Philosophy of Discovery* in the *Athenaeum*:

> When a person colours a map, say the counties in a kingdom, it is clear he must have so many different colours that every pair of counties which have some common boundary *line* – not a mere meeting of two corners – must have different colours. Now, it must have been always known to map-colourers that *four* different colours are enough.

Acquaintance of cartographers with the sufficiency of four colors appears to have been a silly invention of De Morgan—there is no evidence that cartographers (then, or now for that matter!) knew about it or needed to minimize the number of colors, since juxtaposition of colors and addition of textures created sufficient representations for scores of additional colors. While De Morgan did not advance the solution of the 4CC at all, he single handedly popularized it and assured its long life. He managed to make the mathematician of the day, Arthur Cayley, hooked on the problem so much that 18 years after the *Athenaeum* article, Cayley remembered the problem that he was still unable to solve. As reported in [Cay1] and [Cay2], during the June 13, 1878 meeting of the London Mathematical Society, Cayley asked:

> Has a solution been given of the statement that in colouring a map of a country, divided into counties, only four distinct colours are required, so that no two adjacent counties should be painted in the same colour?

Cayley also published a two-page article [Cay3] on this question. Did his choice of the publication, *Proceedings of the Royal Geographical Society and Monthly Record of Geography*, suggest that Cayley believed De Morgan on the usefulness of the 4CC for mapmakers? Perhaps, not, but the coincidence adds a touch of humor to our story. In the paper, Cayley showed that it suffices to prove the 4CC for *trivalent* maps, as they are called now (i.e., maps in which three regions meet at every vertex):

> ...if in any case the figure includes four or more areas meeting in a point (such as the sectors of a circle), then if (introducing a new area) we place at the point a small circular area, cut out from and attaching itself each of the original sectorial areas, it must according to the theorem be possible with four colours only to colour the new figure; and this implies that it must be possible to colour the original figure so that only three colours are used for the sectorial areas.

---

[8] Locations of both [DeM2], [DeM3] come from N. L. Biggs [Big], who analyses De Morgan's contribution to the 4CC and the Separation Axiom.

[9] First found by Bertha Jeffreys in 1979 [JefB].

Finally, Cayley tried to explain at length the difficulty of proving the 4CC by a straightforward induction, and that is all he was able to do.

However, Arthur Cayley stated twice – in the course of two pages – that he had "failed to obtain a proof" of the 4CC. These statements by one of the great mathematicians of his time must have stirred interest in the 4CC. Professionals and amateurs alike jumped on the opportunity to make Cayley out as "a stupid animal," as De Morgan put it in his letter quoted above.

The proof was very soon published in a prestigious journal by Alfred Bray Kempe, a 30-year-old London barrister (lawyer) and avid amateur mathematician, an expert on *linkages*.[10] We will look at his work in Chapter 19.

18.2 A Touch of Historiography

It is very surprising that for over 100 years the confusion reined in the history of the 4CC, one of the most popular problems in the history of mathematics. Truth and fiction were alternating like positive and negative parts of a sin curve. Without presenting here a complete historiography of the problem, I would just mention that countless times Möbius–Weiske's puzzle was mixed up with the 4CC, and consequently credit for the 4CC was often given to Möbius. It has been happening even in recent times. For example, as late as in 1958, the great geometer H. S. M. Coxeter wrote [Cox1]:

> The 4-color theorem was first mentioned by Möbius.

However, there were, authors who presented the problem's history without fantasy and "invention." For example, Alfred Errera was about right in his December 1920 doctoral thesis [Err]:

> Cayley attributed the exposition of the map theorem [sic] to De Morgan, whereas Frederic Guthrie claimed, in 1880, that his brother Francis Guthrie had demonstrated [it] some thirty years earlier.

In 1965 Kenneth O. May summarizes the 4CC's history very well [May], and apparently, is the first to quote De Morgan's letter:

> A hitherto overlooked letter from De Morgan to Sir William Rowan Hamilton.

May then goes on to quote De Morgan's October 23, 1852 letter and Hamilton's reply from the monumental three-volume edition *Life of Sir William Rowan Hamilton*, 1882–1889, written by Hamilton's close friend, the Rev. Robert Perceval Graves [Grav]. Volume 3 includes Hamilton's correspondence with De Morgan, and the letter of our interest, De Morgan to Hamilton of October 23, 1852 appears on pages 422–423. Graves was pressed for space – he wrote (vol. 3, p. v):

---

[10] His 1877 book *How to Draw a Straight Line* was published again 100 years later by the National Council of Teachers of Mathematics, with a funny (for 1977) statement on the copyright page: "Alfred Bray Kempe, 1849–" indicating Kempe's long life indeed.

The … larger portion of the volume [673 pp. long] consists of a selection from a very extensive correspondence between Sir W. R. Hamilton and Professor Augustus De Morgan. … The quantity of material was so great that I have had to exclude matter that possessed inherent value, either because it was in subject unsuited to this work, or because, being mathematical, the investigations carried on were too abstruse or too extended. The general reader will perhaps complain that I have introduced more than enough of mathematical investigation; but he will, I hope, withdraw the complaint when he calls to mind that it was as scientific men that the writers corresponded, that it would be unjust to them if their correspondence as printed should not retain this character, and that the mathematical discussion did in fact most often afford suggestion to the play of thought which, passing beyond the boundaries of science, prompted the wit and the learned and pleasant gossip which the readers will enjoy.

Thus, May knows that Graves condensed letters—in fact, Graves used quotation marks to show in practically every letter that he published only selections and not complete letters. Graves favored "pleasant gossip" indeed. For example, in De Morgan's letter of our interest, he keeps in De Morgan's trying "a fine pen with which to write in books," and "Having given the nibbler a fair trail, I now resume my ordinary pen." But Graves—and consequently May—omit all De Morgan's mathematical drawings, illustrating first ever thoughts on the 4CC, and they omit an important phrase. As a fine historian, May should have looked at these important letters in manuscript at the place where Hamilton spent his life, the place that sponsored Graves' biography of Hamilton, which appeared in "Dublin University Series," – Trinity College Dublin. He would have found there the 1900 *Catalog of the Manuscripts in the Library of Trinity College, Dublin* compiled by T. K. Abbott, where I read ([Abb], p. v):

> In 1890 the Rev. Robert P. Graves presented [Trinity College, Dublin] a collection of mss, which had belonged to Sir W. R. Hamilton, including his correspondence with Sir John Herschel, Professor De Morgan, and others.

In fact, the two letters of our interest, catalogued as 668 and 669 (De Morgan's and Hamilton's respectively) are contained in the group TCD MS 1493 of Hamilton–De Morgan correspondence manuscripts, which were donated to the Trinity College Dublin in 1900, as Stuart Ó Seanór, Assistant Librarian of the Manuscripts Department at Trinity disclosed to me in a letter on March 21, 1997 [OSe]:

> TCD MS 1493 was presented by J R H O'Regan of Marlborough, Wilts in 1900 (a descendant of Hamilton's through his daughter Helen) just in time to be mentioned in T K Abbott's *Catalogue of the manuscripts in the library of Trinity College Dublin* published that year…
>
> Graves' three volume biography of Hamilton or other writings of his may reveal that Hamilton corresponded with De Morgan and even citation of them might date from before the papers were in a library.
> Le meas
>   Stuart Ó Seanór [signed]

By now you must be wondering: which important phrase is missing in Graves and May; I will put it in italic:

He says that if a figure be any how divided and the compartments differently colored so that figures with any portion of common boundary line are differently colored – four colors may be wanted but not more – *the following is his case in which four are wanted [.]*

In 1976 the missing phrase was restored by Norman L. Biggs, E. Keith Lloyd, and Robin J. Wilson in their wonderful textbook of graph theory through its history [BLW]. To do that, the authors clearly had to see the manuscript letter or its photocopy. Unfortunately, they misread a word while transcribing the missing phrase, and the wrong word appeared in all editions of their book [BLW] as follows:

... the following is the [sic] case in which four <u>are</u> wanted [.]

In the manuscript one can clearly see the word "his" where the authors of [BLW] put the second "the." The difference is subtle but important: "the following is the case" would have indicated that De Morgan showed to Hamilton his own counterexample.[11] In fact, De Morgan wrote "the following is *his* [i.e., student's] case," i.e., De Morgan conveyed an example that four colors are wanted which Francis Guthrie devised and passed on to De Morgan through his brother Frederick![12]

Having established that at least one example and drawing in the De Morgan's letter belonged to Francis Guthrie, I wonder whether all arguments and drawings in the letter were Francis's as well – after all, De Morgan did not have much time to ponder on the problem, as he wrote his letter the very day Frederick asked him for a proof!

We have thus established that De Morgan's contemporaneous account agreed with Frederick Guthrie's 1880 recollection: Frederick presented to De Morgan not only the 4CC, but also his brother's proof, albeit "not altogether satisfactory to himself [i.e., to Francis]," as Frederick put it.

In 1976 the history of the 4CC was enriched by discoveries by John Wilson [WilJ], and in 1979 by Bertha S. Jeffreys of Cambridge, England [JefB], who found additional examples of De Morgan's writings about 4CC.

18.3 Creator of the 4 CC, Francis Guthrie

It is fascinating for me to read old newspapers: yes, they became worthless one day after their publication. But for a reader a century later, they are a treasure trove of

[11] The authors of [BLW] misread another word as well: they quote De Morgan as "I am not sure of the [sic] convolutions," whereas De Morgan wrote "I am not sure of all convolutions," which makes more sense.

[12] This section had been written in early 1990s. As I have just noticed during proof-reading, a decade later, in 2002, the third author of [BLW] and celebrated expositor Robin J. Wilson corrected this mistake in his popular engaging book *Four Colors Suffice*, Princeton University Press, Princeton. He surely noticed the mistake independently from me, as we have not discussed it during our meetings.

the life's interests, people's aspirations. They allow us to "touch" the distant culture, and to breathe in its air.

I am looking at *Cape Times* of Monday, October 23, 1899. In the column of my interest first comes *The America Cup*:

> The possession of the America Cup was decided to-day, when the Columbia won her second race against the Shamrock by five minutes. The Cup therefore remains in America.

All important for people of the day *Ship's Movements* come next:

> The Clan Macpherson left Liverpool for Algon Bay on Thursday morning.
> The Pombroks Castle arrived at Plymouth at two on Thursday afternoon.
> The Spartan left St. Vincent last night.

Francis Guthrie. Courtesy of John Webb and Mathematics Department, University of Cape Town

Following these 1.75-inch long reports, I see something that must have mattered to the folks of the Colony of South Africa a great deal: a 22-inch long column *The Late Professor Guthrie*. Let us read a bit of it together [Gut1]:

There has just passed away from us a man who has left a greater mark upon our Colonial life than will be readily recognized by many who did not come into contact with him; or by some who have been taught by this age of self-advertisement to suppose that no good work can be done in modesty and retirement. Professor Francis Guthrie, L.L.B., B.A., whose death on the 19th. . .we briefly announced in Saturday's issue, was born in London in 1831.

We can learn much about Francis Guthrie from this eulogy [Gut1], and from [Gut2] and [Gut3].

Born on January 22, 1831 in London, Francis Guthrie received his B.A. degree with first class honors from the University College, London. He then earned L.L.D., a law degree, and for some time was a consulting barrister in Chancery practice. In 1861 Guthrie left the old world and accepted an appointment at the newly established Graaff-Reinet College in the Colony of South Africa. Following his resignation in 1875 and a brief visit to England, in 1876 Guthrie was appointed to the Chair of Mathematics in the South African College, Cape Town (presently called the University of Cape Town), from which he retired after 22 years on January 31 of 1899. Several months later, on October 19, 1899, Guthrie died in Claremont, Cape Town.

Professor Guthrie was universally liked by his peers. He served on the University Council, 1873–1879, and was Secretary of Senate during 1894. He was an early member of South African Philosophical Society (now the Royal Society of South Africa) and of its Council, a member of the Meteorological Commission, and for many years the Examiner of the Cape University.

His several publications cover mathematics (none on the 4CC), meteorology, and his true passion, botany. Guthrie and his lifelong friend Harry Bolus were pioneers in the study of *ericas* of Southern Africa. In 1973 Harry Bolus discovered a new genus on the summit (altitude 6,500 feet) of the Gnadouw-Sneeuwbergen near Graaff-Reinet. Bolus named it in honor of his friend *Guthriea capensis*.

I am compelled to return to *Cape Times* [Gut1], as it conveys the life of the frontier unknown to most of us through personal experience, and shows a side of Francis Guthrie that is not widely known. Guthrie was a pioneer of the frontier. He discovered not only the 4CC, but also routes for the railroad that determined the future of his region of South Africa:

In 1871-2-3, when the agitation for railway extension was at its height and the battle of the routes was being fought, Professor Guthrie ardently espoused the Midland cause. The problem of that day was to show the Government and Parliament how, if a railway were made to Graaff-Reinet, it could get over the Sneeuwberg Mountains to the northwards. Some case had to be made out before the Government would sanction even a flying survey. Professor Guthrie, in a company with the late Charles Rubidge and some others, climbed the mountains, aneroid in hand, in search of the most available pass. Their efforts had for immediate result the construction of Forth Elizabeth and

Graaff-Reinet line; and it is a tribute to the accuracy of those early amateur railway explorers that the recent extension of that line to Middelburg follows very nearly the route over the Lootsberg which they had suggested as the most feasible. The people of Graaff-Reinet were not ungrateful, and a public banquet and laudatory addresses showed their appreciation of the efforts of Professor Guthrie and his colleagues.

This remarkable 22-inch long eulogy ends with unattributed poetic lines, which I traced to James Shirley (1596–1666):

Only the actions of the just
Smell sweet, and blossom in the dust.

18.4 The Brother

While we are moving through the Victorian history of the problem, I can offer you something mathematical to do as well. Frederick Guthrie (1833–1886), by 1880 a Professor of Chemistry and Physics at the School of Science, Kensington and the younger brother of the 4CC creator Francis, in his letter quoted above [GutFr], created and solved a three-dimensional analog of the 4CC that Francis allegedly neglected:

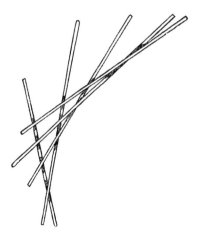

I have at various intervals urged my brother to complete the theorem in three dimensions, but with little success.

It is clear that, at all events when unrestricted by continuity of curvature, the maximum number of solids having superficial contact each with all is infinite. Thus, to take only one case n straight rods, one edge of whose projection forms the tangent to successive points of a curve of one curvature, may so overlap one another that, when pressed and flattened at their points of contact, they give $n - 1$ surfaces of contact.

Thus, Frederick Guthrie posed and solved the following problem:

Problem 18.1 (Frederick Guthrie,1880). Is there a positive integer n such that n colors suffice for proper coloring of any Euclidean three-dimensional map?

Frederick Guthrie continued:

How far the number is restricted when only one kind of superficial curvature is permitted must be left to be considered by those more apt than myself to think in three dimensions and knots.

Guthrie's words are not precise. It seems to me that he posed the following problem:

Problem 18.2 (Frederick Guthrie, 1880) What is the minimum number of colors required for proper coloring of any Euclidean three-dimensional map if each monochromatic set is convex?

I am compelled to allow you time to ponder on an alternative solution of Problem 18.1 and a solution of Problem 18.2. We will return to them in Chapter 20.

19
Victorian Comedy of Errors
and Colorful Progress

19.1 Victorian Comedy of Errors

This period in the history of The Four4-Color Conjecture (4CC) plays itself out like a Victorian version of Shakespeare's *The Comedy of Errors*. Judge for yourself!

Alfred Bray Kempe's proof of the 4CC was announced on July 17, 1879 in *Nature* [Kem1]. The proof itself was published later the same year in the *American Journal of Mathematics Pure and Applied* [Kem2], as Kempe writes (p. 194), "at the request of the Editor-in-Chief," i.e., James J. Sylvester.[13] In accordance with mapmakers' myth of De Morgan–Cayley, Kempe chose the title *On the Geographical Problem of the Four Colours*.[14] The proof was an unqualified success. While Kempe was elected a Fellow of The Royal Society based on this work on linkages, the coloring success must have been a factor for Cayley, Sylvester and others to nominate him for the honor.

Simplifications and variations appeared: first one by William E. Story, Associate Editor in Charge of the *American Journal of Mathematics Pure and Applied* ([Sto]. Story's paper immediately followed the Kempe's article [Kem2]. Simplifications then came from Kempe himself [Kem3] and [Kem4]. New "series of proofs of the theorem that four colours suffice for a map" by Peter Guthrie Tait followed [Tai1], [Tai2], [Tai3].

Popularity of the Four-Color Theorem (4CT) became so great that in late 1886 the Head Master of Clifton College somehow learned about it and ... offered the problem as a "Challenge Problem" to his students:

---

[13] Being Jewish, the famous British mathematician James Joseph Sylvester had to leave the Land of strictly religiously controlled Oxford and Cambridge, for the New World, where he became the first professor of mathematics in the just founded Johns Hopkins University, to the great benefit of the young American mathematics. But that is another story.

[14] We read in [Kem2], with amusement, an expansion of the De Morgan-Cayley myth: "...it has been stated somewhere by Professor De Morgan [must be a reference to Athenaeum [DeM4]] that it has long been known to map-makers as a matter of experience – an experience however probably confined to comparatively simple cases – that *four* colors will suffice in any case."

A. Soifer, *The Mathematical Coloring Book*,
DOI 10.1007/978-0-387-74642-5_19, © Alexander Soifer 2009

In colouring a plane map of counties, it is of course desired that no two counties which have a common boundary should be coloured alike; and it is found, on trial [sic], that four colours are always sufficient, *whatever the shape or number* of the counties or areas may be. Required, a good proof of this. Why *four* ?[15] Would it be true if the areas are drawn so as to cover a whole sphere?

In the funniest turn of this story, the Head Master warned the contestants that "no solution may exceed one page, 30 lines of MS., and one page of diagrams!" Published on January 1, 1887 in the *Journal of Education* [Head1], the challenge attracted a solution from such an unlikely problem solver as The Bishop of London, whose "proof" [Head2] was published in the same journal on June 1, 1889.

Then came along the 29-year-old Percy John Heawood—who spoiled the party! Almost with regret for his own discovery [Hea1], Heawood apologetically writes:

The present article does not profess to give a proof of this original Theorem [i.e., 4CT]; in fact its aims are rather destructive than constructive, for it will be shown that there is a defect in the now apparently recognized proof.

Yes, 11 (eleven!) years after the Kempe's 1879 publication [Kem2], Heawood discovered a hole in the proof (as well as in the two later versions of Kempe's proof [Kem3] and [Kem4]). Moreover, Heawood constructed an example showing that Kempe's argument as it was, could not work. There was a constructive side to Heawood's paper, in spite of his assurance to the contrary: he showed that Kempe's argument actually proves that *five* colors suffice for coloring any map.

In a gentlemanly way, Heawood informed Kempe first, and Kempe was the one who reported Heawood's findings to the London Mathematical Society at its Thursday, April 9, 1891 meeting, while "Major P. A. MacMahon, R.A., F.R.S., Vice-President, in the Chair" [Kem5]:

Mr. Kempe spoke on the flaw in his *proof* "On the Map-colour Theorem," which had recently been detected by Mr. P. J. Heawood, and showed that a statement by the latter at the close of his paper failed. He further stated that he was unable to solve the question to his satisfaction.

The authors of [BLW] researched publications of the period at hand. They reported that they found "no complimentary references to Heawood in the popular journals, and no record of honors granted to him." Heawood's work [Hea1] and his consequent papers dedicated to map coloring were certainly major contributions, and deserved more recognition. As it were, Heawood's work [Hea1] remained almost unnoticed and unquoted by his contemporaries. Long after 1890, we can still find papers giving credit to A. B. Kempe and Peter Guthrie Tait for proving 4CT (see, for example [DR]).

While giving credit to Kempe, Tait offered his own "proofs." It appears that the belief in Kempe's proof was extrapolated by the contemporaries to the belief in Tait's proofs: I was unable to find any contemporaneous refutation of Tait's "proofs."

---

[15] "Why four?" was a great question. Even today, when we have two proofs of the 4CT (see Chapters 21 and 22), we still do not really know the answer to this innocent question.

Tait described his strategy as follows [Tai1]:

> The proof of the elementary theorem is given easily by induction; and then the proof
> that four colours suffice for a map follows almost immediately from the theorem, by
> an inversion of the demonstration just given.

This is true: Tait found a nice proof that his "elementary theorem" was equivalent
to 4CT. The trouble is, it was not so "elementary," and moreover its proof was not
"given easily by induction."

The Bishop of London erred too: he mistakenly believed that the Möbius–
Weiske's problem IV-1 was equivalent to 4CT. Many years later, in 1906, the direct
refutation of his "proof" was published by John C. Wilson [wilJC]. Both De Morgan
and Cayley, nearly half a century earlier, knew that 4CC was much more than a
mere fact that five countries in a map cannot be mutually adjacent. Obviously, the
Headmaster of Clifton College and the Bishop of London did not.

True to its genre, our *Comedy of Errors* has a happy end. Alfred Bray Kempe
eventually becomes the President of the London Mathematical Society. Frederick
Temple, our Bishop of London, reaches the highest religious title of the Archbishop
of Canterbury.

The great Russian poet Aleksand Pushkin ends his "Fairytale about the Gold
Cockerel" ("Сказка о золотом петушке") with the words: "*A fairytale is a lie,
but with a hint, a lesson for a good lad.*"[16] Accordingly, our Victorian Comedy
of Errors leaves us plenty of valuable and enjoyable mathematics. Bright ideas of
Kempe, Tait, and Heawood are alive and well. Get your paper and pencil ready: in
this chapter and the next we will look at our British Victorian inheritance. As the
Bard put it:

All's well that ends well!

19.2 2-Colorable Maps

Let us now look at some of the Victorian problems. To simplify the excursion, we
will translate the Victorian problems into today's jargon. I suggest we start with a
warm-up.

Problem 19.1 Prove that a map formed in the plane by finitely many circles can be
2-colored (Fig. 19.1).

Proof We partition regions of the map into two classes (Fig. 19.2): those contained
in an even number of circles (color them gray), and those contained in an odd number of circles (leave them white). Clearly, neighboring regions got different colors
because when we travel across their boundary line, the parity changes. ∎

---

[16] Translated from Russian by Maya Soifer. The original rhymed Russian text is:
"Сказка ложь, да в ней намек!
добрым молодцам урок."

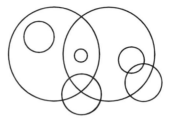

Fig. 19.1 A map formed by circles

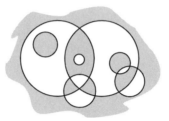

Fig. 19.2 2-Coloring of a map formed by circles

I am sure you realize that the shape of a circle is of no consequence. We can replace circles in Problem 19.1 by their continuous one-to-one images, called *simple closed curves*, because the Jordan Curve Theorem holds for them all:[17]

Jordan Curve Theorem 19.2 A simple closed curve in the plane divides the plane into two regions (inside and outside).

Problem 19.3 Prove that a map formed in the plane by finitely many simple closed curves is 2-colorable.

We can replace simple closed curves by straight lines, or a combination of the two:

Problem 19.4 Prove that a map formed in the plane by finitely many straight lines is 2-colorable (Fig. 19.3).

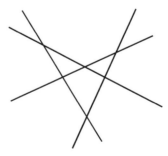

Fig. 19.3 A map formed by straight lines

---

[17] see its proof, for example, in [BS]

An inductive proof is well-known,[18] but as is usually the case with proofs by induction, it does not provide an insight. I found a "one-line" proof that takes advantage of similarity between simple closed curves and straight lines.

Proof Attach to each line a vector perpendicular to it (Fig. 19.4). Call the half-plane *inside* if it contains the vector, and *outside* otherwise. Repeat the proof of problem 19.1 word-by-word to complete the proof (Fig. 19.5). ∎

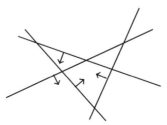

Fig. 19.4 Inventing vectors

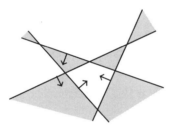

Fig. 19.5 2-Coloring of a map formed by straight lines

Problem 19.5 Prove that a map formed in the plane by finitely many simple closed curves and straight lines is 2-colorable.

So what is common between simple closed curves and straight lines? What allows a 2-coloring to exist? Each vertex in the maps above is a result of the intersection of two or more curves or lines, and therefore, has an even degree! This fact first appears in print on the last page of the 1879 paper by A. B. Kempe in which he attempts to prove 4CC [Kem2].

Kempe's Two-Color Theorem 19.6 (A. B. Kempe, 1879, [Kem2]) A map is 2-colorable if and only if all its vertices have even degree.

Let us take another look at the map M formed by circles in Fig. 19.1. We can construct the *dual graph* $G(M)$ of the map M as follows: we represent every region by a vertex (think of the capital city), and call two vertices adjacent if and only if

---

[18] See, for example, [DU].

the corresponding two countries are adjacent, i.e., have a common boundary (not just a point or finitely many points).[19] The dual graph $G(M)$ of the map M of Fig. 19.1 is presented in Fig. 19.6 (I bent and stretched edges to make the graph look aesthetically pleasing).

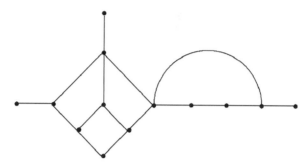

Fig. 19.6 The dual graph of a map from Fig. 19.1

Observe: the dual graph $G(M)$ of any map M is planar: we can draw its edges through common boundaries of adjacent regions so that the edges will have no points in common except the vertices of the graph.

Now the problem of coloring maps can be translated into the language of coloring vertices of planar graphs. But, this problem is not new to us: we have already solved it as Problem 12.2. Let us repeat it here:

Kempe's Two-Color Theorem 19.7 (In Graph–Theoretical Language) The chromatic number $\chi(G)$ of a graph G does not exceed 2 if and only if G contains no odd cycles.

19.3 3-Colorable Maps

It is natural to give a name to the smallest number of colors required to color a map M; let us call it the *chromatic number of a map M*, or *face chromatic number* and denote it by $\chi_2 (M)$.

We have an abundance of maps of chromatic number 2 around us: maps created by circles, straight lines, simple closed curves (Problems 19.1–19.5). Square grid delivers us an example of large periodic map of chromatic number 2: just recall the chessboard coloring. Can you think of a way of creating large periodic maps of

[19] The idea of the dual graph of a map was one of the first ideas of graph theory: Leonard Euler used it in 1736 to solve the Problem of Bridges of Königsberg. The language of maps was universally used by the first researchers of the 4CC. Yet, it is interesting to notice that while Kempe used the language of maps in the main body of his 1879 paper [Kem2], he did describe the construction of the dual graph on the last page of this paper.

chromatic number 3? You have already seen a couple of such constructions in this book, but in a totally unrelated context.

Problem 19.8 Find the chromatic number of the hexagonal map created by the old Chinese lattice in Fig. 6.7.

Solution: Behold (Fig. 19.7):

Fig. 19.7 3-Coloring of an ancient Chinese lattice from Fig. 6.7.

Problem 19.9 Find the chromatic number of the map in Fig. 6.6, which is formed by octagons and squares.

Solution: Behold (Fig. 19.8).

What is special about the maps in Problems 19.8 and 19.9 that make their chromatic number to be 3? Is it the fact that they are *cubic*, i.e., each vertex of these maps has degree 3? Or is it due to an even number of neighbors of every region?

A. B. Kempe [Kem2] repeats Cayley's argument that we can convert any map M into the trivalent map M', such that

$$\chi_2(M) \leq \chi_2(M').$$

Kempe writes:

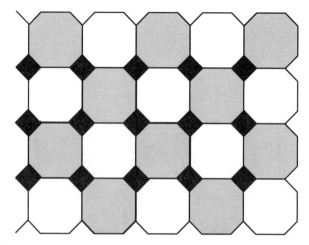

Fig. 19.8 3-Coloring of Soifer's tiling of the plane from Fig. 6.6.

I should show that the colours could be so arranged that only three should appear at every point of concourse [i.e., vertex of the map of degree at least 3]. This may readily be shown thus: Stick a small circular patch, with a boundary drawn round its edge, on every point of concourse, forming new districts. Colour this map [M']. Only three colours can surround any district, and therefore the circular patches. Take off the patches and colour the uncovered parts the same colour as the rest of their districts. Only three colours surrounded the patches, and therefore only three will meet at the points of concourse they covered.

Our maps in Problems 19.8 and 19.9 are cubic, and for cubic maps an even number of neighbors is the key indeed:

Kempe's Three-Color Theorem 19.10 (A. B. Kempe, 1879, [Kem2]) A cubic map M has face chromatic number 3 if and only if the boundary of each of its regions consists of even number of edges.[20]

Let us translate the Three-Color Theorem into the language of graphs by going to the dual graph $G = G(M)$ of the map M. Of course, since M is a trivalent map, all regions of G are bounded by triangles (i.e., three-cycles). A plane graph, where all regions are bounded by three-cycles, is called a *triangulation*.

Kempe's Three-Color Theorem 19.11 (*Three-Color Theorem for Graphs*) Let G be a connected plane triangulation. Then the following three assertions are equivalent:

(a) the chromatic number $\chi(G)$ of G satisfies the inequality $\chi(G) \leq 3$;

(b) the face chromatic number $\chi_2(M)$ of G satisfies the inequality $\chi_2(M) \leq 2$;

(c) the degree of every vertex of G is even.

---

[20] Kempe states only the sufficient condition, but the necessary condition is easier to prove, and was most likely known to him.

Proof Kempe does not prove his statement. The proof presented here is a substantially simplified version of a cycle of problems from the 1952 Russian book by Evgenii B. Dynkin and Vladimir A. Uspensky [DU].

$(a) \Rightarrow (b)$. Since $\chi(G) = 3$, we can label each vertex of G with one of the colors a, b or c. For every face we have one vertex of each of the colors a, b or c. Take a face F; if the direction of going around its vertices $a \to b \to c$ is clockwise, then we color F red, otherwise we color F blue. It is easy to see that any two adjacent faces are thus assigned different colors.

$(b) \Rightarrow (a)$. Let G be face 2-colored red and blue. For every edge xy of G we assign one direction (out of possible two: $x \to y$ or $y \to x$), such that when we travel along the assigned direction, the red triangle is on our right (and thus a blue triangle is on our left). Obviously, for any two vertices v, w of G there is a directed path from v to w, and while the length of such a path (i.e., the number of edges in it) is not unique, its length modulo 3 is *unique*.

Assuming we proved this uniqueness (see next paragraph for the proof), the rest is easy. Let us call our three colors 0, 1, and 2. Pick a vertex v and color it 0; then for any vertex w of G we select one directed path P from v to w, and the remainder upon division of the length $l(P)$ of P by 3 determines the color we assign to w. This guarantees that the adjacent vertices are assigned different colors (do you see why?), and the implication $(b) \Rightarrow (a)$ is proven.

Proof of Uniqueness Let us first prove that the length $l(P)$ of any closed directed path P is divisible by 3. Assume it is not, then among all directed closed paths of length not divisible by 3, there is one of *minimum* length l, call it P'. P' has no self-intersections, as otherwise it could be shortened (can you see how?) in contradiction to its minimum length. P' partitions the plane into two areas: the inside and the outside. We combine the outside into one region O, and as the result get a new map M_1, all regions of which are already colored red and blue, except the region O (Fig. 19.9).

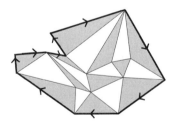

Fig. 19.9

If the loop P' has the clockwise direction, then all triangles bordering on P' are colored red; otherwise they are all colored blue. Since in either case all triangles bordering on P' are assigned the same color, say, red, we can complete the 2-coloring of the map M_1 by assigning the outside region O the opposite color—blue.

Every edge belongs to the boundary of one red and one blue regions, therefore, the total numbers of edges on the boundaries of all red and all blue regions are equal.

Thus, we get the following equality:

$$3r = 3b + l,$$

where r and b denote the numbers of red and blue triangles respectively (and l, as you recall, is the length of P'). This equality contradicts the fact that l is not divisible by 3.

Assume now that there are two directed paths P_1 and P_2 from a vertex v to a vertex w, such that their lengths are l_1 and l_2 give different remainders upon division by 3. Let P_3 be a path of length l_3 from w to v. Then we get two different *closed* paths $P_1 + P_2$ and $P_1 + P_3$ of lengths $l_1 + l_2$ and $l_1 + l_3$ respectively. Therefore, in view of the above both integers $l_1 + l_2$ and $l_1 + l_3$ are divisible by 3. But then the number

$$(l_1 + l_2) - (l_1 + l_3) = l_2 - l_3$$

is divisible by 3, and the desired uniqueness is proven.

$(b) \Leftrightarrow (c)$. This is precisely the 2-Color Theorem we have discussed above. ∎

Kempe's 3-Color Theorem, as can be easily seen, has the following corollary:

Corollary 19.12 (P. J. Heawood, 1898, [Hea2]) Let G be a connected planar graph G. Then the following assertions are equivalent:

(a) the chromatic number $\chi(G)$ of G satisfies the inequality $\chi(G) \le 3$;

(b) G can be embedded (as a subgraph) into a triangulation graph G' such that degree of every vertex of G' is even.

In his 1993 survey [Ste] Richard Steinberg describes the history of the 3-color problem and its state at the time of his writing. In this otherwise wonderful historical work, Steinberg dismisses Alfred B. Kempe in a number of unjustified ways:

> The most notorious paper in the history of graph theory: the 1879 work by A. B. Kempe [Kem2] that contains the fallacious proof of the Four Color Theorem...
> Kempe's language is somewhat unclear – he was a barrister by profession.

Pierre de Fermat was a "barrister by profession" too. Does a mere fact that professionals are paid for services, make them necessarily superior to amateurs? And when an amateur turns professional (which happens every day), does his language improve overnight?

Yes, Kempe's language is not as precise as our present standards require. But the same can be said of Tait and Heawood, yet Steinberg quotes approvingly Gabriel Dirac's passionate but illogical argument in defense of Heawood's writing:

> Most of the assertions stated in [Hea2] are not actually proved, only made plausible, but they have since been proved rigorously by other writers, which indicates [sic] that Heawood was in possession of the necessary proofs but did not choose to include them.

As we have seen, Kempe's last page of [Kem2] contained a number of observations, including both the Two-Color Theorem and the Three-Color Theorem that are listed without proof, as "two special cases" of map coloring. I believe that Kempe knew the proofs, but omitted them possibly because his main, if not the only goal was to prove the much more complex Four-Color Theorem.

Tomas L. Saati, in his 1967 title, calls the attempt "The Kempe Catastrophe" [Saa1]. I cannot disagree more. As we will see in the next chapter, Alfred B. Kempe did not succeed in his goal, but what a fine try it had been, far exceeding anything his celebrated professional predecessors De Morgan and Cayley achieved in years of toying with the 4CC! Moreover, both known today successful assaults of 4CC used Kempe's approach in their foundation. Kempe came up with beautiful ideas; his chain argument was used many times by fine twentieth century professionals— Dénes König in his 1916 work on the chromatic index of bipartite graphs, and Vadim Vizing (Chapter 16) in his famous 1964 chromatic index theorem.

For his important work on linkages, the contemporaries elected A. B. Kempe (1849–1922) a Fellow of the Royal Society (1881) and President (1892–1894) of the London Mathematical (that is: *Mathematical*) Society. Kempe was knighted in 1913.

19.4 The New Life of the Three-Color Problem

In the first half of the twentieth century it seemed that the Three-Color Problem had been settled in the Victorian Age. Since the late 1940s and the 1950s, we have witnessed the accelerating explosion of results on the relationship between the chromatic number of a graph and its small cycles (please, see the discussion of it in Chapter 12). Examples of triangle-free graphs were in the mathematical air. Only one word, planar, needed to be added for revisiting the Three-Color Problem, and seeking a deeper understanding of what causes a map to be 3-colorable.

The first significant step of this new era of 3-colorable graphs was made by the German mathematician Herbert Grötzsch in 1958 [Grö].

Grötzsch's Theorem 19.13 A triangle-free planar graph is 3-colorable.

In order to demonstrate that the restriction to planar graphs cannot be omitted, Grötzsch constructed the graph we discussed in Chapter 12 (Fig. 12.8). His theorem, however, allowed an improvement, which was delivered by the celebrated geometer (and *Geombinatorics'* editor from its inception in 1991 to present) Branko Grünbaum [Grü] in 1963.

Grünbaum's Theorem 19.14 A planar graph with at most three 3-cycles is 3-colorable.[21]

---

[21] A lemma used in the proof of Grünbaum's theorem was corrected and proved by Valeri A. Aksionov in 1974 [Aks].

This result is of course best possible, as K_4, a graph with four 3-cycles shows. Is there a life after the best possible result?

In mathematics—of course! As Valeri A. Aksionov and Leonid S. Mel'nikov observed [AM], "Grünbaum put forth the question which determined the direction of further research." Grünbaum defined *the distance between triangles* of a graph as the length of the shortest path between vertices of various pairs of triangles. He conjectured that if this distance is at least 1, then the planar graph is 3-colorable. Ivan Havel, who constructed a counterexample to Grünbaum's conjecture, posed and refuted his own conjecture (with distance at least 2), and in the end posed a more restrained question in 1969 [Hav].

Havel's Open Problem 19.15 Does there exist an integer n such that if the distance between any pair of triangles in a planar graph G is at least n, then G is 3-colorable?

Havel's problem is still open. According to Baogang Xu (e-mail of May 10, 2007), it is known that if such an n exists, it is at least 4.

Meanwhile Richard Steinberg reasoned as follows: the restrictions on 3-cycles have been settled; but what if we were to impose no restrictions on 3-cycles, but instead limit 4- and 5-cycles. In his 1975 letter to the Russian mathematicians V. A. Aksionov and L. S. Mel'nikov, Steinberg posed his now well-known and still open problem [Ste].

Steinberg's Open Problem 19.16 Must a planar 4- and 5-cycle-free planar graph be 3-colorable?

The further research on Three-Color Problem was inspired by Havel's and Steinberg's open problems, and often by a combination of both of them. The explosion of recent results is so great that the field is surely in need of a new comprehensive survey, like the one Richard Steinberg authored in 1993. I am grateful to the Chinese mathematician Baogang Xu for navigating me through the labyrinth of the current state of the problem. Let us look at the explosion of 3-coloring results.

Abbott–Zhou's Theorem 19.17 ([AZ], 1991) A planar graph without cycles of lengths 4 to 11 is 3-colorable.

Sanders–Zhao and Borodin's Theorem 19.18 ([SZ], 1995; [Bor], 1996) A planar graph without cycles of lengths 4 to 9 is 3-colorable.

Borodin–Glebov–Raspaud–Salavatipour's Theorem 19.19 ([BGRS], 2005) A planar graph without cycles of lengths 4 to 7 is 3-colorable.

Luo–Chen–Wang's Theorem 19.20 ([LCW], 2007) A planar graph without cycles of lengths 4, 6, 7 and 8 is 3-colorable.

Chen–Raspaud–Wang's Theorem 19.21 ([CRW], 2007) A planar graph without cycles of lengths 4, 6, 7 and 9 is 3-colorable.

In 2003, Oleg V. Borodin and André Raspaud [BR] started a direction that combined Steinberg's and Havel's problems.

Borodin–Raspaud's Theorem 19.22 ([BR], 2003) A planar graph without 5-cycles and triangles of distance less than 4 is 3-colorable.

They also formulated two conjectures stronger than the (still open) positive answer to Steinberg's problem 19.16. The authors called them "Bordeaux 3-color conjectures" – I will add the authors' names to give them credit. By *intersecting* (*adjacent*) *triangles* the authors mean those with a vertex (an edge) in common.

Bordeaux 3-Color Borodin–Raspaud's Conjecture 19.23 ([BR], 2003) A planar graph without 5-cycles and intersecting triangles is 3-colorable.

Bordeaux 3-Color Borodin–Raspaud's Strong Conjecture 19.24 ([BR], 2003) A planar graph without 5-cycles and adjacent triangles is 3-colorable.

A proof of Conjecture 19.24 in the positive would imply the validity of Conjecture 19.23 and the positive answer to Steinberg's Problem 19.16.
Baogang Xu has just improved Borodin–Raspaud's result 19.22.

Xu's Theorem 19.25 ([Xu2], 2007) A planar graph without 5-cycles and triangles of distance less than 3 is 3-colorable.

In a significant improvement of Borodin et al. Theorem 19.19, Xu proved the strongest result to date in the direction of proving Bordeaux 3-Color Borodin–Raspaud's Conjecture 19.24:

Xu's Theorem 19.26 ([Xu1], 2006) A planar graph without adjacent triangles and 5- and 7-cycles is 3-colorable.

Two more very recent results have been obtained in the direction of Steinberg's and Havel's problems.

Lu-Xu's Theorem 19.27 ([LX], 2006) A planar graph without cycles of lengths 5, 6, and 9 and without adjacent triangles is 3-colorable.

Xu's Theorem 19.28 ([Xu3], submitted) A planar graph without cycles of lengths 5, 6 and without triangles of distance less than 2 is 3-colorable.

The international group of mathematicians Oleg V. Borodin, Aleksey N. Glebov, both from Russia, Tommy R. Jensen from Denmark, and André Raspaud from France [BGJR] recently put a new twist on the 3-color oeuvre.

Borodin–Glebov–Jensen–Raspaud's Theorem 19.29 ([BGJR], 2006) A planar graph without triangles adjacent to cycles of lengths 3 to 9 is 3-colorable.

The authors have also formulated an attractive conjecture.

Borodin–Glebov–Jensen–Raspaud's Conjecture 19.30 ([BGJR], 2006) A planar graph without triangles adjacent to cycles of lengths 3 or 5 are 3-colorable.

It is fascinating to see how the seemingly lesser known cousin of the celebrated 4CC has flourished so beautifully and became an exciting area of mathematical inquiry, even after 4CC was settled!

20

Kempe–Heawood's Five-Color Theorem and Tait's Equivalence

20.1 Kempe's 1879 Attempted Proof

I am compelled to present here Alfred Bray Kempe's attempted proof of 4CC. As you recall from Chapter 19, that proof contained an oversight, that was found a "mere" 11 years later by Percy John Heawood. Why then do I choose to present the unsuccessful attempt here? First of all, because of beautiful ideas Kempe invented. Secondly, because it is not so easy to notice a flaw right away. Thirdly, because P. J. Heawood did not have to do much to salvage Kempe's ideas and show that, in fact, *they (i.e., Kempe's ideas!)* prove the Five-Color Theorem. And finally, because just like their contemporaries underestimated the work of Heawood, my contemporaries often underestimate contributions of Kempe.

And so it comes. Fasten your seat belts, I challenge you to find Kempe's oversight!

I will translate both the theorem and Kempe's proof into the usual nowadays language of dual graphs. The authors of *The Four-Color Problem* [SK], the first ever book on the subject, Thomas L. Saaty and my friend Paul C. Kainen, write (p. 7):

> The notion of dual graph mentioned above was introduced by Whitney (1931) and used to give an elegant characterization of when a graph is planar.

In fact, the notion of dual graph appears on the last page of A. B. Kempe's 1879 paper [Kem2], as I mentioned in the footnote after Theorem 19.6 in the previous chapter, and in 1736 Leonard Euler had already used it. Kempe reinvented the notion, but did not do much with it (what can one do with a promising notion that is introduced too late, on the last page of the paper!). We will use it here to make Kempe's attempted proof easier to read. We will also rearrange Kempe's proof.

The Four Color Theorem for Graphs 20.1 Chromatic number of any planar graph does not exceed 4.

Attempted proof by Alfred Bray Kempe. First Kempe presents his brilliant chain argument, then he rediscovers Euler's formula 20.2, and uses it to find the graph theory's first set of unavoidable configurations (Tool 20.3 and the equivalent Tool 20.3'), as it is called today. We'll do the latter two first.

A. Soifer, *The Mathematical Coloring Book*,
DOI 10.1007/978-0-387-74642-5_20, © Alexander Soifer 2009

Euler's Formula for Maps 20.2 For any map M in the plane, the following equality holds:

$$R + V = E + 2,$$

where R, V and E are number of regions, vertices and edges of M respectively.

Hint. You can add edges to M as necessary, until you get a triangulation $T(M)$, such that the Euler formula holds for M if and only if it holds for $T(M)$; and then use induction. Let me not present here the complete proof: too many books have already done so. ∎

Kempe's Tool 20.3 (A. B. Kempe, 1879, [Kem2]). Any planar map contains a vertex of degree at most 5.

Proof We can assume without loss of generality that each face is incident to at least three edges, for otherwise we can insert some vertices of degree 2 to remedy such a situation.

We will argue by contradiction. Assume that the desired statement does not hold for a planar graph G, i.e., all V vertices of G have degree at least 6. Let R and E stand for the numbers of regions and edges of G respectively. Since every edge is incident to two vertices, and to two regions, we get $6V \geq 2E$ and $3R \geq 2E$, or $V \geq \frac{1}{3}E$ and $R \geq \frac{2}{3}E$. Then by Euler's formula 20.2, we get:

$$\frac{2}{3}E + \frac{1}{3}E \geq E + 2,$$

which is absurd. ∎

I enjoyed the idea of translating Kempe's attempt into the contemporary terminology of unavoidable sets of reducible configurations that I found in Douglas R. Woodall's paper [Woo2]. I will present Kempe's attempted proof here in this language, for this would better prepare you for the next chapter, where we will discuss Appel and Haken's proof of 4CT.

A configuration C is called *reducible* if the minimal (in terms of the number of vertices) counterexample G to 4CT cannot contain C, i.e., G can be *reduced* to a smaller counterexample.

A finite set S of configurations is called *unavoidable* for a certain class Φ of maps if every map from Φ contains at least one element of S. Tool 20.3 could be reformulated in a language of unavoidable configurations:

Kempe's Tool in Current Terms 20.3' The set of four configurations in Fig. 20.1 is unavoidable, i.e., at least one of them appears in any non-trivial plane triangulation.

Kempe's argument: Kempe is set out to prove that the four configurations in Fig. 20.1 form an unavoidable set of reducible configurations.

Assume that there is a planar graph that is not 4-colorable. Then among all planar non 4-colorable graphs there is a graph, call it G, of minimum order (i.e., minimum

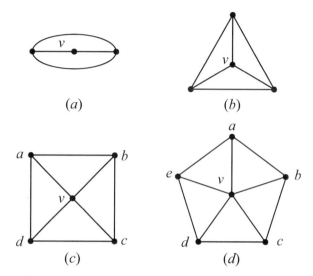

Fig. 20.1 Kempe's unavoidable set of configurations

number of vertices). Embed G in the plane, and add edges, if necessary, to make a triangulation T out of G. T is not 4-colorable as it is of the same order as G, but $T - v$ is 4-colorable for any vertex v. Fix a vertex v, and color $T - v$ in four colors. According to tool 20.3', T contains one of the four configurations listed in Fig. 20.1.

1. If T contains a configuration (a) or (b), the 4-coloring of $T - v$ can be easily extended to a 4-coloring of T: just assign the vertex v a color not used on the vertices adjacent to v. A contradiction, therefore the assumption that T is a minimal counterexample to 4CT is false, and thus T can be reduced. Configurations (a) and (b) are reducible.

2. Let T contain a configuration (c). We will look at three subcases.

2a. If no more than three colors have been used to color the vertices a,b,c and d, we can extend the 4-coloring of $T - v$ to a 4-coloring of T: just assign the vertex v a color not used on the vertices adjacent to v.

2b. Assume now that the vertices a,b,c and d are assigned four different colors: following Kempe's taste let these colors be red, blue, green, and yellow respectively. Consider a subgraph T_{RG} of $T - v$ that is formed by all red and green vertices of $T - v$, with all edges connecting these vertices (we call T_{RG} a subgraph *induced* by the red and green vertices). If the vertices a and c belong to different components of T_{RG}, we interchange colors, red and green, in the component that contains the vertex c. As a result, we get a new 4-coloring of $T - v$, but in this coloring both vertices a and c are colored red. Thus we can extend the 4-coloring of $T - v$ to a 4-coloring of T: just color the vertex v green!

2c. Let us now assume that both vertices a and c belong to the same component of T_{RG}, i.e., there is, what we call now the *Kempe chain* C_{RG} in T_{RG} that connects a and c (Fig. 20.2).

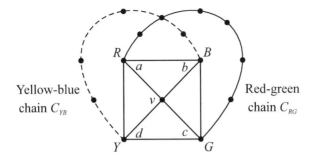

Fig. 20.2 Kempe's chains at work

Consider a subgraph T_{YB} of $T - v$ induced by all yellow and blue vertices of $T - v$. Since the chain C_{RG} separates the vertices b and d, they must lie in different components of T_{YB}. Therefore, we, interchange colors, yellow and blue, in the component of T_{YB} that contains the vertex b. As a result, we get a new 4-coloring of $T - v$, but in this coloring both vertices b and d are colored yellow. Thus we can extend the 4-coloring of $T - v$ to a 4-coloring of T: just color the vertex v blue.

We have thus proved in all cases that T is 4-colorable. A contradiction, therefore the assumption that T is a minimal counterexample to 4CT is false, and thus T can be reduced. Configuration (c) is reducible.

3. Let finally T contain a configuration (d). We will consider three subcases.

3a. If no more than three colors have been used to color the vertices a,b,c,d and e, we can extend the 4-coloring of $T - v$ to a 4-coloring of T: just assign the vertex v a color not used on the vertices adjacent to v.

3b. Assume now that the vertices a,b,c,d and e are assigned four different colors: following Kempe's choice, let these colors be red, blue, yellow, green, and blue respectively. Consider subgraphs T_{RY} and T_{RG} of $T - v$ that are induced by all its red-and-yellow, and red-and-green vertices respectively. If the vertices a and c belong to different components of T_{RY}, or a and d belong to different components of T_{RG}, we interchange colors in the component that contains the vertex a. As the result, we get a new 4-coloring of $T - v$, such that the color red is not assigned to any of the vertices a,b,c,d and e. Thus we can extend the 4-coloring of $T - v$ to a 4-coloring of T: just color the vertex v red.

3c. Let us now assume that vertices a and c belong to the same component of T_{RY}, and a and d belong to the same component of T_{RG}, i.e., there is a Kempe chain C_{RY} in T_{RY} that connects a and c, and a Kempe chain C_{RG} in T_{RG} that connects a and d (Fig. 20.3).

Consider subgraphs T_{BG} and T_{BY} of $T - v$ induced by all its blue-and-green and blue-and-yellow vertices respectively. The vertex b must lie in a component of T_{BG} that is different from those to which d and e belong; and e lies in a component of T_{BY} that is different from those to which b and c belong. We, therefore, interchange colors, blue and green, in the component of T_{BG} that contains b; and blue and yellow, in the component of T_{BY} that contains e. As a result, b becomes green and e yellow.

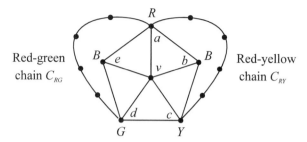

Fig. 20.3 Kempe's chains at work

Thus, we can extend the 4-coloring of $T - v$ to a 4-coloring of T: just color the vertex v blue.

We have proved in all cases that T is 4-colorable. A contradiction, therefore the assumption that T is a minimal counterexample to the 4CT is false, and thus T can be reduced. Configuration (d) is reducible. ∎

The Four-Color Theorem has thus been proven, or has it? In hindsight we know that is has not been. Have *you* noticed the hole? Try finding it on your own before reading the next subsection, in which I will play hide-and-seek and reveal where the hole is.

20.2 The Hole

The hole occurs in the subcase 3c. Everything Kempe does in the neighborhood of the vertex v is fine! He does get rid of color blue among the vertices adjacent to v, and therefore is able to assign blue to v.

However, while interchanging 2 colors in one component (as was done in the subcase) 2c does create an allowable coloring of $T - v$, in the subcase 3c Kempe interchanges coloring in *two* components. Moreover, he interchanges colors in components of T_{BG} and T_{BY} that *share a color* (blue). Thus, there is no guarantee that what he gets in the outset is an allowable coloring of $T - v$ (i.e., everywhere in the graph adjacent vertices are assigned different colors). Thus, Kempe's attempted proof has a hole.

20.3 The Counterexample

In fact, Percy John Heawood was not only first to find the above hole: he constructed a map such that if one follows Kempe's argument, two adjacent regions would get the same color assigned to them. Tomas L. Saati did not just translate Heawood's example into the language of graph theory, but also added niceties of symmetries to his graph [Saa2, p. 9]. My assistant Phillip Emerich and I added further niceties of regular hexagons and pentagons to Saati's graph—see Fig. 20.4 for our embedding.

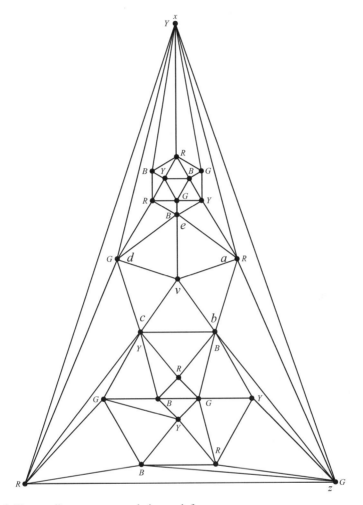

Fig. 20.4 Heawood's counter example in graph form

Letters R, B, Y, and G, stand for colors red, blue, yellow and green, respectively. As a result of Kempe's re-coloring, the adjacent vertices, x and z end up with the same color assigned to them.

Heawood's counterexample is a graph of order 25. While reading Kempe's attempted proof, I found a counterexample of order just 9 that refutes Kempe's proof as written by him.

Problem 20.4 Construct a counterexample to Kempe's attempted proof of order not greater than 9.

Behold (Fig. 20.5).

In fact, I believe that this is the smallest such counterexample:

Conjecture 20.5 For any graph of order less than 9, Kempe's argument works.

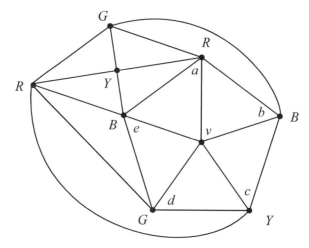

Fig. 20.5 A small counter example

20.4 Kempe–Heawood's Five-Color Theorem

In his 1890 paper P. J. Heawood [Hea1] pointed out that Kempe's argument actually proves that five colors suffice. When we use five colors, there is no need to simultaneously interchange colors in two Kempe chains, and thus Kempe's chain argument works. Do verify the proof of the Five-Color Theorem on your own.

I believe that the name often used today for this result, "The Heawood 5-Color Theorem," is unfair. While Heawood was first to formulate and prove the theorem, he merely adjusted an ingenious argument created by Kempe. It therefore is only fair to name the result after both inventors. I have little doubt Heawood would have agreed!

Kempe–Heawood's Five-Color Theorem 20.6 Five colors suffice to color any map in the plane.

20.5 Tait's Equivalence

Not only did Augustus De Morgan, but also Arthur Cayley contributed to spreading the word about the 4CC. Peter Guthrie Tait is clear about it [Tai1, p. 501]:

> Some years ago, while I was working at knots, Professor Cayley told me of De Morgan's [sic] statement that four colours had been found by experience [sic] to be sufficient for the purpose of completely distinguishing from one another the various districts on a map.

When in 1880 Alfred B. Kempe published yet another sketch of a proof similar to his original attempt [Kem5], Tait was apparently inspired to enter the map-coloring arena. In 1 year, 1880, he published a paper [Tai1], withdrew and replaced it with a one-page "abstract" [Tai2], which he expanded to an article [Tai3]. These papers contain some amusing statements, for example [Tai3, p. 657]:

> The difficulty in obtaining a simple proof of this theorem originates in the fact that it is not true without limitation.

One can paraphrase it to say, "It is difficult to prove what is not true." Indeed, very much so! However, the Tait papers, also contain brilliant observations, such as what we call Tait's Equivalence (Problem 20.8 below). Let us start our Tait-Review with his inductive attempt of proving 4CC.

The Four-Color Theorem 20.7 Every map in the plane is 4-colorable.

Tait's Attempted Proof ([Tai3]): Proof by induction in the number of regions.
For a map with one region 4CC holds.
Assume it holds for any map with less than n regions, i.e., any map with less than n regions is 4-colorable.
Given a map M with n regions, by Kempe's Tool 20.3, M contains a region R bounded by at most five edges. If R is bounded by two or three edges, erase one of them, say e. The resulting map can be 4-colored by inductive assumption. Now reinstate e. At most three colors are forbidden for coloring R (one per each neighbor), and we, therefore, use the remaining color for R.
Let R be bounded by four edges, and adjacent regions clockwise are R_1, R_2, R_3 and R_4. [At least one of the two pairs of the opposite regions R_1 and R_3, R_2 and R_4, is non-adjacent; let R_2 and R_4 be the non-adjacent regions.] We erase a pair of opposite edges e_2 and e_4 that separate the regions R_2, R and R_4 (Fig. 20.6).[22] The resulting map can be 4-colored by the inductive assumption.
Now reinstate e_2 and e_4. At most three colors are forbidden for coloring R (*because R_2 and R_4 are assigned the same color!*), and we, therefore, can use the

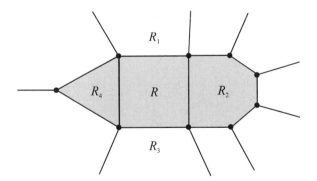

Fig. 20.6 Tait's attempted proof of the Four-Color Theorem

---

[22] Tait in [Tai3, p. 660] wrote: "*either pair* of opposite sides of a four-sided region may be erased, and afterwards restored." This choice can cause a problem if the opposite regions are adjacent, hence I had to correct Tait's attempt by adding the previous sentence in brackets.

remaining color on *R*. *Please, observe: no Kempe chain argument was used in this case, and the proof is much shorter than in Kempe's attempt.*

Let finally *R* be bounded by five edges, erase a pair of non-adjacent edges, say e_1 and e_2. Here Tait suddenly stops and writes:

> But when we erase any two non-adjacent sides of a five-sided district, a condition is thereby imposed on the nomenclature of the remaining lines, with which I do not yet see how generally to deal.

Of course, Tait knew that he could continue his proof by 4-coloring the resulting map, which can be done by inductive assumption; then reinstate e_1 and e_2, and use Kempe chain argument, as in [Kem2] or [Kem5]. *He did not! Why?*

The only plausible explanation, in my opinion, is that *Tait at the very least had doubts about the validity of Kempe's argument in the last case, if not realized the existence of the hole—10 years prior to Heawood's work* [Hea1]. ∎

With mathematical Olympiad-like brilliance, Tait proves the following fabulous equivalence. A fine statement meets as fine a proof. Enjoy!

The dual graph of a planar triangulation graph is a planar graph, whose all vertices have degree 3. If all vertices of a graph have the same degree 3, we say that the graph is *regular of degree 3*, or simply a *3-regular graph*.

Tait's Equivalence, Graph Version 20.8 (Tait, 1880). A planar 3-regular graph can be (vertex) 4-colored if and only if it can be edge 3-colored.

Proof [Tai2], [Tai3]. Let vertices of a planar 3-regular graph *G* be 4-colored in colors *a*, *b*, *c* and *d*. We then color edges in colors *x*, *y*, and *z* as follows: an edge is colored *x* if it connects vertices colored *a* and *b*, or *c* and *d*; an edge is colored *y* if it connects vertices colored *a* and *c*, or *b* and *d*; and an edge is colored *z* if it connects vertices colored *a* and *d*, or *b* and *c*. We can easily verify that a proper edge coloring is thus obtained, i.e., no adjacent edges are assigned the same colors. In view of symmetry, it suffices to show it for the edges incident to a vertex colored *a*, which is demonstrated in Fig. 20.7.

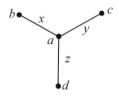

Fig. 20.7 Proof of Tait's equivalence

For the proof of the *converse* statement, Tait adds points and edges to make degrees of every vertex even. Instead, I will subtract (remove) edges, which makes the argument more transparent.

Let edges of a planar 3-regular graph G be 3-colored in colors x, y, and z. Look at the subgraph G_{xy} of G induced by all edges colored x and y.[23] Every cycle of G_{xy} must be even, as it alternates edges colored x and y. Therefore, by Kempe's Two-Color Theorem (problem 19.7), the vertices of G_{xy} (which comprise precisely *all* vertices of G) can be 2-colored in colors, say A and B. Similarly, we create the subgraph G_{yz} of G induced by all edges colored y and z, and color all its vertices in 2 colors, say 1 and 2. We thus assigned every vertex of G one of the following *four pairs* of colors: *A1, A2, B1,* or *B2*. It is easy to verify (do) that we have ended up with the proper vertex 4-coloring of G! ∎

The Tait equivalence can also be formulated in the dual language of maps:

Tait's Equivalence, Map Version 20.9 (Tait, 1880, [Tai2], [Tai3]). A map whose underlying graph is 3-regular, can be (face) 4-colored if and only if it can be edge 3-colored.

20.6 Frederick Guthrie's Three-Dimensional Generalization

Have you– found your own solution of Frederick Guthrie's Problems 18.1 and 18.2? As you know, he generalized his brother's 4CP to the three-dimensional Euclidean space and proved that no finite number of colors suffices.

Problem 20.10 For any positive integer n, there is a three-dimensional map that cannot be colored in n colors (so that regions having a common boundary – and not merely finitely many points, are assigned different colors).

In fact, unlike Möbius–Weiske's puzzle, for any positive integer n, there are n solids such that every two have a common boundary surface.

Second Solution. This solution appears in the 1905 paper of the Austrian mathematician and puzzlist Heinrich Tietze [Tie1]. As Frederick Guthrie before him, Tietze showed that in the three-dimensional space we can easily construct $n + 1$ mutually adjacent solids. Just put $n + 1$ long enough parallelepipeds, numbered 1 through $n + 1$ on a plane; then put $n + 1$ more parallelepipeds that are perpendicular to first ones on top; and combine into one solid two parallelepipeds that are labeled with the same number (Fig. 20.8). ∎

Granted, this puzzle was easy to solve. However, according to Tietze, the German mathematician Paul Stäckel, who also solved the above problem, posed the same question for *convex* solids (something that I believe Frederick Guthrie posed first, but not in very precise words – see the end of Chapter 18). Heinrich Tietze solved this harder problem in the same 1905 paper [Tie1].

---

[23] I substantially simplified here Tait's language without changing his ideas. He talks about converting every triangular face into a four-sided one by inserting one new vertex per face inside an edge. He then throws the edges with inserted vertices away, which is equivalent to keeping precisely edges without insertions. These kept edges are then 2-colored.

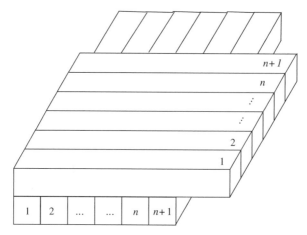

Fig. 20.8 Tietze's argument in the 3-space

Tietze's Theorem 20.11 (H. Tietze, 1905) For any positive integer n, there are n convex solids such that every two have a common boundary surface.

Thus, map coloring in three dimensions did not provide as lasting a fun as has the two-dimensional variety of map coloring.

21
The Four-Color Theorem

> *The most famous conjecture of graph theory or*
> *perhaps of the whole mathematics, the four colour*
> *conjecture, became recently the theorem of Appel*
> *and Haken.*
>
> – Paul Erdős, 1979 [24]

> *Four-colour problem, the as yet unsolved [sic]*
> *problem of proving as a mathematical theorem that*
> *on any plane map only four colours are needed to*
> *give different colours to any regions that have a*
> *common boundary.*
>
> – Oxford English Dictionary, June **2007** [sic] Edition[25]

The year was 1976. I read a notice about a meeting of the Moscow Mathematical Society in disbelief: the topic was the proof of the Four-Color Conjecture (4CC) just obtained by two Americans, whose names did not sound familiar to me, but certainly were destined to enter the history of mathematics, perhaps, history of culture.

So it happened: Kenneth Appel and Wolfgand Haken of the University of Illinois, with the aid of John Koch and some 1200 hours of fast main frame computing, converted Francis Guthrie's 4CC into 4CT, the Four-Color Theorem.

Four-Color Theorem 21.1 (K. Appel, W. Haken and J. Koch [AH1-4]) Every planar map is four-colorable.

However concisely, in this chapter we will look, at the roots of this result and the ideas of Appel–Haken's proof as presented by the authors in their monograph [AH4].

---

[24] [E81.16], published in 1981 in the premier issue of *Combinatorica*; received by the editors on September 15, 1979.

[25] [OED]. *The New Encyclopedia Britannica* did better: "The four-colour problem was solved in 1976 by a group [sic] of mathematicians at the University of Illinois, directed by Kenneth Appel and Wolfgang Haken."

A. Soifer, *The Mathematical Coloring Book*,
DOI 10.1007/978-0-387-74642-5_21, © Alexander Soifer 2009

Appel and Haken's work grew from the 1879 approach discovered by Alfred B. Kempe (discussed in Chapter 20), improved in 1913 by George D. Birkhoff of Harvard University, and brought into the realm of possibility by Heinrich Heesch of the University of Hanover through his committed work over the long years 1936–1972.

Birkhoff found new reducible configurations [Bir], larger then those of Kempe. Heesch built on the work of his predecessors and developed a theory of reducible configurations [Hee1]:

> An investigation of the concepts of reduction has been attempted in the author's "Untersuchungen zum Vierfarbenproblem" (Mannheim, 1969, Chapter I), where the concepts of A-, B-, C-, or D-reducible configurations are developed from the work of A. Errera, G. D. Birkhoff, and C. E. Winn.

Heesch was first to utilize computer in his pursuit [Hee2]:

> The D- or the C-reducibility of a configuration can be recognized much better by computing than by such direct calculations as have been given by the authors up to now.

Above all technical contributions, Heinrich Heesch envisioned and conjectured the existence of a finite set of unavoidable reducible configurations. Appel and Haken paid their tribute to Heesch on the very first lines of their major paper that preceded their great announcement [AH0]:

> This work has been inspired by the work of Heesch [Hee1], [Hee2] on the Four Color Problem, especially his conjecture [Hee1, p. 11, paragraph 1, and p. 216] that there exists a finite set S of 4-color reducible configurations such that every planar map contains at least one element of S. (This conjecture implies the 4CC but is not implied by it.) Furthermore, in 1970 Heesch communicated an unpublished result...which he calls a finitization of the Four-Color Problem.

In 1969 Heesch also pioneered a brilliant idea of *discharging* in search for unavoidable sets of configurations [Hee1]. This book paved the way to computer-aided pursuits of reducibility.[26]

Heesch's role is hard to overestimate. In addition to the credits I have enumerated above, Heesch personally influenced Haken and shared with him much of unpublished ideas. In Appel and Haken's own words [AH2]:

> Haken, who had been a student at Kiel when Heesch gave his talk, communicated with Heesch in 1967inquiring about the technical difficulties of the project of proving Heesch's conjecture and the possible use of more powerful electronic computers.
>
> In 1970 Heesch communicated to Haken an unpublished result which he later referred to as a finitization of the Four-Color Problem, namely that the first discharg-

---

[26]Looking back, it seems surprising that in his 1972 42-page survey [Saa2] of various approaches to the 4CC, T. L. Saati did not even mention the name of Heinrich Heesch.

ing step..., if applied to the general case, yields about 8900 z-positive configurations (most of them not containing any reducible configurations) which he explicitly exhibited...

Heesch asked Haken to cooperate on the project and, in 1971, communicated to him several unpublished results on reducible configurations.

To understand how the discharging works, let us look at the following simple example that I found in Douglas R. Woodall's papers [Woo2], [Woo3] where he gives credit to K. Appel and H. Haken for it.

Problem 21.2 (K. Appel and H. Haken) The set of five configurations in Fig. 21.1 is unavoidable, i.e., at least one of them appears in any plane triangulation.

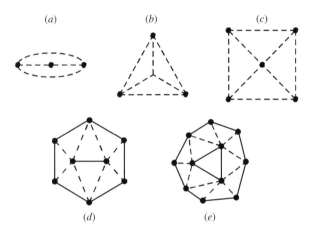

Fig. 21.1 An unavoidable set of configurations

Proof

1. Observations: We will argue by contradiction. Assume that there is a plane triangulation G that contains none of the configurations from Fig. 21.1. We can make the following observations:

Observation A; G has no vertices of degree less than 5, because G contains no configurations (a), (b), and (c).

Observation B: Every vertex v of degree 5 in G has at least three neighbors of degree 7 or greater; for otherwise v would have at least one neighbor of degree 5 and hence G would contain the forbidden configuration (d), or v would have at least three neighbors v_1, v_2, v_3 of degree 6. What is wrong with the latter, you may ask? In the latter case at least 2 of the 6-valent neighbors of v, say v_1 and v_2, must be neighbors of each other (in the *triangulation* G the neighbors of v are connected to each other in a closed path, Fig. 21.2), and thus the forbidden configuration (e) is contained in G.

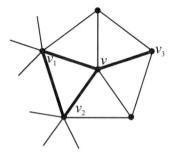

Fig. 21.2

Observation C; Every vertex v of degree 7 has at most three neighbors of degree 5, for otherwise two of its 5-valent neighbors would be neighbors of each other (it is similar to the argument in observation B above: prove it on your own), which would mean precisely that G contains a configuration (d).

Observation D; Every vertex v of degree $i \geq 8$ has at most $\left[\frac{i+1}{2}\right]$ neighbors of degree 5, where for a real number r the symbol $[r]$ denotes the maximum integer such that $[r] \leq r$. The proof of this observation is similar to the proof of observation C above (try it on your own).

2. **Charging**: To each vertex of G of degree i we assign an *electrical charge* equal to $6 - i$. This means that vertices of degree 5 receive unit charge, vertices of degree 6 get zero charge, vertices of degree 7 receive charge equal negative one, etc. In his paper [Kem2], Kempe derives the following equality as a corollary of his rediscovering the Euler's formula (Problem 20.2):

$$\sum_{i=2}^{\Delta} (6 - i)V_i = 12,$$

where V_i stands for the number of vertices of degree i. In Heesch's language of electrical charges, this equality means precisely that the sum of charges of all vertices in G, i.e., the *total charge* in G is equal to a positive 12 units.

3. **Discharging**: Let us now do *discharging*, i.e., redistribution of charge among the vertices without changing the total charge of G. The crux of such a proof is to find the discharging that "works" for the set of configurations in question, which in our case is presented in Fig. 21.1, i.e., it brings us the desired contradiction. Let us transfer $\frac{1}{3}$ of the charge from each vertex of degree 5 to each of its neighbors of degree 7 or greater.

As a result, every vertex of degree 5 ends up with zero or negative charge, because such a vertex has at least three neighbors of degree 7 or greater (see observation B above). Vertices of degree 6 will remain with zero charge, as they are unaffected by discharging. A vertex v of degree 7 would not end up with a positive charge, because v has at most three neighbors of degree 5, each contributing charge $\frac{1}{3}$ to v (observation C). And finally, a vertex of degree $i \geq 8$ that started with a charge

$6 - i$, in view of observation D, can end up with the charge at most

$$6 - i + \frac{1}{3}\left[\frac{i+1}{2}\right] < 0$$

Thus, we end up with *no* vertices of position charge, which contradicts the total charge remaining the positive 12. ∎

Did you like the mathematical Olympiad-like discharging argument? Then you would enjoy proving on your own the following result first obtained without discharging in 1904 by Paul August Ludwig Wernicke from Göttingen University, who in the same year defended his doctorate under the great Hermann Minkowski.

Problem 21.3 (Wernicke, 1904, [Wer]) Prove that the set of five configurations in Fig. 21.3 is unavoidable.

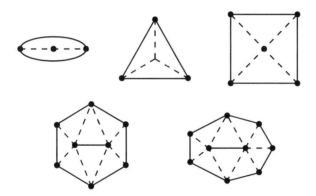

Fig. 21.3 An unavoidable set of configurations

Let us now look at the other critical aspect of Appel–Haken proof: *reducibility*. Appel and Haken used so-called C- and D-reducibilities introduced by Heesch as vast extensions of the technique used by Kempe. In fact, it suffices to restrict ourselves to configurations with vertices of degree 5 and greater since Kempe showed that vertices of lesser degree cannot occur in a minimal counterexample. The authors give an example [AH4]:

Assume that the planar triangulation Δ is the minimal counterexample to the 4CC, which contains, for example, a configuration C of Fig. 21.4(*a*). (Legend in Fig. 21.4(*d*) shows how to read the degrees of the vertices of the configuration from the diagram in Fig. 21.4.) Then the graph $\Delta - C$ obtained from Δ by removing C and edges connecting C to the rest of Δ, must be four-chromatic. A contradiction would be obtained, if we show that every 4-coloring of $\Delta - C$ can be extended to a 4-coloring of Δ.

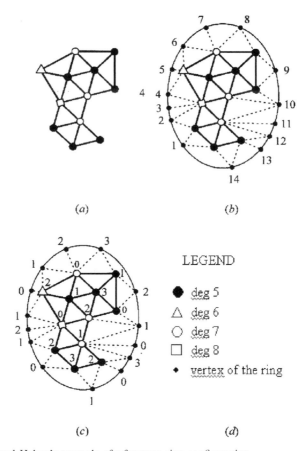

Fig. 21.4 Appel-Haken's example of a fourteen-ring configuration

Appel and Haken repeatedly use good humor in praising the use of computer. Here is one example (numbers 0, 1, 2, 3 on the ring in Fig. 21.4(c) indicate the four colors we are using):

> ... if one were lucky, one might be able to show a fourteen-ring configuration D-reducible with only a few years of careful work. There are obviously some slackers who would not be fascinated by such a task. Such people, with an immorally low tolerance for honest hard work, tend to program computers to do this task. In fact, they find it ideally suited to computers, which are fast, meticulous, and not able to complain about the boring aspects of the work.

Yet, of course, their solution required an enormous amount of both manual and computer work. The crux of Appel–Haken proof was to find such a set of configurations that was *both* unavoidable and consisted of reducible configurations, the so-called *unavoidable set of reducible configurations*. In one of the early, 1978 analyses of the proof [Woo3], Douglas R. Woodall assessed this critical part as follows:

> Discharging procedure and the unavoidable set of configurations were modified every time a configuration in the set turned out not to be C-reducible (or was not quickly proved to be C-reducible) It is clear that these progressive modifications relied on a large number of empirical rules, which enabled an unwanted configuration to be excluded from the unavoidable set at the expense of possibly introducing one or more further configurations. Appel and Haken carried out about 500 such modifications in all. They continued until they had excluded

> i. Every configuration that contained one of three "reduction obstacles"–features that Heesch had discovered, by trial and error, to prevent configurations from being C-reducible;
> ii. Every configuration of ring size 15 or more; and
> iii. Every configuration that was not proved to be reducible fairly quickly (in particular, within 90 minutes on an IBM 370–158 or 30 minutes on a 370–168).

> By the time they had finished (iii), they had constructed an unavoidable set all of whose configurations had been proved reducible; they had therefore proved the theorem. Probably they had excluded from the unavoidable set many configurations that are actually C-reducible but it turned out to be quicker to exclude any configuration that was *not quickly* proved reducible, and to replace it by one or more other configurations, than to carry the analysis of any one configuration to its limit.

> The empirical rules, upon which these progressive modifications were based, were discovered in the course of a lengthy process of trial and error with the aid of a computer, lasting over a year. By the end of this time, however, Appel and Haken had developed such a feeling for what was likely to work (even though they could not always explain why) that they were able to construct the final unavoidable set without using the computer at all. This is the crux of their achievement. Unavoidable sets had been constructed before, and configurations had been proved reducible before, but no-one before had been able to complete the monumental task of constructing an unavoidable set of reducible configurations.

It was a great achievement of Appel and Haken, for they reduced the infinity of various maps to the *finite* set of unavoidable reducible configurations, which needed to be checked. The difficulty was, the set was very large, at first consisting of 1936 configurations. This reduction was a mathematical achievement, and it allowed the use of computer (surely, with infinitely many cases, computer would have been useless!). The enormous computer verification used over 1200 hours of main frame computing of the time (IBM-360 and IBM-370). By 1989, when Appel and Haken produced the 741-page book [AH4] presenting their solution, they reduced the number of configurations to 1476. Such a surprising resolution of the famous problem, both in its volume of work and in use of computing, was bound to cause controversy, and it promptly has. The Appel–Haken–Koch proof of 4CT was a cultural event: it prompted debates and reassessment in many fields of human endeavor, particularly in mathematical and philosophical circles. In the next chapter we will look at the debate and some striking views it has inspired, as well as at the new proof of 4CT and the old but still most promising Hadwiger's Conjecture.

In a phone interview in the fall 1991 (before October 14, 1991), Wolfgang Haken shared with me brief details of his life: born on June 21, 1928 in Berlin; obtained his doctorate from the University of Kiel in 1953; came to the United States in 1962; started to work on 4CC in 1968; came up with first ideas of his own in October 1970. Kenneth Ira Appel was born on October 8, 1932 in Brooklyn, New York; and got his doctorate from the University of Michigan in 1959. During the interview, Haken accepted my invitation to write his view of the Appel–Haken accomplishment entitled, on my recommendation, in Alexandre Dumas' style, "Fifteen Years Later." I offered to include here his complete unedited essay, but no text has ever been received.

The second epigraph, from the Oxford English Dictionary [OED] shows how very little attention is paid to mathematics: Oxford failed to notice even by 2007 the now 31-year-old solution of one of two most famous problems in the multi-millennial history of mathematics (the other, of course, being Fermat's Last Theorem)!

In conclusion, I must quote from the March-2005 unpublished, but web-posted paper by the Microsoft-Cambridge, UK researcher from the Programming Principles and Tools Group, Georges Gonthier [Gon]. With a deep insight of someone who has verified a 4CT proof and came up with a "machine proof," he assesses contributions of the players to the first successful assault of 4CC:

> Although Heesch had correctly devised the plan of the proof of the Four Colour Theorem, he was unable to actually carry it out because he missed a crucial element: computing power. The discharge rules he tried gave him a set R containing configurations with a ring of size 18, for which checking reducibility was beyond the reach of computers at the time. However, there was hope, since both the set of discharge rules and the set R could be adjusted arbitrarily in order to make every step succeed.
>
> Appel and Haken cracked the problem in 1976 by focusing their efforts on adjusting the discharge rules rather than extending R, using a heuristic due to Heesch for predicting whether a configuration would be reducible (with 90% accuracy), without performing a full check. By trial and error they arrived at a set R, containing only configurations of ring size at most 14, for which they barely had enough computing resources to do the reducibility computation. Indeed the discharging formula had to be complicated in many ways in order to work around computational limits. In particular the formula had to transfer arity between non-adjacent faces, and to accommodate this extension unavoidability had to be checked manually. It was only with the 1994 proof by Robertson et al. that the simple discharging formula that Heesch had sought was found.

We will discuss the Robertson–Sanders–Seymour–Thomas proof and its Gonthier's verification in the next chapter.

22
The Great Debate

Computers are useless.
They can only give you answers.

– Pablo Picasso

To reject the use of computers as what one may call
"computational amplifiers" would be akin to an
astronomer refusing to admit discoveries made by
telescope.

– Paul C. Kainen, 1993[27]

I would be much happier with a computer-free proof
of the four color problem, but I am willing to accept
the Appel–Haken proof – beggars cannot be
choosers.

– Paul Erdős, 1991[28]

Interest in the 4CC seems not to be high
in the math literature because it is now
thought to have been proven or something.

– Thomas L. Saaty, 1998[29]

22.1 Thirty Plus Years of Debate

"Thirty years later," as the *Three Musketeers*'s author Alexandre Dumas would have said, the controversy surrounding the Appel and Haken solution is amazingly alive and well. Even when the extraordinary in many respects Appel and Haken's proof was just announced, the President of the Mathematical Association of America Lynn Arthur Steen was very careful [Ste]: he did not write that the conjecture had been

---

[27] [Kai].

[28] Letter to A. Soifer [E91/8/14ltr].

[29] E-mail to A. Soifer of April 13, 1998.

A. Soifer, *The Mathematical Coloring Book*,
DOI 10.1007/978-0-387-74642-5_22, © Alexander Soifer 2009

proven, but instead used the word "verified" in describing the most important mathematical event of that summer.

The proof was met with a considerable amount of confusion in the mathematical community due to the authors' extensive use of computer. This was the first computer-aided solution of a major, celebrated mathematical problem. As such it naturally raised mathematical, philosophical, and psychological questions. In Table 22.1 I put together a "representative" collection of reactions – take a long look at it, then join me for a discussion.

Table 22.1 Reflections on the 4CT

Steen	1976	[Stee]	The four color conjecture...was verified [sic] this summer...
Appel & Haken	1977	[AH1]	Our proof of the four-color theorem suggests that there are limits to what can be achieved in mathematics by theoretical methods alone.
Gardner	1980	[Gar3]	The proof is an extraordinary achievement.... To most mathematicians, however, the proof of the four-color conjecture is deeply unsatisfactory.
Halmos	1990	[Hal]	By an explosion I mean a loud noise, an unexpected and exciting announcement, but not necessarily a good thing. Some explosions open new territories and promise great future developments; others close a subject and seem to lead nowhere. The Mordell conjecture . . . is of the first kind; the four-color theorem of the second.
Erdős	1991	[E91/8/14ltr]	I would be much happier with a computer-free proof of the four-color problem, but I am willing to accept the Appel–Haken proof — beggars cannot be choosers.
Graham	1993	[Hor]	The things you can prove may be just tiny islands, exceptions, compared to the vast sea of results that cannot be proven by human thought alone.
Kainen	1993	[Kai]	To reject the use of computers as what one may call "computational amplifiers" would be akin to an astronomer refusing to admit discoveries made by telescope.
Hartsfield & Ringel	1994	[HR]	Appel and Haken proved it by means of computer program. The program took a long time to run, and no human can read the entire proof, because it is too long.
Jensen & Toft	1995	[JT]	Does there exists a short proof of the four-color theorem... in which all the details can be checked by hand by a competent mathematician in, say, two weeks?
Graham	2002	[Gra4]	Computers are here to stay. There are problems for which computer helps; there are problems for which computer may help; and there are problems for which computer will never help.

The confusion of mathematicians is so clear when we read the words of Martin Garner in his celebrated *Scientific American* column [Gar3]:

> The proof is an extraordinary achievement. . . . To most mathematicians, however, the proof of the four-color conjecture is deeply unsatisfactory.

Which is it, "an extraordinary achievement" or "deeply unsatisfactory?" Surely these terms are mutually exclusive! Paul Halmos, who chose to sum up the twentieth century contribution to mathematics a decade too early (and thus missed a lot, Fermat's Last Theorem, for example), wrote in 1990 [Hal]:

> By an explosion I mean a loud noise, an unexpected and exciting announcement, but not necessarily a good thing. Some explosions open new territories and promise great future developments; others close a subject and seem to lead nowhere. The Mordell conjecture. . . is of the first kind; the four-color theorem of the second.

A loud noise that leads nowhere? It suffices to observe that much of graph theory had been invented through the 124 years of attempts to settle 4CC. Nora Hartsfield and Gerhard Ringel [HR] paint this historic event as absolutely routine, unworthy of a debate:

> Appel and Haken proved it by means of computer program. The program took a long time to run, and no human can read the entire proof, because it is too long.

However, there were, those who gave the event much thought. In 1978, the philosopher Thomas Tymozcko of Smith College illustrated the arrival of computer proofs with a brilliant allegory [Tym]:

> Let us consider a hypothetical example which provides a much better analogy to the appeal to computers. It is set in the mythical community of Martian mathematicians and concerns their discovery of the new method of proof "Simon says." Martian mathematics, we suppose, developed pretty much like Earth mathematics until the arrival on Mars of the mathematical genius Simon. Simon proved many new results by more or less traditional methods, but after a while began justifying new results with such phrases as "Proof is too long to include here, but I have verified it myself." At first Simon used this appeal only for lemmas, which, although crucial, were basically combinatorial in character. In his later work, however, the appeal began to spread to more abstract lemmas and even to theorems themselves. Oftentimes other Martian mathematicians could reconstruct Simon's results, in the sense of finding satisfactory proofs; but sometimes they could not. So great was the prestige of Simon, however, that the Martian mathematicians accepted his results; and they were incorporated into the body of Martian mathematics under the rubric "Simon says."
>
> Is Martian mathematics, under Simon, a legitimate development of standard mathematics? I think not; I think it is something else masquerading under the name of mathematics. If this point is not immediately obvious, it can be made so by expanding on the Simon parable in any number of ways. For instance, imagine that Simon is a religious mystic and that among his religious teachings is the doctrine that the morally good Martian, when it frames the mathematical question justly, can always see the correct answer. In this case we cannot possibly treat the appeal "Simon says" in a purely mathematical context. What if Simon were a revered political leader like Chairman Mao? Under these circumstances we might have a hard time deciding where

Martian mathematics left off and Martian political theory began. Still other variations on the Simon theme are possible. Suppose that other Martian mathematicians begin to realize that Simonized proofs are possible where the attempts at more traditional proofs fail, and they begin to use "Simon says" even when Simon didn't say! The appeal "Simon says" is an anomaly in mathematics; it is simply an appeal to authority and not a demonstration.

The point of the Simon parable is this: that the logic of the appeals "Simon says" and "by computer" are remarkably similar. There is no great formal difference between these claims: computers are, in the context of mathematical proofs, another kind of authority. If we choose to regard one appeal as bizarre and the other as legitimate, it can only be because we have some strong evidence for the reliability of the latter and none for the former. Computers are not simply authority, but warranted authority. Since we are inclined to accept the appeal to computers in the case of the 4CT and to reject the appeal to Simon in the hypothetical example, we must admit evidence for the reliability of computers into a philosophical account of computer-assisted proofs. . .

The conclusion is that the appeal to computers does introduce a new method into mathematics.

Tymoczko is right: Appel–Haken–Koch's proof changed the meaning of the word "proof" by letting in a reliable experiment as allowable means, by taking away the absolute certainty we cherished so much in the mathematical proof. Thomas L. Saaty and Paul C. Kainen, whose great timing allowed them to publish in 1977 the first book ever on The Four-Color Problem that included a discussion of its solution, were first to observe the substantial but inevitable trade-off of the acceptance of such a proof [SK, end of part one]:

To use the computer as an essential tool in their proofs, mathematicians will be forced to give up hope of verifying proofs by hand, just as scientific observations made with a microscope or telescope do not admit direct tactile confirmation. By the same token, however, computer-assisted mathematical proof can reach a much larger range of phenomena. There is a price for this sort of knowledge. It cannot be absolute. But the loss of innocence has always entailed a relativistic world view; there is no progress without the risk of error.

In the essay [Kai] written in 1993 especially for *Geombinatorics*, Paul C. Kainen elaborates further on the above allegory:

To reject the use of computers as what one may call "computational amplifiers" would be akin to an astronomer refusing to admit discoveries made by telescope.

This is certainly an elegant metaphor. However, one cannot argue with Tymoczko's warning about keeping the order right – we have accepted the legitimacy of the use of computers first, and only based on this acceptance we have claimed the existence of the formal proof [Tym]:

Some people might be tempted to accept the appeal to computers on the ground that it involves a harmless extension of human powers. In their view the computer merely traces out the steps of a complicated formal proof that is really out there. In fact, our only evidence for the existence of that formal proof presupposes the reliability of computers.

As Tymoczko rightly observes, the timing of Appel–Haken–Koch work was favorable for the acceptance of their proof [Tym]:

I suggest that if a "similar" proof had been developed twenty-five years earlier, it would not have achieved the widespread acceptance that the 4CT has now. The hypothetical early result would probably have been ignored, possibly even attacked (one thinks of the early reaction to the work of Frege and of Cantor). A necessary condition for the acceptance of a computer-assisted proof is wide familiarity on the part of mathematicians with sophisticated computers. Now that every mathematician has a pocket calculator and every mathematics department has a computer specialist that familiarity obtains. The mathematical world was ready to recognize the Appel–Haken methodology as legitimate mathematics.

In their 1978 essay [WW] Douglas R. Woodall and Robin Wilson state that "there is no doubt that Appel and Haken's proof is a magnificent achievement which will cause many mathematicians to think afresh (or possibly for the first time) about the role of the computer in mathematics." Yet, they share concern with Paul Halmos and many others:

The length of Appel and Haken's proof is unfortunate, for two reasons. The first is that it makes it difficult to verify. . . .. The other big disadvantage of a long proof is that it tends not to give very much understanding of why the result is true. This is particularly true of a proof that involves looking at a large number of separate cases, whether or not it uses the computer.

Paul Erdős put the state of 4CT most aptly in his 1991 letter to me [E91/8/14ltr]:

I would be much happier with a computer-free proof of the four color problem, but I am willing to accept the Appel-Haken proof – beggars cannot be choosers.

So what are we mathematicians to do? The answer, in the form of a question, came from the Danish graph theorists Tommy R. Jensen and Bjarne Toft in their book of open coloring problems of graph theory [JT]:

Does there exists a short proof of the four-color theorem. . . in which all the details can be checked by hand by a competent mathematician in, say, two weeks?

22.2 Twenty Years Later, or Another Time – Another Proof

Twenty years later, when the familiarity with and trust in computing have dramatically improved, a new team of players came on 4CT stage: leading graph theorists Neil Robertson and Paul Seymour and their young students and colleagues Daniel Sanders and Robin Thomas. This reminded me the Hollywood film, *Seven Brides for Seven Brothers*. Only here we had *Four Mathematicians for Four Colors*. Four on four, they had to be able to handle 4CT, and handle they did.

In their work on graph theory, the authors thought that in a sense the validity of $H(6)$ (Hadwiger's Conjecture for 6 – we will formulate it later in this chapter)

depended upon the validity of 4CT. Thus, they felt compelled to either verify Appel–
Haken proof or find their own. They decided that the latter was an easier task.

The "four musketeers" undoubtedly realized that "20 years later," as Alexandre
Dumas used to say (precisely the title of Alexandre Dumas' sequel to the famed
Three Musketeers), they would have to get a much better proof than the original one
by Appel and Haken for otherwise they would be asked "why did you bother?" The
remarkable thing is that these athors have achieved just such a proof!

I first learned about it in February 1993 at Florida Atlantic University from
Ronald L. Graham, who also forwarded to me the e-mail announcement of the forthcoming March 24, 1994 DIMACS talk by Paul Seymour, who at the time worked at
Bellcore. I asked Paul's coauthor Neil Robertson for the details. His May 9, 1994
reply [Rob1] due to its medium, e-mail, concisely and instantly summarized what
he thought was most important about the new proof:

> We have a new proof, along the same lines as the AHK[30] proof, relying more on the
> computer, and so more reliable. The unavoidable set is in the area of 600 configurations
> ($<= 638$), and we get a quadratic algorithm. Dan wrote a nice article about this for
> SIAM (I think). Seymour, Thomas, Sanders and I are involved. With a slightly larger
> unavoidable set the overall proof becomes very simple (apart from the calculations) as
> we avoid almost all degeneracies by using D-reduction and reducers for C-reduction
> from the single edge contraction minors of the given configuration. Will forward to
> you a copy of Dan's article.

One important point to notice here is that the proof relies more (not less!) on
the computer than Appel–Haken–Koch's proof, and it makes the proof more (not
less) reliable due to its clear separation of human and machine tasks. The size of the
unavoidable set of reducible configurations was substantially reduced from Appel
and Haken's 1476 to 638, but even greater improvements were in the much simpler
discharging procedure. Later that same day, Neil forwarded to me Daniel Sander's
summary, entitled, in a word,

NEWPROOFOFTHEFOURCOLORTHEOREM.

I have got to share with you parts of this announcement summary, as it includes the
authors' assessment of the Appel–Haken proof and comparisons of the two proofs.
I will add my comments as footnotes:

> ... Before and after Appel and Haken, many claims have been made to prove 4CT with
> the aid of a computer, none of which held up to the test of time. But Appel and Haken's
> proof has stood; for 18 years. Why? Some may say that the proof is inaccessible.
> It is so long and complicated; has anyone actually read every little detail? At least
> two attempts were made to independently verify major portions of Appel and Haken's
> proof [AH2, AH3], which yielded no significant problems. Appel and Haken [AH4]
> published a more complete (741 pages) version of their proof 5 years ago, but many
> remain hesitant.

---

[30] Appel, Haken and Koch.

The author of this paper, together with Neil Robertson, Paul Seymour, and Robin Thomas, announces a new proof of 4CT. The proof uses the same techniques as that of Appel and Haken: discharging and reducibility. The new proof, however, makes improvements in the complexity of the arguments. Hopefully these improvements will help people to better understand and appreciate Appel and Haken's method.

To describe the improvements in more detail requires a discussion of the discharging method. Simple reductions show that one need[s] only [to] consider plane triangulations of minimum degree 5.

An easy manipulation of Euler's formula gives the following equality for these graphs: $\sum_{v \in V(G)} (6 - \deg(v)) = 12.$[31] This value $6 - \deg(v)$ has come to be known as the charge of v. The vertices of degree 5 are the only vertices of positive charge. The vertices of degree at least 7 have negative charge, and are known as major vertices.

The discharging method is to locally redistribute the positive charge from the vertices of degree 5 into the major vertices. The sum of the new charges will equal the sum of the old charges, and thus the[re] will be a vertex which has its new charge positive, known as an overcharged vertex.

The structure of the graph close to an overcharged vertex is determined by the rules that were used to discharge the vertices of degree 5. Each possible structure that can yield an overcharged vertex must be examined [to] find within it some *configuration* that is reducible (provably cannot exist in minimal counterexample to 4CT). Thus there are the two steps of the proof of 4CT.

Discharging: defining a set of discharging rules which in turn gives a list of configurations that a plane *triangulation of* minimum degree 5 must have.

Reducibility: showing that no minimal counterexample to 4CT can contain any of these configurations.

The two forms of reducibility that Appel and Haken use are known as C-reducibility and D-reducibility. The idea of D-reducibility is that no matter what coloring the ring (border of the configuration) has, it can be changed by Kempe chains (swapping the colors of an appropriate 2-colored subgraph) into a coloring that extends into a coloring of the configuration. C-reducibility is the same idea, except with first replacing the *configuration* [b]y a smaller configuration, thus restricting the possible colorings of the ring. Bernhart (see [GS]) found a new form of reducibility which can show some configurations reducible that D- and C-reducibility cannot.

Although we were able to produce six configurations which were reducible by the block count method, these configurations turned out not to be needed.

The new proof still uses only D- and C-reducibility, which were clearly defined by Heesch [Hee1] based upon ideas of Birkhoff [Bir].

The primary discharge rule that Appel and Haken use is the following:

A vertex x of degree 5 originally has a charge of 1. Send a charge of 1/2 from x to each major neighbor of x. Unfortunately, this simple rule is not enough to prove 4CT. It yields a list of configurations, but not all of them are reducible. So, for each non-reducible configuration, they define secondary discharge rules, which move the charge around a bit more.

These new rules produce the need for even more rules, and so on, but eventually the process stopped with a list of 1476 reducible configurations. The total number

---

[31] This formula is due to A. B. Kempe, 1879 [Kem2].

of secondary discharge rules that they used was 486. A better primary discharge rule permits improvement in both of these areas.[32]

Here is the primary discharge rule used in the new proof. Imagine each vertex x of degree 5 expelling its positive charge equally in each of the five directions around it. Thus x will send 1/5 to each of its neighbors. The major vertices have a negative charge that attracts this positive charge that was expelled. Thus if the neighbor y of x is major, it absorbs this 1/5. If the neighbor y is not major, the charge just keeps going, splitting half to the left and half to the right. Let p and r be the common neighbors of x and y, to the left and right the edge xy.

The left 1/10 rotates counterclockwise through the neighbors of p, while the right 1/10 rotates clockwise through the neighbors of r. If $\deg(p) \geq 8$, its attraction is so great that the 1/10 doesn't make it to the next neighbor; this charge gets absorbed by p. Otherwise, the 1/10 rotates until it reaches a major neighbor of p, unless $\deg p = 7$, and it has rotated through four neighbors; in this case p absorbs it. Similarly for r. Using this primary rule, only 20 secondary discharge rules are necessary to produce a list of 638 reducible configurations...

The largest size ring that Appel and Haken use is a 14-ring; their original list had 660 14-rings. The list of 638 mentioned above contains 161 14-rings. It is not known whether 14-rings can be avoided altogether, but at least 12-rings appear to be necessary...

Totally automating the discharge analysis allowed us to try several heuristics on how to make these choices. Having the discharge analysis automated also hinders the possibility of errors creeping in; a human error was found in Appel and Haken's discharge analysis (its correction can be found in [AH4, p. 24]).

Recently, Appel and Haken [AH4] have proven a quartic algorithm to 4-color planar graphs using their list of 1476 reducible configurations... we have found a quadratic algorithm to 4-color planar graphs...

The reducibility and discharging programs that were used to complete the new proof of 4CT will soon be available by anonymous ftp. The total amount of computer time required to prove 4CT on a Sun Spark 10 is less than twenty-four hours.

About 2 years later, on February 19, 1996 I attended Paul Seymour's plenary talk at the International Southeastern Conference on Combinatorics, Graph Theory and Computing at Florida Atlantic University. I knew that the new proof was superior to the original one in a number of aspects. Yet, I was wondering what compelled the authors to look for another computer-aided proof of 4CT. Paul Seymour addressed it right in the beginning of his talk, as I was jotting down notes:

> It was difficult to believe it [Appel–Haken's proof]: you can't check it. First you need a computer. Second, non-computer part is awful: it contains hundreds of pages of notes. You can't understand. You are not quite sure that the theorem is true. Nobody checked the proof. This is a bit scary.
>
> We assumed 4CT is true in earlier work, so we had to have a sure proof. General framework is the same, but details are better.

---

[32] Better discharge rule prompted a reduction of secondary discharge rules from 486 in Appel and Haken to just 20, which resulted is a much more accessible ideologically proof and a 43-page paper [RSST] vs. the 741 book [AH4] of Appel and Haken.

The talk ended with questions and Paul Seymour's answers:

Erdős: Is there going to be a normal proof?
Seymour: I don't have any reason to think it is impossible. I try it from time to time.
Soifer: What are approaches to "normal proof?"
Seymour: I am not going to tell you my wrong proofs. Start with a triangle, flip it
   over, put a rubber band around, look for a smaller set of reducible configurations.
Soifer: Why did you use Sun Microsystems Workstation in your solution?
Seymour: This is what I have in my office.
Soifer: How long does it take to verify your proof?
Seymour: Computer can verify the proof in 5 minutes; 6 months by hand.

On May 25, 1995, when the paper [RSST] with the new proof was submitted, it
consisted of just 43 pages—a vast improvement over the 741 + 15-page monograph
[AH4]. However, with great advantages of the new proof (Table 22.2), let us not
forget, that Appel and Haken discovered a proof first. And let us remember that
"the most notorious paper in the history of graph theory: the 1879 work by A. B.
Kempe [Kem2] that contains the fallacious proof of the Four Color Theorem" [Ste],
"The Kempe Catastrophe" [Saa1] – paved the way!

Table 22.2 Comparison of the Two 4CT Proofs

	Appel–Haken–Koch	Robertson–Sanders Seymour–Thomas
Number of secondary discharging rules	486	20
Number of unavoidable configurations	1476	638
Computer time to prove	1200 hours	24 hours
Computer time to verify	Not available	5 minutes
Speed of graph coloring algorithm	Quartic	Quadratic
Number of pages in the final publication	741	43

March 2005 brought a new development in the 4CT saga, when Georges Gonthier
of Miscrosoft-Cambridge, UK, produced "a formal proof of the famous Four Color
Theorem that has been fully checked by the Coq proof assistant." "It's basically
a machine verification of our proof," wrote Paul Seymour in his January 17, 2008
e-mail to me. Let us give the podium to Georges Gonthiers for the assessment of his
work in the unpublished but posted on the web paper [Gon]:[33]

> We took the work of Robertson et al. as our starting point, reusing their optimized
> catalog of 633 reducible configurations, their cleverly crafted set of 32 discharge rules,

---

[33] *The Economist* (April 2–8, 2005) reported "Dr. Gonthier says he is going to submit his paper to a
scientific journal in the next few weeks." This, however, has not happened.

and their branch-and-bound enumeration of second neighborhoods [RSST]. However, we had to devise new algorithms and data structures for performing reducibility checks on configurations and for matching them in second neighborhoods, as the C integer programming coding tricks they used could not be efficiently replicated in the context of a theorem prover, which only allows pure, side effect free data structures (e.g., no arrays). And performance *was* an issue: the version of the Coq system we used needed three days to check our proof, whereas Robertson et al. only needed a three hours. . .ten years ago! (Future releases of Coq should cut our time back to a few hours, however.)

We compensated this performance lag in part by using more sophisticated algorithms, using multiway decision diagrams (MDDs) for the reducibility computation, concrete construction programs for representing configurations, and tree walks over a circular zipper to do configuration matching. This sophistication was in part possible because it was backed by formal verification; we didn't have to "dumb down" computations or recheck their outcome to facilitate reviewing the algorithms, as Robertson et al. did for their C programs [RSST].

Even with the added sophistication, the program verification part was the easiest, most straightforward part of this project. It turned out to be much more difficult to find effective ways of stating and proving "obvious" geometrical properties of planar maps. The approach that succeeded was to turn as many mathematical problems as possible into program verification problems.

In the concluding section, "Looking ahead," Gonthier sees his success as a confirmation that the "programming" approach to theorem proving may be more effective than the traditional "mathematical" approach, at least for researchers with computer science background:

As with most formal developments of classical mathematical results, the most interesting aspect of our work is not the result we achieved, but how we achieved it. We believe that our success was largely due to the fact that we approached the Four Colour Theorem mainly as a *programming* problem, rather than a *formalization* problem. We were not trying to replicate a precise, near-formal, mathematical text. Even though we did use as much of the work of Robertson et al. as we could, especially their combinatorial analysis, most of the proofs are largely our own.

Most of these arguments follow the generate-and-test pattern exposed in Chapter 4. We formalized most properties as computable predicates, and consequently most of our proof scripts consisted in verifying some particular combination of outcomes by a controlled stepping of the execution of these predicates. In many respects, these proof scripts are closer to debugger or testing scripts than to mathematical texts. Of course this approach was heavily influenced by our starting point, the proof of correctness of the graph colouring function. We found that this programs-as-proof style was effective on this first problem, so we devised a modest set of tools (our tactic shell) to support it, and carried on with it, generalizing its use to the rest of the proof. Perhaps surprisingly, this worked, and allowed us to single-handedly make progress, even solving subproblems that had stumped our colleagues using a more orthodox approach.

We believe it is quite significant that such a simple-minded strategy succeeded on a "higher mathematics" problem of the scale of the Four Colour Theorem. Clearly, this is the most important conclusion one should draw from this work. The tool we used to support this strategy, namely our tactic shell, does not rely on sophisticated technology of any kind, so it should be relatively easy to port to other proof assistants (including the newer Coq). However, while the tactic shell design might be the most obvious byproduct of our work, we believe that it should have wider implications on the interface design of proof assistants. If, as this work seems to indicate, the "programming" approach to theorem proving is more effective than a traditional "mathematical" approach, and given that most of the motivated users of poof assistants have a computer science background and try to solve computer-related problems, would it not make interface of a proof assistant more similar to an program development environment, rather than strive to imitate the appearance of mathematical texts?

22.3 The Future that commenced 65 Years Ago: Hugo Hadwiger's Conjecture

There remained a number of conjectures that, if proved, would imply the 4CT. Hugo Hadwiger posed the most prominent of these conjectures in 1943 [Had3].

An *edge contraction* of a graph G consists of deleting an edge and "gluing" together (i.e., identifying) its incident vertices. We say that a graph G is *contractible* to a graph H if H can be obtained from G by a sequence of edge contractions. In this case H is called a *contraction* of G, and G is said to be *contractible* to H.

We can view the Hadwiger conjecture $H(n)$ as a series of conjectures, one for every positive integer n.

The Hadwiger's Conjecture $H(n)$ 22.1 [Had3] Every connected n-chromatic graph G is contractible to K_n.

The truth of the conjecture $H(n)$ for $n < 5$ has been proven in 1952 by G. A. Dirac [Dir]. But it is the case $H(5)$ that proved to be particularly important. Why? Because the following equivalence takes place:

Theorem 22.2 $H(5)$ is equivalent to the 4CT.

$H(5) \Rightarrow 4CT$. Proof in this direction is very simple. Given $H(5)$, assume G is a planar graph that is not 4-colorable. But then by $H(5)$, G is contractible to K_5, which is absurd since any contraction of the planar G must be planar as well.

$4CT \Rightarrow H(5)$. Proof is this direction is more involved: here is its sketch. Let G be a 5-chromatic graph, not contractible to K_5, of the minimum order with this property. Then it can be shown that G is 4-connected. In 1937, before Hadwiger formulated his conjecture, K. Wagner [Wag] showed that a 4-connected graph not contractible to K_5, is planar. Thus, G is a planar graph of chromatic number 5, which contradicts 4CT. ∎

The most surprising result was published in 1993 by Neil Robertson, Paul D. Seymour and Robin Thomas. They proved that $H(6)$ is also equivalent to 4CT!

Theorem 22.3 ([RST], 1993) The following statements are equivalent:

(a) 4CT;
(b) $H(5)$;
(c) $H(6)$.

The authors [RST] comment in the abstract:

We show (without assuming the 4CC) that every minimal counterexample to Hadwiger's conjecture [for 6] is "apex", that is, it consists of a planar graph with one additional vertex. Consequently, the 4CC implies Hadwiger's conjecture [for 6], because it implies that apex graphs are 5-colourable.

Right after his plenary talk on February 19, 1996, at the Conference on Combinatorics, Graph Theory and Computing in Boca Raton, Florida, I asked Paul Seymour about his and Robertson result about the relationship between $4CT$ and Hadwiger's Conjecture. His reply was:

I believe that all of them (Hadwiger's Conjectures for various n) are equivalent. We have a result that if 4CC is true, then for every n there is $f(n)$ such that for Hadwiger's Conjecture to be true, it suffices to check graphs of order not exceeding $f(n)$.

Conjecture 22.4 (Paul D. Seymour) All Hadwiger's conjectures for various $n \geq 5$ are equivalent, and equivalent to $4CT$.

It seems plausible that a computer-free proof of the 4CT (shouldn't it exist!) will come as a consequence of a (computer-free) proof of the Hadwiger's conjecture $H(n)$ for some $n > 4$.

Another way of finding a short computer-free proof was suggested by Appel and Haken [AH1]:

Of course, a short proof of the 4-color theorem may some day be found, perhaps, by one of those bright high school students.

I would love that, amen!

23
How Does One Color Infinite Maps? A Bagatelle

How does one measure fun in mathematics? Certainly not by the length of exposition. This is a short chapter, a bagatelle. I hope nonetheless that you will enjoy it.

We know (4CT) that every finite map on the plane is 4-colorable. What about maps with infinitely many countries? This sounds like a natural question, which I have heard from various people at various times. In particular, Peter Winkler, then Director of Fundamental Mathematics Research at Bell Labs and now professor of mathematics at Dartmouth, asked me this question on October 11, 2003 right after my talk at Princeton-Math. The 4-colorability of infinite maps follows from 4CT due to de Bruijn–Erdős' Compactness Theorem 26.1. Let us record it formally.

Infinite Map-Coloring Theorem 23.1 Every map with infinitely many countries is 4-colorable.

Proof Given an infinite map M. As we know, we can translate the problem of coloring M into the problem of coloring the planar graph $G(M)$. Since by 4CT every finite subgraph of $G(M)$ is 4-colorable, $G(M)$ is 4-colorable as well by De Bruijn–Erdős' Compactness Theorem 26.1. ∎

In late December 2004 I was giving talks at the Mathematical Sciences Research Institute in Berkeley, California. There my old friend Prof. Gregory Galperin showed me his proof of Theorem 23.1. His proof was longer and worked only for countable maps. Nevertheless, I have got to share it with you here because of its striking beauty. Plus, of course, Galperin's proof—unlike the proof above—does not require the axiom of choice in its full force.

Countable Map-Coloring Theorem 23.2 Every map with countably many countries is 4-colorable.

Proof by G. Galperin [Gal] Given a countable map M. Enumerate the countries of the map by positive integers: $1, 2, \ldots, n, \ldots$. Let integers 1, 2, 3, and 4 be the names of the colors to be used.

Let n be a positive integer. Take the first n countries. By 4CT there is a 4-coloring of the submap consisting of these n countries. Let the colors assigned to these countries be $a_1, a_2, \ldots, a_n$ respectively (of course, each a_i is equal 1, 2, 3, or 4). We represent this coloring by a number x_n in its decimal form: $x_n = 0.a_1 a_2 \ldots a_n$.

A. Soifer, *The Mathematical Coloring Book*,
DOI 10.1007/978-0-387-74642-5_23, © Alexander Soifer 2009

As we do this for each positive integer n, we end up with the sequence $S = \{x_1, x_2, \ldots, x_n, \ldots\}$ of real numbers. Since S in bounded, $S \subset [0, 1]$, by the Bolzano–Weierstrass theorem it contains a convergent subsequence S':

$$S' = \left\{x_{i_1}, x_{i_2}, \ldots, x_{i_n}, \ldots\right\},$$

where $i_1 < i_2 < \ldots < i_n < \ldots$ Let the limit point of S' be y, which in decimal form looks like

$$y = 0.y_1 y_2 \ldots y_n \ldots$$

It is easy to prove that the sequence $y_1, y_2, \ldots, y_n, \ldots$ delivers a (proper) 4-coloring of the respective regions $1, 2, \ldots, n, \ldots$ Indeed, since all the decimal digits of the x_i were 1, 2, 3, or 4, the same must be true about the decimal digits $y_1, y_2, \ldots, y_n, \ldots$ of the limit point y. Two neighboring regions could not be assigned the same color by this rule, for otherwise they would have been assigned the same color already in a coloring that we decoded by one of the x_{i_n}! ∎

24

Chromatic Number of the Plane Meets Map Coloring: Townsend–Woodall's 5-Color Theorem

In Chapter 8, I described Douglas R. Woodall's attempt to obtain a result on the chromatic number of the plane under an additional condition that monochromatic sets are closed or simultaneously divisible into regions [Woo1]. Six years after his publication, Stephen P. Townsend found a logical mistake in Woodall's proof, constructed a counterexample showing that Woodall's proof cannot work, and went on to discover his own proof of the following major result.

Townsend–Woodall's 5-Color Theorem 24.1 [Tow2]. Every 5-colored planar map contains two points of the same color until distance apart.

This implies result 8.1:

Townsend–Woodall's Theorem 24.1' The chromatic number of the plane under map-type coloring is 6 or 7

In this chapter, I will give you the story of the proof and the proof itself.

24.1 On Stephen P. Townsend's 1979 Proof

This story must remind the readers of the famed Victorian Affair, which we discussed in Chapters 19 and 20. To sum it up, in 1879 Alfred B. Kempe published a proof of the 4-Color Map-Coloring Theorem, in which 11 years later Percy J. Heawood found an error and constructed a counterexample to demonstrate the irreparability of the hole. Heawood salvaged Kempe's proof as the 5-Color Theorem, but the 4CC had to wait nearly another century for its proof.

Our present story started with Douglas R. Woodall's 1973 publication, in which 6 years later Steven P. Townsend found an error and constructed a counterexample to demonstrate the irreparability of the hole. So far the two stories are so close! However, unlike its Victorian counterpart, Townsend went on to prove Woodall's statement, and so I thought the new story had a happy end—until February 11, 2007, when I asked Stephen about "the story of the proof." The surprising reply reached me by e-mail on February 20, 2007:

A. Soifer, *The Mathematical Coloring Book*,
DOI 10.1007/978-0-387-74642-5_24, © Alexander Soifer 2009

Story of the proof

I first became interested in the plane-colouring problem in 1977 or 1978. At that time I was a lecturer in the Department of Mathematics at the University of Aberdeen, having just completed my doctoral thesis (in Numerical Analysis). I had read an article that listed some of the unsolved problems in Combinatorics at that time, and this one caught my attention.

I was totally unaware of Douglas's 1973 proof, which was both my folly and my good fortune. Folly, in that I should have conducted a more exhaustive literature search before devoting time to the problem. Good fortune in that had I been aware of Douglas's paper I would not have spent any time on the problem; I certainly would not have had the temerity to check Douglas's proof for accuracy. It should be noted that I was a numerical analyst, not a combinatorialist, so my awareness of the field of combinatorics was somewhat limited, in spite of brushing shoulders at Aberdeen with some eminent contributors to the field.

It was not until I had completed the proof, and was considering what references to include, that I came upon Douglas's paper. I was both devastated and puzzled. The puzzlement came from my intimate knowledge of the difficulties of certain aspects of the proof and the fact that Douglas seemed to have produced a proof that circumnavigated these difficulties. So it was with an attitude of "how did he manage this?" that I went through his proof and consequently spotted the error.

A colleague at Aberdeen, John Sheehan, whom I'm sure you will have come across, encouraged me nonetheless to submit my proof for publication, but including a reference to Douglas's work. The rest I think you know.

Yes, Stephen Townsend was lucky, for not only was he the first to produce a proof—but he also rediscovered the statement of the result on his own, albeit after Woodall's publication—and this Townsend's rediscovery was a necessary condition for finding the proof.

However, Townsend's good luck, ran into a wall, when the *Journal of Combinatorial Theory*'s Managing Editor and the distinguished Ramsey theorist Bruce L. Rothschild wrote to Townsend on April 3, 1980:

The Journal of Combinatorial Theory – Series A is now trying very hard to reduce its large backlog, and we ask all our referees to be especially attentive to the question of the importance of the papers. In this case the referee thought that the result was not of great importance. In view of our backlog situation then, we are reluctant to publish the paper. However, since it does correct an error in a previously published paper, we would like to have a very short note about it. Perhaps, you would be willing to do the following: Write a note pointing out the error, stating the theorem (Theorem 1) (without proof) used to get around the trouble, and that the theorem must be used with care to get around the problem.

Stephen P. Townsend had satisfied the Editor (what choice did he have!), and produced a 2-page proof-free note [Tow1], which was published the following year. This is where the story was to end in 1981.

No blame should be directed at Douglas R. Woodall, we all make mistakes (except those of us who do nothing). The mistake notwithstanding, Woodall's 1973 paper has remained one of the fine works on the subject. Moreover, he was the first to alert me of his mistake and Townsend's 2-page note. "I am a fan of your 1973 paper," I wrote to Woodall in the October 10, 1993 e-mail, in which I called [Tow1] "the Townsend's addendum." The following day Woodall replied as follows:

> I will put a reprint in the post to you today, together with a photocopy of Townsend's "addendum," as you so tactfully describe it. (The fact is, I boobed, and Townsend corrected my mistake.)

However, regret is in order about the decision by the *Journal of Combinatorial Theory Series A* (*JCTA*). While they apparently (and correctly) assessed Woodall's paper as being "of great importance" (an impossible test if one interprets it literally), they denied its readers—and the world—the pleasure and the profit of reading Townsend's proof of the major result.

I have corrected *JCTA*'s quarter-a-century old mistake, when I published Townsend's work in the April 2005 issue of *Geombinatorics* [Tow2]. Townsend's work was preceded by my historical introduction [Soi25], a version of which you have just read. I ended that introduction with the words I would like to repeat here: It gives me a great pleasure to introduce and publish Townsend's proof. In my opinion, it *is* of great importance—judge for yourselves!

It pains me to see that most researchers in the field are still unaware of Woodall's mistake and Townsend's proof. It suffices to look at the major problem books to notice that: not only the 1991 book by Croft–Falconer–Guy [CFG], but even the recent 2005 book by Brass–Moser–Pach [BMP] give credit to Woodall and do not mention Townsend! I hope this chapter will inform my colleagues of the correct credit and of Townsend's achievement.

Stephen Phillip Townsend was born on July 17, 1948 in Woolwich, London, England. He received both graduate degrees, Master's (1972) and doctorate (1977) from the University of Oxford. Townsend has been a faculty first in the department of mathematics (1974–1980) and then in the department of computer science (1982–present) at the University of Aberdeen, Scotland. Since 1995 he has also been Director of Studies (Admissions) in Sciences. In addition to publications in mathematics, Steven's list of publications includes "Women in the Church–Ordination or Subordination?" (1997).

24.2 Proof of Townsend–Woodall's 5-Color Theorem

In this chapter, I will present Stephen P. Townsend's proof. As you now know, it first appeared in 2005 in *Geombinatorics* [Tow2]. However, when I was preparing this chapter, I asked Stephen to improve the exposition, make his important proof more accessible to the reader not previously familiar with topology, and include plenty of drawings to help the reader to visualize the proof. He did it, quite brilliantly. Thus, presented below exposition of the proof has been written by Professor Townsend especially for this book in 2007.

He starts with a few basic definitions from general (point set) topology.

Definitions A pair of points in E^2 unit distance apart having the same color is called a *monochrome unit*.

Let S and T be subsets of E^2. S is said to *subtend T at unit distance* if T is the union of all unit circles centered on points in S.

Let A be any closed, bounded doubly connected set in E^2 containing a circle of unit radius. If the removal of any point in A renders it simply connected then such a point is called a *cut point* of A. If A has no cut points, its interior A^0 is said to be a *unit annulus*. If A has a finite number of cut points (which must occur on a circle of unit radius) then A^0 is said to be a *finitely disconnected unit annulus* (Fig. 24.1).

A *planar map* (Fig. 24.2) is an ordered pair M(S,B) where S is a set of mutually disjoint bounded finitely connected open sets (*regions*) in E^2 and B is a set of simple closed curves (*frontiers*) in E^2 satisfying

   i. the union of the members of S and B forms a covering of E^2;
  ii. there exists a one-to-one function F : S → B such that b = F(s), s ∈ S, is the *exterior boundary* of s;
 iii. the *boundary* of s ∈ S is the union of F(s) and at most a finite number of other members of B, which are the *interior boundaries* of s.

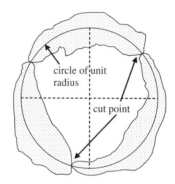

Fig. 24.1

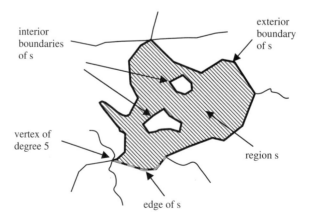

Fig. 24.2

A point on the boundary of s is called a *boundary point* of s. A boundary point, which lies on the boundary of k regions, $k \geq 3$, is called a *vertex of degree k*. A closed subset of a frontier $b \in B$, which is bounded by two vertices and contains no other vertices, is called an *edge* of each region for which b is part of the boundary. Two regions are *adjacent* if their boundaries contain a common edge or a common frontier.

The above definition is more general than the usual definition of a planar map, which requires each region $s \in S$ to be simply connected, and requires each frontier $b \in B$ to contain at least two vertices.

An *r-coloring* of a planar map is a function of $C_r : E^2 \rightarrow \{c_1, c_2, \ldots, c_r\}$ where C_r is constant over each region in S and where a boundary point is given the color of one of the regions in the closure of which it lies.

Initial Observations: To prove that an r-colored map must contain a monochrome unit it is sufficient to examine only those r-colored maps satisfying

  (i) each region has no interior boundaries, i.e., its closure does not contain the closure of any other region;
  (ii) different regions of the same color have no common boundary points.

This is best understood by observing that every r-colored map with no monochrome units may be simplified to an r-colored map with no monochrome units satisfying (i) and (ii) above as follows.

(a) For each region s with interior boundaries, remove these boundaries and assimilate into s all regions whose closures are contained in the closure of s.

(b) Remove any edges common to adjacent regions of the same color.

(c) For each vertex $\underline{v}$ which is a boundary point of two non-adjacent regions of the same color, choose $\varepsilon > 0$ sufficiently small and describe an ε-neighborhood whose closure contains $\underline{v}$ and whose intersection with each of the two regions is non-null, coloring this ε-neighborhood the same color as the two regions, and thus forming one new region incorporating the original two and the ε-neighborhood. (Fig. 24.3.)

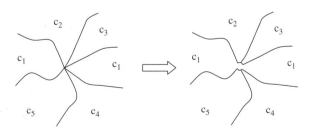

Fig. 24.3

Note that a consequence of (ii) is that we do not need to consider vertices of degree greater than r in an r-colored map. A sequence of theorems now follows, concluding with the main result that *every 5-colored planar map contains a monochrome unit*. Here is an outline of the proof:

1. we show that every 4-colored planar map contains a monochrome unit;
2. we show that every 5-colored planar map containing a vertex of degree 3 contains a monochrome unit;
3. we show that every 5-colored planar map without a monochrome unit must contain a vertex of degree 3;
4. for 2 and 3 both to be true, every 5-colored planar map must contain a monochrome unit.

The Proof: Townsend presents the proof in stages through five theorems.

Theorem 24.2 Let A^0 be a finitely disconnected unit annulus (Fig. 24.1) for which a circle of unit radius contained in its closure, A, has at least one arc of length greater than $\pi/3$ containing no cut points of A. Then any 2-coloring of A^0 contains a monochrome unit.

Outline of Proof The basic argument is as follows (Fig. 24.4).

1. We assume that A^0 is 2-colored and contains no monochrome unit.
2. Points x and y can be selected from A^0, so that they are differently colored and as close together as we want.
3. The points x and y can also be chosen so that (a) x is unit distance from at most one cut point of A, and (b) y is unit distance from no cut points of A.

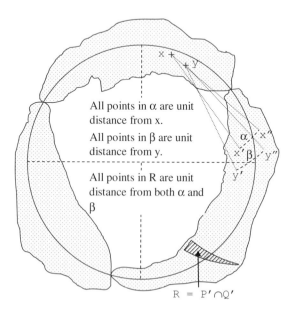

Fig. 24.4

4. Point x subtends an arc α of finite length in A^0, each point of which is unit distance from x, and consequently the opposite color to x. Similarly y subtends an arc β in A^0 which is the opposite color to y.
5. Arc α subtends a two-dimensional region, each point of which is unit distance from a point on α. This region intersects A^0 in a band P' of finite width, each point of which must be the same color as x. A similar region subtended at unit distance by arc β intersects A^0 in a band Q', each point of which is the same color as y.
6. Points x and y can be chosen to lie sufficiently close together to make $R = P' \cap Q'$ non-null.
7. But points in R must simultaneously have the color of x and the color of y, which is impossible. Consequently, the initial two assumptions are incompatible.

The proof hinges on our ability to construct arcs α and β that each does not intersect a cut point of A. This will be true if x is unit distance from at most one cut point of A, and y is unit distance from no cut points of A.

Tool 24.3 Let γ be any simple arc of length L in A^0 with the following properties:

- γ contains at least two points unit distance apart;
- γ contains at most M points, each unit distance from exactly one cut point of A;
- all other points in γ are unit distance from no cut points of A;
- γ is 2-colored with no monochrome units.

Then given $\varepsilon > 0$ there exists an ε-neighbourhood in γ containing a point of each color, one of which is unit distance from no cut points of A and the other of which is unit distance from at most one cut point of A.

Proof Let d(x,y) be the straight line distance between two points x and y on γ, and let δ (x,y) be the distance along γ between x and y.

By assumption there exist two points x_1 and y_1 in γ, not both the same color, with $d(x_1, y_1) = 1$. Let $\varepsilon > 0$ be given. The following algorithm uses the method of bisection to prove the lemma (Fig. 24.5).

1. If $M > 1$ then from the M points in γ that are unit distance from exactly one cut point of A, select the two that are closest together measuring along γ. Let h be the distance between them along γ.
2. If $h < \varepsilon$ then set $\varepsilon = h$.
3. Set $i = 1$.
4. Let w_i be the point in γ mid-way (by arc-length) between x_i and y_i.

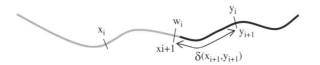

Fig. 24.5

5. If the colors of w_i and x_i are not the same then put $x_{i+1} = x_i$ and $y_{i+1} = w_i$ otherwise put $x_{i+1} = w_i$ and $y_{i+1} = y_i$.
6. If $\delta(x_{i+1}, y_{i+1}) \geq \varepsilon$ increase i by 1 and re-cycle from 4.
7. Points x_{i+1} and y_{i+1} satisfy the requirements.

The algorithm terminates in not more than n cycles, where n is the smallest integer such that $\varepsilon 2^n > L$. ∎

Proof of Theorem 24.2 Let A^0 be 2-colored with no monochrome units. Let N be the number of cut points of A. Let C be a circle of unit radius contained in A. By assumption, C has at least one arc of length greater than $\pi/3$ containing no cut points of A; hence C has an arc containing no cut points of A, whose end points are unit distance apart. There are at most 2N points on C in A^0 that are unit distance from a cut point of A. Some of these may be unit distance from two different cut points of A, but none can be unit distance from more than two cut points of A. By following a path sufficiently close to C it is possible to construct a simple closed curve that, apart form the cut points of A, lies entirely within A^0, which contains at most 2N points in A^0 that are unit distance from a cut point of A, and that contains no points in A^0 that are unit distance from more than one cut point of A. (This curve can merely trace the path of C for the most part, deviating only to bypass any points on C in A^0 that are unit distance from two different cut points of A.) There exists an infinite family Γ of such simple closed curves, for each of which there is an arc of finite length containing two points unit distance apart not separated by a cut point (Fig. 24.6). This must be so since C has two such points, and we can choose the members of Γ to be as close to C as required. For any given $\varepsilon > 0$, this arc contains an ε-neighborhood in which lies a point of each color, one of which is unit distance from at most one cut point of A, and the other of which is unit distance from no cut points of A (by Tool 24.3).

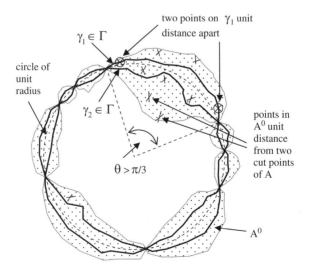

Fig. 24.6

Let γ_1 and γ_2 be members of Γ. Let x and y be two differently colored points in an ε-neighborhood on γ_1 such that x is unit distance from at most one cut point of A and y is unit distance from no cut points of A.

In A^0 there exists an arc α of unit radius and centre x which intersects γ_1 at x' and γ_2 at x'' and no point of which is a cut point of A. (If x is unit distance from one cut point of A then the arc α can be constructed on the other side of x from this cut point.) Arc α cannot be the same color as x, so must be the same color as y. Similarly there exists an arc β in A^0 of unit radius and centre y which intersects γ_1 at y' and γ_2 at y'' and no point of which is a cut point of A. Arc β must be the same color as x.

Let P and Q be sets subtended at unit distance by α and β respectively. P and Q are finitely disconnected unit annuli, each having one cut point at x and y respectively, and each intersecting A^0 in a band of finite width between γ_1 and γ_2 Let these bands be P' and Q' respectively. All points in P' must be the same color as x, and all points in Q' the same color as y. Q' may be considered to be the image of P' under a homeomorphism T which depends on $|x-y|$. Defining $d(P', Q') = \sup\{|p\text{-}T(p)| : p \in P'\}$ we have $d(P', Q') \to 0$ as $|x\text{-}y| \to 0$; in this sense we say $P' \to Q'$ as $|x\text{-}y| \to 0$. There must then exist $\varepsilon > 0$ such that for $|x\text{-}y| < \varepsilon$, $P' \cap Q' \neq 0$. But all points in $P' \cap Q'$ must simultaneously be colored the same as x and y, which is impossible. Consequently the original assumptions are incompatible, and so if A^0 is 2-colored it must contain a monochrome unit. ∎

Using this result it is possible to exclude two configurations from any 4-coloring of E^2 without monochrome units, and show as a natural consequence that any 4-colored map in E^2 contains a monochrome unit.

Theorem 24.4 Let E^2 be 4-colored. If for some distinct points x and y there exist two simple arcs with endpoints x and y, each, excepting the endpoints, being monochrome but not both the same color, then E^2 contains a monochrome unit.

Proof Let the two simple arcs be γ and δ. If $|x - y| > 1$ then both γ and δ contain a monochrome unit.

Assume $|x - y| \leq 1$. Then the intersection of the sets subtended at unit distance by γ and δ (excluding the endpoints) is a finitely disconnected unit annulus with at most two cut points (Fig. 24.7). This annulus is 2-colored at most, since it cannot contain the colors of γ and δ, and a circle of unit radius contained in its closure has an arc of length greater than $\pi/3$ containing no cut points, and so by Theorem 24.2 the annulus contains a monochrome unit. ∎

Theorem 24.5 If a 4-coloring of E^2 contains two differently colored, bounded, open connected monochrome sets with a common boundary of finite length, then E^2 contains a monochrome unit.

Proof Let G and F be two such sets, and let x and y be two distinct points on the common boundary. Because the closure of G is a simply connected Jordan region, there is a simple arc γ with endpoints x and y which, apart from its endpoints, lies in G. There exists a similar arc δ in F. By theorem 24.4 E^2 contains a monochrome unit. ∎

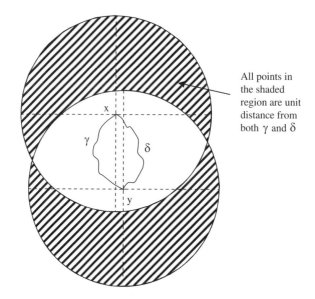

All points in
the shaded
region are unit
distance from
both γ and δ

Fig. 24.7

Corollary Every 4-colored planar map contains a monochrome unit.

A similar result involving three sets can be proved for 5-colorings of E^2, and again the consequence is that every 5-colored planar map contains a monochrome unit, but this requires a careful proof.

Theorem 24.6 If a 5-coloring of E^2 contains three disjoint, differently colored, bounded, open, connected, monochrome sets each having a common boundary with each of the other two, and all three having one common boundary point, then E^2 contains a monochrome unit.

Proof Let v be the boundary point common to all three sets and let a_1, a_2, and a_3 respectively be boundary points common to each pair of sets. We assume these points are distinct and are chosen to be not more than one unit from each other. There are simple closed curves γ_1 colored c_1 containing v, a_1, and a_2; γ_2 colored c_2 containing v, a_1, and a_3; and γ_3 colored c_3 containing v, a_2, and a_3, where in each case the coloring refers to every point on the curve with the possible exception of the points v, a_1, a_2, and a_3 (Fig. 24.8). Let P be the intersection of the sets subtended at unit distance by γ_1, γ_2, and γ_3 excepting the points v, a_1, a_2, and a_3. P is either a unit annulus or a finitely disconnected unit annulus with at most three cut points. (A necessary condition for such a cut point to exist is that a set boundary incident to v is an arc of a circle of unit radius; if the cut point exists then it lies at the centre of this circle.) P satisfies the requirements of Theorem 24.2, and since it is 2-colored (viz. not c_1, c_2 or c_3) it must contain a monochrome unit. ∎

Corollary Every 5-colored planar map containing a vertex of degree 3 contains a monochrome unit.

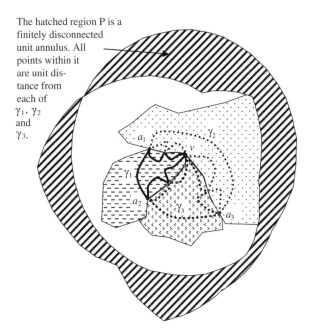

The hatched region P is a finitely disconnected unit annulus. All points within it are unit distance from each of γ_1, γ_2 and γ_3.

Fig. 24.8

Theorem 24.7 Every 5-colored planar map contains a monochrome unit.

Proof We show (i) that every 5-colored planar map with no monochrome units contains a vertex of degree 3 or 4 and (ii) that every such map containing a vertex of degree 4 also contains a vertex of degree 3.

i. Let v be any vertex in a 5-colored planar map, and assume that this has degree 5. Assume that the map has no monochrome units.

Let γ be the boundary of one of the regions which has v as a boundary point. Let a be a point on γ that lies on an edge connected to v. Let b be a point on γ that lies on the other edge connected to v (Fig. 24.9). Let c be a point on the edge connected to v that is on the opposite side of va to b. Let d be a point on the edge connected to v that is on the opposite side of vb to a.

There is a simple closed curve γ_1 passing through v, a, and b all the points of which, except possibly v, a, and b, are colored c_1. There is a simple closed curve γ_2 passing through v, a, and c all the points of which, except possibly v, a, and c, are colored c_2. And there is a simple closed curve γ_3 passing through v, b, and d all the points of which, except possibly v, b, and d, are colored c_3. Let T_2 be the intersection of the sets subtended at unit distance by γ_1 and γ_2 and let T_3 be the intersection of the sets subtended at unit distance by γ_1 and γ_3 (In Fig. 24.9 T_2 is the hatched region and T_3 is the grey region).

We consider two cases. The first is when the angle θ subtended at v by a line from a to b (through the region enclosed by γ) is greater than π. The interiors

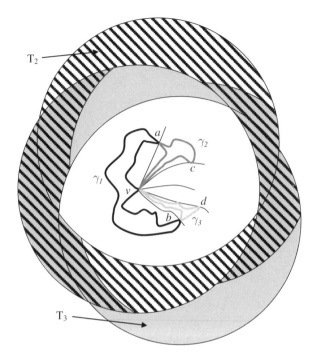

Fig. 24.9

of T_2 and T_3, T_2^0 and T_3^0 respectively, are unit annuli with no cut points, and so by Theorem 1 cannot be 2-colored. T_2^0 must contain regions colored c_3, c_4, and c_5, and T_3^0 must contain regions colored c_2, c_4, and c_5. The interior of $T_1 = T_2 \cup T_3$ is a 4-colored unit annulus with no cut points.

There is a vertex in T_1^0. To prove this, assume it is not so. Then there must be edges in T_1^0 that do not intersect each other in T_1^0, each of which intersects both the interior and the exterior boundary of T_1. Any such edge, e, must cross both T_2^0 and T_3^0. This means that the regions on either side of e must be colored c_4 and c_5. Consequently T_1^0 is a 2-colored unit annulus, containing no cut points.

The second case is when the angle θ is not greater than π. It is clear, since v is a vertex of degree 5, that the region enclosed by γ may be chosen such that θ is not less than $2\pi/5$. Let a_1 be a point between v and a on the edge on which a lies. Similarly let b_1 be a point between v and b on the edge on which b lies. Choose curve γ_1 so that it passes through a_1 and b_1 as well as v, a, and b and so that all of its points, except possibly v, a, a_1, b_1, and b, are colored c_1. Similarly choose γ_2 to pass through a_1 as well as v, a, and c and γ_3 to pass through b_1 as well as v, b, and d.

Now each of T_2^0 and T_3^0 is a finitely disconnected unit annulus with at most one cut point (Fig. 24.10). The single cut point in T_2^0, say p, only occurs in the event that v, a, and a_1 lie on the circle of unit radius centre p. Similarly the single cut point in T_3^0, say q, only occurs in the event that v, b, and b_1 lie on the circle

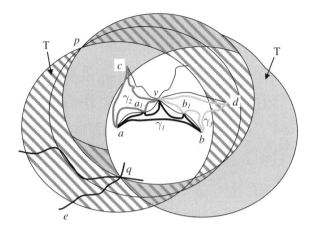

Fig. 24.10

of unit radius centre q. for $T_3{}^0$. The interior of $T_1 = T_2 \cup T_3$ is a 4-colored finitely disconnected unit annulus with at most one cut point. This cut point only occurs in the event that p and q are coincident, and all of v, a, a_1, b_1, and b, lie on the same circle of unit radius. If one of p and q lies on the exterior boundary of T_1 and the other lies on the interior boundary then the length of the arc of the unit circle centre v passing through p and q is θ radians, and this means the distance between p and q is greater than one.

As before we assert there is a vertex in $T_1{}^0$. To prove this, assume it is not so. Then there must be edges in $T_1{}^0$ that do not intersect each other in $T_1{}^0$, each of which intersects both the interior and the exterior boundary of T_1. Any such edge, e, must cross both $T_2{}^0$ and $T_3{}^0$ except in the case that e passes through p and remains entirely within T_3 until it reaches the opposite boundary of T_1, or e passes through q and remains entirely within T_2 until it reaches the opposite boundary of T_1. Note that such an edge e cannot pass through both p and q, since this would imply the existence of a monochrome unit in one of the regions on either side of e. Apart from these exceptional edges every edge in $T_1{}^0$ must separate and regions colored c_4 or c_5. This means that $T_1{}^0$ contains a 2-colored finitely disconnected unit annulus, containing at most two cut points.

Clearly there is a circle of unit radius in T_1 which has an arc of length greater than $\pi/3$ containing no cut points of $T_1{}^0$. Therefore, by Theorem 24.2 $T_1{}^0$ contains a monochrome unit. This is a contradiction of the initial assumption, consequently there must be a vertex in $T_1{}^0$, and since $T_1{}^0$ is 4-colored this vertex is of degree 4 at most.

ii. We show that every 5-colored planar map with no monochrome units containing a vertex of degree 4 also contains a vertex of degree 3.

Suppose v is a vertex of degree 4 in a 5-colored planar map. Let c_1, c_2, c_3, and c_4 be the colors of the four regions of which v is a boundary point. Let a, b,

c, and d be points on the four edges incident to v. Let a_1, b_1, c_1, and d_1 be points on the edges between a and v, b and v, c and v, and d and v respectively. Assume that the map has no monochrome units.

There exists a simple closed curve γ_1, defined in the closure of the region colored $c1$, that passes through v and four of the edge points defined above, and such that every point in γ_1, except possibly v and the four edge points, is colored $c1$. Similarly, there exist simple closed curves γ_2, γ_3, and γ_4, each of which contains v and four of the edge points, the points on each curve being colored c_2, c_3, and c_4 respectively except possibly v and the edge points. Let the order of the γ_i be chosen such that γ_2 and γ_4 have only the point v in common (Fig. 24.11).

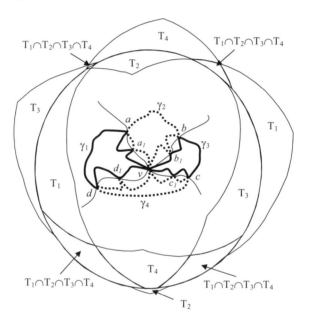

Fig. 24.11

Let T_i, $i = 1$, 2, 3, 4, be the intersection of sets subtended at unit distance by γ_j, $j = 1..4$, $j \neq i$. Set T_i is 2-colored with colors c_i and c_5. Define $T = \cup T_i$. The interior of T, T^0, is a unit annulus with centre v, possibly finitely disconnected with at most two cut points (Fig. 24.11).

Every point within T^0 that is on a boundary of a region of the planar map is a boundary point of at most three regions. Suppose none of these boundary points is a vertex. Then there must exist edges that pass from the interior boundary to the exterior boundary of T which pass through either both of T_1 and T_3 or both of T_2 and T_4. It is possible for an edge to cut T and only cut one of T_1 and T_3 or one of T_2 and T_4, but such an edge must intersect the unit circle centre v at one of four points, these points being cut points (if they exist) of the finitely disconnected annuli which are the interiors of $T_1 \cup T_2$, $T_3 \cup T_4$, $T_1 \cup T_4$, and

$T_2 \cup T_3$. There must be edges crossing T which intersect the circle of unit radius centered on v at points other than these four cut points. (If not then there is an arc of the circle of unit radius centered on v, of length greater than or equal to $\pi/2$ that lies in or on the boundary of a region of the map. But then this region must contain a monochrome unit.) An edge crossing both T_1 and T_3 (or both T_2 and T_4) must separate regions with different colors. But the only color common to both T_1 and T_3 (or both T_2 and T_4) is c_5. We have arrived at a contradiction. Hence, there must be vertices in T^0, and these are of degree 3.

Now, by the corollary of Theorem 24.6 our 5-colored map contains a monochrome unit! ■

V

Colored Graphs

25
Paul Erdős

*I hope several [of my] results will survive for
centuries, but we will see*
— Paul Erdős[1]

*Paul Erdős' contributions to mathematics cannot be
measured through his papers alone. Over the years
he has traveled extensively among the mathematical
centers of the globe. Like the bumblebee, flying from
flower to flower transmitting pollen, Paul Erdős has
created an enormous cross-pollinization effect in
mathematics. An Erdős visit to a mathematical
center is marked by intense work. Mathematicians
gather round and discuss the current problems in
their various fields. The resulting interplay of ideas
is exhausting and highly productive.*
— Joel H. Spencer [Sp1]

*The early involvement of Paul Erdős in problem
solving at high school level had a strong influence on
his own life-work, and to this day he can make the
young feel close to him. This closeness to the young
is determined also by another factor: the human side
of Erdős, his warmth and compassion, his love of
youth, his strong sense of justice, unspoilt and at
times childishly naïve.*
—Martha Sved [Sve1]

---

[1] Talk at the Keszthely 1993 Conference, dedicated to Paul Erdős's 80th birthday.

A. Soifer, *The Mathematical Coloring Book*,
DOI 10.1007/978-0-387-74642-5_25, © Alexander Soifer 2009

25.1 The First Encounter[2]

Once upon a time I came to Budapest for a congress.

I called a Hungarian friend of my friend, who offered me his hospitality:

– Would you like me to show you something—he asked—a place to buy Hungarian souvenirs, or perhaps, a disco to meet beautiful Hungarian girls?

The offer sounded attractive. But the many legends I have heard about Erdős came to mind, and I replied:

– Is Erdős a real person? Is he in Budapest?
– Of course, he is a real person, but he can be anywhere in the world on any given day.
– Well, said I, you offered to grant me one wish. My choice is to meet Erdős.

The following day he told me that I was lucky: Paul Erdős was in town and willing to see us. We found Paul in his huge office with high ceilings at the Alfréd Rényi Institute of Mathematics on Reáltanoda 13–15 speaking with two Russian mathematicians, the father and the son Stechkins. I joined them. No language was known to all, but every two had a language in common: Erdős and the Stechkins spoke German, Erdős and I used English, and the Stechkins and I knew Russian.

The Russians soon left. Without looking at me, Paul said:

– Let x sub 1, . . . , x sub n be n points in the plane no three on a line. . .

Paul Erdős and Alexander Soifer, the first meeting, August 1988, Rényi Institute of Mathematics, Budapest

---

[2] This section is based on my essay published in *Geombinatorics* in 1993 [Soi28].

The problem was beautiful. But did he ask himself or me? Did he want me to solve it right there? Did he want me to offer him a problem in return? I gave him the most difficult problem of that year's Colorado Mathematical Olympiad, that I created and none of some 1,000 participating students solved:

– Five points lie inside a triangle of area one...

To my disbelief, he solved it!

The next day I came back. I had an idea but still no solution for his problem. My embarrassment disappeared when Paul said:

– This is an open problem, and I offer...dollars for its [first] solution.

The few meetings with Paul during the congress affected my life and started our very special friendship. The idea for the book *Problems of pgom Erdős* occurred to me right then, in August 1988, when for the first time I was listening to Paul presenting "some of my favorite problems." Erdős's problems were legendary, and as true legends, they were passed from person to person, and sometimes changed in the process to become something else, not intended by the author. Right after the talk, I asked Paul to write such a book, but he replied, "why don't *you* write a book of my problems?" I disagreed: "I envision it as a book of problems *you* have posed with *your* commentaries for each of the problems." Paul accepted the challenge, but in a couple of years of limited progress, Ron Graham suggested to Paul and me to unite our efforts. The book is not finished yet, but it would be my highest priority upon the completion of this book.

In a few months we were already working together in Colorado Springs, accompanied by the gentle "noise" of Mozart (Beethoven was too much of a distraction), and taking walks in the Garden of the Gods, an old Indian sacred ground full of remarkable red vertical rocks.

A list of mathematicians inspired by Paul Erdős may go on for longer than the list of his 1,600 publications. Trajectories of his travels probably added up to a set dense at every point on our globe. Paul signed reprints for me with mysterious sequence of letters after his name:

Erdős Pal, pgom, ld, ad, ld, cd.

Paul explained:

pgom = poor great old man
ld   = living dead (i.e., over 60 years old)
ad   = archaeological discovery (>65)
ld   = legally dead (>70)
cd   = counts dead (>75)

"Two Thinkers": Paul Erdős, Colorado Springs, December 28, 1991. Photograph by A. Soifer

In July 1993 in Keszthely I reminded Paul that "the emergency" of adding another pair of initials has arrived. Paul thought for a moment, and then declared: "nd, nearly dead."

It is impossible to overestimate how much the field of this book owes to Paul Erdős. With the intuition of a genius, Paul saw the beauties to be had in what became known as *Ramsey Theory*, and lead our way to this Garden of Eden.

25.2 Old Snapshots of the Young[3]

Martha (Waksberg) Svéd, a member of the legendary Budapest circle of young student-mathematicians that included Paul Erdős, Paul Turan, Tibor Gallai, Esther Klein, George Szekeres, and others, had known Erdős as few did. For Paul's 80th birthday, she wrote her fabulous, lyrical reminiscences especially for *Geombinatorics* [Sve2]. This subsection is all hers: Martha Svéd recollects:

Yes, E. P., this is the name: initials for Erdős Pal, Hungarian form of the name Paul Erdős, name by which we, old Hungarian friends called him and still refer to him. This is the year when he is 80 years YOUNG. At this point I recall his Cambridge lecture attended by the two of us, G. [George] and M. Sved in 1959. Ahead of those formulae about the secrets of primes, Hebrew words appeared on the blackboard:

זקנה לא נעימה

---

[3] First published in *Geombinatorics* [Sve2].

(This lecture was held just after an extended stay in Israel.) The translation is: "Old Age Is Not Pleasant," referring to himself. His greeting words to us (having last seen each other in Budapest in 1938) were: "SAM SURRENDERED.".

It is now necessary to give an Erdős dictionary for those who are not initiated. Since the set of those whose Erdős number is 1, has measure, there will be a large subset not needing such a dictionary, hence they may skip it. I add here for the uninitiated the definition of the "Erdős number."[4] It is the number of links in the chain leading to the origin, E. P. himself. The aristocrats are those whose name has appeared together with E. P. in at least one publication, hence can boast of number 1. My own number is 2, but without great claims of merit. The thanks for it must go to George Szekeres for a single joint publication with him. G. Szekeres holds number 1 with high multiplicity. Since G. Sved and I have lived in Australia for more than 50 years and only for a few years in the same city as G.&E. Szekeres, my mathematical contacts have been locals, so my Erdős number would have been hard to trace. This is why I am proud and happy for being asked to add my lines to this celebratory volume. I should add here that some conjecture is floating around: if you have joint publications with at least three coauthors then your Erdős number is finite, (though it could be distressingly large!). This is my reason to leave the mathematical bits to others, restricting myself to reminiscences about our great and faithful friend whose letters still begin: "G. and M. Sved", followed by a paragraph about his own personal jaunts across four continents, with the last paragraph beginning "Let n points. . ." I try to translate (not adequately) his Hungarian self description of "not being a university professor but a world professor of mathematics," the traveling missionary, to whom "Sam surrendered" in 1959.

Now to the Dictionary, apologizing for its incompleteness and haphazard ordering:

- **Epsilon**: small, negligible in some respects, but the word has also another meaning, an endearing one: child generally. When talking about the offsprings of his friends, the Epsilon could be quite grown up, perhaps having Epsilons of his own, i.e., epsilon squares.
- **Omega**: large, many.
- **Trivial**: of a person: mean, uncaring, unjust etc., hence Triviality.
- **Victory**: solution of a problem found. However, once during a hike he sang out: "Victory! I lost my wallet" ???
- **Fascism**: the nastiest swearword Paul can think of, when he is clumsy and drops or mislays something.
- **Boss**: wife or girlfriend.
- **Slave**: husband or boyfriend.
- **Captured**: snared into marriage or long-term relationship.
- **Liberated slave**: divorced man.
- **Sam**: United States of America.

---

[4] Casper Goffman must have been first to define Erdős Numbers in print in 1969 [Gof]. Apparently the concept was born in the 1960s and Paul Erdős himself did not know about it before 1968.

Paul Erdős with the epsilon, Isabelle Soifer, Kalamazoo, Michigan, June 4, 1992. Photograph by A. Soifer

- **Joe**: the late Soviet Union, abbreviated name of Joseph Stalin.
- **Cured**: passed away, "cured" of the illness of life.

I am now in the position of being able to explain that greeting in Cambridge in 1959. While E. P. is not "patriotic" in the nationalistic sense, he could never deny his Hungarian identity. During the oppressive communist regime of Hungary in the 1950s he visited Budapest and was probably the only person who was allowed to leave freely. While having lived in the States permanently after World War II, he did not acquire US citizenship, moreover refused to sign the "loyalty oath" of the McCarthy era during the "cold war." He was not expelled for this, but was warned that he would not get a return visa when leaving. Nevertheless when invited to an international mathematical conference in Holland in 1954, he took it up. His return visa was refused for years. A letter went then to President Eisenhower, signed by the greatest names of American Mathematics. This letter pointed out what the loss of Paul Erdős meant to their country. There was now a whole generation of young-sters growing up, entirely missing his inspiring influence. This was a loss which US mathematics could not afford. This worked. E. P. was given the visa, for a limited period at that time, but by now he is welcome with open arms at any time when he wishes to enter and spend a very short or a long period there. In fact, he has now two main "bases." One is in Budapest at the Mathematical Research Institute of the Hungarian Academy, where all his publications are kept and which serves him as a home when in Hungary, since he has not entered his own flat since the death of his mother at an advanced age. She had been his faithful companion and secretary through all his travels. His flat is now used by visiting mathematician friends. The other base is at Bell Laboratories, where his friend Ronald Graham looks after the business matters of his life.

I go back now to the early days to write reminiscences. The name Erdős Pal was well-known to us, Esther Szekeres (then Klein) and myself before we met him at the university. All of us were frequent problem solvers of our loved Hungarian magazine, the *School Journal for Mathematics and Physics*.

The beginnings of this journal date to the turn of the century, in a great period of prosperity, liberalism, and culture. World War I and years in the era that followed washed it all away. The mathematical school journal was revived in 1925, to go into oblivion again late during World War II. Nazism, taking hold of Hungary killed not only the journal, but also its editor, Andrew Farago together with his family. There were also number of mathematicians, most of them young and of great promise who became victims of fascism. A plaque at the Research Institute of Mathematics commemorates their names. The school journal came to vigorous life again after this second war, also new journals for secondary students were born around the world to inspire the young, but the human victims could not be brought back to life.

Esther Klein (Szekeres) was my classmate and best friend in the final 4 years of our secondary schooling. We had an exceptional mathematics teacher, R. Rieger. That day in the beginning of year 1925 pictures still clearly in my mind, when our teacher appeared in class with the first issue of the revived school journal in his hand. He pointed out the two sets of problems, aimed at two levels, at the lower and higher grades of schooling. Both Esther and I became problem solvers working completely independently of each other, continuing until the time when we left school and became university students, a privilege meted to a very restricted number of Jewish youngsters.

Esther to whom I shall refer from now on as Eps, (for, as coined by E. P. twisting the petting name used by her mother) remained really faithful to mathematics during our university years but I became somewhat wavering. My loss of dedication was only partly due to the early "capture" of each other with G. Sved whose mathematics was rich but not "pure," being an engineer. The case with G. Szekeres was different. He studied chemical engineering, to satisfy his father and in 1928 was still a leather manufacturer. Yet the real love of this other George was always mathematics, with some theoretical physics thrown in, as his later contributions to relativity theory show. Nevertheless he completed his course successfully, and worked for some years in the leather industry, in Hungary first, then during the early war years in Shanghai. He found time during engineering studies to join our little circle. His name together with those of T. P. (Paul Turán) and G. T.(Tibor Gallai) and of course, E. P. known to us through that journal which published not only the names of successful problem solvers, specially printing (with some editing) the best solutions, but supplied also at the end of each school year the photos of the most frequent contributors of solutions.

T. P. (Paul Turán) was in the same year as ourselves, but E. P. and G. T. younger than the three of us appeared on the scene 2 years later, E. P. as mover and shaker, G. T. as sharp critic. E. P. seldom graced the lecture room with his presence. I am not sure now whether he was even enrolled, like the rest of us for secondary teachers training, running in conjunction with our courses in mathematics and physics. He certainly missed the fifth and final year required, consisting of teaching practice in one of the schools officially prescribed and ending with the examination in philosophy and theory of education. In that fifth year of academic education he was already in Manchester with a scholarship, working as a post doctoral fellow with Mordell, having gained his Ph. D. with L. Fejer being his supervisor. His doctoral thesis was based on results obtained on the distribution of prime numbers. Still as an undergraduate he obtained new results, (elementary proof of Csebisev's theorem). To this day primes are one of his prime concerns.

However, I must go back now to those earlier years, when E. P. was holding court, in the students' common room, or being one of our crowd in our City Park, where we were tackling problems set in the then new and by now classic collection of problems in analysis and number theory by G. Polya and G. Szego. In those years with him at university, Esther and I had to take charge of him to ensure that required enrolment and semester end formalities were satisfied. Our rewards were rich. Our group held together strongly then and in some later years, with some of us already graduates (though not holding teachers' appointments, with our teaching work being confined to private tutoring). We shared hikes at our charming hills, near our city and continued our mathematical meetings at a site around the Anonymous Statue in our city park.

Paul's parents adopted all his young friends. The Erdős home became our second home. The parents, Louis and Anne were both mathematics teachers, but in our time only Louis was active as a school teacher. His mother was sacked in 1919, in the days following the upheavals after the war. Louis, who had been on active war service and had returned after long years in Russian camps for war prisoners, could not be dismissed. Fascism (the word had not yet come into existence) in those days was "mild" in comparison with what came at the end of World War II. Louis, outstanding as a teacher, a man of wisdom, vision combined with a sense of humor, was a delightful company for us. Taking breath in that warm and stimulating atmosphere created by Anne and Louis nurturing Paul was a gift. During the years after 1939, E. P. was in the US and we in Australia, and were able to keep some contact. I was shaken when I read the news in one of Paul's letters that his father died a natural death during those war years. Then I was comforted by the thought that he was spared the dangers, degradations, and humiliations to be meted out at the end of that war.

Since E. P. and his mathematics, (the two are being inseparable), form the pivot of such large collection of mathematicians (with their bosses or slaves), the number of stories surrounding him together with those histories of mathematical problems solved or still in states of conjecture is also of an impressive multitude, I want to add here my own story about a problem I witnessed at birth.

Paul calls it "the happy [end] problem." My friend Eps, not much after her return from Göttingen, in those days the world centre of mathematics, posed the following question: given 5 points, no 3 collinear, in the plane, conjecture: it is always possible to select 4 to form the vertices of a convex quadrilateral. It was a problem of unusual flavor, but my own waverings did not point in that direction. All the more were E. P.

and Gy. Szekeres aroused. As Gy. S. confessed later, his attraction to the problem was sparked by the person proposing it. Actually Eps found the proof and efforts to generalize began. They resulted in the first Erdős–Szekeres joint publication to appear about 2 years later. The authors were not aware at the time that they solved and extended an old theorem by Ramsey. The significance of this publication was that it yielded lifetime results for the Trio involved: Eps and Gyu, (pronounce Dew, and I am not giving here a linguistic lecture to explain this) became the couple of mathematicians bearing the name, G. and E. Szekeres; P. Erdős and G. Szekeres became lifetime cooperators, though in fields different from that first joint paper; E. P. became the originator of a new field in mathematics: combinatorial geometry, one of the new chapters created by him.

The youth of E. P. is of lifetime duration. His approach to problems is "elementary," his best working pals are the young; his games, hobbies, and relaxations do not belong to the world of old, and ignoring social conventions are those of a child. He has remained the Peter Pan of mathematics.

26
De Bruijn–Erdős's Theorem and Its History

26.1 De Bruijn–Erdős's Compactness Theorem[5]

They were both young. On August 4, 1947 the 34-year-old Paul Erdős, in a letter to the 29-year-old Nicolaas Govert de Bruijn of Delft, The Netherlands, offered the following conjecture [E47/8/4ltr]:

Let G be an infinite graph. Any finite subset of it is the sum of k independent sets (two vertices are independent if they are not connected). Then G is the sum of k independent sets.

Paul added in parentheses "I can only prove it if $k = 2$". In his 5-page August 18, 1947 reply [Bru1], de Bruijn reformulated the Erdős conjecture in a way that is very familiar to us today:

Theorem: Let G be an infinite graph, any finite subgraph of which can be k-colored (that means that the nodes are coloured with k different colours, such that the two connected nodes have different colours). Then G can be k-coloured.

Following a nearly three-page long transfinite induction proof of the "Theorem," de Bruijn observed [Bru1]:

> I am sorry that this proof takes so much paper; its idea, however, is simple. Perhaps, you do not call it a proof at all, because it contains "Well ordering", but we can hardly expect to get along without that.

This was an insightful observation, for de Bruijn and Erdős relied on the Axiom of Choice or equivalent (like Well-Ordering Principle or Zorn's Lemma) very heavily. When in early 2004 Professor de Bruijn received from me a reprint of Shelah–Soifer 2003 paper (to be discussed in Chapter 46) which analyzed what happens with the de Bruijn–Erdős Theorem in the absence of the Axiom of Choice, de Bruijn replied to me on January 27, 2004 as follows [Bru7]:

---

[5] I am infinitely grateful to N.G. de Bruijn for providing me with copies of his correspondence with Paul Erdös.

A. Soifer, *The Mathematical Coloring Book*,
DOI 10.1007/978-0-387-74642-5_26, © Alexander Soifer 2009

About the axiom of choice, I remember a conversation with Erdős, during a walk around 1954. I told him that I hated the axiom of choice, and that I wanted to do analysis without it, maybe except for the countable case. He was surprised, and said: but you were always so good at it. Indeed, I had loved transfinite induction, just because it worked exactly the same way as ordinary induction.

This invaluable de Bruijn's e-mail also contained the conclusion of the story of the de Bruijn–Erdős Theorem [Bru5]:

Erdős and I did not take any steps to publish the k-coloring theorem. In 1951 I met Erdős in London, and from there we went together by train to Aberdeen, which took a full day. It was during that train ride that he told me about the topological proof of the k-coloring theorem. Not long after that, he wrote it up and submitted it for publication. I do not think I had substantial influence on that version.

Let us look at a proof of this celebrated theorem, which we have formulated without proof and used in chapter 5.

De Bruijn–Erdős's Compactness Theorem 26.1 ([BE2], 1951). An infinite graph G is k-colorable if and only if every finite subgraph of G is k-colorable.[6]

In what follows, we will need a few definitions from set theory.

Given a set A; any subset R of the so-called Cartesian product $A \times A = \{(a_1, a_2) : a_1, a_2 \in A\}$ is called a *binary relation* on A. We write $a_1 R a_2$ to indicate that the ordered pair (a_1, a_2) is an element of R.

Poset, or *partially ordered set*, is a set A together with a particularly "nice" binary relation on it, i.e., a relation that satisfies the following three properties:

1. Reflexivity: $a \leq a$ for all $a \in A$;
2. Anti-symmetry: If $a \leq b$ and $b \leq a$ for any $a, b \in A$, then $a = b$;
3. Transitivity: If $a \leq b$ and $b \leq c$ for any $a, b, c \in A$, then $a \leq c$.

A *chain*, or *totally ordered set*, is a poset that satisfies a fourth property:

4. Comparability: For any $a, b \in A$, either $a \leq b$ or $b \leq a$.

Let A be a set with a partial ordering $\leq$ defined on it, and B a subset of A. An *upper bound* of B is an element $a \in A$ such that $b \leq a$ for every $b \in B$.

Let $\leq$ be a partial ordering on a set A, and $B \subseteq A$. Then, we say that $b \in B$ is a *maximal element* of B if there exists no $x \in B$ such that $b \leq x$ and $x \neq b$.

In 1935 Max Zorn (1906, Germany-1993, USA) introduced the following important tool, which he called *maximum principle*. (It was shown by Paul J. Campbell that, in fact, a number of famous mathematicians—Hausdorff, Kuratowski, and Brouwer—preceded Zorn, but Zorn's name got as attached to this tool as, say, Amerigo Vespucci's name to America.)

---

[6] This theorem requires the Axiom of Choice or equivalent.

Zorn's Lemma 26.2 If S is any non-empty partially ordered set in which every chain has an upper bound, then S has a maximal element.

During the summer of 2005, I supervised at the University of Colorado, a research month of Dmytro (Mitya) Karabash, who had just completed his freshman year at Columbia University, and asked to come and work with me. One of my assignments for him was to prove the de Bruijn–Erdős Theorem 26.1, and then to write the solution as well. After going through several revisions, Mitya produced a fine proof, which follows here, slightly edited by me.[7]

Proof of Theorem 26.1 by D. Karabash: We say that G has the property P and write $P(G)$ if every finite subgraph of G is k-colorable. For a graph G we write $G = (V, E)$, where V is the vertex set and E is the edge set of G. Now let S be the set of all graphs with the property P which are obtained from G by an addition of edges, i.e., $S = \{(V, F) | E \subseteq F \text{ and } P(V, F)\}$.

Let S be partially ordered by the inclusion of edge sets. Observe that for every chain A_i in S, its union $A = (V, \bigcup_i E(A_i))$ is also in S (here $E(A_i)$ stands for the edge set of the graph A_i). Indeed, every finite subgraph F of A must be contained in some A_i (because F is finite) and therefore F is k-colorable. Since A has property P, A is in S, as desired.

We have just proved that in S every chain has an upper bound. Therefore, by *Zorn's Lemma*, S contains a maximal element, call it M. Since M is in S, M has property P; since M *is* maximal, no edges can be added to M without violating property P.

We will now prove that *non-adjacency* (here to be denoted by the symbol $\neg adj$) is an equivalence relation on M, i.e., for every a, b, $c \in V(M)$, if $a \ \neg adj \ b$ and $b \ \neg adj \ c$, then $a \ \neg adj \ c$. Let us consider all finite subgraphs of M that contain a and b, and all k-colorings on them. Since $a \ \neg adj \ b$, there must be a subgraph M_{ab} for which the colors of a and b are the same for all k-colorings of this subgraph, for otherwise we could add the edge ab to M with preservation of property P and attain a contradiction to M being a maximal element of S. Construct a subgraph M_{bc} similarly. The subgraph $M_{ab} \cup M_{bc}$ is finite and thus k-colorable. $M_{ab} \cup M_{bc}$ contains subgraphs M_{ab} and M_{bc}, therefore by construction of M_{ab} and M_{bc}, any coloring of $M_{ab} \cup M_{bc}$ must have pairs (a, b) and (b, c) colored in the same color. Thus, a and c have the same color for all k-colorings of the subgraph $M_{ab} \cup M_{bc}$ and therefore a is not adjacent to c.

From the fact that the non-adjacency is an equivalence relation on M, we conclude that the edge-complement M' of M is made of some number of disjoin complete graphs K_i because in M' *adjacency* is an equivalence relation. Therefore $a \in K_i$, $b \in K_j$, $i \neq j$ implies $a \ \neg adj \ b$ in M' or equivalently $a \ adj \ b$ in M.

---

[7] You can also read the original proof in [BE2]; a nice proof by L. Pósa in the fine book [Lov2] by László Lovász; and a clear insightful proof of the countable case in the best introductory book to Ramsey Theory [Gra2] by Ronald L. Graham.

Suppose there is more than k disjoint complete subgraphs K_i in M'. Then pick $k+1$ vertices, all from distinct $V(K_i)$. Since all of the vertices are located in distinct $V(K_i)$, they must all be pairwise non-adjacent in M' and thus form a complete graph M_{k+1} on $k+1$ vertices in M. We obtained a finite subgraph M_{k+1} of M which is not k-colorable, in contradiction to M having property P. Therefore, M' consists of at most k complete subgraphs $V(K_i)$, $i = 1, \ldots, k$. Now we can color each subgraph $V(K_i)$ in a different color. Since no two vertices of an $V(K_i)$ are adjacent in M, this is a proper k-coloring. Since G is a subgraph of M, G is k-colorable, as desired. ∎

Corollary 26.3 Compactness Theorem 5.1 is true.

The proof of Theorem 26.1 is much more powerful than you may think. It works not only for graphs, but even for their important generalization—hypergraphs. Permit me to burden you with a few definitions.

As you recall from chapter 12, a *graph* $G = G(V, E)$ is a non-empty set V (of vertices) together with a family E of 2-element subsets (edges) of V. If we relax the latter condition, we will end up with a hypergraph.

A *hypergraph* $H = H(V, E)$ is a non-empty set V (of *vertices*) together with a family E of subsets (*edges*) of V each containing *at least two* elements. Thus, an edge e of H is a subset of V; its elements are naturally called *vertices of the edge e* (or *vertices incident with e*).

Let n be a positive integer. We would say that a hypergraph H is *n-colored*, if each vertex of H is assigned one of the given n colors. If *all* vertices of an edge e are assigned the same color, we call e a *monochromatic edge*.

The *chromatic number* $\chi(H)$ of a hypergraph H is the smallest number of colors n for which there is an n-coloring of H without monochromatic edges.

A hypergraph $H_1 = H_1(V_1, E_1)$ is called a *subhypergraph* of a hypergraph $H = H(V, E)$, if $V_1 \subseteq V$ and $E_1 \subseteq E$.

Compactness Theorem for Hypergraphs 26.4 The chromatic number $\chi(H)$ of a hypergraph H is equal to the maximum chromatic number of its finite subhypergraphs.

Proof Repeat word-by-word the proof of Theorem 26.2 (just replace "graph" by "hypergraph"). ∎

26.2 Nicolaas Govert de Bruijn

Ever since 1995, I have exchanged numerous e-mail messages—and sometimes letters—with the Dutch mathematician N. G. de Bruijn. His elegant humor, openness in expressing views even on controversial issues, and his eyewitness accounts of post W.W.II events in Holland made this correspondence most fascinating and enjoyable for me. We also shared interest in finding out who created the conjecture on monochromatic arithmetic progressions, which was proven by B. L. van der Waerden (see chapter 34 for the answer). Yet, for years I have been asking Professor de Bruijn to share with me his autobiography to no avail. For a long while, I did not even know what "N. G." stood for. On October 29, 2005, I tried to be a bit more specific in my e-mail. I wrote:

> May I ask you to describe your life – and any participation in political affairs – during the occupation, May 1940–1945, and during the first post war years, 1945 up to your Sep-1952 appointment to replace Van der Waerden at Amsterdam?

De Bruijn understood my maneuver, but provided the desired reply on November 1, 2005 [Bru12]:

> You are asking for an autobiography in a nutshell.
>
> I was born in 1918 [on July 9th in Den Haag], so I just left elementary school in 1930 when the great depression broke out. I managed to finish secondary school education in 4 years (the standard was 5 or 6). After that, I could not get any job, and could not get any financial support for university education. I used my next two years (1934–1936) to study mathematics from books, without any teacher. I passed the examinations that qualified me as a mathematics teacher in all secondary schools in the Netherlands. But there weren't any jobs. Yet I had some success: I could get a small loan that enabled me to study mathematics and physics at Leiden University. In the academic year 1936–1937 I attended courses in physics and astronomy, and in 1937–1938 courses in mathematics on the master's degree level. That was all the university education I had. The most inspiring mathematician in those days at Leiden was H. D. Kloosterman.
>
> In 1939 I was so lucky to get an assistantship at Delft Technical University. It didn't pay very much, but it left me plenty of time to get involved in various kinds of mathematical research. It was quite an inspiring environment, and actually it was the only place in the Netherlands that employed mathematical assistants (Delft had about 8 or 9 of them). In 1940 the country was occupied, and from then on the main problem was to avoid being drawn into forced labour in Germany. In that respect my assistantship was a good shelter for quite some time.
>
> All the time I lived with my parents in The Hague, not so safe as it seemed. We were hiding a Jewish refugee (a German boy, a few years younger than me), who assisted my brother in producing and distributing forbidden radio material, like antennas that made it possible to eliminate the heavy bleep-bleep-bleep that the Germans used in the wavelengths of the British Radio. And later, when radios were forbidden altogether, my brother built miniature radios, hidden in old encyclopedia volumes. All this activity ended somewhere in the beginning of 1944 when our house was raided by the *Sicherheitspolizei*. My brother and his Jewish assistant where taken into custody, but by some strange coincidence they came back the next day. Nevertheless they had to leave to a safer place, where both of them survived the war. A few months later, I got my first real job. It was at the famous Philips Physical Labs at Eindhoven. The factory worked more or less for German war production, just like most factories in the country, but the laboratories could just do what they always did.
>
> Four months later, Eindhoven was occupied by the allied armies, in their move towards the battle of Arnhem. From then on we were cut off from the rest of the country, where people had a very bad time.
>
> So this was about my life during the war. Compared to others, I had been quite lucky. I had even managed to get my doctorate at the [Calvinist] Free University, Amsterdam [March 1943], just a few weeks before all universities in the country

were definitely closed (Leiden University had already been closed in 1940, because of demonstrations against the dismissal of Jewish professors).

In 1946 I got a professorship at Delft Technical University. I had to do quite elementary teaching, leaving me free to do quite some research, mainly in analytical number theory. It got me into correspondence with Erdős, and around 1948 he visited us at Delft.

In 1951 I made a mathematical trip abroad for the first time in my life. There I had contact with Erdős too. We had a long train ride together from London to Edinburgh.

In 1952 I got that [Van der Waerden's] professorship at Amsterdam, at that time the mathematical Mecca of the Netherlands. I stayed there until 1960, when I got my professorship at Eindhoven Technological University, where I retired in 1984. After that, I always kept a place to work there.

I think this is all you wanted to know.

In fact, on November 1, 2005 I asked for a few additional details:

I know you are one of the most modest men. Yet, I would think you were not just an observer when your family hid a Jewish boy, and your brother did activities not appreciated by the occupiers. Would you be so kind to share with me your role is these activities during 1940–1945? What were the names of your brother and his Jewish-German assistant? What was the difference in age between you and your brother?

Two days later, my questions were answered [Bru13]:

I hardly ever participated in my brother's activities. At most three times I delivered an antenna or a radio to some stranger. My brother was a year and a half older than I. His name was Johan.

The Jewish boy's name was Ernest (Ernst) Goldstern. He was born 24 December 1923 (in Muenchen, I believe). His family came to Holland in the late 1930's, where Ernst just completed his secondary school education in Amsterdam. He lived with us in The Hague from 1940 to 1944. I helped him to study advanced mathematics, which he could use after the war. He went into Electrical Engineering and got his degree in Delft. He died 19 January 1993. Johan died in 1996.

On July 9, 2008 N. G. de Bruijn is turning 90—Happy Birthday, Nicolaas!

27
Edge Colored Graphs: Ramsey and Folkman Numbers

27.1 Ramsey Numbers

In this chapter we will see that no matter how edges of a complete graph K_n are colored in two or, more generally, finitely many colors (each edge in one color), we can guarantee the existence of the desired monochromatic subgraph as long as we choose n to be large enough.

Naturally, when we talk about edge colored graphs, we call a subgraph *monochromatic* if all its edges are assigned the same color.

Frank Harary told me that he was once asked to suggest problems for the W. L. Putnam Mathematical Competition, and he suggested to use a problem that had already existed in the mathematical folklore:

Problem 27.1 (W. L. Putnam Mathematical Competition, March 1953). Prove that no matter how the edges of the complete graph K_6 are colored in two colors, there is always a monochromatic triangle K_3.

Proof Let v_0 be a vertex of K_6, whose edges are colored red and blue. Then v_0 is incident with at least three edges of the same color, say, red (Fig. 27.1).

If any two of the vertices v_1, v_2, v_3, say, v_1 and v_2, are connected by a red edge, then we are done: v_0, v_1, v_2 is a red monochromatic triangle. Otherwise all three edges v_1v_2, v_2v_3, and v_3v_1 are blue, and we are done as well. ∎

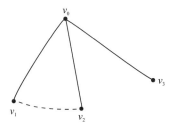

Fig. 27.1

A. Soifer, *The Mathematical Coloring Book*,
DOI 10.1007/978-0-387-74642-5_27, © Alexander Soifer 2009

Problem 27.2 Show that in the statement of Problem 27.1, 6 is best possible number, i.e., there is a way to color the edges of K_5 in two colors without creating any monochromatic triangles.

Solution: Behold (Fig. 27.2):

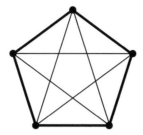

Fig. 27.2 2-colored K_5 without monochromatic triangles

For positive integers m and n, the *Ramsey number* $R = R(m, n)$ is the smallest positive integer such that *any* red and blue edge coloring of the complete graph K_R contains a red monochromatic K_m or a blue monochromatic K_n.

Problems 27.1 and 27.2 together prove, for example, that

$$R(3, 3) = 6.$$

You do not need more than definitions to prove the following two equalities.

Problem 27.3 For any two positive integers m and n

$$R(m, n) = R(n, m).$$

Problem 27.4 For any positive integer n

$$R(2, n) = n.$$

When and who coined the term "Ramsey number"? The publication search readily proves that it did not exist in print before 1966. "Ramsey number" makes its first appearance in January 1966 in the remarkable Ph.D. thesis *Chromatic Graphs and Ramsey's Theorem* by James (Jim) G. Kalbfleisch [Ka2] at the University of Waterloo, Ontario, Canada. He proved a good number of new upper and lower bounds, uniqueness of certain colorings and the exact value $R(3, 6) = 18$ (which was also proven independently by G. Kéry). Kalbfleisch may have been first to use computer programs in aid of his Ramsey numbers research. Kalbfleisch was also first to use "Ramsey number" term in print (his thesis, as was typical in mathematics in North America, was not published), in his 1966 paper [Ka3], submitted for publication on February 26, 1966. Nearly half a year later, on July 13, 1966, Jack E. Graver and James Yackel's paper [GY] was communicated by Victor Klee.

Both papers, [Ka3] and [GY], proudly displayed the new term "Ramsey number" in their titles. The term took hold, and since then was used in an enormous number of publications.

As to Jim Kalbfleisch, following a number of fine Ramsey number related publications, he "defected" to statistics. Kalbfleisch served the University of Waterloo for 37 years, 1963–2000, as a student, professor, dean of mathematics, academic vice-president and provost. On December 31, 2000, he retired at 60 to enjoy his artistic hobbies. *Daily Bulletin* reported on Thursday, October 5, 2000:

> "There's never a good time to go," he [Kalbfleisch] said, but after 14 years, "I feel the need for a break." He said he is looking forward to a chance to get back to stained glass work (his long-time hobby) and enjoy music, bridge, and some travel that isn't just for business.

The following year Kalbfleisch was awarded the title "Provost Emeritus," a rare distinction indeed.

I would like to compute a few Ramsey numbers with you. For this we will need the following simple but quite useful tool.

Basic Tool 27.5 For any graph G with p vertices v_1, v_2, ..., v_p and q edges, $\deg v_1 + \deg v_2 + \cdots + \deg v_p = 2q$.

The following Ramsey numbers were first found in 1955 by Robert E. Greenwood of the University of Texas and Andrew M. Gleason of Harvard.

Problem 27.6 (R. E. Greenwood and A. M. Gleason, [GG]) Prove that $R(3, 4) = 9$.

Proof Let the edges of a complete graph K_9 be colored red and blue. We will consider two cases.
Case 1: Assume there is a vertex, say v_0, of K_9 that is incident with at least four red edges (Fig. 27.3). Then should any two of the vertices v_1, v_2, v_3, v_4 be connected by a red edge, we get a red triangle. Otherwise we get a blue monochromatic K_4 on the vertices v_1, v_2, v_3, v_4.

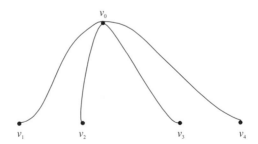

Fig. 27.3

Case 2: Every vertex of K_9 is incident with at least five blue edges. The nine vertices of K_9 with all blue edges form a graph G. The degree of each vertex of G may not be equal to five because we would get an odd $5 \cdot 9 = 45$ in the left side of the equality

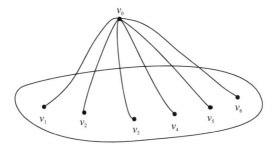

Fig. 27.4

of Tool 27.5 with an even $2q$ in the right side. Therefore, at least one vertex, say v_0, of K_9 is incident with at least six blue edges (Fig. 27.4).

By Problem 27.1 applied to the complete graph K_6 on the vertices v_1, v_2, ..., v_6, K_6 contains a monochromatic triangle K_3. If K_3 is red, we are done. If K_3 is blue, then the three vertices of K_3 plus v_0 form a blue monochromatic graph K_4, and we are done again.

Thus, we proved the inequality $R(3, 4) \leq 9$.

Figure 27.5 shows all red edges of K_8. The edges that are not drawn are colored blue. It is easy to verify (do) that this 2-coloring of the edges of K_8 creates neither a red monochromatic K_3, nor a blue monochromatic K_4. ∎

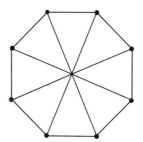

Fig. 27.5

Problem 27.7 (R.E. Greenwood and A. M. Gleason, [GG]) Prove that $R(4, 4) = 18$.

Proof First we will prove the inequality $R(4, 4) \leq 18$.

Let the edges of a complete graph K_{18} be colored red and blue, and v_0 be a vertex of K_{18}. Since v_0 is incident with 17 edges, by the Pigeonhole Principle v_0 must be incident with at least 9 edges of the same color.

If these 9 edges are red, we apply the equality $R(3, 4) = 9$ of Problem 27.6 to the 9-element set $S = \{v_1, v_2, ..., v_9\}$. If S contains a blue monochromatic K_4, we are done. If S contains a red monochromatic triangle T, then T together with v_0 and three red edges between them comprise a red monochromatic K_4.

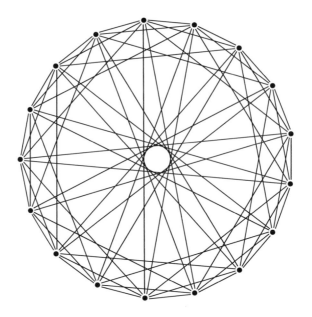

Fig. 27.6

If the nine edges are blue, we apply the equality $R(4, 3) = 9$ to the 9-set $S = \{v_1, v_2, \cdots, v_9\}$ and reason similarly to the above "red" case. Thus, the inequality $R(4, 4) \leq 18$ is proven.

Now we have to prove that $R(4, 4) > 17$. Figure 27.6 shows all red edges of the red–blue edge coloring of K_{17} (all missing edges are blue). It is easy to verify (do, and use symmetry) that our K_{17} contains no monochromatic K_4. ∎

You can now solve on your own the following couple of problems, of which the first one gives a rare exact value of a Ramsey number, while the second problem is just an exercise.

Problem 27.8 (R.E. Greenwood and A.W. Gleason, [GG]) Prove that $R(3, 5) = 14$.

Problem 27.9 Prove that $R(4, 5) \leq 32$ and $R(5, 5) \leq 64$.

In fact, the problem of calculating $R(4, 5)$ has been settled completely by Brendan D. McKay of the Australian National University and Stanisław P. Radziszowski of the Rochester Institute of Technology, originally of Poland.

Result 27.10 (B. D. McKay and S. P. Radziszowski, [MR4]). $R(4, 5) = 25$.

This remarkable result took years of computing to achieve, with the happy end taking place right in front of my eyes. I attended Stanisław Radziszowski's talk in early March of 1993 at the Florida Atlantic University conference. During the talk he mysteriously remarked that the value of $R(4, 5)$ may be established very soon. Imagine, in a matter of days I received his e-mail, announcing the birth of the result up to a hundredth of a second (this is what computer-aided communication delivers):

From: MX%"spr@cs.rit.edu"
To: ASOIFER
Subj: R(4, 5) = 25
From: spr@cs.rit.edu (Stanisław P Radziszowski)
Message-ID: <9303191824.AA22893@rit.cs.rit.edu>
Subject: R(4, 5) = 25
To: jackkasz@utxvm.cc.utexas.edu, asoifer@happy.uccs.edu,
    goldberg@turing.cs.rpi.edu
Date: Fri, 19 Mar 1993 19-MAR-1993 11:24:29.37 (EST)

$$R(4, 5) = 25$$

Brendan D. McKay, Australian National University
Stanisław P. Radziszowski, Rochester Institute of Technology
The Ramsey number R(4,5) is defined to be the smallest n such that every graph on n vertices has either a clique of order 4 or an independent set of order 5. We have proved that R(4, 5) = 25. Previously it was only known that R(4,5) is one of the four numbers 25–28. Our proof is computational.

For integers s,t define an (s,t,n)-graph to be an n-vertex graph with no clique of order s or independent set of order t. Suppose that G is a (4,5,25)-graph with 25 vertices. If a vertex is removed from G, a (4,5,24)-graph H results; moreover, the structure of H can be somewhat restricted by choosing which vertex of G to remove. Our proof consists of constructing all such structure-restricted (4,5,24)-graphs and showing that none of them extends to a (4,5,25)-graph. In order to reduce the chance of computational error, the entire computation was done in duplicate using independent programs written by each author. The fastest of the two computations required about 3.2 years of cpu time on Sun workstations.

A side result of this computation is a catalogue of 350866 (4,5,24)-graphs, which is likely to be most but not all of them.

We wish to thank our institutions for their support. Of particular importance to this work was a grant from the ANU Mathematical Sciences Research Visitors Program.

– bdm@cs.anu.edu.au and spr@cs.rit.edu; March 19, 1993.

Imagine how quickly the amount of computation increases in these "small" Ramsey numbers: "The fastest of the two computations required about 3.2 years of cpu time on Sun workstations," and "A side result of this computation is a catalogue of 350,866[8] (4,5,24)-graphs"!

What about the value of the next Ramsey number, $R(5, 5)$? In the historical summary included in [MR5], we see that the lower bound of $R(5, 5)$ increased slowly from 38 (Harvey Leslie Abbott in his impressive 1965 Ph.D. thesis [Abb]) to 42 (Robert W. Irving, 1974 [Irv2]), to finally 43 (Geoffrey Exoo, 1989 [Ex4]): Exoo produced a K_5-free 2-coloring of the edges of K_{42}.

I included an easy upper bound in Problem 27.10 just as an exercise—already in 1965 J. G. Kalbfleish [Ka1] knew better when he came up with 59. The first half of the 1990s saw a rapid improvement due to the works by Brendan D. MacKay and

---

[8] Later in 1993 this number grew to 350,904.

Stanisław P. Radziszowski: 53 (1992), 52 (1994), 50 (1995, an implication of the $R(4, 5)$ result above), and finally 49 (1995, [MR5]).

Thus, today's world records in lower and upper bound competitions for the value of $R(5, 5)$ are due to Geoffrey Exoo, Brendan McKay and Stanisław Radziszowski, respectively:

Best Bounds 27.11 ([Ex4], [MR5]). $43 \leq R(5, 5) \leq 49$.

And when the great expert of lower bounds Geoffrey Exoo and the great experts of upper bounds Brendan McKay and Stanisław Radziszowski agree that there is evidence for a "strong conjecture," we'd better listen and record:

McKay–Radziszowski–Exoo's Conjecture 27.12 [MR5]

$$R(5, 5) = 43.$$

It may take decades or even a century to settle this number—when done, we will see whether the three authors of the conjecture are right. In fact, Paul Erdős liked to popularly explain the difficulties of this problem [E94.21]: "It must seem incredible to the uninitiated that in the age of supercomputers $R(5, 5)$ is unknown. This, of course, is caused by the so-called combinatorial explosion: there are just too many cases to be checked." He even made up a joke about it, which I have heard during his talks in a few different variants:

> Suppose aliens invade the earth and threaten to destroy it in a year if human beings do not find $R(5, 5)$. It is, probably, possible to save the earth by putting together the world's best minds and computers. If, however, the invaders were to demand $R(6, 6)$, the human beings might as well attempt a preemptive strike without even trying to ponder the problem.[9]

Ever since 1994 Stanisław Radziszowski has maintained and revised 11 times a major 60-page compendium of "world records" in the sport of small Ramsey numbers [Radz1]. This is an invaluable service to the profession. I will present here only Table 27.1 of all known non-trivial classic 2-color small Ramsey numbers and their best lower and upper bounds. Where lower and upper bounds do not coincide, they both are listed in the appropriate cell. The cells below the main diagonal are left empty because filling them in would be redundant due to symmetry of the Ramsey function $R(m, n) = R(n, m)$, (Problem 27.3). As Table 27.2, I am presenting only a part of Radziszowski's Reference Table for Table 27.1—see the rest in his compendium [Radz1] readily available on the Internet. You will find there a wealth of other fascinating small Ramsey-related world records, Ramsey numbers (understood broader than here), Ramsey numbers inequalities, and a bibliography of 452 referenced items.

In the standard text on *Ramsey Theory* [GRS2, pp. 89–90], a tiny "Table 4.1" of known values and bounds is presented, accompanied by quite a pessimistic prediction:

---

[9] Alternative versions appear in [E93.20] and [E94.21].

Table 27.1 World records in classical 2-color small Ramsey numbers

k \ l	3	4	5	6	7	8	9	10	11	12	13	14	15
3	6	9	14	18	23	28	36	40	46	52	59	66	73
								43	51	59	69	78	88
4		18	25	35	49	56	73	92	97	128	133	141	153
				41	61	84	115	149	191	238	291	349	417
5			43	58	80	101	125	143	159	185	209	235	265
			49	87	143	216	316	442		848		1461	
6				102	113	127	169	179	253	262	317		401
				165	298	495	780	1171		2566		5033	
7					205	216	233	289	405	416	511		
					540	1031	1713	2826	4553	6954	10581	15263	22116
8						282	317				817		861
						1870	3583	6090	10630	16944	27490	41525	63620
9							565	580					
							6588	12677	22325	39025	64871	89203	
10								798					1265
							23556		81200			

Table 27.2 References for a part of Table 1

k \ l	4	5	6	7	8	9	10
3	GG	GG	Kéry	Ka2	GR	Ka2	Ex5
				GY	MZ	GR	RK2
4	GG	Ka1	Ex9	Ex3	Ex15	Ex17	HaKr
		MR4	MR5	Mac	Mac	Mac	Mac
5		Ex4	Ex9	CET	HaKr	Ex17	Ex17
		MR5	HZ1	Spe3	Spe3	Mac	Mac
6			Ka1	Ex17	XXR	XXER	Ex17
			Mac	Mac	Mac	Mac	Mac

Table 4.1 gives all known exact bounds [values] and some upper and lower bounds on the function R. It is unlikely that substantial improvement will be made on this table.

Just compare their Table 4.1 to the Table 27.1 here, and you would agree with me that the researchers in small Ramsey numbers have dramatically exceeded expectations of the authors of [GRS2] in a short span of 17 years. We have a race here: combinatorial explosion vs. improvements in computers and computational methods. It seems that computers and mathematicians in this field have held their own and gained some!

What would happen if we were to color edges of a complete graph K_n in more than two colors? Can we then guarantee the existence of, say, a monochromatic triangle K_3? Yes, we can.

Problem 27.13 (R.E. Greenwood and A. M. Gleason, 1955, [GG]) Prove that for any positive integer r there is a positive integer $n(r)$ such that any r-coloring of edges of a complete graph $K_{n(r)}$ contains a monochromatic triangle K_3.

Proof We will prove this statement by induction. For $r = 1$ (i.e., one-color edge coloring), we can certainly choose $n = 3$: one-colored edges of K_3 form a monochromatic triangle. The statement for $r = 2$ has been proven as Problem 27.1.

Assume that for a positive integer r there is $n(r)$ such that any r-coloring of edges of the complete graph $K_{n(r)}$ contains a monochromatic triangle K_3. We need to find the value of the function $n(r + 1)$ such that any $(r + 1)$-coloring of edges of a complete graph $K_{n(r+1)}$ contains a monochromatic triangle K_3.

Let us define the value of the function $n(r + 1)$ as

$$n(r + 1) = (r + 1)(n(r) - 1) + 2.$$

Assume that the edges of $K_{n(r+1)}$ are $(r+1)$-colored, and v_0 is a vertex of $K_{n(r+1)}$. Since v_0 is incident with $(r + 1)(n(r) - 1) + 1$ edges, by the Pigeonhole Principle there is a color, say color A, such that v_0 is incident with $n(r)$ edges of color A (Fig. 27.7).

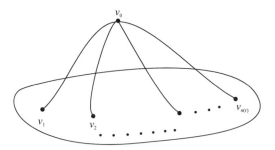

Fig. 27.7

If any two of the vertices $v_1, v_2, \ldots, v_{n(r)}$, are connected by an A-colored edge, these two vertices plus v_0 form an A-colored monochromatic triangle. Otherwise, we have a complete graph $K_{n(r)}$ on vertices $v_1, v_2, \ldots, v_{n(r)}$, whose edges are r-colored. By the inductive assumption, $K_{n(r)}$ contains a monochromatic triangle.

Note that in fact it is easy to prove by induction that $n(r) \le \lfloor r!e \rfloor + 1$, where e is the base of the natural logarithms, $e = 2.718281828459045 \ldots$ ∎

We computed some particular Ramsey numbers and looked at ideas of proofs. Surprisingly, they are fairly recent. Even more surprising to me is that general existence results came first. The foundation for this beautiful direction in mathematics, now called *Ramsey Theory*, was laid by the young British mathematician Frank P. Ramsey. We will discuss his impressive work and short life in chapters 28 and 30 respectively. Here I will only formulate particular cases, graph-theoretic diagonal versions of Ramsey's Theorems.

Ramsey Theorem, Infinite Diagonal Graph Version 27.14 Every complete infinite graph with 2-colored edges contains a complete infinite monochromatic subgraph.

Ramsey Theorem, Finite Diagonal Graph Version 27.15 For any positive integers n and k, there is an integer $R(n; k)$ such that if $m > R(n; k)$ and the edges of a complete graph K_m are k-colored, then K_m contains a complete monochromatic subgraph K_n.

These theorems should sound familiar to you. We have solved some particular cases of Problem 27.15 earlier in this chapter, and even found the values of $R(n; k)$, which for $k = 2$ we will simply denote as $R(n, n)$.[10] For example, Problems 27.1 and 27.2 show that $R(3, 3) = 6$; Problem 27.6 gives us $R(3, 4) = 9$; Problem 27.13 demonstrates that $R(3; k)$ exists for any positive integer k.

Instead of demonstrating 27.15, we will prove a stronger pair of results, 27.16 and 27.17, obtained in the early 1933 and published 2 years later by two young unknown Hungarian university students, Pal (Paul) Erdős and Gjörgy (George) Szekeres.

Problem 27.16 (P. Erdős and G. Szekeres, [ES1]) Assume that the Ramsey number $R(m, n)$ exists for every pair of positive integers m and n. Then for any integers $m \geq 2$ and $n \geq 2$

$$R(m, n) \leq R(m - 1, n) + R(m, n - 1).$$

Proof Let $L = R(m - 1, n) + R(m, n - 1)$. We have to prove precisely that if the edges of a complete graph K_L with L vertices are colored red and blue, K_L contains a K_m with all red edges or a K_m with all blue edges. Indeed, let v_0 be a vertex of K_L whose edges are colored red and blue. We consider two cases and use an approach that proved successful in Problems 27.7 and 27.8.

Case 1: Let v_0 be incident with at least $R(m - 1, n)$ red edges. Then by definition of $R(m - 1, n)$, the vertex set $S = \{v_1, v_2, \ldots, v_{R(m-1,n)}\}$ contains a blue monochromatic K_n (and we are done), or a red monochromatic K_{m-1}. In the latter case, K_{m-1} together with v_0 and $m - 1$ red edges connecting them, form a red monochromatic K_m.

Case 2: Let v_0 be incident with less than $R(m - 1, n)$ red edges. Since v_0 is incident with $L - 1 = R(m - 1, n) + R(m, n - 1) - 1$ edges, each colored red or blue, we see that in this case v_0 is incident with at least $R(m, n - 1)$ blue edges.

By the definition of $R(m, n - 1)$, the vertex set $S = \{v_1, v_2, \ldots v_{R(m,n-1)}\}$ contains a red monochromatic K_m (and we are done), or a blue monochromatic K_{n-1}. In the latter case, K_{n-1} together with v_0 and $n - 1$ blue edges connecting them, form a blue monochromatic K_n. ∎

Problem 27.17 (P. Erdős and G. Szekeres, [ES]). For every two positive integers m and n, the Ramsey number $R(m, n)$ exists, moreover,

---

[10] Please, do not overlook the significant difference between $R(n, k)$ and $R(n; k)$.

$$R(m, n) \leq \binom{m + n - 2}{m - 1}.$$

Proof We will use induction on $k = m + n$. We have the equality when one of the numbers m, n equals 1 or 2 and the other is arbitrary (see Problems 27.4 and 27.3, and observe that $R(1, n) = 1$). Therefore the inequality is true for $k \leq 5$, and we can assume that $m \geq 3$ and $n \geq 3$.

Assume that $R(m - 1, n)$ and $R(m, n - 1)$ exist and that

$$R(m - 1, n) \leq \binom{m + n - 3}{m - 2}$$

and

$$R(m, n - 1) \leq \binom{m + n - 3}{m - 1}.$$

Then by Problem 27.17 and Pascal binomial equality, we get

$$
\begin{aligned}
R(m, n) &\leq R(m - 1, n) + R(m, n - 1) \\
&\leq \binom{m + n - 3}{m - 2} + \binom{m + n - 3}{m - 1} = \binom{m + n - 2}{m - 1},
\end{aligned}
$$

as desired. We are done. $R(m, n)$ exists and satisfies the required inequality. ∎

In the same paper Paul Erdős and Gjörgy Szekeres also proved in similar spirit the Monotone Subsequence Theorem, which we will discuss in Chapter 29.

What can we learn about large Ramsey numbers if we could only compute some small Ramsey numbers? Nothing at all as far as the exact values are concerned. We can, however, aspire to estimate their growth, strive for asymptotics. This is precisely what interested Paul Erdős the most. Paul traces the developments in this direction at the 1980 Graph Theory conference at Kalamazoo [E81.20]: "it is well-known that

$$c_1 n 2^{n/2} < R(n, n) < c_2 \binom{2n - 2}{n - 1}, \quad (c_2 < 1)."$$

He reports an improvement in the upper bound a few years later [E88.28]:

$$c_1 n 2^{n/2} < R(n, n) < c_2 \binom{2n}{n} \Big/ (\log n)^\varepsilon \qquad (*)$$

Every time Erdős speaks on this subject, he offers the same important conjecture, which still remains open today:

Erdős's $100 Conjecture 27.18

$$\lim_{n \to \infty} R(n, n)^{1/n} = c.$$

Paul adds [E88.28] "I offer 100 dollars for a proof [of this conjecture] and 10,000 dollars for a disproof. I am sure that [the conjecture] holds." He continues with the problem of determining the limit in Conjecture 27.18.

Erdős's $250 Problem 27.19 Determine c in Conjecture 27.18.

Paul gives a hint too [E88.28]: "$\sqrt{2} \le c \le 4$ follows from (*), perhaps $c = 2$?" Let us record it formally.

Erdős's Open Problem 27.20 Prove or disprove that

$$\lim_{n \to \infty} R(n, n)^{1/n} = 2.$$

These problems matter a great deal to Paul Erdős, for he repeats these problem in his many problem talks and papers, for example [E81.20], [E88.28], [E90.23], and [E93.20]. He even offers, rare for Erdős, *unspecified* compensation [E90.23]:

Any improvement of these bounds [$\sqrt{2} \le c \le 4$] would be of great interest and will receive an "appropriate" financial reward. ("Appropriate" I am afraid is not the right word, I do not have enough money to give a really appropriate award.)

He is pessimistic about finding the asymptotic formula any time soon [E93.20]:

An asymptotic formula for $R(n, n)$ would of course be very desirable, but at the moment this looks hopeless.

Yet, Erdős poses a number of other problems related to the Ramsey numbers' asymptotic behavior. Let me mention here just two examples.

Erdős's Open Problem 27.21 ([E91.31]). Is it true that for every $\varepsilon > 0$ and $n > n_0(\varepsilon)$

$$R(4, n) > n^{3-\varepsilon}?$$

In fact, probably

$$R(4, n) > cn^3 / (\log n)^{\alpha}.$$

Erdős's Open Problem 27.22 ([E91.31]). Is it true that

$$R(n+1, n+1) > (1+c) R(n, n)?$$

In fact it is not even known that

$$R(n+1, n+1) - R(n, n) > cn^2.$$

In the mid 1990s, when I was looking for the author of the term *Ramsey Theory* (you will find out the answer in chapter 30), on February 19, 1996 in Baton Rouge the famous graph theorist Frank Harary half-wrote, half-dictated to me a letter [Har3], which is most relevant to this section on Ramsey numbers, and so I will transcribe it here in its entirety (see facsimile of the opening lines in Fig. 27.8):

To A Soifer   19 Feb 96
In 1965, I looked into the Ramsey Nos. of C_4 and $2K_2$ [two copies of K_2] and found (proved) their values are 6 and 5 resp.
    Before then only ramsey nos. of complete graphs had been studied; e.g.

$r(K_3) = 6$ (folklore)
$r(K_4) = 18$ (A. Gleason + R. Greenwood)
$r(K_5) = ?$ ($100 from FH for 1st exact solution)

I called $r(C_4) = 6$ and $r(2K_2) = 5$ generalized ramsey nos. for graphs.
    In November 1970, V. Chvatal defended his Ph.D. thesis at U Waterloo on Ramsey nos. of hypergraphs. Erdős was visiting professor at the same time at Waterloo. He saw me drinking tea and grabbed my elbow saying "You must hear this doctorate defense, as Chvatal is brilliant." The next night Chvatal invited me to dinner at his house, and I proposed a series of papers to him. He accepted gladly and we had a good time writing them. I told Erdős that this was part of my big research project on Ramsey Theory.

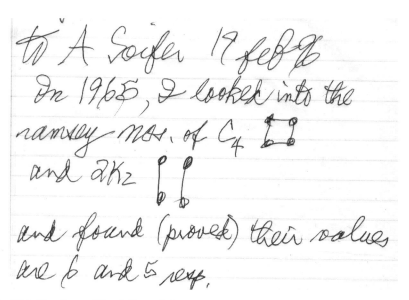

Fig. 27.8 First lines of Frank Harary's Letter

I saw that he [Chvatal] and I would be able to carry out my research project of calculating the Ramsey Numbers of all the small G [graphs]. We wrote a series of papers
"Generalized Ramsey theory for graphs" I, II, III, and maybe IV. I then continued the
series to XVII. I referred early to this as the study of Ramsey theory for graphs.

Thus, Frank Harary and Václav Chvátal introduced the term *Generalized Ramsey
Theory for Graphs* and started an impressive series of papers under this title. They
generalized the notion of Ramsey number by including in the study the existence
of subgraphs other than complete graphs. This is a flourishing field today, and I
refer you to Radziszowski's compendium [Radz1] and Section 5.7 of the monograph
[GRS2] for a summary of many of the achievements of this promising direction of
research. The authors of [GRS2, p. 138] write:

> A major impetus behind the early development of Graph Ramsey theory was the hope
> that it would eventually lead to methods for determining larger values of the classical
> Ramsey numbers $R(m, n)$. However, as so often happens in mathematics, this hope has
> not been realized; rather, the field has blossomed into a discipline of its own. In fact,
> it is probably safe to say that the results arising from Graph Ramsey theory will prove
> to be more valuable and interesting than knowing the exact value of $R(5, 5)$ [or even
> $R(m, n)$].

I do not know how one measures value. It seems that the relationship between
theoretical and numerical directions of inquiry has been, as so often happens in
mathematics, a marriage made in heaven. Numerical results provided a foundation
for theoretical generalizations and asymptotics, while theoretical results allowed
to dramatically reduce sprawling explosion of computation thus making numerical
results possible. Moreover, numerical results can contain beauties both in mathematical arguments and in extreme graphs they uncover—just look at the graph in
Fig. 27.9 below! I hope the rest of this section will illustrate my point of view. We
will look at one related train of thought and the direction it has inspired, Folkman

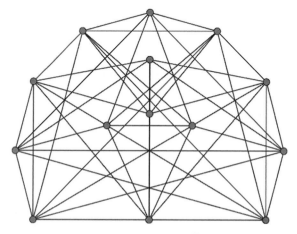

Fig. 27.9 Unique 14-vertex bi-critical $F_v(3, 3; 4)$ graph[11]

numbers. We will see that Paul Erdős was very interested in small Folkman numbers, and the authors of the above quotation have contributed their energies and results to this cause as well.

27.2 Folkman Numbers

In 1967 Paul Erdős and András Hajnal took the first step in a typically subtle Erdős style: they posed a particular problem.

Problem 27.23 (Erdős–Hajnal [EH2]) Construct a graph G which does not contain K_6 such that every 2-coloring of its edges contains a monochromatic K_3.

This time Erdős and Hajnal added a more general conjecture:

Erdős–Hajnal's Conjecture 27.24 [EH2]. For every positive integer r there is a graph G which contains no K_4 such that every r-coloring of its edges contains a monochromatic K_3.

In the same year of 1967, Jon H. Folkman [Fol] generalized Erdős–Hajnal conjecture for the case of two colors. Folkman, a winner of the 1960 William Lowell Putnam Mathematical Competition and University of California Berkeley graduate before joining Rand Corporation, tragically left this world in 1969. He was 31 years old. Before I formulate Folkman's theorem in the contemporary terminology, I need to introduce a few terms that have recently become standard.

Given positive integers m,n,l, an *edge Folkman graph* G is a graph without a K_l subgraph, such that if its edges are 2-colored, there will be a subgraph K_m with all edges of color 1 or a subgraph K_n with all edges of color 2.

The *edge Folkman number* $F_e(m,n;l)$ is defined as the smallest positive integer k such that there exists an k-vertex Folkman graph G.

More generally, given positive integers n, m_1, m_2, ..., m_n, l, an *edge Folkman graph* G is a graph without a K_l subgraph, such that if its edges are n-colored, there will be a subgraph K_{m_i} with all edges of color i for at least one value of i, $1 \le i \le n$.

The *edge Folkman number* $F_e^n(m_1, m_2, \ldots, m_n; l)$ is defined as the smallest positive integer k such that there exists a k-vertex Folkman graph G.

In this recent terminology, Folkman's result can be formulated as follows:

Folkman's Theorem 27.25 [Fol] For all positive integers m, n, l; $l > \max(m, n)$, edge Folkman numbers $F_e(m, n; l)$ exist.[12]

Folkman ends [Fol] with a far reaching generalization of Erdős–Hajnal's Conjecture 27.24:

---

[11] This graph is so striking, that I chose it to decorate the cover of the April-2007 issue of *Geombinatorics* XVI(4) that contained Stanisław P. Radziszowski's paper.

[12] Of course, Jon Folkman did not use the term "Folkman number," which seems to have appeared first in 1993, and since has become standard.

Folkman's Conjecture 27.26 [Fol] For all positive integers $m_1, m_2, \ldots, m_n, l$, $l > \max(m_1, m_2, \ldots, m_n)$, edge Folkman numbers $F_e^n(m_1, m_2, \ldots, m_n; l)$ exist.

Vertex Folkman graphs and *vertex Folkman numbers* $F_v^n(m_1, m_2, \ldots, m_n; l)$, are defined similarly for coloring of vertices instead of edges. When $n = 2$, we omit the superscript n.

In his lyrical (a rare quality for mathematical writing) paper [Sp2], Joel Spencer recalls that in 1973, during the Erdős 60th birthday conference in Keszthely,[13] Hungary, the Erdős–Hajnal Conjecture 27.20 was given to the Czech mathematician Jaroslav Nešetřil and his student Vojtěch Rödl, who proved it during the conference. Moreover, they came up with pioneering results so general that they could be considered as principles, not unlike the Ramsey theorem!

A *clique number* $\omega(G)$ of a graph G is the order n of its largest complete subgraph K_n.

Edge Nešetřil–Rödl's Theorem 27.27 [NR]. Given a positive integer n and a graph G, there exists a graph H of the same clique number as G, such that if edges of H are n-colored, H has an edge-monochromatic subgraph isomorphic to G.

Vertex Nešetřil–Rödl's Theorem 27.28 [NR] Given a positive integer n and a graph G, there exists a graph H of the same clique number as G, such that if vertices of H are n-colored, H has a vertex-monochromatic subgraph isomorphic to G.

These remarkable theorems, in our terminology, imply the following:

Corollary 27.29 For any positive integers m, n, the edge Folkman numbers $F_e^n(m, m, \ldots, m; m + 1)$ and vertex Folkman numbers $F_v^n(m, m, \ldots, m; m + 1)$ exist, where m inside the parentheses repeats n times.

Erdős–Hajnal problems, Folkman's paper, and Nešetřil–Rödl theorems inspired a new direction in Ramsey theory as well as new exciting problems on Ramsey-like numbers.

What should these new numbers be called? Several names were used at first: *restricted Ramsey, Erdős–Hajnal, Graham–Spencer* [HN1], [Irv1]. Nešetřil and Rödl [NR] mention such names as "*Galvin–Ramsey*" and "*EFGH*" (which, I guess, stood for Erdős–Folkman–Graham–Hajnal). In 1993, some researchers in these new numbers seem to have started, suddenly and simultaneously, to use the name *Folkman numbers*: Jason I. Brown and Vojtěch Rödl [BR]; Martin Erickson [Eri]. Slowly, through the decade that followed, this name has won out and become standard.

Obviously, if $l > R(m, n)$, where $R(m, n)$ is a Ramsey number, then $F_e(m, n; l) = R(m, n)$. The real challenge in calculating Folkman numbers occurs when $l \leq R(m, n)$, even in the simplest case $F_e(3, 3; l)$. As could be expected, the lower the l is, the harder is the problem (except for trivially small values of l).

---

[13] Spencer misidentifies the town as Balatonfüred; both towns are on lake Balaton.

In 1968 Ronald L. Graham [Gra0] published a solution of the first problem, 27.23. In fact, Graham found the smallest order graph that does the job for $F_e(3, 3; 6)$.

Graham's Result 27.30 [Gra0] The graph $G = K_8 - C_5 = K_3 + C_5$ satisfies the conditions of problem 27.23, i.e.,

$$F_e(3, 3; 6) = 8.$$

Graham was not alone working on this new kind of problem, as he wrote in the conclusion of [Gra0]:

> To the best of the author's knowledge, the first example of a graph satisfying the conditions [27.23] of Erdős and Hajnal was given by J. H. van Lint; subsequently L. Pósa showed the existence of such a graph containing no complete *pentagon* [$l = 5$] and Jon Folkman constructed such a graph containing no complete *quadrilateral* [$l = 4$] (all unpublished).

Paul Erdős's problem, reported in 1971 [GS1, p. 138] in the current terminology, reads simply as:

Paul Erdős's Open Problem 27.31 Compute edge Folkman numbers.

The simplest unknown edge Folkman number was at that time $F_e(3, 3; 5)$. Its first upper bound of 42 was established by M. Schäuble in 1969, which 2 years later was reduced to 23 by Graham and Spencer [GS1], who conjectured that 23 was the exact value. In 1973 Robert W. Irving [Irv1] reduced the upper bound to 18 and thus disproved the Graham–Spencer conjecture. A year earlier the best lower bound of 10 was established by Shen Lin [Lin]. Joel Spencer, in a review of Irving's paper, wrote (MR0321778):[14]

> It is now known that $10 \le F_e(3, 3, 5) \le 18$. The determination of $F_e(3, 3, 5)$ appears to be extremely difficult.

Afterwards the Bulgarian mathematicians took over the problem. In 1979, N. G. Hadziivanov and N. D. Nenov [HN1] reduced the upper bound to 16. A year later Nenov [Nen1] improved the lower bound to 11, and in 1981 he reduced the upper bound to 15 [Nen2]. In 1985 Hadziivanov and Nenov [HN2] increased the lower bound to 12.

In 1999 Stanisław P. Radziszowski, Konrad Piwakowski and Sebastian Urbánski [PRU] increased the lower bound to match the Nenov's upper bound at 15, and thus closed the problem: $F_e(3, 3; 5) = 15$. They also proved that $F_v(3, 3; 4) = 14$, and found a unique bi-critical 14-vertex Folkman graph without a K_4 subgraph, such that any vertex 2-coloring contains a monochromatic triangle K_3 (Fig. 27.9). They, as well as some of their predecessors, observed that by adding a new vertex adjacent to all 14 vertices of this graph, they get a 15-vertex Folkman graph without a K_5 subgraph, and such that any edge 2-coloring contains a monochromatic triangle K_3.

---

[14] He used the greek α in place of not yet established Folkman symbol.

In 1974 in Prague, Paul Erdős gave a talk of a special kind [E75.33]:

I discuss some of the problems which occupied my collaborators and myself for a very long time. I tried to select those problems which are striking and which are not too well known.

One of the striking problems addressed there dealt with the next Folkman's number, $F_e(3, 3; 4)$. In 1975 we knew very little about it, and Paul Erdős summarized the state of the problem as follows:

Folkman's upper bound for $F_e(3, 3, 4)$ is enormous, it is much bigger than $10^{10^{10^{10^{10^{10}}}}}$, the same holds for the bound of Nešetřil and Rödl.

Erdős then offered an unusual, mathematically defined price, max(100 dollars, 300 Swiss francs) for a specific bound.

Paul Erdős's max ($100, 300 SF) Problem 27.32 [E75.33]. Prove or disprove the inequality $F_e(3, 3, 4) < 10^{10}$.

A dozen years later, in 1986 Frankl and Rödl [FR2] came close, within a factor of 100 from Paul Erdős's conjectured upper bound: they used a probabilistic proof to show that $F_e(3, 3; 4) \leq 10^{12}$. Soon after, in 1988, Joel Spencer in a paper proudly called *Three Hundred Million Points Suffice* [Sp3] squeezed out of the probabilistic approach a better bound: $F_e(3, 3; 4) < 3 \times 10^8$. A mistake found in Spencer's proof by Mark Hovey of MIT prompted Spencer in 1989 [Sp4] to increase his bound to $F_e(3, 3; 4) < 3 \times 10^9$ and change the title to *Three Billion Points Suffice*, which miraculously was still within Paul Erdős's limit for the cash prize.

Spencer's Upper Bound 27.33 ([Sp3], [Sp4], 1988–1989)

$$F_e(3, 3, 4) < 3,000,000,000.$$

The first lower bound is a consequence of Lin's 1972 results [Lin], $10 \leq F_e(3, 3; 5) \leq F_e(3, 3; 4)$. In the recently published *Geombinatorics* work [RX] Stanisław P. Radziszowski and the Chinese mathematician Xiaodong Xu remark that the analysis in the above cited 1999 result $F_e(3, 3; 5) = 15$ [PRU] allows to devise a better lower bound: all 659 15-vertex graphs that have no K_5 subgraph and in every 2-coloring of edges contain a monochromatic K_3, have a subgraph K_4, hence $16 \leq F_e(3, 3; 4)$. This new 2007 paper [RX] contains a computer-free proof that $18 \leq F_e(3, 3; 4)$, and a further computer-aided improvement to $19 \leq F_e(3, 3; 4)$, which is currently the best-known lower bound.

Radziszowski–Xu's Lower Bound 27.34 ([RX], 2007)

$$19 \leq F_e(3, 3; 4).$$

So, the state of the problem today is this:

$$19 \leq F_e(3, 3; 4) < 3,000,000,000.$$

As you can see, the gap between the best known lower and upper bounds of $F_e(3,\ 3;\ 4)$ is enormous! The authors of [RX] report:

> Geoffrey Exoo suggested to look at the well known Ramsey coloring of K_{127} defined by Hill and Irving [HI] in 1982 in order to establish the bound $128 \le R(4,\ 4,\ 4)$.

During Staszek Radziszowski's March 8, 2007 talk at the Florida Atlantic University conference, I hinted that a prize for a dramatic improvement in upper bound would be in order, and the speaker obligated by offering $500. Better yet, in his March 22, 2007 e-mail to me, Staszek offered two $500 prizes, for proof or disproof of the lower bound 50, and the upper bound 127 of $F_e(3,\ 3;\ 4)$. "I believe that both of these bounds are true," he added in the e-mail.

Radziszowski's Double $500 Conjecture 27.35 .

$$50 \le F_e(3, 3; 4) \le 127.$$

In his talk Radziszowski mysteriously hinted that an upper bound around the conjectured 127 may be proved in the year 2013. I hope we will witness this great result!

VI
The Ramsey Principle

28
From Pigeonhole Principle to Ramsey Principle

28.1 Infinite Pigeonhole and Infinite Ramsey Principles

The Infinite Pigeonhole Principle states:

Infinite Pigeonhole Principle 28.1 Let k be a positive integer. If elements of an infinite set S are colored in k colors, then S contains an infinite monochromatic subset S_1.

I am sure you will have no difficulties in proving it.

Do you see anything in common between this simple principle and Infinite Diagonal Graph Version of Ramsey Theorem, 27.14? Both say that if we have enough objects, then we can guarantee the existence of something: in Pigeonhole Principle it is an infinite subset; in Ramsey Theorem 27.14 we get an infinite subset of edges, i.e., the subset of two-element sets of vertices of the graph (since an edge is a pair of vertices). This connection is very close, both results are particular cases of the so called Ramsey Theorem, one result for $r = 1$ and the other for $r = 2$. Let me formulate it here under the new, more appropriate name in my opinion:

Infinite Ramsey Principle 28.2 For any positive integers k and r, if all r-element subsets of an infinite set S are colored in k colors, then S contains an infinite subset S_1 such that all r-element subsets of S_1 are assigned the same color.

I have always felt that something is wrong with the title "Ramsey Theorem." To see that it suffices to read the leader of the field Ronald L. Graham, who in 1983 wrote [Gra2]:

The generic [sic] result in Ramsey Theory is due (not surprisingly) to F. P. Ramsey.

Exactly: a "generic result," compared to much more specific typical examples, such as Schur's Theorem (Chapter 32), Van der Waerden's Theorem (Chapter 33), etc. The Ramsey Theorem occupies a unique place in the Ramsey Theory. It is a powerful tool. It is also a philosophical principle stating, as Theodore S. Motzkin put it, that "complete disorder is an impossibility. Any structure will necessarily contain an orderly substructure".[1] It is, therefore, imperative to call the Ramsey Theorem by a much better fitting name: *The Ramsey Principle.*

The original double-induction proof of the Infinite Ramsey Principle 28.2 by F. P. Ramsey is crystal clear—read it in the original [Ram2]. I choose to present here the proof by Ronald L. Graham from [Gra1]. This is not only a beautiful proof: it demonstrates a method that has worked very productively in the Ramsey Theory. Keep it in your mathematical tool box.

Proof by Ronald L. Graham [Gra1] For $r = 1$ we get the Infinite Pigeonhole principle.

Without loss of generality, we can assume that our infinite set S coincides with the set of positive integers N. (Every infinite set S contains a countable subset equivalent to N, and N is sufficient for us to select the required in the problem subset S_1.)

We first treat the case $r = 2$ since it is easy to visualize. We can identify the two-element subsets of N with edges of the infinite complete graph K_N with the vertex set $N = \{1, 2, \ldots, n, \ldots\}$. Let the edges of K_N be colored in k colors. It is convenient to denote the color of an edge $\{x, y\}$ by $\chi\{x, y\}$.

(1) Consider the edges of the form $\{1, x\}$, i.e., the edges incident with the vertex 1. There are infinitely many of them and only k colors, therefore by the Infinite Pigeonhole Principle infinitely many of these edges $\{1, x_1\}$, $\{1, x_2\}$, $\ldots$, $\{1, x_n\}$, $\ldots$ are assigned the same color, say c_1. Denote $X = \{x_1, x_2, \ldots, x_n, \ldots\}$; and let x_1 be the smallest number in X. Please note that $\chi\{1, x\} = c_1$ for any x in X.

(2) Consider the edges of the form $\{x_1, x\}$ where $x \in X$; i.e., the edges incident with the vertex x_1 with the other endpoint x being an element of the set X. Once again, by the Infinite Pigeonhole Principle infinitely many of these edges $\{x_1, y_1\}$, $\{x_1, y_2\}$, $\ldots$, $\{x_1, y_n\}$, $\ldots$ are assigned the same color, say c_2. Denote $Y = \{y_1, y_2, \ldots, y_n, \ldots\}$; and let y_1 be the smallest number in Y. Note that $\chi\{x_1, y\} = c_2$ for any y in Y.

(3) Consider the edges $\{y_1, y\}$, where $y \in Y$, i.e., edges incident with the vertex y_1 with the other endpoint y being an element of Y. Again, by the Infinite Pigeonhole Principle infinitely many of these edges $\{y_1, z_1\}$, $\{y_1, z_2\}$, $\ldots$, $\{y_1, z_n\}$, $\ldots$ are assigned the same color, say c_3. Denote $Z = \{z_1, z_2, \ldots, z_n, \ldots\}$; and let z_1 be the smallest number in Z. We have $\chi\{y_1, z\} = c_3$ for any z in Z, etc.

We can continue this construction indefinitely. As a result, we get the infinite set $T = \{1, x_1, y_1, z_1, \ldots\}$. It has one key property: *for any two elements t, t' from T the color of the edge $\{t, t'\}$ depends only on the value of $\min\{t, t'\}$. Consequently, our edge coloring χ on T uniquely determines vertex coloring χ^* on T as follows*

$$\chi^*(t) = \chi\{t, t'\} \text{ for } t' > t.$$

Thus, we get the set T colored in k colors. By the Infinite Pigeonhole Principle some infinite subset S_1 of T must be monochromatic under χ^*, i.e., all colors $\chi^*(s)$ for s

---

[1] Quoted from [GRS2].

from S_1 are the same. However, by the definition of χ^* this means precisely that all edges $\{s, s'\}$ of S_1 have the same color under χ. This proves the Infinite Ramsey Principle for $r = 2$.

As an example of the method, we sketch the proof for $r = 3$. The given k-coloring χ of three-element subsets of N uniquely determines k-coloring χ_1 of the two-element subsets of $X' = X \backslash \{1\}$ by $\chi_1\{x, x'\} = \chi\{1, x, x'\}$. By the Infinite Ramsey Principle for $k = 2$, X contains an infinite subset X' monochromatic under χ_1, (i.e., all values $\chi_1\{x, x'\}$ are the same for $x, x' \in Y$), say having color c_1, and the smallest element x_1. Next, the original k-coloring χ uniquely defines k-coloring of two-element subsets of $Y = X' \backslash \{x_1\}$ by $\chi_2\{y, y'\} = \chi\{x_1, y, y'\}$. Once again, by the Infinite Ramsey Principle for $r = 2$, Y contains an infinite subset Y' monochromatic under χ_2, having color c_2, and the smallest element y_1. We next observe that the original k-coloring χ uniquely defines k-coloring χ_3 of two-element subsets of $Z = Y' \backslash \{y_1\}$ by $\chi_3\{z, z'\} = \chi\{y_1, z, z'\}$, etc.

Similarly to the case $r = 2$ above, we end up with the infinite set $T = \{1, x_1, y_1, z_1, \ldots\}$ which by construction has the property that the color of any triple $\{t, t', t''\}$ depends only on $\min\{t, t', t''\}$. Thus, the original k-coloring χ of three-element subsets of T uniquely defines the k-coloring χ^* of the vertices of T as follows

$$\chi^*(t) = \chi(\{t, t', t''\}) \text{ for } t'' > t' > t.$$

By the Infinite Pigeonhole Principle some infinite subset S_1 of T is monochromatic under χ^*. By the definition of χ^* this means that all three-element subsets of S_1 have the same color under χ. We are done for $r = 3$.

The inductive step for the general r follows *exactly* the same lines. ∎

Have you heard of the famous Helly Theorem? I noticed in 1990 that the Helly Theorem and its variations are ready for the marriage to the Infinite Ramsey Principle. This could be a new observation: not just I, but the leading expert Branko Grübaum, a coauthor of the monograph *Helly's Theorem and its Relatives* [DGK] written jointly with Ludwig Danzer and Victor Klee has not heard of such a marriage. Here is a plane version of the Helly Theorem for the case of infinitely many figures.

Helly's Theorem for Infinite Family of Convex Figures in the Plane 28.3 Given an infinite family of closed convex figures in the plane, one of which is bounded. If every three of them have a common point, then the intersection of all figures in the family is non-empty.

We can obtain the following result, for example, by combining the Helly Theorem and the Infinite Ramsey Principle.

Problem 28.4 Let $F_1, F_2, \ldots, F_n, \ldots$ be a family of closed convex figures in the plane, and F_1 be bounded. If among any four figures there are three figures with a point in common, then infinitely many figures of the family have a point in common.

Proof Consider the set $S = \{F_1, F_2, \ldots, F_n, \ldots\}$. We color a three-element subset $\{F_i, F_j, F_k\}$ of S red if $F_i \cap F_j \cap F_k \neq \varnothing$, and blue otherwise. By the Infinite Ramsey Principle S contains an infinite subset S_1 such that all three-element subsets of S_1 are assigned the same color. This color cannot be blue because every four-element subset of S_1 contains a three-element subset $T = \{F_i, F_j, F_k\}$, such that $F_i \cap F_j \cap F_k \neq \varnothing$, i.e., T is colored red. Thus, all three-element subsets of S_1 are red. By the Helly Theorem 28.3, all figures of the infinite subset S_1 have a point in common. ∎

In 1990, Paul Erdős informed me in a letter that a stronger statement was conjectured (he was not sure by whom).

Conjecture 28.5 Given an infinite family of closed convex figures in the plane, one of which is bounded. If among any four figures there are three figures with a point in common, then there is a finite set S (consisting of n points), such that every given figure contains at least one point from S.

Moreover, n is an absolute constant (i.e., it is one and the same for all families that satisfy the above conditions).

Vladimir Boltyanski and I first published this conjecture in 1991 [BS]. 18 years later, on September 26, 2008 while reading the manuscript of the forthcoming new expanded 2009 Springer edition of [BS], Branko Grünbaum, resolved this conjecture in the negative: I am mailing to him the \$25 prize. Grünbaum showed that Conjecture 28.5 does not hold even for the line R.

Grünbaum's Counterexample 28.6 (e-mail to A. Soifer, September 26, 2008). Define the sets as follows:

$$F_0 = \{0\};$$

$$F_n = \{x \in R : x \geq n\}, \text{ for every positive integer } n.$$

Of course, all conditions of Conjecture 28.5 are satisfied, while for any finite set S of reals, there is an integer n that is greater than any number from S. By definition, F_n does not contain any element from S. ∎

On September 29, 2008, I asked Branko, Grünbaum whether he can "save" Conjecture 28.5, and the following day he sent me his saving recipe:

> Yes, I conjecture that Erdos's problem may be resuscitated by requiring two (instead of just one) of the sets to be compact. But I do not see any easy proof.

Grünbaum's Conjecture 28.7 (e-mail to A. Soifer, September 30, 2008). Given an infinite family of closed convex figures in the plane, *two* of which are compact. If among any four figures there are three figures with a point in common, then there is a finite set S (consisting of N points) such that every given figure contains at least one point from S.

28.2 Pigeonhole and Finite Ramsey Principles

Let us take another look at Frank P. Ramsey's pioneering 1930 paper [Ram2]. Having disposed of the infinite case, Ramsey proves the finite one ([Ram2], Theorem B). As a methodology of the new theory, it ought to be elevated to the status of a principle.

Finite Ramsey Principle 28.8 For any positive integers r, n, and k there is an integer $m_0 = R(r, n, k)$ such that if $m > m_0$ and all r-element subsets of an m-element set S_m are colored in k colors, then S_m contains an n-element subset S_n such that all r-element subsets of S_n are assigned the same color.

Proof The Ramsey Principle follows from the Infinite Ramsey Principle 28.2 by de Bruijn–Erdős's Compactness Theorem 26.1.

A clearly written direct proof, without the use of the compactness argument, can be found in the original 1930 paper by F. P. Ramsey [Ram2]; it is also reproduced in full in [GRS2]. ∎

As you surely noticed, the Pigeonhole Principle is a particular case of the Fininte Ramsey Principle for $r = 1$.

Pigeonhole Principle 28.9 Let n and k be positive integers. If elements of a set S with at least $m_0 = (n - 1)k + 1$ elements are colored in k colors, then S contains a monochromatic n-element subset S_1.

Since edges can be viewed as two-element subsets of the vertex set of a graph, by plugging in $r = 2$ in the Finite Ramsey Principle, we get the result encountered in the previous chapter: Finite Diagonal Graph Version Ramsey Theorem 27.15.

It is amazing to me how quickly the news of the Ramsey Principle traveled in the times that can hardly be called the age of information. Ramsey's paper appeared in 1930. Already in 1933 the great Norwegian logician Thoralf Albert Skolem (1887–1963) published his own proof [Sko] of the Ramsey Theorem (with a reference to the Ramsey's 1930 publication!). In 1935 yet another proof (for the graph-theoretic setting) appeared in the paper [ES1] by the two young Hungarians, Paul Erdős and Gjörgy (George) Szekeres. We will look at this remarkable paper in the next chapter.

29
The Happy End Problem

29.1 The Problem

During the winter of 1932–1933, two young friends, mathematics student Paul Erdős, age 19, and chemistry student George (György) Szekeres, 21, solved the problem posed by their youthful lady friend Esther Klein, 22, but did not send it to a journal for a year and a half. When Erdős finally sent this joint paper for publication, he chose J. E. L. Brouwer's journal *Compositio Mathematica*, where it appeared in 1935 [ES1].

Erdős and Szekeres were first to demonstrate the power and striking beauty of the Ramsey Principle when they solved this problem. Do not miss G. Szekeres' story of this momentous solution later in this chapter. In the process of working with Erdős on the problem, Szekeres actually rediscovered the Finite Ramsey Principle before the authors ran into the 1930 Ramsey publication [Ram2].

Erdős–Szekeres's Theorem 29.1 [ES1] For any positive integer $n \geq 3$ there is an integer m_0 such that any set of at least m_0 points in the plane in general position[2] contains n points that form a convex polygon.

To prove Erdős–Szekeres's Theorem, we need two tools.

Tool 29.2 (Esther Klein, Winter 1932–1933) Any 5 points in the plane in general position contain 4 points that form a convex quadrilateral.

In fact, in anticipation of the proof of Erdős–Szekeres's Theorem, it makes sense to introduce an appropriate notation $ES(n)$ for the Erdős–Szekeres function. For a positive integer n, $ES(n)$ will stand for the minimal number such that any $ES(n)$ points in the plane in general position contain n points that form a convex n-gon. Esther Klein's result can be written as

Result 29.3 (Esther Klein). $ES(4) = 5$.

---

[2] That is, no three points lie on a line.

A. Soifer, *The Mathematical Coloring Book*,
DOI 10.1007/978-0-387-74642-5_29, © Alexander Soifer 2009

Condensed Proof (Use paper and pencil.) Surely, $ES(4) > 4$. Given 5 points in the plane in general position, consider their convex hull H.[3] If H is a quadrilateral or a pentagon, we are done. If H is a triangle, the line determined by the two given points a, b inside H does not intersect one of the triangle H's sides de. We get a convex quadrilateral formed by the points a, b, d, and e. ∎

Tool 29.4 (P. Erdős and G. Szekeres, [ES1]) Let $n \geq 3$ be a positive integer. Then n points in the plane form a convex polygon if and only if every 4 of them form a convex quadrilateral.

According to Paul Erdős, two members of his circle E. Makai and Paul Turán established (but never published) one more exact value of $ES(n)$:

Result 29.5 (E. Makai and P. Turán). $ES(5) = 9$.

Erdős mentioned the authorship of this result numerous times in his problem papers. However, I know of only one instance when he elaborated on it. During the first of the two March 1989 lectures Paul gave at the University of Colorado at Colorado Springs, I learned that Makai and Turán found proofs *independently*. Paul said that Makai proof was lengthy, and shared with us Turán's short Olympiad-like proof. Turán starts along Esther Klein's lines by looking at the convex hull of the given 9 points. Let me stop right here and allow you the pleasure of finding a proof on your own.

We are now ready to prove Erdős–Szekeres's Theorem asserting the existence of the function $ES(n)$.

Proof of Theorem 29.1 (*P. Erdős and G. Szekeres*) Let $n \geq 3$ be a positive integer. By the Ramsey Principle 28.8 (we set $r = 4$ and $k = 2$) there is an integer $m_0 = R(4, n, 2)$ such that if $m > m_0$ and all four-element subsets of an m-element set S_m are colored in two colors, then S_m contains a n-element subset S_n such that all four-element subsets of S_n are assigned the same color.

Now let S_m be a set of m points in the plane in general position. We color a four-element subset of S_m red if it forms a convex quadrilateral, and blue if it forms a concave (i.e., non-convex) quadrilateral. Thus, all four-element subsets of S_m are colored red and blue. Hence, S_m contains an n-element subset S_n such that all four-element subsets of S_n are assigned the same color. This color cannot be blue, because in view of Tool 29.2 any five or more element set contains a red four-element subset! Therefore, all four-element subsets of S_n are colored red, i.e., they form convex quadrilaterals. By Tool 29.4, S_n forms a convex n-gon. ∎

I must show you a beautiful alternative proof of Erdős–Szekeres's Theorem 29.1, especially since it was found by an undergraduate student, Michael Tarsi of Israel. He missed the class when the Erdős–Szekeres solution was presented, and had to

---

[3] Convex hull of a set S is the minimal convex polygon that contains S. If you pound a nail in every point of S, then a tight rubber band around all nails would produce the convex hull.

come up with his own proof under the gun of the exam! Tarsi recalls (e-mail to me of December 12, 2006):

> Back in 1972, I took the written final exam of an undergraduate Combinatorics course at the Technion – Israel Institute of Technology, Haifa, Israel. Due to personal circumstances, I had barely attended school during that year and missed most lectures of that particular course. The so-called Erdős-Szekeres Theorem was presented and proved in class, and we have been asked to repeat the proof as part of the exam. Having seen the statement for the first time, I was forced to develop my own little proof.
>
> Our teacher in that course, the late Professor Mordechai Levin, had published the story as an article, I cannot recall the journal's name, the word 'Gazette' was there and it dealt with Mathematical Education.
>
> I was born in Prague (Czechoslovakia at that time) in 1948, but was raised and grew up in Israel since 1949. Currently I am a professor of Computer Science at Tel-Aviv University, Israel.

Proof of Theorem 29.1 *by Michael Tarsi.* Let $n \geq 3$ be a positive integer. By the Ramsey Principle 28.8 ($r = 3$ and $k = 2$) there is an integer $m_0 = R(3, n, 2)$ such that, if $m > m_0$ and all three-element subsets of an m-element subset S_m are colored in two colors, then S_m contains an n-element subset S_n such that all three-element subsets of S_n are assigned the same color.

Let now S_m be a set of m points in the plane in general position labeled with integers 1, 2, ..., m.

We color a three-element set $\{i, j, k\}$, where $i < j < k$, red if we travel from i to j to k in a clockwise direction, and blue if counterclockwise. By the above, S_m contains an n-element subset S_n such that all three-element subsets of S_n are assigned the same color, i.e., have the same orientation. But this means precisely that S_n forms a convex n-gon! ∎

In their celebrated paper [ES1], P. Erdős and G. Szekeres also discovered the Monotone Subsequence Theorem.

A sequence $a_1, a_2, \ldots, a_k$ of real numbers is called *monotone* if it is increasing, i.e., $a_1 \leq a_2 \leq \ldots \leq a_k$, or decreasing, i.e., $a_1 \geq a_2 \geq \ldots \geq a_k$ (we use weak versions of these definitions that allow equalities of consecutive terms).

Erdős–Szekeres's Monotone Subsequence Theorem 29.6 [ES1] Any sequence of $n^2 + 1$ real numbers contains a monotone subsequence of $n + 1$ numbers.

I would like to show here how the Ramsey Principle proves such a statement with, of course, much worse upper bound than $n^2 + 1$. I haven't seen this argument in literature before.

Problem 29.7 Any long enough sequence of real numbers contains a monotone subsequence of $n + 1$ numbers.

Solution. Take a sequence S of $m = R(2, n + 1, 2)$ numbers $a_1, a_2, \ldots, a_m$. Color a two-element subsequence $\{a_i, a_j\}$, $i < j$ red if $a_i \leq a_j$, and blue if $a_i > a_j$. By the Ramsey Principle, there is an $(n + 1)$-element subsequence S_1 with every two-element subsequence of the same color. But this subsequence is monotone! ∎

In [ES1] P. Erdős and G. Szekeres generalize Theorem 29.6 as follows:

Erdős–Szekeres's Monotone Subsequence Theorem 29.8 Any sequence S: a_1, a_2, ..., a_r of $r > mn$ real numbers contains a decreasing subsequence of more than m terms or an increasing subsequence of more than n terms.

A quarter of a century later, in 1959, A. Seidenberg of the University of California, Berkeley, found a brilliant "one-line" proof of Theorem 29.8, thus giving it a true Olympiad-like appeal.

Proof of Theorem 29.8 *by A. Seidenberg* [Sei] Assume that the sequence S : $a_1, a_2, \ldots, a_r$ of $r > mn$ real numbers has no decreasing subsequence of more than m terms. To each a_i assign a pair of numbers (m_i, n_i), where m_i is the largest number of terms of a decreasing subsequence beginning with a_i and n_i the largest number of terms of an increasing subsequence beginning with a_i. This correspondence is an injection, i.e., distinct pairs correspond to distinct terms $a_i, a_j, i < j$. Indeed, if $a_i \le a_j$ then $n_i \ge n_j + 1$, and if $a_i > a_j$ then $m_i \ge m_j + 1$.

We get $r > mn$ distinct pairs (m_i, n_i), they are our pigeons, and m possible values (they are our pigeonholes) for m_i, since $1 \le m_i \le m$. By the Pigeonhole Principle, there are at least $n + 1$ pairs (m_0, n_i) with the same first coordinate m_0. Terms a_i corresponding to these pairs (m_0, n_i) form an increasing subsequence! ∎

Erdős and Szekeres note that the result of their Theorem 29.8 is best possible:

Problem 29.9 ([ES1]) Construct a sequence of mn real numbers such that it has no decreasing subsequence of more than m terms *and* no increasing subsequence of more than n terms.

Proof Here is a sequence of mn terms that does the job:

$$m, m - 1, \ldots 1; 2m, 2m - 1, \ldots, m + 1; \ldots;$$
$$nm, nm - 1, \ldots, (n - 1)m + 1. \quad \blacksquare$$

H. Burkill and Leon Mirsky in their 1973 paper [BM] observe that the Monotone Subsequence Theorem holds for countable sequences as well.

Countable Monotone Subsequence Theorem 29.10 [BM]. Any countable sequence S : $a_1, a_2, \ldots, a_r, \ldots$ of real numbers contains an infinite increasing subsequence or an infinite strictly decreasing subsequence.

Hint: Color the two-element subsets of S in two colors. ∎

The authors "note in passing that the same type of argument enables us to show" the following cute result (without a proof):

Curvature Preserving Subsequence Theorem 29.11 [BM]. Every countable sequence S possesses an infinite subsequence which is convex or concave.

Hint: Recall Michael Tarsi's proof of Erdős–Szekeres Theorem above, and color the three-element subsets of S in two colors! ∎

The results of this section reminded me of the celebrated Helly Theorem.

Helly's Theorem 29.12 Let $F_1 \ldots, F_m$ be convex figures in n-dimensional space R^n. If each $n + 1$ of these figures have a common point, then the intersection $F_1 \cap \ldots \cap F_m$ is non-empty.

In particular, for $n = 2$ we get Helly's Theorem for the plane.

Helly's Theorem for the Plane 29.13 A finite family $F_1 \ldots, F_m$ of convex figures is given in the plane. If every three of them have a non-empty intersection, then the intersection $F_1 \cap \ldots \cap F_m$ of all of these figures is non-empty as well.

The structure of the Helly Theorem appears to me similar to the one of Theorem 29.1. This is why I believe that the Helly Theorem and its numerous beautiful variations are a fertile ground for applications of the powerful tool, the Finite Ramsey Principle 28.8. To the best of my—and Branko Grünbaum's—knowledge this marriage of Helly and Ramsey has not been noticed before. To illustrate it, I have created a sample problem. Its result is not important, but the method may lead you to discovering new theorems.

Problem 29.14 Let m be a large enough positive integer ($m \geq R(3, 111, 2)$ to be precise), and $F_1, \ldots, F_m$ be convex figures in the plane. If among every 37 figures there are 3 figures with a point in common, then there are 111 figures with a point in common.

Hint: The fact that $37 \times 3 = 111$ has absolutely nothing to do with solution: the statement of Problem 29.14 remains true if we replace 37 and 111 by arbitrary positive integers l and n, respectively, as long as $l \leq n$.

Solution: Let $m \geq R(3, 111, 2)$, and $F_1, F_2, \ldots, F_m$ be convex figures in the plane. Consider the set $S = \{F_1 F_2, \ldots, F_m\}$. We color a three-element subset $\{F_i, F_j, F_k\}$ of S red if $F_i \cap F_j \cap F_k \neq \emptyset$, and blue otherwise. By the Finite Ramsey Principle 28.8, there is a 111-element subset S_1 of S such that all its three-element subsets are assigned the same color. Which color can it be? Surely not blue, for among every 37 figures there are 3 figures with a point in common, thus forming a red three-element subset. Thus, all three-element subsets of S_1 are red. Therefore, by the Helly Theorem 29.13 the intersection of all 111 figures of the set S_1 is non-empty. ■

29.2 The Story Behind the Problem

On Paul Erdős's 60th birthday, his lifelong friend George (György) Szekeres gave Paul and us all a present of magnificent reminiscences, allowing us a glimpse into Erdős and Szekeres's first joint paper [ES1]and the emergence of a unique group of unknown young Jewish Hungarian mathematicians in Budapest, many of whom were destined to a great mathematical future. To my request to reproduce these remarkable reminiscences, George Szekeres answered in the March 5, 1992 letter:

Dear Alexander, ... Of course, as far as I am concerned, you may quote anything you like (or see fit) from my old reminiscences in "The Art of Counting"... But of course it may be different with MIT Press, that you have to sort out with them.

György Szekeres and Esther Klein, Bükk Mountains, Northern Hungary, 1938 (shortly after their 1937 marriage), provided by George Szekeres

I am grateful to George Szekeres and the MIT Press for their kind permission to reproduce George's memoirs here. His Reminiscences are sad and humorous at the same time, and warm above all. György Szekeres recalls [Szek]:

It is not altogether easy to give a faithful account of events which took place forty years ago, and I am quite aware of the pitfalls of such an undertaking. I shall attempt to describe the genesis of this paper, and the part each of us played in it, as I saw it then and as it lived on in my memory.

For me there is a bit more to it than merely reviving the nostalgic past. Paul Erdős, when referring to the proof of Ramsey's theorem and the bounds for Ramsey numbers given in the paper, often attributed it to me personally (e.g., in [E42.06]), and he obviously attached some importance to this unusual step of pinpointing authorship in a joint paper. At the same time the authorship of the "second proof" was never clearly identified.

I used to have a feeling of mild discomfort about this until an amusing incident some years ago reassured me that perhaps I should not worry about it too much. A distinguished British mathematician gave a lunch-hour talk to students at Imperial College on Dirichlet's box principle, and as I happened to be with Imperial, I went along. One of his illustrations of the principle was a beautiful proof by Besicovitch of Paul's theorem (2nd proof in [ES1]), and he attributed the theorem itself to "Erdős and someone whose name I cannot remember." After the talk I revealed to him the identity of Paul's coauthor (incidentally also a former coauthor of the speaker) but assured him that no historical injustice had been committed as my part in the theorem was less than ε.

Paul Erdős, early 1930s, Budapest

The origins of the paper go back to the early 1930s. We had a very close circle of young mathematicians, foremost among them Erdős, Turán, and Gallai; friendships were forged which became the most lasting that I have ever known and which out-lived the upheavals of the 1930s, a vicious world war and our scattering to the four corners of the world. I myself was an "outsider," studying chemical engineering at the Technical University, but often joined the mathematicians at weekend excursions in the charming hill country around Budapest and (in summer) at open air meetings on the benches of the city park.

Paul, then still a young student but already with a few victories in his bag, was always full of problems and his sayings were already a legend. He used to address us in the same fashion as we would sign our names under an article and this habit became universal among us; even today I often call old members of the circle by a distortion of their initials.

"Szekeres Gy., open up your wise mind." This was Paul's customary invitation— or was it an order?—to listen to a proof or a problem of his. Our discussions centered around mathematics, personal gossip, and politics. It was the beginning of a desperate era in Europe. Most of us in the circle belonged to that singular ethnic group of European society which drew its cultural heritage from Heinrich Heine and Gustav Mahler, Karl Marx and Cantor, Einstein and Freud, later to become the principal target of Hitler's fury. Budapest had an exceptionally large Jewish population, well over 200,000, almost a quarter of the total. They were an easily identifiable group speaking an inimitable jargon of their own and driven by a strong urge to congregate under the pressures of society. Many of us had leftist tendencies, following the simple reasoning that our problems can only be solved on a global, international scale and socialism was the only political philosophy that offered such a solution. Being a leftist had its dangers and Paul was quick to spread the news when one of our members got into trouble: "A. L. is studying the theorem of Jordan." It meant that following a political police action A. L. has just verified that the interior of a prison cell is not in the same component as the exterior. I have a dim recollection that this is how I first heard about the Jordan curve theorem.

Apart from political oppression, the Budapest Jews experienced cultural persecution long before anyone had heard the name of Hitler. The notorious "numerus clausu" was operating at the Hungarian Universities from 1920 onwards, allowing only 5% of the total student intake to be Jewish. As a consequence, many of the brightest and most purposeful students left the country to study elsewhere, mostly in Germany, Czechoslovakia, Switzerland, and France. They formed the nucleus of that remarkable influx of Hungarian mathematicians and physicists into the United States, which later played such an important role in the fateful happenings towards the conclusion of the second world war.

For those of us who succeeded in getting into one of the home universities, life was troublesome and the outlook bleak. Jewish students were often beaten up and humiliated by organized student gangs and it was inconceivable that any of us, be he as gifted as Paul, would find employment in academic life. I myself was in a slightly better position as I studied chemical engineering and therefore resigned to go into industrial employment, but for the others even a high school teaching position seemed to be out of reach.

Paul moved to Manchester soon after his Ph.D. at Professor Mordell's invitation, and began his wanderings which eventually took him to almost every mathematical corner of the world. But in the winter of 1932/1933 he was still a student; I had just received my chemical degree and, with no job in sight, I was able to attend the mathematical meetings with greater regularity than during my student years. It was at one of these meetings that a talented girl member of our circle, Esther Klein (later to become Esther Szekeres), fresh from a one-semester stay in Göttingen, came up with a curious problem: given 5 points in the plane, prove that there are 4 which form a convex quadrilateral. In later years this problem frequently appeared in student's competitions, also in the *American Mathematical Monthly* (53(1946)462, Problem

E740). Paul took up the problem eagerly and a generalization soon emerged: is it true that out of $2^{n-2}+1$ points in the plane one can always select n points so that they form a convex n-sided polygon? I have no clear recollection how the generalization actually came about; in the paper we attributed it to Esther, but she assures me that Paul had much more to do with it. We soon realized that a simple-minded argument would not do and there was a feeling of excitement that a new type of geometrical problem emerged from our circle which we were only too eager to solve. For me the fact that it came from Epszi (Paul's nickname for Esther, short for ε) added a strong incentive to be the first with a solution and after a few weeks I was able to confront Paul with a triumphant "E. P., open up your wise mind." What I really found was Ramsey's theorem from which it easily followed that there exists a number $N < \infty$ such that out of N points in the plane it is possible to select n points which form a convex n-gon. Of course at that time none of us knew about Ramsey. It was a genuinely combinatorial argument and it gave for N an absurdly large value, nowhere near the suspected 2^{n-2}. Soon afterwards Paul produced his well-known "second proof" which was independent of Ramsey and gave a much more realistic value for N; this is how a joint paper came into being.

I do not remember now why it took us so long (a year and a half) to submit the paper to the *Compositio*. These were troubled times and we had a great many worries. I took up employment in a small industrial town, some 120 kms from Budapest, and in the following year Paul moved to Manchester; it was from there that he submitted the paper.

I am sure that this paper had a strong influence on both of us. Paul with his deep insight recognized the possibilities of a vast unexplored territory and opened up a new world of combinatorial set theory and combinatorial geometry. For me it was the final proof (if I needed any) that my destiny lay with mathematics, but I had to wait for another 15 years before I got my first mathematical appointment in Adelaide. I never returned to Ramsey again.

Paul's method contained implicitly that $N > 2^{n-2}$, and this result appeared some 35 years later [ES2] in a joint paper, after Paul's first visit to Australia. The problem is still not completely settled and no one yet has improved on Paul's value of

$$N = \binom{2n-4}{n-2} + 1.$$

Of course we firmly believe that $N = 2^{n-2} + 1$ is the correct value. ∎

These moving memories prompted me to ask for more. George Szekeres replied on November 30, 1992:

Dear Sasha, ... Marta Svéd rang me some time ago from Adelaide, reminding me of an article that I was supposed to write about the old Budapest times... From a distance of 60 years, as I approach 82, these events have long lost their "romantic" freshness... My memories of those times are altogether fading away into the remote past, even if they are occasionally refreshed on my visits to Budapest. (I will certainly be there to celebrate Paul's 80-th birthday.)

The following year George did come to Hungary, and we met for dinner during the conference dedicated to Paul Erdős's 80th birthday, when George Szekeres and Esther Klein shared with me unique memories of Tibor Gallai, a key Budapest group member. See them in Chapter 42, dedicated to the Gallai Theorem.

29.3 Progress on the Happy End Problem

In May 1960, when Paul Erdős visited George Szekeres in Adelaide, they improved the lower bound of the Happy End problem [ES2].

Lower Bound 29.15 (Erdős and Szekeres [ES2]). $2^{n-2} \le ES(n)$, where $ES(n)$ is the Erdős–Szekeres function, i.e., the smallest integer such that any $ES(n)$ points in general position contain a convex n-gon.[4]

It is fascinating how sure Erdős and Szekeres were of their conjecture. In one of his last, posthumously published problem papers [E97.18], Erdős attached the prize and modestly attributed the conjecture to Szekeres: "I would certainly pay $500 for a proof of Szekeres's conjecture."

Erdős–Szekeres Happy End $500 Conjecture 29.16

$$ES(n) = 2^{n-2} + 1.$$

Their confidence is surprising[5] because the foundation for the conjecture was very thin, just results 29.2 and 29.5:

$$ES(4) = 5,$$
$$ES(5) = 9.$$

Computing exact values of the Erdős–Szekeres function $ES(n)$ proved to be a very difficult matter. It took over 70 years to make the next step. In 2006, George Szekeres (posthumously) and Lindsay Peters, with the assistance of Brendan McKay and heavy computing, have established one more exact value in the paper [SP] written "In memory of Paul Erdős":

Result 29.17 (G. Szekeres and L. Peters [SP]). $ES(6) = 17$.

In his latest surveys [Gra7], [Gra8],[6] Ronald L. Graham is offering $1000 for the first proof, or disproof, of the Erdős–Szekeres Happy End Conjecture 29.16.

---

[4] Erdős and Szekeres actually proved a strict inequality.

[5] In fact, Paul Erdős repeated $500 offer for the proof of the conjecture in [E97.21], but offered there "only 100 dollars for a disproof."

[6] I thank Ron Graham for kindly providing the preprints.

George Szekeres was, of course, correct when he wrote in his 1973 reminiscences above that their 1935 upper bound

$$E S(n) \leq \binom{2n - 4}{n - 2} + 1$$

had not been improved. In fact, it withstood all attempts of improvement until 1997 when Fan Chung and Ronald L. Graham [CG] willed it down by 1 point to

$$E S(n) \leq \binom{2n - 4}{n - 2}.$$

In the process, Chung and Graham offered a fresh approach which started an explosion of improvements. First it was improved by Daniel J. Kleitman and Lior Pachter [KP] to

$$E S(n) \leq \binom{2n - 4}{n - 2} + 7 - 2n.$$

Then came Géza Tòth and Pavel Valtr [TV1] with

$$E S(n) \leq \binom{2n - 5}{n - 2} + 2.$$

These developments happened so swiftly that all three above papers appeared in the same 1998 issue of *Discrete Computational Geometry*! In 2005 Tòth and Valtr came again [TV2] with the best known today upper bound

$$E S(n) \leq \binom{2n - 5}{n - 2} + 1,$$

which is about half of the original Erdős–Szekeres upper bound.

Paul Erdős's trains of thought are infinite—they never end, and each problem gives birth to a new problem, or problems. The Happy End Problem is not an exception. Paul writes about the AfterMath of the Happy End Problem with his vintage humor and warmth [E83.03]:

> Now there is the following variant which I noticed when I was once visiting the Szekeres in 1976 in Sydney, the following variant which is of some interest I think. It goes as follows. $n(k)$ is derived as follows, if it exists. It is the smallest integer with the following property. If you have $n(k)$ points in the plane, no three on a line, then you can always find a convex k-gon with the additional restriction that it doesn't contain a point in the interior. You know this goes beyond the theorem of Esther, I not only require that the k points should form a convex k-gon, I also require that this convex k-gon should contain none of the [given] points in its interior. And surprisingly enough this gives a lot of new difficulties. For example it is trivial that $n(4)$ is again 5, that is no problem. Because if your have a convex quadrilateral, if no point is inside we are happy; if from the five points one of them is inside you draw the diagonal (*AC*, Fig. 29.1):

And you join these (*AE, EC*) and now this convex quadrilateral (*AECD*) contains none of the points. And if you have four points and the fifth point is inside then you take this quadrilateral. This is convex again and has no point in the inside. And Harborth proved that $n(5) = 10$. $f(5)$ was 9 in Esther Klein's problem but here $n(5)$ is 10. He dedicated his paper to my memory when I became an archeological discovery. When you are 65 you become an archeological discovery. Now, nobody has proved that $n(6)$ exists. That you can give, for every t, t points in the plane, no three on a line and such that every convex hexagon contains at least one of the points in its interior. It's perfectly possible that can do that. Now Harborth suggested that maybe $n(6)$ exists but $n(7)$ doesn't. Now I don't know the answer here.

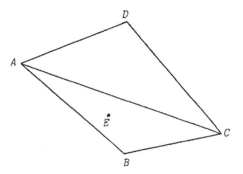

Fig. 29. 1

Indeed, in 1978 Heiko Harborth [Harb] of Braunschweig Technical University, Germany, proved that $n(5) = 10$. In 1983 J. D. Horton [Hort] of the University of New Brunswick, Canada, proved Harborth's conjecture that $n(t)$ does not exist for $t \geq 7$. This left a mystifying gap that is alive and well today:

Open Problem 29.18 Does $n(6)$ exist? If yes, find its value.

This new rich train of thought now includes many cars. I would like to share with you my favorite, the beautiful 2005 result by Adrian Dumitrescu of the University of Wisconsin-Milwaukee.

Dumitrescu's Theorem 29.19 [Dum].[7] For each finite sequence $h_0, h_1, \ldots, h_k$, with $h_i \geq 3$ ($i = 0, \ldots, k$) there is an integer $N = N(h_0, h_1, \ldots, h_k)$ such that any set S of at least N points in general position in the plane contains either

an empty convex h_0-gon (i.e., a convex h_0-gon that contains no points of S in its interior)

or

k convex polygons $P_1, P_1, \ldots, P_1$, where P_i is an h_i-gon such that P_i strictly contains P_{i+1} in its interior for $i = 1, \ldots, k - 1$.

---

[7] Adrian mistakenly credits 1975 Erdős's paper [E72.25] with the birth of the problem about empty convex polygons. In the cited story Erdős clearly dates it to his 1976 visit of the Szekereses.

29.4 The Happy End Players Leave the Stage as Shakespearian Heroes

Paul Erdős named it *The Happy End Problem*. He explained the name often in his talks. On June 4, 1992 in Kalamazoo I took notes of his talk:

> I call it The Happy End Problem. Esther captured George, and they lived happily ever after in Australia. The poor things are even older than me.

This paper also convinced George Szekeres to become a mathematician. For Paul Erdős the paper had a happy end too: it became one of his early mathematical gems and Paul's first of the numerous contributions to and leadership of the Ramsey Theory and, as Szekeres put it, of "a new world of combinatorial set theory and combinatorial geometry."

The personages of The Happy End Problem appear to me like heroes of Shakespeare's plays. Paul, very much like *Tempest's* Prospero, gave up all his property, including books, to be free. George and Esther were so close that they ended their lives together, like Romeo and Juliet. In the late summer 2005 e-mail, Tony Guttmann conveyed to the world the sad news from Adelaide:

> George and Esther Szekeres both died on Sunday morning [August 28, 2005]. George, 94, had been quite ill for the last 2–3 days, barely conscious, and died first. Esther, 95, died an hour later. George was one of the heroes of Australian mathematics, and, in her own way, Esther was one of the heroines.

I always wanted to know the membership in this amazing Budapest group. On May 28, 2000, during a dinner in the restaurant of the Rydges North Sydney Hotel at 54 McLaren Street,[8] I asked George Szekeres and Esther Klein to name the members of their group, so to speak the Choir of the Happy End Production. Esther produced, signed and dated the following list of young participants, of which according to her "half a dozen usually met":

Paul Erdős, Tibor Grünwald (Gallai), Géza Grünwald (Gergör), Esther Klein (Szekeres), Lily Székely (Sag), George (György) Szekeres, Paul Turán, Martha Wachsberger (Svéd), and Endre Vázsonyi.[9]

George Szekeres also told me that night "my student and I proved Esther's Conjecture for 17 with the use of computer." "Which computer did you use?" asked I. "I don't care how a pencil is made," answered George.

---

[8] Esther wrote the list on the letterhead of the hotel.

[9] Mikós Ság and László Molnár occasionally joined the group too.

30
The Man behind the Theory: Frank Plumpton Ramsey

I verified harmony by algebra.
Only then, experienced in science,
I dared to surrender to the bliss of creative dream.
— Aleksandr Pushkin, *Mozart and Salieri*

Knowledge is a correspondence between idea and fact.

— Frank Plumpton Ramsey

30.1 Frank Plumpton Ramsey and the Origin of the Term "Ramsey Theory"

Who was "Ramsey," the man behind the theory named for him by others?

Let us start with the introduction to Ramsey's collected works [Ram3], assembled and edited right after his passing in 1930 by Ramsey's friend and disciple Richard Bevan Braithwaite, then Fellow of King's College and later the Knightbridge Professor of Philosophy at the University of Cambridge, who opens as follows:

> Frank Plumpton Ramsey was born on 22nd February, 1903, and died on 19th January 1930 [a jaundice attack prompted by an unsuccessful surgery]. The son of the President of Magdalene, he spent nearly all his life in Cambridge, where he was successively Scholar of Trinity, Fellow of King's [at 21], and Lecturer in Mathematics in the University [at 23]. His death at the height of his powers deprives Cambridge of one of its intellectual glories and contemporary philosophy of one of its profoundest thinkers. Though mathematical teaching was Ramsey's profession, philosophy was his vocation.

The celebrated British philosopher, Cambridge "Professor of Mental Philosophy and Logic" and Fellow of Trinity College, George Edward Moore wrote the preface for the book [Ram3]:

> He [Ramsey] was an extraordinarily clear thinker: no-one could avoid more easily than he the sort of confusions of thought to which even the best philosophers are liable, and he was capable of apprehending clearly and observing consistently, the subtlest

A. Soifer, *The Mathematical Coloring Book*,
DOI 10.1007/978-0-387-74642-5_30, © Alexander Soifer 2009

distinctions. He had, moreover, an exceptional power of drawing conclusions from a complicated set of facts: he could see what followed from them all taken together, or at least what might follow, in cases where others could draw no conclusions whatsoever. And, with all this, he produced the impression of also possessing the soundest common sense: his subtlety and ingenuity did not lead him, as it seems to have led some philosophers, to deny obvious facts. He had, moreover, so it seemed to me, an excellent sense of proportion: he could see which problems were the most fundamental, and it was these in which he was most interested and which he was most anxious to solve. For all these reasons, and perhaps for others as well, I almost always felt, with regard to any subject that we discussed, that he understood it much better than I did, and where (as was often the case) he failed to convince me, I generally thought the probability was that he was right and I was wrong, and that my failure to agree with him was due to lack of mental power on my part.

Indeed, Ramsey's philosophical essays impress me immensely by their depth, clarity, and common sense—a combination that reminds me the great Michel de Montaigne. Here is my favorite quotation from Ramsey [Ram5, p. 53]:

Knowledge is a correspondence between idea and fact.

Frank P. Ramsey's parents were Arthur Stanley Ramsey and Agnes Mary Wilson. In addition to Magdalene College's presidency, Arthur S. Ramsey was a tutor in mathematics. Frank was the oldest of four children, he had two sisters and a brother, Arthur Michael Ramsey, who much later became The Most Reverend Michael Ramsey, Archbishop of Canterbury (1961–1974). In 1925, Frank P. Ramsey married Lettice C. Baker, and their marriage produced two daughters. It is surprising to find in one family two brothers, Michael, the head of the Church of England and Frank, "a militant atheist," as Lettice described her husband.

The great economist John Maynard Keynes (1883–1946), who was then a Fellow of King's College and a close friend of Frank Ramsey, writes in March 1930 about Ramsey's contribution to economics [Key]:

He [Ramsey] has left behind him in print (apart from his philosophical papers) only two witnesses of his powers – his papers published in the *Economic Journal* on "A Contribution to the Theory of Taxation" in March 1927, and on "A Mathematical Theory of Saving" in December 1928. The latter of these is, I think, one of the most remarkable contributions to mathematical economics ever made, both in respect of the intrinsic importance and difficulty of its subject, the power and elegance of the technical methods employed, and the clear purity of illumination with which the writer's mind is felt by the reader to play about its subject.

Keynes also draws for us a portrait of Ramsey the man (ibid.):

His bulky Johnsonian frame, his spontaneous gurgling laugh, the simplicity of feelings and reactions, half-alarming sometimes and occasionally almost cruel in their directness and literalness, his honesty of mind and heart, his modesty, and the amazing, easy efficiency of the intellectual machine which ground away behind his wide temples and broad, smiling face, have been taken from us at the height of their excellence and before their harvest of work and life could be gathered in.

This portrait reminds me of Frank Ramsey's joking about his size while favoring human emotion over all issues of the universe (February 28, 1925):

Where I seem to differ from some of my friends is in attaching little importance to physical size. I do not feel the least humble before the vastness of the heavens. The stars may be large, but they cannot think of love; and those are qualities which impress me far more than the size does. I take no credit for weighing nearly seventeen stone.

Frank Plumpton Ramsey, aged 18. Reproduced by kind permission of the Provost and Scholars of King's College, Cambridge

By kind permission of the Provost and Scholars of King's College, Cambridge, I can share with you two photographs of the gentle giant, Frank Plumpton Ramsey. As Jacqueline Cox, Modern Archivist of King's College Library advises in her November 21, 1991 letter, "Both photographs come from the J. M. Keynes Papers (ref. JMK B/4). The first is a portrait of him at age the 18 in 1921 [page 283]. The second [page 284] shows him sitting on the ground in the open air reading a book at the age 25 in 1928. The photographers are not indicated, but in the case of the second photograph a note records that it was taken in the Austrian Tyrol in August 1928."

Considering his short life, Ramsey produced an enormous amount of work in logic, foundations of mathematics, mathematics, probability, economics, decision theory, cognitive psychology, semantics, and of course philosophy. Ramsey manuscripts, held in the Hillman Library of the University of Pittsburg, fill seven boxes and number about 1500 pages[10] [Ram5]. Probability fare is worthy of our attention. In his February 27, 1978 BBC radio broadcast (reprinted as an article [Mel] in 1995), Emeritus Professor of Philosophy at Cambridge D. H. Mellor explains:

> The economist John Maynard Keynes, to whom Braithwaite introduced Ramsey in 1921, published his *Treatise on Probability* in August of that year...It did not satisfy Ramsey, whose objections to it – some of them published before he was nineteen – were so cogent and comprehensible that Keynes himself abandoned it.

Frank Plumpton Ramsey, aged 25, Austrian Tyrol, August, 1928. Reproduced by kind permission of the Provost and Scholars of King's College, Cambridge

In fact, the Princeton Professor Emeritus of both Mathematics and Economics Harold W. Kuhn tells me that Keynes decided against continuing with mathematics because Ramsey was so much superior in it. Mellor continues:

> In this paper [Ram4], after criticizing Keynes, Ramsey went on to produce his own theory. This starts from the fact that people's actions are largely determined by what they believe and what they desire – and by strength of those beliefs and desires. The strength of people's beliefs is measured by the so-called 'subjective probability' they

---

[10] In *A Tribute to Frank P. Ramsey* [Har2], Frank Harary writes: "At her home, she [Mrs. Lettice Ramsey, the widow] showed me box upon box of notes and papers of Frank Ramsey and invited me to pore through them. As they dealt mostly with philosophy, I had to decline." As "a tribute," could Prof. Harary have shown more interest and curiosity?

attach to events. . .Subjective utility measures the strength of people's desires just as subjective probability measures the strength of their beliefs.

The problem is how to separate these two components of people's actions. . . One of the things Ramsey's paper did was to show how to extract people's subjective utilities and probabilities from the choices they make between different gambles; and by doing so it laid the foundations for the serious use of these concepts in economics and statistics as well as in philosophy.

It took a long time, however, from this 1926 paper of Ramsey's to bear fruit. Only after the publication in 1944 of a now classic book [NM] by John von Neumann and Oskar Morgenstern, *The Theory of Games and Economic Behavior*, did utility theory begin to catch on and be applied in modern decision theory and games theory. And for many years no one realized how much of it had been anticipated in Ramsey's 1926 paper.

I am looking at the classic 1944 book [NM] Mellor mentions above, written by the two celebrated Institute for Advanced Study and Princeton University members, John von Neumann (1903–1957) and Oskar Morgenstern (1902–1977) respectively, and at its later editions (Fine Library of Princeton-Math is very good). The authors cite many colleagues in the book: Daniel Bernoulli, Dedekind, Kronecker, D. Hilbert, F. Hausdorff, E. Zermelo, G. Birkhoff, E. Borel, W. Burnside, C. Carathéodory, W. Heisenberg, A. Speiser – and even Euclid. One missing name that merits credit the most is that of Frank P. Ramsey. Harold W. Kuhn tells me that in a 1953 letter he asked von Neumann why the latter gave no credit to Ramsey for inventing subjective probability. Indeed, this question and von Neumann's answer are reflected in H. W. Kuhn and A. W. Tucker's 1958 memorial article about von Neumann [KT, pp. 107–108]:

> Interest in this problem as posed [measuring "moral worth" of money] was first shown by F. P. Ramsey [Ram4] who went beyond Bernoulli in that he defined utility operationally in terms of individual behavior. (Once von Neumann was asked [by H. W. Kuhn] why he did not refer to the work of Ramsey, which might have been known to someone conversant with the field of logic. He replied that after Gödel published his papers on undecidability and the incompleteness of logic, he did not read another paper in symbolic logic.[11]

Ramsey's priority was discovered and acknowledged in print by others. In already mentioned D. H. Mellor's broadcast, the philosopher of probability Richard Carl Jeffrey (1926–2002; Ph.D. Princeton 1957; Professor of Philosophy at Princeton 1974–1999) says:

> It was when Leonard Savage, statistician, was working on his book on subjective probability theory, and he wished to find out what if anything the philosophers had to say on the subject, he went to Ramsey article [Ram4] and read it, and he found that what

---

[11] Indeed, von Neumann and Morgenstern probably did not expect Ramsey to publish on a topic far away from the foundations, such as economics, and thus might not have known about Ramsey's pioneering work by the time of the first 1944 edition of their celebrated book. However, new editions, which came out in 1947, 1953, 1961, etc., did not give Ramsey a credit either.

he [Ramsey] had done was to a great extend fairly describable as rediscovering another aspect of Ramsey's work in that article – the foundations of the theory of subjective probability. It was Savage's book, *The Foundations of Statistics*, that was published in 1954, that made subjectivism a respectable sort of doctrine for serious statistician to maintain; and the remarkable thing is that Ramsey in this little paper to the Moral Sciences Club in 1926 has done all of that already.

Indeed, Leonard Jimmie Savage (1917–1971) writes in 1954 [Sav, pp. 96–97]:

Ramsey improves on Bernoulli in that he defines utility operationally in terms of the behavior of a person constrained by certain postulates. . .

Why should not the range, the variance, and the skewness, not to mention countless other features, of the distribution of some function join with the expected value in determining preference? The question was answered by the construction of Ramsey and again by that of von Neumann and Morgenstern.

Richard C. Jeffrey writes [Jef, p. 35]:

This method of measurement [of desirability] was discovered by F. P. Ramsey and rediscovered by von Neumann and Morgenstern, through whose work it came to play its current role in economics and statistics.

More importantly, most of his 1965 book *The Logic of Decision* [Jef] is based on Ramsey's ideas, while one Chapter is simply called *Ramsey's Theory*.

Ramsey's first mathematical paper, *Mathematical Logic* [Ram1] appeared in 1926 in the midst of the *Grundlagenstreit* (Crisis in the Foundations), the confrontation between the two giants, David Hilbert and L. E. J. Brouwer, over the foundations of mathematics. Ramsey, who always addressed the most important issues of his day did not shy away from this one either. However, he did not, take either side. Ramsey did not agree with the intuitionist approach:

Weyl has changed his view and become a follower of Brouwer, the leader of what is called the intuitionist school, whose chief doctrine is the denial of the Law of Excluded Middle, that every proposition is either true or false. This is denied apparently because it is thought impossible to know such a thing *a priori*, and equally impossible to know it by experience. . . *Brouwer would refuse to agree that either it was raining or it was not raining, unless he had looked to see.*

Ramsey did not support Hilbert either:

I must say something of the system of Hilbert and his followers, which is designed to put an end to such skepticism once and for all. This is to be done by regarding higher mathematics as the manipulation of meaningless symbols according to fixed rules. We start with certain symbols called axioms: from these we can derive others by substituting certain symbols called constants for others called variables, and by proceeding from the pair of formulae p, if p then q to the formula q.

Mathematics proper is thus regarded as a sort of game, played with meaningless marks on paper rather like noughts and crosses; but besides this there will be another subject called metamathematics, which is not meaningless, but consists of real assertions about mathematics, telling us what this or that formula can or cannot be obtained from the axioms according to the rules of deduction. . .

Now, whatever else a mathematician is doing, he is certainly making marks on paper, and so this point of view consists of nothing but the truth; but it is hard to suppose it is the whole truth. There must be some reason for the choice of axioms. . .Again, it may be asked whether it is really possible to prove that the axioms do not lead to contradiction, since nothing can be proved unless some principles are taken for granted and assumed to lead to contradiction.

Summing up both Hilbert and Brouwer–Weyl approaches, Ramsey concluded:

We see then that these authorities, great as they are the differences between them, are agreed that mathematical analysis as originally taught cannot be regarded as a body of truth, but is either false or at best a meaningless game with marks on paper.

What was a mathematician to do? Ramsey was in favor of using the Axiom of Infinity. "As to how to carry the matter further, I have no suggestion to make; all I hope is to have made it clear that the subject is very difficult," wrote Ramsey in the end. (4 years later Ramsey would take a finitist view of rejecting the existence of any actual infinity.)

Ramsey came back with a specific approach in his second mathematical paper *On a Problem of Formal Logic* [Ram2], submitted on November 28, 1928, and published posthumously in 1930. This paper gives a clear and unambiguous start to what was later named the *Ramsey Theory*. What is the aim of this work? Fortunately, Ramsey answers this question right in the beginning of this paper:

This paper is primarily concerned with a special case of one of the leading problems of mathematical logic, the problem of finding a regular procedure to determine the truth or falsity of any given logical formula. But in the course of this investigation it is necessary to use certain theorems on combinations which have an independent interest and are most conveniently set out by themselves beforehand.

Indeed, Ramsey solves the problem in the special case, as he promises. However, little does he—or for that matter anyone else—expect that the next year, in 1931 another young genius, the 25-year-old Kurt Gödel will shock the mathematical world by publishing the (Second) Incompleteness Theorem [Göd1] that shows that Hilbert–Ackermann's *Entscheidungsproblem*, "the leading problem of mathematical logic" as Ramsey calls it, cannot have a solution in general case. Ramsey continues:

The theorems which we actually require concern finite classes only, but we shall begin with a similar theorem about infinite classes which is easier to prove and gives a simple example of the method of argument.

Yes, the infinite case here—as often happens—is easier than the finite, but is very well worth of the presentation (in fact, the finite case follows from the infinite by the de Bruijn–Erdős Compactness Theorem, as we have seen in Chapter 28). Later in the paper, Ramsey also observes that his infinite case requires the use of the Axiom of Choice:

Whenever universe is infinite we shall have to assume the axiom of selection.

In fact, some 40 years later, in 1969 Eugene M. Kleinberg [Kle] will prove that Ramsey's Theorem is independent from ZF, the Zermelo–Fraenkel set theory. (More precisely, if ZF is consistent, then Ramsey's theorem is not provable in ZF.)

As we have seen in Chapter 28, Frank P. Ramsey realizes— and clearly states— that his new pioneering method and his "theorems on combinations have an independent interest." Indeed, Ramsey's theorems deliver the principles and the foundation to a new field of mathematics, the *Ramsey Theory*. Now, this requires a certain clarification.

Three Ramsey Theory results appeared before Frank P. Ramsey erected its foundation, and is the reason why I combine these three early results under the name *Ramsey Theory before Ramsey*. They are Hilbert's Theorem of 1892, Schur's Theorem of 1916, and Baudet–Schur–Van der Waerden's Theorem of 1927. These classic results, which we will discuss in great detail in the next part, discovered particular properties of colored integers or colored spaces in particular circumstances. These theorems contributed real "meat" to the Ramsey Theory, real applications of the Ramsey Principle to particular contexts before Ramsey even formulated it!

Ramsey's amazing logical and philosophical gifts allowed him to abstract the idea from any particular context, to formulate his theorems as a *method*, a *principle* of the new theory—a great achievement indeed. Surely, Ramsey fully deserves his name to be placed on the new theory, whose principle he so clearly formulated and proved, but could anyone point out to who and when coined the term *Ramsey Theory*?

We have already seen *Ramsey's Theory of Decision* in Richard C. Jeffrey's 1965 book [Jef]. But we are after *The Ramsey Theory*, a new and flourishing branch of combinatorial mathematics. On July 21, 1995, I posed the question to the leader of the Ramsey Theory, Ronald L. Graham. Here is our brief exchange of the day:

Dear Ron:
Who and when coined the name "Ramsey Theory"?
Yours, Sasha

Sasha,
Beats me! Who first used the term Galois theory?
Ron

On January 22, 1996, I asked Ron again, and received another concise reply the same day:

Dear Sasha,
I would imagine that Motzkin may have used the term Ramsey Theory in the 60's. You might check with Bruce Rothschild at UCLA who should know.

Still the same day I received a reply from Bruce Rothschild:

Dear Alexander,

This is a good question, to which I have no real answer. I do not recall Motzkin using the phrase,[12] though he might have. I also don't recall hearing Rota use it when I was at MIT in the late 60's. My best recollection is that I began using the term informally along with Ron sometime in the very early '70's... But I could be way off here.

Frank Harary was less concise. On February 19, 1996, during a conference in Baton Rouge, Louisiana, he gave me a multi-page statement (you saw it in its entirety in Chapter 27), showing that Frank Harary and Václav Chvátal were the first to introduce the term *Generalized Ramsey Theory for Graphs* in their series of papers that started in 1972. I am looking at the first paper [CH] of the series: Chvátal, Václav, and Harary, Frank, *Generalized Ramsey theory for graphs*. The authors generalize the notion of the Ramsey number by including in the study graphs other than complete graphs. By doing so Harary and Chvatal open a new, now flourishing chapter, *Graph Ramsey Theory*. However, *The Ramsey Theory* as we understand it today stands for so much broader a body of knowledge, including Schur's, Van der Waerden's, and Hales-Jewett's Theorems that it does not completely fit inside Graph Theory. Thus, my search for the true birth of the name continued.

One 1971 survey [GR2], by Ronald L. Graham and Bruce L. Rothschild show a clear realization that a new theory has been born and needs an appropriate new name. Following a recitation of Ramsey's Theorem and Schur's Theorem, the authors write:

> These two theorems are typical of what we shall call a *Ramsey theorem* and a *Schur theorem*, respectively. In this paper we will survey a number of more general Ramsey and Schur theorems which have appeared in the past 40 years. It will be seen that quite a few of these results are rather closely related, e.g., van der Waerden's theorem on arithmetic progressions [Wae2], [Khi4], Rado's work on regularity and systems of linear equations [Rad1], [Rad2], the results of Hales and Jewett [HJ] and others [Garsia, personal communication] on arrays of points and Rota's conjectured analogue of Ramsey's Theorem for finite vector spaces, as well as the original theorems of Ramsey and Schur.

Yes, I agree that the new theory was created by 1971, and the choice of its name was between two deserving candidates: *The Schur Theory*, in honor of the main early contributor Issai Schur and his School (Schur's work was continued by his students Alfred Brauer and Richard Rado); and *The Ramsey Theory*, in honor of Frank P. Ramsey who formulated the principles of the new theory. Soon Graham and Rothschild arrived at the decision, and in their 1974 survey made the first published announcement of their choice [GR3]:

> Recently a number of striking new results have been proved in an area becoming known as RAMSEY THEORY. It is our purpose here to describe some of these. Ramsey Theory is a part of combinatorial mathematics dealing with assertions of a certain type, which we will indicate below. Among the earliest theorem of thus type

---

[12] Motzkin did not use "Ramsey Theory" in his 1960s articles, as I have verified shortly after.

are RAMSEY's theorem, of course, VAN DER WAERDEN's theorem on arithmetic progressions and SCHUR's theorem on solutions of $x + y = z$.

It seems that *The Ramsey Theory* has been shaping throughout the 1970s, and the central engine of this process was new results and the above mentioned surveys. In 1980 the long life of the name was assured when it appeared as the title of the book *Ramsey Theory* [GRS1] by three of the leading researchers of the field, Ronald L. Graham, Bruce L. Rothschild, and Joel H. Spencer. A decade later, the authors produced the second, updated edition [GRS2]. This book has not only assured the acceptance of the name—it has become the standard text in the new field of mathematics. It still remains the standard bearer today.[13]

Now is the time to share a bit of information about the co-creators of the term *Ramsey Theory*, who of course contributed much more than just the name.

Bruce Lee Rothschild was born on August 26, 1941 in Los Angeles. Following his B.S. degree from the California Institute of Technology in 1963, he earned a Ph.D. degree from Yale in 1967 with the thesis *A Generalization of Ramsey's Theorem and a Conjecture of Rota*, supervised by the legendary Norwegian graph theorist Øystein Ore (1899–1968). After 2 years 1967–1969 at MIT, Bruce became a professor at the University of California, Los Angeles, where he continues his work today. In 1972 Graham, Rothschild, and Leeb shared the Polya Prize of SIAM with Hales and Jewett.

Ronald Lewis Graham was born on October 31, 1935 in Taft, California. In 1962, he earned his Ph. D. degree from the University if California, Berkeley with the thesis *On Finite Sums of Rational Numbers*, supervised by Derrick Lehmer. Following decades as Director of Mathematical Sciences at Bell Laboratories, Ron moved South and West, and is now Irwin and Joan Jacobs Professor in the Department of Computer Science and Engineering at the University of California, San Diego. There is much to be said about this unique individual, who besides publishing well over 300 papers and several books, served as President of the American Mathematical Society (AMS), President of the Mathematical Association of America, and since 1996 is the Treasurer of the National Academy of Sciences. In 2003, AMS awarded Graham Steele Prize for lifetime achievement. I can attest to Ron's supreme elegance and depth as author and lecturer, and limitless energy in promoting the Ramsey Theory.

This certainly does not cover Ron's excellence in juggling ("juggling is a metaphor," he likes to say), fluency in Mandarin, friendship with Paul Erdős, etc. See all those on "Ronald Graham's special page" created by Ron's wife and well-known mathematician in her own rights Fan Chung at http://math.ucsd.edu/~fan/ron/.

Ron maintained a room for Paul Erdős in his New Jersey house, and took care of Erdős's finances. In the whole world, Ron knew best where Paul Erdős was on any given day, although as Ron's December 20, 1993 e-mail shows, even his knowledge was imperfect:

---

[13] I have only one problem with this beautiful book: today it sells for a whopping $199 at Wiley, its publisher, and on Amazon.com.

Sasha,
Erdős is staying with me for a while. During the night he is at (908)322–4111. During
the day it's anyone's guess where he will be!
Ron

Now that Paul passed on, Ron has graciously taken upon himself to keep the
tradition going by paying Erdős's prizes for first solutions of Erdős's problems,
but not all of them. So, all those interested in making a living by solving Erdős's
problems, pay attention to the small print in Ron's e-mail of February 12, 2007:

Hi Sasha,
I am willing to pay all the prizes offered by Paul that are listed in the book that Fan and
I wrote: Erdos on Graphs: His Legacy of Unsolved Problems. These we have checked.
The others (e.g., in number theory or set theory) are not (automatically) part of the
offer. I did have to pay $100 last year (the first time for a problem in this book that was
solved) to Jacques Verstraete in Canada!

Best regards,
   Ron Graham

30.2 Reflections on Ramsey and Economics, by Harold W. Kuhn

Why is it that sometimes people work together for decades and still remain
strangers, while in other instances friendship arrives at the first sight? This is a
question for psychologists to ponder. I should only observe that Harold W. Kuhn
and I instantly became friends in early 2003, when I arrived in Princeton-Math., just
as in 1988 when an instant friendship linked Paul Erdős and I. It has always been
intellectually stimulating to discuss any subject with Harold, from mathematics to
the cinema of Michelangelo Antonioni, and from African Art to Pierre Bonnard's
drawings.

Harold William Kuhn was born in Santa Monica, California on July 29, 1925.
Following his B.S. degree in 1947 from the California Institute of Technology, he
earned a Ph.D. degree from Princeton University in 1950, while also serving as
Henry B. Fine Instructor in the Mathematics Department, 1949–1950. Following
a professorship at Bryn Mawr, 1952–1958, Harold has been a Professor of Mathematical Economics at Princeton's two departments, Mathematics and Economics,
becoming Emeritus in 1995. His honors include presidency of the Society for Industrial and Applied Mathematics (1954–55), service as Executive Secretary of the
Division of Mathematics of the National Research Council (1957–1960), John von
Neumann Theory Prize of the Operation Research Society of America (1982; jointly
with David Gale and A. W. Tucker), and Guggenheim Fellowship (1982). It was
Harold Kuhn who nominated John F. Nash Jr. for Nobel Prize (awarded in 1994).

In the fall of 2006, upon my return to Princeton University, I asked myself, who
could best evaluate Frank P. Ramsey's works on economics? It would take an expert
on mathematics and economics. Harold was the only choice, and he most generously
agreed. Best of all, Harold wrote a triptych about F. P. Ramsey, John von Neumann

and John F. Nash Jr., especially for this book. In all that follows in this section, the podium—shall I say, the pages—belong to Harold W. Kuhn. ∎

Although mathematics became the *lingua franca* of 20th century economics, only a handful of mathematicians have exerted a direct and lasting influence on the subject. They surely include Frank Plumpton Ramsey, John von Neumann, and John Forbes Nash Jr. The similarities and differences in their life trajectories are striking. Ramsey died at 26 years of age after an exploratory liver operation following a bout of jaundice, while Nash's most productive period ended when he fell prey to schizophrenia at the age of 30. Von Neumann's original work on game theory and growth models was done before he was 30 years old. For all three, the work in economics appears as a sideline. Ramsey's friend and biographer, Richard Braithwaite has written: "Though mathematical teaching was Ramsey's profession, philosophy was his vocation," without mentioning his contributions to economics at all or including the three papers on economics in the posthumous "complete" works that Braithwaite edited. The contributions of von Neumann to mathematical economics is but one chapter in the seven chapters comprising the memorial issue devoted to von Neumann's research and published as a special issue of the Bulletin of the American Mathematical Society. Regarding Nash, John Milnor considered "...Nash's [Nobel Economics] prize work [to be] an ingenious but not surprising application of well-known methods, while his subsequent mathematical work was much more rich and important."

Ramsey, von Neumann, and Nash came from very different backgrounds and had very different relationships to the economics and the economists of their day. Ramsey, an intimate friend of Bertrand Russell and Wittgenstein, was a Cambridge man by birth. He appears to have been interested in economics from the age of 16 and wrote his first published piece on economics at 18. He had close personal and professional contacts with such well-known economists as John Maynard Keynes, Arthur Pigou, Piero Sraffa, and Roy Harrod. He served as an advisor to the Economic Journal, where Keynes took his counsel most seriously. He was well acquainted with the trends in economic theory of his day.

Von Neumann, the scion of a Jewish banking family in Budapest, had a wide circle of intellectual friends from Budapest, Berlin, and Vienna that included economists such as William Fellner (who was a friend from gymnasium days) and Lord Nicholas Kaldor, who gave von Neumann a reading list in contemporary economics in the 1920s, and who arranged for an English translation of von Neumann's growth model to be published in the Review of Economic Studies in 1945. Thus there is ample evidence that von Neumann was well-informed of the state of economics throughout his life.

The case of John Nash, who grew up in the coal mining and railroad town of Bluefield, West Virginia, is very different. When he came to Princeton to do graduate work in mathematics at the age of 20, he had taken one undergraduate course in economics (on International trade) at Carnegie Tech, taught by an Austrian émigré, Bert Hoselitz. His major contribution on bargaining, which appears to have had its origin in this course, has two boys (Bill and Jack) trading objects such as a whip, a bat, a ball, and a knife. This was the work of a teenager. There is no evidence that Nash had read any contemporary economist outside the required readings of his one undergraduate course. Of course, later in his life, in the period when he was on the faculty at the Massachusetts Institute of Technology, he had contact with Paul Samuelson and Robert Solow, Nobel Prize winners in Economics, who knew of his work in game theory. Nash's only later excursion into economics is a theory of "ideal money," an idea that appears to have been anticipated in part by Friederich Hayek.

Now that game theory has become part of the economist's tool kit, anyone who takes an introductory economics course learns about the contributions of von Neumann and Nash. Ramsey's work, however, is less well-known and the principal reason for this note is to give the reader an appreciation for the contributions of Ramsey to economics. Between the ages of 18 and 29, Ramsey wrote four papers, which we shall discuss in detail below.

(A) "The Douglas Proposals," *The Cambridge Magazine*, Vol. XI, No. 1, January 1922, pp. 74–76.

Ramsey's first work related to economics (A) was published when he was 18. He was no common 18-year-old; here is how Keynes described him: "From a very early age, about 16 I think, his precocious mind was intensely interested in economic problems." The Cambridge Magazine was edited by C. K. Ogden, a Fellow of Magdalene College where Ramsey's father was President, from 1912 to 1922. Ramsey and Ogden met while Ramsey was still a student in his public school, Winchester, and Ogden persuaded him to study the then much-discussed social credit proposals of a certain Major Douglas. I. A. Richards recalled the upshot: "Soon after he'd done the Douglas credit thing, you know, A. S. Ramsey, his father, called up Ogden and said 'What have you been doing to Frank?', and Ogden said 'What's he been doing?'. 'Oh he's written a paper on Douglas Credit which would have won him a Fellowship in any University anywhere in the world instantly. It's a new branch of mathematics'."

Who was this Major Douglas? Briefly, he was one of those crackpots who exist on the fringe of academic economics and whose theories promise a redistribution of wealth that appealed to a large part of the public (including, in Douglas's case, Ezra Pound and T. S. Eliot). Like many of those offering a panacea for the Great Depression, he was also an anti-Semite who invoked the theses expounded in the Protocols of the Elders of Zion in defense of his economic theories.

What was Major Douglas's heresy that Ramsey demolished? It is centered on the so-called A + B "theorem" (called by Keynes "mere mystification"). In producing a good, price is made up of two parts of the cost paid out by the producer: A equals the amount paid out for raw materials and overhead and B equals the sums paid out in wages, salaries, and dividends. According to Douglas, the amount B, paid to the consumers, is never sufficient to buy all of the good, whose cost (and price) is A+B. Therefore, the state should make up the difference through "social credit."

Ramsey first provides a verbal argument that shows that, in a stationary state, the total rate of distribution of purchasing power (taking into account payments originating in intermediate goods) equals the rate of flow of costs of consumable goods. He then writes:

". . .it is possible, using some complicated mathematics to show that the ratio is unity under much wider conditions which allow for changes in the quantity of production, in the rate of wages, in the productivity of labor, and in the national wealth." The "complicated mathematics," other than Ramsey's curiously rigid set of modeling assumptions, consists of the use of "integration by parts," a technique taught to every beginning student of the calculus.

(B)   "A Contribution to the Theory of Taxation," *The Economic Journal*, Vol. XXXVII, March 1927, pp. 47–61.

The young Ramsey assisted A. C. Pigou, who was the successor to Alfred Marshall in the chair of Political Economy at Cambridge, on a number of occasions beginning before 1926. After providing Pigou with a mathematical proposition and examples for two articles, one on credit and one on unemployment, Ramsey assisted Pigou with changes in the third edition of *The Economics of Welfare*, published in 1929. However, it appears that Ramsey's work on taxation (B) was inspired by questions raised in Pigou's *A Study in Public Finance*.

The problem posed by Ramsey in (B) was to find an optimal system of taxation of n commodities so as to raise a given quantity of revenue. For Ramsey in (B), "optimal" means minimizing aggregate sacrifice. Using this objective function, he shows that the production of each commodity should be reduced in the same proportion, thus a system of differential taxation. The mathematics employed is rather standard, namely, optimization under equality constraints using Lagrange multipliers which was taught to mathematicians of this period by treatises such as de la Vallee Poussin's *Cours d'Analyse*. The treatment is careful for the period and Ramsey includes a number of examples of potential applications of his results. Of particular interest is a discussion of the application of income tax to savings, a subject that I believe was part of a larger research agenda that Ramsey had formulated.

(C)   "A Mathematical Theory of Saving," *The Economic Journal*, Vol. XXXVIII, December 1928, pp. 543–549.

Papers (B) and (C) were published in the Economic Journal which Keynes controlled with an iron hand. Keynes wrote of (C) that it "is, I think, one of the most remarkable contributions to mathematical economics ever made, both in respect of the intrinsic importance and difficulty of its subject, the power and elegance of the technical methods employed, and the clear purity of illumination with which the writer's mind is felt by the reader to play about its subject." The article (C) is concerned with the derivation of optimal saving programs under a variety of conditions. Samuelson captures the spirit of the paper in the society in which it was created when he wrote: "Frank Ramsey, living in a happier age and being a Cambridge philosopher assumed society would last forever and seek to maximize the utility of its consumption over all infinite time." A major stumbling block immediately presents itself in that the "utility of its consumption over all infinite time" is an improper integral which, in general, will not have a finite maximum value. Ramsey proposed an elegant device to get around this problem. He assumed that there was a maximum amount of attainable utility (called "bliss") and, instead of maximizing the improper integral he minimized the deviation from bliss over the infinite horizon.

Ramsey then derives a result that is easy to express in common English, namely: "The optimal rate of saving multiplied by the marginal utility of consumption should always equal the difference between bliss and the actual rate of utility enjoyed." The paper contains a derivation of this result by simple verbal reasoning provided by Keynes (which does not apply to the most general cases considered by Ramsey but which does give the non-mathematically adept, the feeling of "understanding the result"). Contemporary mathematical economists will instantly recognize the problem as one to which the calculus of variations applies and, indeed, over 30 years after Ramsey wrote (C) such techniques took over the theoretical models of growth. We can say with real justice that Ramsey was "ahead of his time."

Recently, three economic historians (D. A. Collard, M. Gaspard, and P. C. Duarte) have put forth a very persuasive theory (based largely on unpublished notes of Ramsey that are archived at the University of Pittsburg) that Ramsey's two papers on taxation and savings were not isolated works of a mathematician answering questions put to him by economists but were rather part of an over-arching research program that Ramsey had clearly in mind. If this plausible theory is true, it makes his early death even more tragic.

(D) "Truth and Probability," in R. B. Braithwaite (ed.), *The Foundations of Mathematics and Other Logical Essays*, London: Routledge and Kegan Paul, 1931, pp. 156–198. Reprinted in H. E. Kyburg and H. E. Smokler (eds.) *Studies in Subjective Probability*, New York: Wiley 1964, pp. 61–92.

In modeling the decisions of an individual who chooses an alternative from a set of uncertain outcomes, it has long been the tradition to introduce a numerical function to measure the objective of the individual involved. When von Neumann first formulated "the most favorable result" for a player in a strategic game, he identified "the most favorable result" with "the greatest expected monetary value," remarking that this or some similar assumption was necessary in order to apply the methods of probability theory. While doing so, he was well aware of the objections to the principal of maximizing expected winnings as a prescription for behavior, but wished to concentrate on other problems. The St. Petersburg paradox illustrates in clear terms the fact that the principle of maximizing expected winnings does not reflect the actual preferences of many people.

To resolve this paradox, Daniel Bernoulli suggested that people do not follow monetary value as an index for preferences but rather the "moral worth" of the money. He then proposed a quite serviceable function to measure the moral worth of an amount of money, namely, its logarithm. Whatever the defects of this function as a universal measure of preferences, and they are many, it raises the question of the existence of a numerical index which will reflect accurately the choices of an individual in situations of risk. Interest in this problem was first shown by Ramsey in (D) in which he defined utility operationally in terms of individual behavior. As Mellor has written: "In this paper (D), after criticizing Keynes, Ramsey went on to produce his own theory. This starts from the fact that people's actions are largely determined by what they believe and what they desire, and by strength of those beliefs and desires. The strength of people's beliefs is measured by the so-called subjective probability' they attach to events. . .Subjective utility measures the strength of people's desires just as subjective probability measures the strength of their beliefs. The problem is how to separate these two components of people's actions. One of the things Ramsey's paper did was to show how to extract people's subjective utilities and probabilities from the choices they make between different gambles; and by doing so it laid the foundations for the serious use of these concepts in economics and statistics as well as in philosophy."

The bible of game theory, *The Theory of Games and Economic Behavior* by von Neumann and Morgenstern which confronts similar problems contains no reference to the work of Ramsey. When von Neumann was queried about this omission, he explained it by saying that, after Goedel published his papers on undecidability and the incompleteness of logic, he did not read another paper in symbolic logic. Although his excuse is strengthened by the fact that (D) first appeared in the volume that Braithwaite edited after Ramsey's death, no such excuse exists for Morgenstern, when he wrote *"Some Reflections on Utility"* in 1979 and cites two articles by J. Pfanzagl while overlooking Ramsey's paper (D) and Savage's *The Foundations of Statistics*.

Aside from Ramsey's paper on Major Douglas, which was an exemplary mathematical model refuting errant nonsense, he has clear precedence in four major themes of 20th century economics. The paper on taxation (B) was a source for both public finance theorists and for monetary economists who have characterized inflation as a tax on money holdings and have formulated optimal inflation policies as optimal taxation schemes. The paper on savings (C) has become the touchstone for economists working on growth. The fourth area is the theory of expected utility and decisions under risk which has used in an essential way Ramsey's insights on subjective probability in (D).

I have been a friend of John Nash since he arrived in Princeton in 1948. I knew John von Neumann from 1948 until his death in 1957. I very much regret not having known Frank Ramsey. Given the modernity of his work, it is hard to grasp the fact that he died over 77 years ago.

VII
Colored Integers: Ramsey Theory Before Ramsey and Its AfterMath

History will be written many different ways. Look out, the Chinese are coming, the Chinese are coming and they will write history from their perspective and many things we believe are important facts will not matter to them.
— Thomas L. Saaty[1]

[1] E-mail to A. Soifer, April 13, 1998.

31
Ramsey Theory Before Ramsey: Hilbert's Theorem

A new theory is an attempt to answer new questions, or to shed a new light on old problems. It is not usually born overnight. Before its birth, a new mathematical theory usually grows unnoticed within old and well-established branches of mathematics. Ramsey Theory was not an exception. Its roots go back decades before the 1930 pioneering paper of Frank Plumpton Ramsey saw the light of day after his untimely passing at the age of 26. As far as we know today, the first Ramseyan-type result appeared in 1892 as a little noticed assertion in [Hil]. Its author was the great David Hilbert. In this work Hilbert proved the theorem of our interest merely as a tool for his study of irreducibility of rational functions with integral coefficients.

A set $Q_n (a, x_1, x_2, \ldots, x_n)$ of integers is called an *n-dimensional affine cube* if there exist $n + 1$ positive integers $a, x_1, \ldots, x_n$ such that

$$Q_n (a, x_1, x_2, \ldots, x_n) = \left\{ a + \sum_{i \in F} x_i : \varnothing \neq F \subseteq \{1, 2, \ldots, n\} \right\}.$$

In this chapter and the rest of the book it is convenient to use the symbol $[n]$ for the starting segment of positive integers:

$$[m] = \{1, 2, \ldots, m\}.$$

This theorem, which preceded the Schur (Chapter 32) and the Baudet–Schur–Van der Waerden (Chapter 33) Theorems, reads as follows.

Hilbert's Theorem 31.1 For every pair of positive integers r, n, there exists a least positive integer $m = H(r, n)$ such that in every r-coloring of $[m]$ there exists a monochromatic n-dimensional affine cube.

Proof easily follows from the Baudet–Schur–Van der Waerden Theorem (Chapter 33): the arithmetic progression (AP) $\{a, a + x, a + 2x, \ldots, a + nx\}$ is precisely the cube $Q_n (a, x_1, x_2, \ldots, x_n)$ with $x_1 = x_2 = \ldots = x_n = x$. Of course, this is not Hilbert's proof, for his paper preceded Van der Waerden's paper by 35 years. ∎

A. Soifer, *The Mathematical Coloring Book*,
DOI 10.1007/978-0-387-74642-5_31, © Alexander Soifer 2009

This is a Ramseyan theorem, as it asserts a property invariant under all
r-colorings of a certain set, in this case the initial segment $[m]$ of the set of positive
integers.

Nearly 100 years later, in 1989, Paul Erdős, András Sárközy, and Vera T. Sós published [ESS] a generalization of Hilbert's Theorem. They called it aptly "a density
version" of Hilbert's Theorem.

Density Version of Hilbert's Theorem 31.2 [ESS] For every positive integer n
there is a number $m_0 = H(n)$ such that for any $m > m_0$, $B \subseteq [m]$ with
$|B| > 3m^{1-2^{-n}}$, there exist distinct positive integers $a, x_1, x_2, \ldots, x_n$ such that all
2^n sums forming the n-dimensional affine cube $Q_n(a, x_1, x_2, \ldots, x_n)$ belong to B.

Hilbert's place as one of the world's leading mathematicians at the turn of the
twentieth century had certainly not been won by this result. He did not come back to
Ramseyan style mathematics (unlike Issai Schur, as we will see in the following few
chapters). Still, in the style of this book I would like to briefly describe Hilbert's life.
I refer the reader to Hilbert's celebrated biography by Constance Reid [Reid] and
Herman Weil's paper *David Hilbert and his mathematical work*, contained therein
for a much worthier narration.

Hilbert was born near Königsberg (currently Kaliningrad, Russia) in Wehlau
(currently Znamensk). In 1885 he obtained his Ph.D. degree at the University of
Königsberg under Ferdinand von Lindemann. Following 10 years at Königsberg, he
moved to the University of Göttingen where he remained for the rest of his life.

Hilbert made major contributions to numerous areas of mathematics and physics.
In 1900, at the International Congress of Mathematicians in Paris, he presented a set
of problems, known as The 23 Hilbert's Problems (during the talk he was able to
articulate 10 of them) that profoundly influenced the development of mathematics in
the twentieth century. The problems included questions related to Cantor's Continuum Hypothesis and Zermelo's Axiom of Choice (Problem 1), the provability of the
consistency of axioms for logic (Problem 2), the possibility of the axiomatization of
physics (Problem 6), and The Riemann Hypothesis (Problem 8).

Following Felix Klein, Hilbert made Göttingen the world's premier center of
mathematics. However, he lived to see Göttingen's fall, when almost immediately
following Hitler's early 1933 rise to power: many leaders of mathematics and
physics had to leave the University and the country.

Constance Reid [Reid] conveys how Hilbert must have felt:

Sitting next to the Nazi's newly appointed minister of education [Bernard Rust]
at a banquet, he [Hilbert] was asked, "And how is mathematics at Göttingen now
that it has been freed of the Jewish influence?"

"Mathematics in Göttingen?" Hilbert replied. "There is really none any more."

Hilbert passed away in Göttingen on February 14, 1943.

32

Ramsey Theory Before Ramsey: Schur's Coloring Solution of a Colored Problem and Its Generalizations

32.1 Schur's Masterpiece

Probably no one remembered—if anyone ever noticed—Hilbert's 1892 lemma by the time the second Ramseyan type result appeared in 1916 as a little noticed assertion in number theory. Its author was Issai Schur.

Our interest here lies in the result he obtained during 1913–1916 when he worked at the University of Bonn as the successor to Felix Hausdorff.[2] There he wrote his pioneering paper [Sch]: *Über die Kongruenz $x^m + y^m \equiv z^m$ (mod. p)*. In it Schur offered another proof of a theorem by Leonard Eugene Dickson from [Dic1], who was trying to prove Fermat's Last Theorem. For use in his proof, Schur created, as he put it, "a very simple lemma, which belongs more to combinatorics than to number theory."

Nobody then asked questions of the kind Issai Schur posed and solved in his 1916 paper [Sch]. Consequently, nobody appreciated this result much when it was published. Now it shines as one of the most beautiful, classic theorems of mathematics. Its setting is positive integers colored in finitely many colors. The beautiful solution I am going to present utilizes coloring as well. I have got to tell you how I received this solution (see [Soi9] for more details).

In August 1989 I taught at the International Summer Institute in Long Island, New York. A fine international contingent of gifted high school students for the first time included a group from the Soviet Union. Some members of this group turned out to be Mathematics Olympiads "professionals," winners of the Soviet Union National Mathematical Olympiads in Mathematics and in Physics. There was nothing in the Olympiad genre that they did not know or could not solve. I offered them and everyone else an introduction to certain areas of combinatorial geometry.

---

[2] Both Alfred Brauer [Bra2] and Walter Ledermann [Led] reported 1911 as the time when Schur became an *Extraordinarius* in Bonn, while Schur's daughter Mrs. Hilde Abelin-Schur [Abe1] gave me 1913 as the time her family moved to Bonn. The Humboldt University's Archive contains personnel forms (Archive of Humboldt University at Berlin, document UK Sch 342, Bd.I, Bl.25) filled up by Issai Schur himself, from which we learn that he worked at the University of Bonn from April 21, 1913 until April 1, 1916, when he returned to Berlin.

A. Soifer, *The Mathematical Coloring Book*,
DOI 10.1007/978-0-387-74642-5_32, © Alexander Soifer 2009

We quickly reached the forefront of mathematics, full of open problems. Students shared with me their favorite problems and solutions as well. Boris Dubrov from Minsk, Belarus, told me about a visit to Moscow by the American mathematician Ronald L. Graham. During his interview with the Russian mathematics magazine *Kvant*, Graham mentioned a beautiful problem that dealt with 2-colored positive integers. Boris generalized the problem to *n*-coloring, strengthened the result and proved it all! He gave me this generalized problem for the Colorado Mathematical Olympiad.

This problem was the celebrated Schur Theorem of 1916, rediscovered by Boris, with his own proof that was more beautiful than Schur's original proof, but which was already known. Paul Erdős received this proof from Vera T. Sós, and included it in his talk at the 1970 International Congress of Mathematicians in Nice, France [E71.13]. Chances of receiving a solution of such a problem during the Olympiad were very slim. Yet, the symbolism of a Soviet kid offering an astonishingly beautiful problem (and solution!) to his American peers was so great that I decided to include this problem as an additional Problem 6 (Colorado Mathematical Olympiad usually offers 5 problems).

Schur's Theorem 32.1 ([Sch]) For any positive integer n there is an integer $S(n)$ such that any n-coloring of the initial positive integers array $[S(n)]$ contains integers a, b, c of the same color such that $a + b = c$.

In this case we call a, b, c a *monochromatic solution* of the equation $x + y = z$. In fact, Schur proved by induction that $S(n) = n!e$ would work.[3]

Proof of Schur's Theorem Let all positive integers be colored in n colors $c_1, c_2, \ldots, c_n$. Due to Problem 27.13, there is $S(n)$ such that any n-coloring of edges of the complete graph $K_{S(n)}$ contains a monochromatic triangle K_3.

Construct a complete graph $K_{S(n)}$ with its vertices labeled with integers from the initial integers array $[S(n)] = \{1, 2, \ldots, S(n)\}$. Now color the edges of $K_{S(n)}$ in n colors as follows: let i and j, $(i > j)$, be two vertices of $K_{S(n)}$, color the edge ij in precisely the color of the integer $i - j$ (remember, all positive integers were colored in n colors!). We get a complete graph $K_{S(n)}$ whose edges are colored in n colors. By Problem 27.13, $K_{S(n)}$ contains a triangle ijk, $i > j > k$, whose all three edges ij, jk, and ik are colored in the same color (Fig. 32.1).

Fig. 32.1

---

[3] Here e stands for the base of natural logarithms.

Denote $a = i - j$; $b = j - k$; $c = i - k$. Since all three edges of the triangle ijk are colored in the same color, the integers a, b, and c are colored in the same color in the original coloring of the integers (this is how we colored the edges of $K_{s(n)}$). In addition, we have the following equality:

$$a + b = (i - j) + (j - k) = i - k = c$$

We are done! ∎

The result of the Schur Theorem can be strengthened by an additional clever trick in the proof.

Strong Version of Schur's Theorem 32.2 For any positive integer n there is an integer $S^*(n)$ such that any n-coloring of the initial positive integers array $[S^*(n)]$ contains distinct integers a, b, c of the same color such that $a + b = c$.

Proof Let all positive integers be colored in n colors $c_1, c_2, \ldots, c_n$. We add n more colors $c'_1, c'_2, \ldots, c'_n$ different from the original n colors and construct a complete graph $K_{S(2n)}$ with the set of positive integers $\{1, 2, \ldots, S(2n)\}$ labeling its vertices (See the definition of $S(2n)$ in the proof of Theorem 32.1). Now we are going to color the edges of $K_{S(2n)}$ in $2n$ colors.

Let i and j, $(i > j)$, be two vertices of $K_{S(2n)}$, and c_p be the color in which the integer $i - j$ is colored, $1 \le p \le n$ (remember, all positive integers are colored in n colors $c_1, c_2, \ldots, c_n$). Then we color the edge ij in color c_p if the number $\left\lfloor \frac{i}{i-j} \right\rfloor$ is even, and in color c'_p if the number $\left\lfloor \frac{i}{i-j} \right\rfloor$ is odd (for a real number r, the symbol $\lfloor r \rfloor$, as usual, denotes the largest integer not exceeding r).

We get a complete graph $K_{S(2n)}$ whose edges are colored in $2n$ colors. By Problem 27.13, $K_{S(2n)}$ contains a triangle ijk, $i > j > k$, whose all three edges ij, jk, and ik are colored in the same color (Fig. 32.1).

Denote $a = i - j$; $b = j - k$; $c = i - k$. Since all three edges of the triangle ijk are colored in the same color, from the definition of coloring of edges of $K_{S(2n)}$ it follows that in the original coloring of positive integers, the integers a, b, and c were colored in the same color. In addition we have

$$a + b = (i - j) + (j - k) = i - k = c.$$

We are almost done. We only need to show (our additional pledge!) that the numbers a, b, c are all distinct. In fact, it suffices to show that $a \neq b$. Assume the opposite: $a = b$ and c_p is the color in which the number $a = b = i - j = j - k$ is colored. But then

$$\left\lfloor \frac{i}{i-j} \right\rfloor = \left\lfloor 1 + \frac{j}{i-j} \right\rfloor = 1 + \left\lfloor \frac{j}{i-j} \right\rfloor = 1 + \left\lfloor \frac{j}{j-k} \right\rfloor,$$

i.e., the numbers $\left\lfloor \frac{i}{i-j} \right\rfloor$ and $\left\lfloor \frac{j}{j-k} \right\rfloor$ have different parity, thus the edges ij and jk of the triangle ijk must have been colored in different colors. This contradiction to the fact that all three edges of the triangle ijk have the same color proves that $a \neq b$. Theorem 32.2 is proven. ■

32.2 Generalized Schur

It is fitting that the Schur Theorem was generalized by one of Schur's best students—Richard Rado. Rado calls a linear equation

$$a_1x_1 + a_2x_2 + \ldots + a_nx_n = b \qquad (*)$$

regular, if for any positive integer r, no matter how all positive integers are colored in r colors, there is a monochromatic solution of the equation $(*)$. As before, we say that a solution $x_1, x_2, \ldots, x_n$ is *monochromatic*, if all numbers $x_1, x_2, \ldots, x_n$ are colored in the same color.

For example, the Schur Theorem 32.1 proves precisely that the equation $x + y - z = 0$ is regular. In 1933 Richard Rado, among other results, found the following criterion:

Rado's Theorem 32.3 (A particular case of [Rad1]) Let E be a linear equation $a_1x_1 + a_2x_2 + \ldots + a_nx_n = 0$, where all $a_1, a_2, \ldots, a_n$ are integers. Then E is regular if and only if some non-empty subset of the coefficients a_i sums up to zero.

For example the equation $x_1 + 3x_2 - 2x_3 + x_4 + 10x_5 = 0$ is regular because $1 + 3 - 2 = 0$.

Problem 32.4 (*trivial*) Schur Theorem 32.1 follows from Rado's Theorem.

Richard Rado found regularity criteria for systems of homogeneous equations as well. His fundamental contributions to and influence on Ramsey Theory is hard to overestimate. I have just given you a taste of his theorems here. For more of Rado's results read his papers [Rad1], [Rad2], and others, and the monograph [GRS2]. Instead of a formal biographical data, I prefer to include here a few passages about Richard Rado (1906, Berlin—1989, Henley-on-Thames, Oxfordshire) written by someone who knew Rado very well—Paul Erdős—from the latter's paper *My joint work with Richard Rado* [E87.12]:

> I first became aware of Richard Rado's existence in 1933 when his important paper *Studien zur Kombinatorik* [Rado's Ph.D. thesis under Issai Schur] [Rad1] [4] appeared. I thought a great deal about the many fascinating and deep unsolved problems stated in this paper but I never succeeded to obtain any significant results here and since I have to report here about our joint work I will mostly ignore these questions. Our joint work extends to more than 50 years; we wrote 18 joint papers, several of them jointly with A. Hajnal, three with E. Milner, one with F. Galvin, one with Chao Ko, and we have a book on partition calculus with A. Hajnal and A. Mate. Our most important work is

---

[4] Two years later, Rado obtained his second Ph.D. degree at Cambridge under G. H. Hardy.

undoubtedly in set theory and, in particular, the creation of the partition calculus. The term partition calculus is, of course, due to Rado. Without him, I often would have been content in stating only special cases. We started this work in earnest in 1950 when I was at University College and Richard in King's College. We completed a fairly systematic study of this subject in 1956, but soon after this we started to collaborate with A. Hajnal, and by 1965 we published our GTP (Giant Triple Paper - this terminology was invented by Hajnal) which, I hope, will outlive the authors by a long time. I would like to write by centuries if the reader does not consider this as too immodest...

I started to correspond with Richard in late 1933 or early 1934 when he was a German [Jewish] refugee in Cambridge. We first met on October 1, 1934 when I first arrived in Cambridge from Budapest. Davenport and Richard met me at the railroad station in Cambridge and we immediately went to Trinity College and had our first long mathematical discussion...

Actually our first joint paper was done with Chao Ko and was essentially finished in 1938. Curiously enough it was published only in 1961. One of the reasons for the delay was that at that time there was relatively little interest in combinatorics. Also, in 1938, Ko returned to China, I went to Princeton and Rado stayed in England. I think we should have published the paper in 1938. This paper "Intersection theorems for systems of finite sets" became perhaps our most quoted result.

It is noteworthy to notice how differently people see the same fact. For Richard Rado, Schur's Theorem was about monochromatic solutions of a homogeneous linear equation $x + y - z = 0$, and so Rado generalized the Schur Theorem to a vast class of homogeneous linear equations (Rado's Theorem 32.3) and systems of homogeneous linear equations [Rad1]. Three other mathematicians saw Schur's Theorem quite differently. This group consisted of Jon Folkman, a young Rand Corporation scientist; Jon Henry Sanders, the last Ph.D. student of the legendary Norwegian graph theorist Øystein Ore at Yale (B.A. 1964 Princeton University; Ph.D. 1968, Yale University); and Vladimir I. Arnautov, a 30-year-old Moldavian topological ring theorist. For the three, the Schur Theorem spoke about monochromatic sets of symmetric sums

$$\{a_1, a_2, a_1 + a_2\} = \left\{ \sum_{i=1,2} \varepsilon_i a_i : \varepsilon_i = 0, 1; \varepsilon_1 \varepsilon_2 \neq 0 \right\}.$$

Consequently, the three proved a different —from Rado's kind —Schur's Theorem generalization and paved the way for further important developments. I see therefore no choice at all but to name the following fine theorem by its three inventors. This may surprise those of you accustomed to different attributions. I will address these concerns later in this chapter.

Arnautov–Folkman–Sander's Theorem 32.5 ([San], [Arn]) For any positive integers m and n there exists an integer $AFS(m, n)$ such that any m-coloring of the initial integers array $[AFS(m, n)]$ contains an n-element subset $S \subset [AFS(m, n)]$ such that the set $\left\{ \sum_{x \in F} x : \varnothing \neq F \subseteq S \right\}$ is monochromatic.

Problem 32.6 (*trivial*) Show that both Hilbert's Theorem 31.1 and Schur's Theorem 32.1 follow from Arnautov–Folkman–Sander's Theorem 32.5.

In their important 1971 paper [GR1] Ron Graham and Bruce Rothschild, having vastly generalized some theorems of the Ramsey Theory (beyond the scope of this book), conjectured that the word "finite" in reference to the subset S in Arnautov–Folkman–Sander's Theorem 32.5 can be omitted. Paul Erdős gave a high praise to their conjecture at his 1971 talk in Fort Collins, Colorado, published in 1973 [E73.21]:

> Graham and Rothschild ask the following beautiful question: split the integers into two classes. Is there always an infinite sequence so that all the finite sums $\sum \varepsilon_i a_i$, $\varepsilon_i = 0$ or 1 (not all $\varepsilon_i = 0$) all belong to the same class? . . . This problem seems very difficult.

Surprisingly, the proof in the positive came in soon. In the paper submitted in 1972 and published in 1974 [Hin], Neil Hindman proved Graham–Rothschild's conjecture. While Graham–Rothschild's conjecture asked for an initial generalization for two colors (probably to test waters before diving into the general case), Hindman proved the result for any finite number of colors, thus fully generalizing Arnautov–Folkman–Sander's Theorem.

Hindman's Theorem 32.7 (Hindman [Hin]) For any positive integer n any n-coloring of the set of positive integers N contains an infinite subset $S \subseteq N$ such that the set $\left\{ \sum_{x \in F} x : \varnothing \neq F \subset S; |F| < \aleph_0 \right\}$ is monochromatic.

Let us now go back and establish the most appropriate credit for Theorem 32.5. It is called Folkman–Rado–Sanders' Theorem in [GRS1], [Gra2] and [EG]; and Folkman's Theorem in [Gra1] and [GRS2]. Most of other authors have simply copied attribution from these works. Which credit is most justified? In one publication only [Gra2], Ronald L. Graham gives the date of Jon Folkman's personal communication to Graham: 1965. In one publication only [Gra1], in 1981 Graham publishes Folkman's proof that uses Schur–Baudet–Van der Waerden's Theorem (see Chapters 33 and 35). Thus, Folkman merits credit. In the standard text on Ramsey Theory [GRS2], I find an argument for giving credit to Folkman alone, disagreeing with the first edition [GRS1] of the same book:

> Although the result was proved independently by several mathematicians, we choose to honor the memory of our friend Jon Folkman by associating his name with the result.

Jon H. Folkman left this world tragically in 1969 at the age of 31. He was full of great promise. Sympathy and grief of his friends is understandable and noble. Yet, do we, mathematicians, have the liberty to award credits? In this case, how can we deny Jon Henry Sanders credit, when Sanders' independent authorship is absolutely clear and undisputed (he could not have been privy to the mentioned above personal communication)? Sanders formulates and proves Theorem 32.5 in his 1968 Ph.D. dissertation [San]. Moreover, Sanders proves it in a different way from Folkman: he does not use Schur–Baudet–Van der Waerden's Theorem, but instead generalizes Ramsey's Theorem to what he calls in his dissertation "Iterated Ramsey Theorem" [San, pp. 3–4].

Vladimir I. Arnautov's discovery is even more striking. His paper is much closer in style to that of Schur's classic 1916 paper, where Schur's Theorem appears as a useful tool, "a very simple lemma," and is immediately used for obtaining a number-theoretic result, related to Fermat's Last Theorem. Arnautov formulates and proves Theorem 32.5, but treats it as a useful tool and calls it simply "lemma 2" (in the proof of lemma 2, he uses Schur–Baudet–Van der Warden's Theorem). He then uses lemma 2 and other Ramseyan tools to prove that every (not necessarily associative) countable ring allows a non-discrete topology. This brilliant paper was submitted to *Doklady Akademii Nauk USSR* on August 22, 1969, and on September 2, 1969 was recommended for publication by the celebrated topologist Pavel S. Aleksandrov.[5] We have no choice but to savor the pleasure of associating Aknautov's name with Theorem 32.5.

What about Rado, one may ask? As Graham–Rothschild–Spencer [GRS2] observe, Theorem 32.5 "may be derived as a corollary of Rado's theorem [Rad1] by elementary, albeit non-trivial, methods."[6] In my opinion, this is an insufficient reason to attach Rado's name to Theorem 32.5. Arnautov, Folkman, and Sanders envisioned a generalization in the direction different from that of Rado, and paved the way for Graham–Rothschild's conjecture proved by Hindman. In fact, Erdős came to the same conclusion in 1973 [E73.21] when he put Rado's name in parentheses (Erdős did not know about Arnautov's paper, or he would have definitely added him to the authors of Theorem 32.5):

> Sanders and Folkman proved the following result (which also follows from earlier results of Rado [Rad1]).

32.3 Non-linear Regular Equations

A number of mathematicians studied regularity of non-linear equations. The following problem was posed by Ronald L. Graham and Paul Erdős circa 1975 (Graham estimates it as "has been opened for over 30 years" in his 2005 talk published as [Gra7]), and still remains open today, as Graham reports in [Gra7], [Gra8], where he offers $250 for the first solution:

Open $250 Problem 32.8 (R. L. Graham and P. Erdős, 1975) Determine whether the Pythagorean equation $x^2 + y^2 = z^2$ is partition regular, i.e., whether for any positive integer k, any k-coloring of the set of positive integers contains a non-trivial monochromatic solution x, y, z of the equation.

"There is actually very little data (in either direction) to know which way to guess," Graham remarks [Gra7], [Gra8]. However, I recall the following story.

---

[5] *Doklady* published only papers by full and corresponding members of the Academy. A non-member's paper had to be recommended for publication by a full member of the Academy.

[6] Theorem 32.5 also follows from Graham and Rothschild's results published in 1971 [GR1].

In May of 1993 in a Budapest hotel, right after Paul Erdős's 80th-Birthday Conference in Keszthely, Hungary, Hanno Lefmann from Bielefeld University, Germany, told me that he and Arie Bialostocki from the University of Idaho, Moscow, generated by computer, with an assistance of a student, a coloring of positive integers from 1 to over 60,000 in two colors that forbade monochromatic solutions x, y, z of the equation $x^2 + y^2 = z^2$. This could be a basis for conjecturing a negative answer to Problem 32.8, but of course the problem remains open and is still awaiting new approaches.

Let us roll back a few years. Inspired by the old K. F. Roth's conjecture (published by Erdős already in 1961 [E61.22], Problem 16, p. 230), Paul Erdős, András Sárközy, and Vera T. Sós proved in 1989 a number of results and posed a number of conjectures [ESS]. I would like to present here one of each.

Erdős–Sárközy–Sós's Theorem 32.9 [ESS, Theorem 3] Any k-coloring of the positive integers, $k \leq 3$, contains a monochromatic pair x, y such that $x + y = z^2$, for infinitely many integers z.

The authors then posed a conjecture:

Erdős–Sárközy–Sós's Conjecture 32.10 [ESS, Problem 2] Let $f(x)$ be a polynomial of integer coefficients, such that $f(a)$ is even for some integer a. Is it true that for any k-coloring of positive integers for some b (for infinitely many b) the equation $x + y = f(b)$ has a monochromatic solution with $x \neq y$?

On the first reading you may be surprised by the condition on $f(a)$ to be even for some integer a. However, you could, easily construct a counterexample to the Erdős–Sárközy–Sós's Conjecture 32.10 if this condition were not satisfied. Indeed, let $f(x) = 2x^2 + 1$, and color the integers in two colors, one for even integers and another for the odd.

In 2006 Ayman Khalfalah, professor of engineering in Alexandria, Egypt, and Endre Szemerédi [KSz] generalized Theorem 32.9 to all k.

Khalfalah–Szemerédi's Theorem 32.11 [KSz] For any positive integer k there exists $N(k)$, such that any k-coloring of the initial segment of positive integers $[N(k)]$ contain a monochromatic pair x, y such that $x + y = z^2$, for an integer z.

Khalfalah and Szemerédi also proved Conjecture 32.10.

Khalfalah–Szemerédi's Generalized Theorem 32.12 [KSz] Given a positive integer k and a polynomial with integer coefficients $f(x)$ such that $f(a)$ is even for some a, there exists $N(k)$, such that any k-coloring of the initial segment of positive integers $[N(k)]$ contains a monochromatic pair x, y, $x \neq y$, such that $x + y = f(z)$, for some integer z.

Endre Szemerédi is a witty speaker, with humor reminiscent to that of Paul Erdős, which he displayed on April 4, 2007 when he presented these results at the Discrete Mathematics Seminar at Princeton-Math.

While these results were a step forward, non-linear regular equations remain a little studied vast area of Ramsey Theory. It deserves its own Richard Rado!

33
Ramsey Theory before Ramsey: Van der Waerden Tells the Story of Creation

> *It is like picking apples from a tree. If one has got an apple and another is hanging a little higher, it may happen that one knows: with a little more effort one can get that one too.*
> — B. L. van der Waerden [Wae18]

> *A thing of beauty is a joy for ever.*
> — John Keats, *Endymion*

The third result in Ramsey Theory before Ramsey was proven by Bartel Leendert van der Waerden in 1926 and published a year later.

Arithmetic Progressions Theorem 33.1 (Van der Waerden, 1927, [Wae2]). For any k, l, there is $W = W(k, l)$ such any k-coloring of the initial segment of positive integers $[W]$ contains a monochromatic arithmetic progression of length l.

B. L. van der Waerden proved this pioneering result while at Hamburg University and presented it the following year at the meeting of *D.M.V., Deutsche Mathematiker Vereinigung* (German Mathematical Society) in Berlin. The result became popular in Göttingen, as the 1928 Russian visitor of Göttingen A. Y. Khinchin noticed and later reported [Khi1], but its publication [Wae2] in an obscure Dutch journal hardly helped its popularity. Only Issai Schur and his two students Alfred Brauer and Richard Rado learned about and improved upon Van der Waerden's result almost immediately (details in Chapter 35); and somewhat later, in 1936, Paul Erdős and Paul Turán commenced density considerations related to Van der Waerden's result [ET] (more in Chapter 35). Only after the World War II, when Khinchin's book *Three Pearls of Number Theory* came out in Russian in 1947 [Khi1] and again in 1948 [Khi2], in German in 1951 [Khi3], and in English in 1952 [Khi4], the result became a classic, and has remained one of the most striking "pearls" of mathematics. In his April 5, 1977 reply to N. G. de Bruijn's compliment, Van der Waerden wrote: "Your praise 'A thing of beauty is a joy forever' pleases me." Let me second de Bruijn: the praise (coined by John Keats, by the way) is well deserved!

A. Soifer, *The Mathematical Coloring Book*,
DOI 10.1007/978-0-387-74642-5_33, © Alexander Soifer 2009

Now that the success of Khinchin's booklet had made the result classic, the latter merited a special attention and commentary by its solver. Van der Waerden obligated, and in 1954 published an essay *Der Beweis der Vermutung von Baudet* with a more expressive English title *How the Proof of the Baudet's Conjecture Was Found* in a later published translation. This essay has appeared four times in German: twice in 1954 [Wae13], [Wae14], in 1965 [Wae16], posthumously in 1998 [Wae26]; and once in English in 1971 [Wae18]. It is not just invaluable as a historical document. The essay delivers a vibrant portrait of mathematical invention in the making. Van der Waerden presents all critical ideas of the proof in the most clear and engaging way. Thanks to the permission granted to me by Professor B. L. van der Waerden in his letter [Wae24] and the permission by Academic Press, London, I am able to bring this delightful essay [Wae18] to you here instead of presenting a formal "dehydrated" proof of the result.

Enjoy, as Bartel Leendert van der Waerden recalls:

Once in 1926, while lunching with Emil Artin and Otto Schreier, I told them about a conjecture of the Dutch mathematician Baudet:

If the sequence of integers 1, 2, 3, ... *is divided into two classes, at least one of the classes contains an arithmetic progression of l terms*:

$$a, a + b, \ldots a + (l - 1)b,$$

no matter how large the length l is.

After lunch we went into Artin's office in the Mathematics Department of the University of Hamburg, and tried to find a proof. We drew some diagrams on the blackboard. We had what the Germans call "*Einfälle*": sudden ideas that flash into one's mind. Several times such new ideas gave the discussion a new turn, and one of the ideas finally led to the solution.

One of the main difficulties in the psychology of invention is that most mathematicians publish their results with condensed proofs, but do not tell us how they found them. In many cases they do not even remember their original ideas. Moreover, it is difficult to explain our vague ideas and tentative attempts in such a way others can understand them.[7] To myself I am accustomed to talk in short hints which I alone can understand. Explaining these hints to others requires making them more precise and thus changing their nature.

In the case of our discussion of Baudet's conjecture the situation was much more favorable for a psychological analysis. All ideas we formed in our minds were at once put into words and explained by little drawings on the blackboard. We represented the integers 1, 2, 3 ... in the two classes by means of vertical strokes on two parallel lines. Whatever one makes explicit and draws is much easier to remember and to reproduce than mere thoughts. Hence, this discussion between

---

[7] And when mathematicians attempt to be subjective, and better express themselves and include the emergence of their results, most of journal editors, these priests of gloom and doom, would mercilessly cut manuscripts to bring them to an 'objective' and relentless theorem–proof style. –A.S.

Artin, Schreier, and myself offers a unique opportunity for analyzing the process of mathematical thinking.

It was clear to us from the very beginning that the case $l = 2$ is trivial. One need not even consider the infinite sequence of integers; it is sufficient to consider the three integers 1, 2, 3. If they are divided into two classes, one of the classes contains a pair of numbers (in arithmetic progression).

The next case we considered was $l = 3$. In this case, too, it is not necessary to consider all integers: it suffices to take the integers from 1 to 9. The numbers 1 to 8 can be divided, in several ways, into classes without obtaining an arithmetic progression of 3 terms in one class, e.g. like this:

$$12 \quad 56 \qquad \text{in the first class}$$
$$34 \quad 78 \qquad \text{in the second class.}$$

However, in any one of these cases, the number 9 cannot escape. If we put it into the first class, we have the progression 1 5 9, and if we put it into the second class, we get the progression 7 8 9, and so on in all other possible cases. I had observed this already the day before.

Next, Schreier asked if at all Baudet's conjecture is true for a certain value of l, is it always possible to find an integer $N(l)$ such that the conjecture holds already for the segment

$$1\,2\,3 \dots N(l),$$

in the sense that every division of this segment into 2 classes yields an arithmetic progression of length l in 1 class?

Schreier himself found the answer: it was Yes. If Baudet's conjecture holds for a fixed value of l, it is possible to find an N such that the conjecture holds already for the segment 1 2 ... N. This was proved by a well-known procedure from set theory, the "diagonal procedure". The argument is as follows.

If no such N existed, then for every N there would be a division D_N of the numbers from 1 to N into 2 classes such that no class contains an arithmetic progression of length l. Thus one could obtain an infinite sequence

$$D_1 \, D_2 \dots$$

of such divisions. The number 1 lies, in every one of these divisions, in one of the 2 classes. Hence it happens an infinity of times that 1 is in the same (first or second) class, and an infinite sequence D_1', D_2', ... exists such that in all these divisions 1 is in the same class, say in class number i_1 ($i_1 = 1$ or 2).

In the divisions D_2', D_3', ... the number 2 belongs to one of the two classes. Hence, by the same argument, an infinite subsequence D_2'', D_3'', ... exists such that 2 is always in the same, i_2th class.

And so on. For every n one obtains a subsequence of divisions

$$D_n^{(n)}, \quad D_{n+1}^{(n)}, \quad \ldots$$

such that in all these divisions the integers 1, 2, ..., n are always in the same classes:

1 in class i_1

2 in class i_2

. . .

n in class i_n.

Next, one can form a "diagonal division" $D\,D$ of all integers 1, 2, 3, ... in which 1 lies in class i_1, 2 in class i_2, and so on. In this division, the number n lies in the same class as in the division $D_n^{(n)}$, hence the name "diagonal procedure."

In this division $D\,D$ no arithmetic progression of length l could exist in which all terms belong to the same class. For if it existed, it would exist already in $D_n^{(n)}$, i.e., in one of the original divisions. But we have assumed Baudet's conjecture to be true for the sequence of integers 1 2 3 ... and for this particular value of l. Thus we obtain a contradiction.

From this point onward, we tried to prove the "strong conjecture", as we called it, for a finite segment from 1 to $N(l)$, i.e., we tried to find a number $N(l)$ having the desired property. For $l = 2$ and $l = 3$ such numbers had been found already:

$$N(2) = 3, \quad N(3) = 9.$$

So we tried to go from $l - 1$ to l. For this induction proof, the replacement of the original conjecture by a stronger one is a definite advantage, as Artin rightly remarked. If one can assume for $l - 1$ the existence of a finite bound $N(l - 1)$, one has a chance to find a proof for the next number l.

Next, Artin observed: If the strong conjecture is true for 2 classes and for all values of l, it must be true for an arbitrary number of classes, say for k classes. To prove this assertion, he first proposed k to be 4. The 4 classes can be grouped into 2 and 2. This gives us a rough division of the integers into 2 big classes, every big class consisting of 2 smaller classes. In one of the big classes an arithmetic progression of $N(l)$ terms exists. The terms of this progression can be numbered from 1 to $N(l)$. These numbers are now divided into two smaller classes, and hence in one of the smaller classes an arithmetic progression of length l exists. Thus, if the strong conjecture is true for 2 classes, it is also true for 4 classes. By the same argument one finds that it also holds for 8 classes, etc., hence, for any number of classes $k = 2^n$. But if it holds for $k = 2^n$, it also holds for every $k \leq 2^n$, because we may always add a few empty classes. Hence, if Baudet's conjecture holds for 2 classes it also holds, even in the strong form, for an arbitrary number of classes.

We now tried to prove the "strong conjecture" for arbitrary k and l by induction form $l - 1$ to l. This means: we tried to find a bound $N = N(l, k)$ such that, if the integers from 1 to N are divided into k classes, one of the classes contains an arithmetic progression of length l.

Artin expected—and he proved right—that the generalization from 2 to k classes would be an advantage in the induction proof. For, he argued, we might now try to prove the strong conjecture for an arbitrary *fixed* value of k and for length l under the induction hypothesis that it holds for *all* k and for length $l - 1$. This means: we have a very strong induction hypothesis to start with, which is a definite advantage.

Following the line indicated by Artin, we now tried to prove Baudet's conjecture for 2 classes and for progressions of length l, assuming the strong conjecture to hold for all k for progressions of length $(l - 1)$.

Next, Artin had another very good idea. If the integers 1, 2, ... are divided into 2 classes, blocks of (say) 3 successive integers are automatically partitioned into $2^3 = 8$ classes. For each of the 3 numbers within the block can lie in the first or second class, and this gives us 8 possibilities for the whole block. Now the blocks of 3 successive integers can be numbered: block number n consists of the integers n, $n + 1$, $n + 2$. If the blocks are partitioned into 8 classes, their initial numbers n are also partitioned into 8 classes, and to this partition we can apply the induction hypothesis. Thus we obtain the following result: among sufficiently many successive blocks we can find an arithmetic progression of $(l - 1)$ blocks all in the same class. The pattern of the distribution of integers over the 2 classes in the first block will be repeated, exactly as it is, in the other $(l - 2)$ blocks.

The same holds for blocks of arbitrary length m, each consisting of m successive numbers

$$n, n + 1, \ldots, n + m - 1.$$

The number of classes for those blocks is 2^m. Again one can obtain arithmetic progressions of $(l-1)$ blocks in the same class, with exact repetition of the pattern in the first block. Moreover, if the blocks are long enough, we can also find arithmetic progressions of $(l - 1)$ integers within each block.

In the simplest case $l = 2$ the conjecture is certainly true for all k, for if the integers from 1 to $k + 1$ are divided into k classes, there must be two integers in one of the classes. This is Dirichlet's "box principle."[8] if $k + 1$ objects are in k boxes, one of the boxes contains two of them. A very useful principle in Number Theory.

Thus, starting with the obvious case $l = 2$, we tried to treat the case of 2 classes and $l = 3$ (although this case had been dealt already by an enumeration of all possible cases). We represented the integers in the 2 classes by small vertical strokes on two parallel lines, as in Fig. 33.1.

Among 3 successive integers there are always 2 in the same class, by the induction hypothesis, i.e., in this case by the "box principle." Now consider a block of 5

---

[8] In the USA, it is usually called the Pigeonhole Principle.

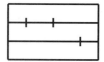

Fig. 33.1

successive integers. Among the first three there are two in the same class; this gives us an arithmetic progression of length 2. The third term of this progression still lies within the block of 5. If it is in the same class as the first two terms, we have in this class a progression of length 3, as desired. Therefore we may suppose that the third term lies in the other class, and we have, within every block of 5, a pattern like the one of Fig. 33.1.

I was drawing such blocks on the blackboard, and thought: There are $2^5 = 32$ classes of blocks of 5, hence among 33 successive blocks of 5 there must be 2 blocks in the same class. In the first of these blocks a pattern like the one in Fig. 33.1 exists, and in the second block of 5 this pattern is exactly repeated (Fig. 33.2).

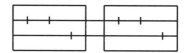

Fig. 33.2

What we wanted to construct were progressions of length 3. Hence I drew one more block at the same distance from the second block as the second from the first, and I drew three strokes in the third block in the same position as the strokes in the first and second block (Fig. 33.3).

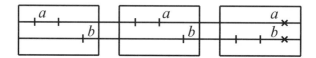

Fig. 33.3

The third of these strokes represents an integer, which may be in the first or second class. If it is in the first, we have in this class an arithmetic progression $a\,a\,a$ (Fig. 33.3). If it is in the second class, we have in this class a progression $b\,b\,b$. Hence we have in any case within the block of integers from 1 to $5 + 32 + 32 = 69$, an arithmetic progression of 3 terms in one class.

After having found this proof in the special case $k = 2$ and $l = 3$, I explained it to Artin and Schreier. I felt sure that the same proof would work in the general case. They did not believe it, and so I proceeded to present the proof in the next higher case $k = 3$, $l = 3$.

Instead of considering blocks of $3 + 2 = 5$, I now considered blocks of $4 + 3 = 7$ successive integers. Since the first four numbers of such a block are distributed among 3 classes, two of them must belong to the same class. The third term of the arithmetic progression starting with these two terms still belongs to same block of 7. If the third term lies in the same class, we have a progression of length 3 in this class. Hence we may suppose the third term to lie in another class. Thus we obtain, in every block of seven, a pattern like the one in the first small block of Fig. 33.4.

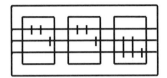

Fig. 33.4

The blocks of 7 are partitioned into 3^7 classes. Hence among $3^7 + 1$ successive blocks of 7 there are two belonging to the same class. In the first block we have three integers in arithmetic progression, two of which belong to the same class, and this pattern repeats itself in the second block. If the second block is shifted once more over the same distance, one contains 3 blocks forming an arithmetic progression of blocks, as shown in Fig. 33.4.

In the third block, I drew 3 strokes in positions corresponding to the 3 strokes in the first or second block, and I considered the possibilities for the third of these strokes. If it falls into the first or second class, we have an arithmetic progression of length 3 in the same class, by the same argument as before; but now the third stroke can escape into the third class. Thus we obtain the pattern drawn in Fig. 33.4.

We have such a pattern in every large block of $3^7 + 3^7 + 7 = h$ successive integers. Now the large blocks of h are divided into 3^h classes. Hence among $3^h + 1$ successive large blocks there are two belonging to the same class. Drawing the small blocks within the large ones, I obtained the picture of Fig. 33.5.

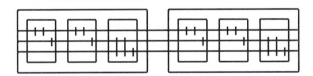

Fig. 33.5

Now shifting the second large block over the same distance, and considering the third stroke in the third small block in the third large block, I showed that it cannot escape anymore. If it lies in the first class, there is a progression $a\ a\ a$ in the first class. If it lies in the second class, there is a progression $b\ b\ b$ in that class, and if in the third class, a progression $c\ c\ c$ in that class (Fig. 33.6).

After this, all of us agreed that the same kind of proof could be given for arbitrary k. However, Artin and Schreier still wanted to see the case $l = 4$.

As before, I first considered the case of 2 classes. For this case I had already proved that among sufficiently many, say n, successive integers there is a progression

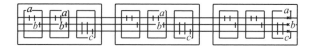

Fig. 33.6

of 3 terms in the same class. We may suppose n to be odd. The distance between the first and the last term of the progression $(n - 1)$ at most, hence the difference between two successive terms is $\frac{1}{2}(n - 1)$ at most. Now consider the fourth term of the same progression. All four terms lie within a block of

$$g = n + \tfrac{1}{2}(n - 1)$$

successive integers. If the fourth term belongs to the same class as the other three, we are satisfied. Suppose it lies in the other class; then we have the pattern of Fig. 33.7.

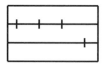

Fig. 33.7

In every block of g successive integers, such a pattern must occur. Now the blocks of g are divided into 2^g classes. Hence among sufficiently many, say $N(3, 2^g)$ blocks of length g, there are three blocks in arithmetic progression belonging to the same class. The pattern in the first block is exactly repeated in the second and third block (Fig. 33.8).

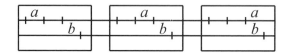

Fig. 33.8

Adding a fourth block to this progression, I easily obtained a progression $a\,a\,a\,a$ in the first or $b\,b\,b\,b$ in the second class.

Now it was clear to every one of us that the induction proof from $(l-1)$ to l works for arbitrary l and for any fixed value of k. Hence if Baudet's strong conjecture is true for length $(l - 1)$ and all k, it is also true for l and any k. Since it is true for $l = 2$, its truth follows quite generally.

Analyzing this record, one can clearly distinguish a succession of sudden ideas, which gave the discussion a new turn every time.

1. The first was Schreier's idea of restricting oneself to a finite segment from 1 to N. This idea was fundamental to the whole proof.

2. The second idea was: to try an induction from $l - 1$ to l. This was quite a natural idea, because the case $l = 2$ was obvious and the case $l = 3$ could be solved by enumerating all possible cases.
3. Artin proved: if the strong conjecture is true for 2 classes, it is also true for 4 classes. In his proof another idea was implicit, viz.: if the conjecture is true for a segment of all integers from 1 to N it is also true for any arithmetic progression of length N

$$a, a + b, \ldots, a + (N - 1)b$$

because the terms of this progression can be numbered by the integers 1 to N. This is also a central idea in the proof.
4. Next, Artin said: in an induction it is always an advantage to have a strong induction hypothesis to start with. Therefore let us start with the assumption that the conjecture holds for progressions of length $(l - 1)$ and for *all* k, and try to prove the conjecture for progressions of length l and for *one* value of k, say $k = 2$. Thus the plan for the proof was devised.
5. The next idea, which also came from Artin, was of decisive importance. He said: we can apply the induction hypothesis not only to single integers, but also to blocks, for they too are divided into classes. Thus we are sure that whole blocks are repeated $(l - 1)$ times.
6. After this, it was only natural to consider progressions of $(l - 1)$ integers within the blocks, and to try to extend these progressions of length $(l-1)$ to progressions of length l. The simplest non-trivial case is $l = 3$, and thus I was led, quite naturally, to consider patterns like the one of Fig. 33.2.
7. This pattern still does not contain a progression of length 3 in one class. Therefore, it was necessary to extend the progression of length 2 occurring in the second class in Fig. 33.2 to a progression of length 3. Hence I extended the pattern of Fig. 33.2 by drawing the third block of Fig. 33.3, and I considered the third term of the progression $b\,b\,b$. As soon as attention was focused upon this term, it was clear that it cannot escape from forming an arithmetic progression of length 3 in the first or second class.

This final idea was accompanied by a feeling of complete certainty. I felt quite sure that this method of proof would work for arbitrary k and l. I cannot explain this feeling; I can only say that the mathematicians often have such a conviction. When a decisive idea comes to our mind, we feel that we have the whole proof we are looking for: we have only to work it out in detail.

However, I can explain, to a certain extent, why Artin and Schreier did not feel so sure. They saw only the result: the presence of the progression $a\,a\,a$ in the first class or $b\,b\,b$ in the second one, but I had discovered a method for finding such progressions, and I was convinced that this method would work in higher cases as well.

It is like picking apples from a tree. If one has got an apple and another is hanging a little higher, it may happen that one knows: with a little more effort one can get that one too. The man standing next to me only sees that I have just got the first apple, and he is in doubt whether I can get the other too, but I myself have not only got the apple, but I also have a feeling of the movement that enabled me to pick it.

The feeling that a method of proof can be carried over to the other cases is still sometimes deceptive. Often the higher cases offer additional difficulties. Still, feelings of this kind are extremely useful in mathematical research.

Finding the proof of Baudet's conjecture was a good example of teamwork. Each of the three of us contributed essential ideas. After the discussion with Artin and Schreier I worked out the details of the proof and published it in *Nieuw Archief voor Wiskunde* **15**, p. 212 (1927). (Interesting applications and generalizations of the theorem proved in my paper were given by Richard Rado[9]).

A. J. Khinchin included the theorem among his "Three Pearls of the Theory of Numbers" (1952) and published a proof due to M. A. Lukomskaja, which is in all essentials the same as mine, the only difference being that in her proof the blocks are required to be non-overlapping. ∎

Van der Waerden, assisted by Emil Artin and Otto Schreier, actually proved a "strong conjecture" as they called the result:

Van der Waerden's Theorem Strong Version 33.2 For all positive integers n and r there exists an integer $W = W(n, r)$ such that if the initial set of integers $[W] = \{1, 2, \ldots, W\}$ is colored in r colors, then there exists a monochromatic n-term arithmetic progression.

It is natural to inquire whether the finality of n and of r is essential. Prove first that the finality of the length n of the guaranteed arithmetic progression is essential:

Problem 33.3 Color the set of all positive integers in two colors in a way that forbids infinite monochromatic arithmetic progressions.

Of course, the finality of the number of colors is essential, for otherwise we can color each integer in its own color and thus exclude even length two arithmetic progressions. However, Paul Erdős and Ronald L. Graham proved a nice "consolation" result [EG]. We will say that a sequence is *representative* if each term is colored in different color.

Theorem 33.4 (Erdős–Graham, [EG]) Any coloring of the positive integers in infinitely many colors contains arbitrarily long monochromatic or representative arithmetic progressions.

Hint: Peek at Szemerédi's Theorem in Chapter 35—and use it. ∎

I have been unable to explain why the leader of the new field, Ramsey Theory, Paul Erdős almost universally quoted Van der Waerden's result as addressing only the case of two colors (see, for example, [E57.13], [E61.22], [E71.13], [E73.21], [E76.35], [E81.16], [E80.03], [E83.03], [E85.33], [E89.27], etc.,). Is it because Van

[9] R. Rado: Studien zur Kombinatorik, Ph.D-Thesis Berlin 1931, *Math. Zeitschr.* **36**, p. 424. Verallgemeinerung eines Satzes von van der Waerden, *Sitzungsber. preuss. Akad.*, Berlin 1933, p. 589. Note on Combinatorial Analysis, *Proc. London Math. Soc.* (2) **48**, p. 122.

der Waerden's paper opens with the Baudet's Conjecture for two colors or because Erdős wanted, as he often did, to gain insight into the simplest case first (and then forgot about the general case)?

Besides Issai Schur and his former Ph. D. students Alfred Brauer and Richard Rado, and Erdős and Turán, nobody seemed to have appreciated and furthered Van der Waerden's proof soon after Van der Waerden's publication. There was, however, one exception—a pair of mathematicians, who published on Van der Waerden's proof very shortly after its publication. Their result was cited in Erdős and Graham's fine 1980 problem book [EG] as "an easy consequence of Van der Waerden's Theorem":

Problem 33.5 [KM] If $A = \{a_1, a_2, \ldots\}$ is an increasing infinite sequence of integers with $a_{k+1} - a_k$ bounded, then A contains arbitrarily long arithmetic progressions.

The great surprise is: [KM] was published by the two Japanese mathematicians Sôichi Kakeya and Seigo Morimoto in 1930, much earlier than even Erdős and Turán's paper! How did they get a hold of the little-read Dutch journal where Van der Waerden published his result just 3 years earlier? The authors misspelled the name of Baudet everywhere, even in the title: *On a Theorem of MM. Bandet [sic] and van der Waerden*. But they were first to recognize that credit is due to both, Baudet for creating the conjecture, and Van der Waerden for proving it. Without the conjecture, Van der Waerden would have had nothing to prove!

In the next chapter I will present my research on the authorship of the conjecture that Van der Waerden proved, and my take on attributing credit for the result. Meanwhile, I invite you to prove Kakeya–Morimoto's statement 33.5. You will find my proof appended to the end of Chapter 35.

34
Whose Conjecture Did Van der Waerden Prove? Two Lives Between Two Wars: Issai Schur and Pierre Joseph Henry Baudet

As far as your advice to leave priority matter . . . alone, it is my opinion that the tiniest moral matter is more important than all of science, and that one can only maintain the moral quality of the world by standing up to any immoral project
                                                      – L. E. J. Brouwer[10]

What amazes us today is, of course, that no one in Hamburg (including Schreier and Artin) had known about Schur's work. In that connection we must realize that the kind of mathematics involved in the [Baudet-Schur] conjecture was not mainstream, and that combinatorics was not a recognized field of mathematics at all.
                                                      – Nicolaas G. de Bruijn[11]

34.1 Prologue

Bartel L. van der Waerden credited "Baudet" [sic] with conjecturing the result about monochromatic arithmetic progression. Decades later, Van der Waerden gave a most insightful story of the birth of his proof, which I have reproduced for you in Chapter 33. As I enumerated there, the "Story of Creation" appeared four times in German: twice in 1954 [Wae13], [Wae14], in 1965 [Wae16], posthumously in 1998 [Wae26]; and once in English in 1971 [Wae18]. In these publications Van der Waerden extended the credit for the conjecture to "the Dutch mathematician Baudet" (still without first name or initials). Biographers of Van der Waerden faithfully followed him with crediting Baudet for the conjecture (see [Fre1], [FTW], [Per] and [Bru1]).

---

[10] February 24, 1929 letter to H. Hahn, quoted from [Dal2], p. 651.

[11] E-mail to A. Soifer, January 5, 2004.

A. Soifer, *The Mathematical Coloring Book*,
DOI 10.1007/978-0-387-74642-5_34, © Alexander Soifer 2009

On the other hand, Ronald L. Graham, Bruce L. Rothschild, and Joel H. Spencer in their definitive monograph [GRS1], [GRS2] cited Alfred Brauer [Bra2], [Bra3] in taking credit for the conjecture away from Baudet and giving it to Issai Schur. Schur was also credited by Hillel Furstenberg in his pioneering paper [Fur1]. Consequently, many mathematicians uncritically quoted or simply copied credit from [GRS1], [GRS2] or [Fur1].

False attributions are never pleasant. One may wonder, however, why the authorship of *this* conjecture is so extremely important that I most thoroughly researched it, and am dedicating this whole chapter to my findings. This is so because we have here, for the third time in the history of mathematics,[12] a totally new Ramseyan type of question, quite uncommon in mathematics of the time: "if a system is partitioned arbitrarily into a finite number of subsystems, then at least one subsystem possesses a certain specified property."[13] It was a major achievement indeed to envision and conjecture such a result. But whose achievement was it, Baudet's or Schur's? And who was "Baudet" anyway? My early investigative reports appeared in mid 1990s [Soi10], [Soi11], and [Soi12]. Let us look at the more complete evidence that I have been able to assemble to date.

34.2 Issai Schur

Germany has surely been one of the best countries in preserving documents through all the cataclysms that have befallen on this land. Issai Schur's personnel file, and personnel forms it contains are preserved in the Archive of the University Library of the Humboldt University at Berlin.[14] Let us make good use of them.

Issai Schur was born on January 10, 1875, in the Russian city of Mogilyov (presently in Belorussia) in the family of the merchant Moses Schur and Golde Landau. Being a Jew, Issai could not enroll in any Russian university. At 13 he went to live with his older sister in Libau, Russia (now Latvia), in order to attend the German language Nicolai-Gymnasium (1888–1894). That prepared him for entering a German university in 1894. In Berlin, on September 2, 1906, Issai Schur married Regina Malka Frumkin, born on January 8, 1881 in Kowno (Kaunas). She was a medical doctor, also Jewish, and apparently an émigré from Russia. Issai Schur, who originally filled the personnel form in his hand, likely in 1916 (it was later updated, probably by clerks), on the line "Arian" promptly put "*nicht*" for himself and "*nicht*" for his wife. The happy and lasting marriage produced two children, Georg (named in honor of Schur's mentor, the celebrated algebraist F. Georg Frobenius), born on July 25, 1907, and Hilde, born on March 15, 1911.

---

[12] First two being Hilbert's Theorem of 1892 [Hil] and Schur's Theorem of 1916 [Sch1] – see Chapters 31 and 32.

[13] Leon Mirsky in [Mir], in reference of Schur's Theorem.

[14] Archive of Humboldt University at Berlin, documents UK-Sch 342, Bd. I, Bl. 1, 1R, and 2R, 3.

Young Issai Schur, Courtesy of Hilde Abelin-Schur

Issai Schur gave most of his life to the University of Berlin, first as a student (1894–1901; Ph.D. in Mathematics and Physics summa cum laude, November 27, 1901), then as a *Privatdozent* (1903–1909), *ausserordentlischer Professor* (equivalent to an associate professor, December 23, 1909–April 21, 1913 and again April 1, 1916–April 1, 1919) and *Ordinarius* (equivalent to a full professor, April 1, 1919–September 30, 1935).[15] On April 1, 1921, Schur was appointed to the *Ordinarius* chair of Prof. Dr. Schottky with a very respectable compensation: 16200 marks base salary; plus local, adjustment and family allowances; plus 5000 marks for lecturing the minimum of 8 hours a week. Schur spent 3 years at the University of Bonn, 1913–1916, the only period he spent away from Berlin. These years are important to our story, and we will look at them in the next section.

Issai Schur was elected to a good number of academies of sciences. He was a legendary lecturer. Schur's student and friend Alfred Theodor Brauer (Ph. D. under Schur 1928) recalls [Bra2] that the number of students in Schur's elementary num-

---

[15] Archive of Humboldt University at Berlin, documents UK-Sch 342, Bd. I, Bl. 4.

Issai Schur, the collection of his daughter, Hilde Abelin-Schur

ber theory courses often exceeded 400, and during the winter semester of 1930 even exceeded 500. Brauer would know, for as Schur's Assistant, he had to grade home works of all those students! Walter Ledermann, who estimates to have taken about 500 lectures from Schur, writes [Led1] that "Schur's lectures were exceedingly popular. I remember attending his algebra course which was held in a lecture theatre filled with about 400 students." Ledermann adds in his 2000 interview [Led2]:

> I was absolutely captivated by Schur. I wrote about 300 lectures in fair copy in cloth bound book which I had until quite recently, running to something like 2000 pages of Schur's lectures.

Hitler's appointment as *Reichskanzler* by President Paul von Hindenburg on January 30, 1933 changed this idyllic life. Schur's former student Menahem Max Schiffer recalls in his talk at the 4th Schur conference in May 1986 at Tel Aviv University, which was consequently published [Schi]:

> Now, the year 1933 was a decisive cut in the life of every German Jew. In April of that year [April 7, 1933 to be precise] all Jewish government officials were dismissed, a boycott of Jewish businesses was decreed and anti-Semitic legislation was begun. When Schur's lectures were cancelled there was an outcry among the students and professors, for Schur was respected and very well liked. The next day Erhard Schmidt started his lecture with a protest against this dismissal and even Bieberbach, who later made himself a shameful reputation as a Nazi, came out in Schur's defense. Schur went on quietly with his work on algebra at home.

Issai Schur (left) and Edmund Landau, the collection of his daughter, Hilde Abelin-Schur

Ledermann gives vivid details [Led2]:

When Hitler finally came to power, all the Jewish faculty were dismissed instantly, including Schur who was not allowed to come even to the library any more.

However Erhardt Schmidt, who was the decent sort of German, found that in the regulations of the Nazis there was a clause to say that these dismissals would not apply to two types of non-Aryan:

1. those who had fought in the First World War in the German army on the front, and
2. those who had during the First World War held a position making them German/Prussian civil servants.

The first of these applied to Alfred Brauer who had been a soldier, and yes, he was badly wounded, and the second applied to Schur because in 1916 he was an extraordinary professor at Bonn, so had effectively become a Prussian civil servant.

So, Schmidt applied this clause. He went to Goebbels and said, "You must abide by your own law and reinstate Schur for this reason", and he was reinstated. He could then come to the University but he was not allowed to lecture. For supervision of my dissertation, I had to go to his house. It was nice to meet with him, he lived in a suburb of Berlin, to see him and his wife and talk not only about mathematics but also about the Jews. He said, "I can read the English Times which is still allowed", all the other papers were taken over by the Nazis. I cannot bear this. And then the time came for me to have my exam, the oral, and he was allowed to come to take this examination in mathematics for one hour. Also, a co-examiner was expected to come. They did not normally ask questions but would take a record, more like a secretary. This co-examiner was, unfortunately, none other than Bieberbach, who appeared in Nazi uniform, brown shirt and swastika. He came and sat down to take notes about what Schur was asking me. But I must say he was quite fair. He didn't interfere and I got a very good result.

Hindenburg negotiated with Hitler exemptions from the April 7, 1933 Restoration of Professional Civil Service Law for those Jews who fought for Germany in World I, those who lost a father or a son in the war and those who entered their civil service jobs (university professors included) before the start of the war. Schur held a civil service (university) appointment before the war, and thus fell under the exemption. Nevertheless, by the order UI No. 6362 of the Prussian Minister for Science, Art and Public Education of on April 29, 1933, Schur and 18 other faculty were "relieved of their duties effective immediately" – yes, immediately, as was customary for the Nazi orders.[16] At that time a representative of the English Jewish Emergency Council visited Schur. We are lucky to have had a witness at the meeting—Schiffer reports [Schi]:

> The lady asked Schur whether and where he wanted to go, because for a man of his reputation all doors would be open. But Schur responded that he did not intend to go; for he did not want to enable the Nazis to say, that many Jewish professors just left for better jobs. Besides, there were many younger colleagues which needed help much more urgently, and he would not take away their chances. He would stick it out in Berlin, for the craze of the Hitlerites could not last long.

I believe in Schur's generosity and genuine care for younger colleagues. Yet, there had to be more to his refusal to leave Germany earlier, as early as the Nazis had come to power in the early 1933. In 1995 Schur's former student Walter Ledermann, Professor at the University of Sussex, UK, sent me his 1983 reprint [Led1], where he introduced additional reasons for the unfortunate Schur's decision to stay in Germany:

> When the storm broke in 1933, Schur was 58 years of age and, like many German Jews of his generation, he did not grasp the brutal character of the Nazi leaders and their followers. It is an ironic twist of fate that, until it was too late, many middle-aged Jews clung to the belief that Germany was the land of Beethoven, Goethe and Gauss rather than the country that was now being governed by Hitler, Himmler, and Goebbels. Thus Schur declined the cordial invitations to continue his life and work in America or Britain. There was another reason for his reluctance to emigrate: he had already once before changed his language, and he could not see his way to undergoing this transformation a second time.
>
> So he endured 6 years of persecution and humiliation under the Nazis.

On October 7, 1933 the Prussian Minister for Science, Art and Public Education, by the order UI No. 8831, "canceled the suspension for Dr. Mittwoch and Dr. Schur, *Ordinarius* professors on the Philosophical Faculty," imposed by the previous order, effective – of course – immediately.[17] The legal exemption had worked; Walter Ledermann, whom we quoted above, and Alfred Brauer [Bra2] credited Erhard Schmidt's efforts for the success. Consequently, Schur was able to carry out some of his duties but not all (no lecturing, for example) and not for long. Issai Schur

---

[16] Archive of Humboldt University at Berlin, documents UK-Sch 342, Bd. I, Bl. 23 and Bl. 23R.

[17] Archive of Humboldt University at Berlin, document UK-Sch 342, Bd.I, Bl.24.

was a famous professor, the pride of his University and of his profession. Yet no achievement was high enough for a Jew in the Nazi Germany. Following 2 years of pressure and humiliation, Schur, faced with imminent expulsion, "voluntarily" asked for resignation on August 29, 1935. On September 28, 1935, Reichs- and Prussian Minister of Science, Instruction and Public Education, replied on behalf of *Der Fürer und Reichskanzler*, i.e., Adolf Hitler himself (see facsimile on p. 327):[18]

> *Fürer* and *Reichskanzler* has relieved you from your official duties in the Philosophical *Facultät* of the University of Berlin effective at the end of September 1935, in accordance with your August 29 of this year request.

As Henrik Hofer of the Humboldt University Library reports [Hof], Schur was the last Jewish professor to lose his job at the University of Berlin. Only a few of his closest friends had the courage to visit him, recalls Schiffer, and retells one such visit, about which he learned from Schur himself [Schi, p. 180]:

> When he complained to [Erhard] Schmidt about the Nazi actions and Hitler, Schmidt defended the latter. He said, "Suppose we had to fight a war to rearm Germany, unite with Austria, liberate Saar and the German part of Czechoslovakia. Such a war would have cost us half a million young men. But everybody would have admired our victorious leader. Now, Hitler has sacrificed half a million of Jews and has achieved great things for Germany. I hope some day you will be recompensed but I am still grateful to Hitler." So spoke a great scientist, a decent man, and a loyal friend. Imagine the feelings of a German Jew at that time.

Clearly, Erhard Schmidt, who, as we have seen, helped Schur after the latter's initial dismissal, held extreme nationalistic aspirations for Great Germany, *Deutschland über Alles*. Schmidt acknowledged the brutal sacrifice of half a million of Jews, including his friend Schur, and Schmidt was willing to sacrifice half a million young German men for "great things for Germany." How low had morality fallen in the Third Reich, if these were the views of "a decent man" (Schiffer's words), Erhardt Schmidt!

One very special 1936 visitor of Issai Schur, Paul Erdős recalls on the pages *Geombinatorics* [E95.32]:

> Schur was of the Russian Jewish origin. He always viewed himself as a German, and he was greatly attracted by the German culture. The horrible degeneration of Nazism was a great disappointment and a personal tragedy to him.

Menahem Schiffer lists Schur's numerous honors that were stripped away:

> He [Schur] was a member of many distinguished academies and learned societies; for example, the Prussian, Bavarian, Saxonian Academies of Science and many more. He had been ejected from each of them.

A document published in 1998 by the authors of [BFS] sheds light on one of these expulsions. Schur had been a member of the Prussian Academy of Sciences

---

[18] Archive of Humboldt University at Berlin, document UK-Sch 342, Bd.I, Bl.25.

Der Reichs=
und Preußische Minister
für Wissenschaft, Erziehung
und Volksbildung

I p Schur.2 a

Es wird gebeten, dieses Geschäftszeichen und den
Gegenstand bei weiteren Schreiben anzugeben.

Berlin W 8, den
Unter den Linden 4
Fernsprecher: A 1 Jäger 0030
Postscheckkonto: Berlin 14402
Reichsbank-Giro-Konto
— Postfach —

Verw. Dir.
b. d. Univ. Berlin
Eing. – 4. OKT 1935

Der Führer und Reichskanzler hat Sie auf Ihren Antrag

vom 29.August d.Js. mit Ablauf des Monats September 1935

von den amtlichen Verpflichtungen in der Philosophischen

Fakultät der Universität Berlin entbunden.

Ich übersende Ihnen anbei die hierüber ausgefertigte

Urkunde.

(Unterschrift)

An Herrn Professor Dr. Issai S c h u r in Berlin-Schmar-
gendorf, Rumlaerstr.14 – Einschreiben –.

————————

Abschrift zur Kenntnis und weiteren Veranlassung.

Jn Vertretung

gez. K u n i s c h

Beglaubigt.

Ministerial-Kanzleisekretär.

An

den Herrn Verwaltungsdirektor
bei der Universität Berlin

hier C 2.

Letter relieving Issai Schur from his duties at the University of Berlin. Courtesy of the Archive of the Humboldt University at Berlin

ever since his election in 1921. The Academy was going to publish works of Weierstrass – what can be political about it? The editorial board was to routinely sign off on the publication on a "*Zirkular*". Let us look together at this document

[BFS, p. 26]. The first two lines seem routine and are written by Erhard Schmidt and Issai Schur respectively (I am translating the lines from German here):

Seen – 11.3.38 Erhard Schmidt
Seen 12/3/38 Schur

Here comes Bieberbach, the founder of the racist doctrine of *German Mathematics* (that he opposed to *Jewish Mathematics*) and writes right below Schur and clearly hinting at Schur's presence:

Bieberbach 29.3.38
I am surprised that Jews still belong to the Academic Commissions. B.

In his turn Theodore Vahlen, a long-term Nazi and anti-Semite, a mathematician and official in the Ministry of Education for University Affairs, who was in charge of hiring professors, agrees with Bieberbach:

Seen Vahlen 30.3.38
I request change. V.

The great Max Plank, a near 80-year-old icon of science, comes last and writes:

Planck 3.4.38
I will settle the affair. Planck

And settle Planck did. Just 4 days later, on April 7, 1938, Schur resigned from all Commissions of the Prussian Academy of Sciences. How does one assess Planck's role? Nazi collaboration, a pedantic fulfillment of his duties as Secretary of the Academy, or a desire to dismiss Schur gentler than someone else, like Vahlen or Bieberbach would have done? We will never know for sure which one(s) of these motivations prompted Planck's actions. I for one deeply regret that whatever the motive, Planck carried out the Nazi's dirty business. There were – had to be – other options. For example, Planck could have resigned from his Secretary's position, or from the Academy itself. Meanwhile the pressure on Issai Schur continued, and later that year he resigned from the Academy itself.

On November 15, 1938, Issai Schur applied for a foreign passport, which was needed for leaving Germany. On January 14, 1939, the *Reichsminister* for Science, Instruction and Public Education stated[19] that he "no longer objected to the issuance of a foreign passport for Dr. Schur" in view of "vulnerable health of Dr. Schur." He even approved paying Schur his emeritus remuneration through the date of Schur's departure.

On February 2, 1939, in the midst of the Gestapo's "personal interest" in him,[20] depressed and sick, Schur had to leave, better said, run away from Germany to Switzerland. Incredibly, the *Reichsminister* for Science, Instruction and Public Education believed that he could order Issai Schur where to live after Schur left Germany

---

[19] Archive of Humboldt University at Berlin, document UK-Sch 342, Bd.I, Bl. 47 and Bl. 47R.

[20] [Bra2]

and apparently when to come back to the Third Reich, for on February 24, 1939, he issued the following order number W T Schur 4:[21]

> I hereby authorize the change of permanent address for the Emeritus Prof. Dr. Issai Schur of the University of Berlin, residing in Berlin-Schmargendorf, Ruhlaer Str. 4, first to Switzerland and thereafter to Palestine starting February 1st 1939 until the end of March 1941.[22]

Schur's wife of 33 years, Med. Dr. Regina Frumkin-Schur, joined him there in March. They stayed in Bern for a few weeks with their daughter Hilde Abelin-Schur and her family. Broken mentally, physically, and financially, the Schurs moved on to Palestine.

While in Palestine, without means, Schur had to sell his only valuables, scientific books and journals, to the Institute for Advanced Study, Princeton, where his former student and friend Alfred Brauer was Herman Weyl's assistant, charged with library acquisitions. This book transfer must have been painful for both Schur and Brauer. Schiffer recalls one 1939 episode that shows how infinitely professional Schur was:

> [Schur] agreed to give a lecture at the Hebrew University and this I will never forget. He spoke about an interesting inequality in polynomial theory with the customary clarity and elegance. Suddenly, in the middle of his talk he sat down, bent his head and was silent. We, in the audience, did not understand what was going on; we sat quietly and respectfully. After a few minutes he got up and finished his talk in his usual manner.
>
> I was sitting next to a physician from the Hadassa Hospital who had come to see this famous man. He was quite upset; after the lecture he told me that Schur had obviously had a heart attack and he could not understand the self-discipline which had enabled Schur to finish his talk. That was the man Schur, for you!

Schiffer informs that Schur eventually got better, wrote several research papers, supervised a number theoretic work of Theodore S. Motzkin, and "started interacting with younger men at the Mathematics Institute." Issai Schur died from yet another heart attack in Tel Aviv right on his 66th birthday on January 10, 1941.[23]

The list of Issai Schur's Ph. D. students, who became world-class mathematicians, is amazing. It includes Heinz Prüfer (1921), Richard Brauer (1925), Eberhard Hopf (1926), Alfred Brauer (1928), Bernhard Neumann (1932), Hans Rohrbach (1932), Wilhelm Specht (1932), Richard Rado (1933), and Helmut Wielandt (1935). The list of successful mathematicians, who were Schur's undergraduates or were influenced by him in other significant ways, is too numerous to be included here. Schur with his teacher and a student produced one of the most remarkable succession lines in the history of modern algebra: Ferdinand Georg Frobenius–Issai Schur–Richard Dagobert Brauer.

---

[21] Archive of Humboldt University at Berlin, document UK-Sch 342, Bd.I, Bl. 53 and Bl. 53R.

[22] Of all people, I should not be surprised, for when I was leaving another bastion of tyranny, the Soviet Union, in 1978, I too was told where to go and where to live.

[23] For more details see [Bra2], [Schi], [Led1] and [Soi10].

34.3 Argument for Schur's Authorship of the Conjecture

Issai Schur made major contributions to various areas of mathematics.[24] Our interest here lies in the result he obtained during 1913–1916 when he worked at the University of Bonn as the successor to the celebrated topologist Felix Hausdorff.[25] There he wrote his pioneering paper [Sch] containing, as he put it, "a very simple lemma, which belongs more to combinatorics than to number theory." We proved Schur's Theorem in Chapter 32. Here I would like to formulate it again for your convenience:

Schur's Theorem 34.1 (Schur, [Sch]) Let m be a positive integer and $N > m!e$. If integers $1, 2, \ldots, N$ are m-colored then there are integers a, b, and c of the same color such that $a + b = c$.

Schur's Theorem gave a new birth to this novel way of thinking (likely nobody remembered Hilbert's 1892 lemma by 1916), a new direction in mathematics, now called the Ramsey Theory.

Leon Mirsky writes [Mir] on the occasion of the centenary of the birth of Issai Schur,

> We have here a statement of the type: 'if a system is partitioned arbitrarily into a finite number of subsystems, then at least one subsystem possesses a certain specified property.' To the best of my knowledge, there is no earlier result which bears even a remote resemblance to Schur's theorem. It is this element of novelty that impresses itself so forcibly on the mind of the reader.

Mirsky continues:

> After writing his paper, Schur never again touched on the problem discussed there; and this is in itself something of a mystery. For the strongest impression one receives on scanning his publications is the almost compulsive striving for comprehensiveness. There are few isolated investigations; in algebra, in analysis, in the theory of numbers, Schur reverts again and again to his original questions and pursues them to the point of where one feels that the last word has been spoken... Why, then, did he not investigate any of the numerous questions to which his Theorem points so compellingly? There is no evidence to enable us to solve the riddle. (Footnote: As will emerge from the discussion below, Professor Rado, if anyone, should be able to throw light on the mystery – and he tells me that he cannot.)

---

[24] For details see [Bra2] and [Led1].

[25] Both Alfred Brauer [Bra3] and Walter Ledermann [Led] reported 1911 as the time when Schur became an *Extraordinarius* in Bonn, while Schur's daughter Mrs. Hilde Abelin-Schur [Abe1] gave 1913 as the time her family moved to Bonn. The Humboldt University's Archive contains personnel forms (Archive of Humboldt University at Berlin, document UK Sch 342, Bd.I, Bl. 1–3) filled up by Issai Schur himself, from which we learn that he worked at the University of Bonn from April 21, 1913 until April 1, 1916, when he returned to Berlin.

The latter Mirsky's statement, apparently backed by Richard Rado was echoed in the standard text on Ramsey Theory [GRS2, p. 70], thus becoming a universal view on this matter: "Schur never again touched on this problem."

I have solved the Mirsky's "*mystery*," and my findings contradict Mirsky's and Graham–Rothschild–Spencer's conclusion. I will show in this section that the new Ramseyan mathematics, discovered by Issai Schur in his 1916 paper, remained dear to his heart for years to come. He thought about this new mathematics himself, and he passed his interest on to a number of his students: Hildegard Ille, Alfred Brauer, and Richard Rado.

As we have seen in Chapter 33, the third classic result of Ramsey Theory was published by B. L. van der Waerden in 1927, in which he presented "*Proof of a Baudet's Conjecture*" [Wae2]. The credit to Baudet for the conjecture remained unchallenged (and unsubstantiated), until 1960 when Alfred Brauer (1894–1985) made his sensational revelations.

"I remember Alfred [Brauer]," told me Mrs. Abelin-Schur, the daughter of Issai Schur [Abe2], "he was Assistant of my father, and I was then a little girl." An Assistant, a doctoral student (Ph.D. in 1928), a colleague (*Privatdozent* at the University of Berlin), coauthor and a friend through the difficult years of the Nazi rule, Alfred Brauer had unique knowledge of Issai Schur. Away from Germany for over 20 years, he returned to Berlin in 1960 to pay tribute to his teacher. His moving talk about Issai Schur given at the Humboldt University of Berlin on November 8, 1960, appeared in print in 1973 as an introduction [Bra3] to the three-volume set of Schur's collected works that Brauer edited jointly with another former Schur's Ph. D. student Hans Rohrbach. This talk offered a wealth of information about Schur. In particular, it revealed that Issai Schur, inspired by E. Jacobsthal's results about quadratic residues,[26] came up with the following two conjectures:

Conjecture 34.2 For any positive integer k and any large enough prime p, there is a sequence of k consecutive quadratic residues modulo p.

Conjecture 34.3 For any positive integer k and any large enough prime p, there is a sequence of k consecutive quadratic non-residues modulo p.

As was the case with Schur's Theorem of 1916 [Sch] (Chapter 32), a search for a proof of number-theoretic Conjectures 34.2 and 34.3 led Schur to conjecture a "helpful lemma":

Conjecture 34.4 For any positive integer k there is $N = N(k)$ such that the set of whole rational numbers 1, 2, ..., N, partitioned into two classes, contains an arithemetic progression of length k in one of the classes.

Alfred Brauer described circumstances of Schur's discovery that his Conjecture 34.4 had been proven:

---

[26] If the congruence $x^n \equiv a \pmod{m}$ has a solution for x, then a is called an *n-th power residue modulo m*. In particular, 2nd power residues are called *quadratic*.

Many years passed, but neither Schur nor other mathematicians, who were familiar with this conjecture, were able to prove it. One day in September of 1927 my brother [i.e., Richard Brauer, Ph.D. in 1925 under Schur] and I were visiting Schur, when [John] von Neumann came unexpectedly. He was participating in the meeting of the D.M.V.[27] and came to tell Schur that at the meeting Van der Waerden, using a suggestion by Artin, gave a proof of the combinatorial conjecture and was going to publish it under the title '*Beweis einer Baudetschen Vermutung.*' Schur was very pleased with the news, but a few minutes later he became disappointed when he learned that his conjecture about sequences [i.e., Conjectures 34.2 and 34.3 above] was not proven yet... It would have made sense if Schur were to propose a change in the title of van der Waerden's publication or an addition of a footnote in order to indicate that this was an old conjecture of Schur. However, Schur was too modest for that.

Paul Erdős, a man of an incredible memory for events, told me that in everything concerned with Schur, Alfred Brauer was by far the most reliable source of information. Paul also provided an additional confirmation of Schur's authorship of the conjecture. In the course of our long conversation,[28] that commenced at 7:30 PM on Tuesday March 7, 1995 in Boca Raton, Florida, during the traditional combinatorics conference's "jungle party," Paul told me that he heard about Schur's authorship of the conjecture from Alfred Brauer. Independently he heard about it from Richard Brauer. Finally, Schur's authorship was confirmed to Erdős by Erich Rothe, who obtained the information from his wife and Schur's former student Hildegard Rothe (born Hildegard Ille; Ph.D. in 1924 under Schur). As I am writing these lines, I am looking at a yellow lined sheet that Paul tore out of his notebook and next to his mathematical texts wrote for me "*Hildegard Ille*," so that I would remember her name when I get to write about it. Thank you, Paul!

I believe you will agree with me that we have produced as rigorous a proof as a historical endeavor allows that Issai Schur had the conjecture and created it independently from anyone else.

The historical research of this section shows for the first time that Issai Schur had been the most instrumental leader in the development of "Ramsey Theory before Ramsey" (I did not know that myself until the completion of this research). Started with his 1916 theorem (Chapter 32), Schur's interest in not-yet-born Ramsey Theory continued with the conjecture on arbitrarily long monochromatic arithmetic progressions in finitely colored integers. Right after Van der Waerden's publication, Issai Schur produced, as we will see in Chapter 35, *Generalized Schur's Theorem*, which generalized at the same time both Schur's and Baudet–Schur–Van der Waerden's theorems. With Schur's guidance, Schur's former student Alfred Brauer proved a Ramseyan result of his own (Chapter 35). Schur offered Ramseyan type problems to his doctoral student Hildegard Ille. Under Schur's guidance, Richard Rado generalized Schur's and Baudet–Schur–Van der Waerden's theorems in his doctoral

---

[27] *Deutsche Mathematiker-Vereinigung*, German Mathematical Society – the Annual September 18–24,1927 meeting took place in Bad Kissingen in Bavaria.

[28] Audio recorded by me.

dissertation and important consequent publications. In fact, Rado contributed to Ramsey Theory, perhaps, more than anyone.

As proof on the Schur's pudding, we will observe in Chapter 35 that Schur appears to have been the first to raise the problem of arbitrarily long arithmetic progressions of primes in 1920–1930, before Paul Erdős took charge of leading the development of Ramsey Theory. Erdős recalls on the pages of *Geombinatorics* [E95.32] on the occasion of Schur's 120th birthday in 1995:

> I first heard about Schur when I was a student from an old Hungarian algebraist Michael Bauer, who advised me to write to Schur about my results on prime numbers in arithmetic progressions. Schur was the first foreign mathematician with whom I corresponded. I wrote him my elementary proofs on some of my results on prime numbers in arithmetic progressions, which Schur liked very much, the results were published in Math. Zeitschrift in 1935.

In fact, the Ramseyan baton from Schur to Erdős may have been passed at their 1936 meeting in Berlin. "I was told that Schur sometimes referred to me as the sorcerer from Budapest," Paul recalled fondly in our conversations and in print [E95.32].[29] Amazingly, I found the eyewitness's reminiscences of this Erdős's visit to Berlin when Hilde Brauer, the widow of Schur's Assistant and friend Alfred Brauer, gave me a copy of her wonderful unpublished memoirs [BraH]. She married Alfred on August 19, 1934, and as a new "mathematical wife" from the Schur's circle, met Erdős during his Berlin visit:

> The latter [Paul Erdős], who was a child prodigy, surprised me at his first visit when he was barely twenty with curious interest for all details in bringing up a baby. He called all children epsilons, but knew all the names of his friends' babies.

I have got to mention one more of Schur's activities, in which he in a sense pre-dates Erdős. Schur's former student Richard Brauer (February 10, 1901–April 17, 1977) writes in his February 1977 introduction to his 3-volume collected papers [BraR] that appeared posthumously in 1980:

> He [Schur] conducted weekly problem hours, and almost every time he proposed a difficult problem. Some of the problems had already been used by his teacher Frobenius, and others originated with Schur. Occasionally he mentioned a problem he could not solve himself. One of the difficult problems was solved by Heinz Hopf and also by my brother Alfred and myself. We saw immediately that by combining our methods, we could go a step further than Schur. Our joint paper [BBH] in the list below originated this way.

Issai Schur had a great interest in creating problems and conjectures, and disseminating them on a regular basis, weekly, starting at least in 1920. Ramseyan-style problems and conjectures must have been part of this Schur's oeuvre. Paul Erdős, who took over the leadership of the Ramsey Theory, also had, as we all know, a

---

[29] Erdős was not only mathematically, but also personally attached to Issai Schur and his wife. "I several times visited his widow. In 1965 I visited her in Tel-Aviv with my mother," writes Paul [E95.32].

great interest in problem posing. He created his first open problem in 1931. In 1957, Paul commenced his celebrated "some of my favorite unsolved problems" papers.

This inquiry into the life of Issai Schur was possible due to most valuable help from Issai Schur's daughter Hilde Abelin-Schur; the widow of Alfred Brauer, Hilde Brauer; Schur's former student, Walter Ledermann; Paul Erdős; and Henrik Hofer and the Archive of Humboldt University of Berlin.

34.4 Enters Henry Baudet II

In 1995 when I presented the above argument for Issai Schur's credit in an essay [Soi10] written on the occasion of his 120th birthday, I specifically included a historically significant disclaimer:

> Nothing presented here excludes the possibility that Baudet created the conjecture independently from Schur. N. G. de Bruijn [Bru3], clearly understanding the rarity of Ramseyan ideas at the time, hypothesizes that Baudet was inspired by the 1916 paper [Sch] of Schur to independently create the conjecture. Perhaps, in the future historians would shed light on the question whether Baudet was an independent from Schur author of the second [counting Hilbert-1892, third] conjecture in the history of Ramsey Theory. Until then the conjecture ought to be rightfully called Schur's.

When my essay [Soi10] appeared, I learned from N. G. de Bruijn about the existence of P. J. H. Baudet's son, Henry Baudet, or as he sometimes called himself Henry Baudet II, and sent him a copy of my paper. I sowed an essay and harvested a fury. Henry Baudet the son (his full name Ernest Henri Philippe Baudet) was born on January 29, 1919 in Scheveningen: He was 76 at the time) when he replied to me in style all of his own:

> I write to you in my own English, which is far from good but it might be better than your own French or Dutch.

He then offered a counterexample[30] to Schur's 1916 theorem, and questioned Brauer's assessment of Schur as follows: "too modest seems hardly possible and hardly believable, considering the revolutionary essence of the theorem or the conjecture." Henry was clearly upset with my putting in doubt his father's credit. In my August 30, 1995 letter, I admitted that indeed "my French and Dutch are far inferior" to his fine English and offered Henry to publish in *Geombinatorics* his essay challenging my proof of Schur's authorship of the conjecture, if he so desired. I also offered Henry Baudet II to join me in the investigation of whether his father Henry Baudet I had created the conjecture independently from Schur.

Henry offered to help with documents upon his return to Holland from his summer home in Bourgogne, France. In addition to being a history professor, Henry was The Historian of the Delft Technical University, and the last Ph. D. student

---

[30] Schur's Theorem survived; Henry simply misunderstood it.

of the legendary Johan Huizinga of *The Waning of the Middle Ages* fame. From letter to letter I was promoted from "Professor Soifer" to "Alexander," to "Sasha." Our correspondence for the ensuing year was very intense: we exchanged some 30 letters. My family and me then paid a 5-day visit to Henry and his wife Senta Govers Baudet in their centuries-old stone house in the medieval village *Corpoyer-la Chapelle* (population 26) in Bourgogne, France.[31] As I am writing these lines, I am holding in front me a copy of Henry's book *Mon Village en France* warmly inscribed to my (then) wife Maya and I by the author on August 1, 1995. Later that year we also visited the Baudet family in their *Oegstgeest* house in the outskirts of Delft.

I learned about Henry and Senta helping Jews in the Netherlands occupied by Germany during the long 5 years 1940–1945. Henry recalls [Bau5]:

> I myself, finally, started studying history at Leiden University but this was interrupted when the Germans, during the war, closed the University. Somehow, nevertheless, I could remain in touch with my professors, at least in the beginning. Of course the German occupation made life extremely difficult, and this every year more and more. Resistance was a new activity we had to learn; hiding Jews was a daily concern and hiding ourselves was another. We lost many friends but somehow or other I got through myself (though my wife, then my girlfriend, then 17 years [old], got temporarily into jail for helping Jewish classmate to escape – she (I mean: her Jewish girlfriend) lives in Dallas now and we see each other and call each other by telephone).

In fact, Senta's name is inscribed in Yad Vashem, Israel, as she was awarded the title of a *Righteous Among the Nations*, granted to non-Jews who risked their lives to save Jews during the Holocaust. Senta helped her Jewish friend Liny L. Yollick escape from the Netherlands by lending her identification card. The escape was successful, but silly Liny sent the card back with a boy who ended up being caught by the Germans. On June 27, 1942 Senta was imprisoned by the Germans and spent a week in jail, interrogated day and night. Only her consistent denial of loaning the card to Liny, had finally convinced the jailers.[32] This was but one episode of the young family's participation in the resistance. Henry and Senta risked their lives, on numerous occasions by helping Jews hide or escape. They had to hide from the Germans, who occasionally came to look for them.

With Henry's help, I was able to successfully investigate the question whether Baudet I had earned the credit that Van der Waerden so nonchalantly had given him.

My dear friend Henry Baudet II was one of the most charming people I have ever met in my life. He passed away on December 16, 1998. In 2003, Delft Technical University created the *Henry Baudet Institute*, dedicated to the history of design, one of his many interests.

---

[31] Both Henry Baudet II and his son Remy Baudet, a wonderful violinist (music, even more than mathematics and chess, was a family tradition for generations), looked so Gascogne that they could play Alexandre Dumas père's D'Artagnan without any make-up.

[32] I thank Yad Vashem, The Holocaust Martyrs' and Heroes' Remembrance Authority, for providing me copies of the relevant documents.

34.5 Pierre Joseph Henry Baudet

B. L. van der Waerden gave Baudet credit in his 1927 paper [Wae2], which in fact
was called *Beweis einer Baudetschen Vermutung* (i.e., *Proof of a Baudet Conjecture*). We do not find Baudet's initials in Van der Waerden's paper. Indeed, Van der
Waerden did not even know that at the time of his publication Baudet had been dead
for 6 years. As is often the case with young and brilliant mathematicians, Van der
Waerden was probably not interested in the history of the problem he solved, nor
in the identity of the author of the conjecture. In reply to my questions, Van der
Waerden wrote on April 24, 1995 [Wae25]:[33]

> 1. *I heard of "Baudet's Conjecture" in 1926.*
> 2. *I never met Baudet.*
> 4. *I never met Schur.*
> 5. *I never heard about Schur's [1916] result.*

However, when, Van der Waerden published a *detailed* story of the emergence of
his proof in German in 1954 and in English in 1971 (presented in Chapter 33), he
became not only a celebrated mathematician, but also a famous historian of science,
author of the well-known book *Science Awakening* [Wae15] and numerous historical
articles. Sometimes he was deservedly harsh towards other historians [Wae15]:

> How frequently it happens that books on the history of mathematics copy their assertions uncritically from other books, without consulting the sources! How many fairy
> tales circulate as "universally known truths"!

Yet as a historian Van der Waerden did not investigate the authorship of the conjecture that became *his* classic theorem. Biographers of Van der Waerden faithfully
followed the master and credited Baudet with the conjecture, ignoring (or being
ignorant of) Brauer's reminiscences [Bra3], and provided no independent historical
analysis (see [Frel], [FTW], [Per], etc.,).

I thought that in all likelihood someone, sometime during the long years between
1927 and 1971, must have mentioned to Van der Waerden Brauer's assertion that in
his celebrated paper Van der Waerden had proved Schur's conjecture. Apparently,
no one did before me, as you can see from Van der Waerden's March 9, and April
4, 1995, replies [Wae23,24] to my inquiry:[34]

> *Dear Professor Soifer: Thank you for informing me that 'Baudet's conjecture' is in
> reality a conjecture of Schur. I did not know this.*

While Van der Waerden's acceptance of my argument for Schur's credit was
important, it contributed nothing to the question whether Baudet created the conjecture independently of Schur. As I wrote [Soi10], "Perhaps, in the future, historians
will shed light on the question whether Baudet was an author, independent of Schur."
This future has come: let us take a look at Baudet's role in our saga.

---

[33] See the facsimile of this letter in chapter 38 on page 429.

[34] See the facsimile of [Wae23] in this chapter on page 337.

Van der Waerden, March 9, 1995 letter to Alexander Soifer

It appears that Alfred Brauer was first to speak about Baudet on November 8, 1960 [Bra3] (see also recent English translation [LN]) ever since Baudet's obituaries appeared in 1921 ([Schuh] and 1922 [Arr]). Since Brauer knew first hand that Schur created the conjecture (and, I gather, assumed it to be unlikely that two people could independently come up with it), he attempted to "prove" that Baudet *did not* create the conjecture independently by showing how the conjecture got from Schur to Baudet: [35]

> Baudet at that time was an unknown student at Göttingen, who has later made no mathematical discoveries. On the other hand, at this time Schur's friend Landau was a professor at Göttingen, who obviously knew the conjecture, and used to offer unsolved conjectures as exercises to every mathematician he met. It was therefore highly probable that Baudet learned the conjecture directly or indirectly [from Landau].

Brauer repeated his assertions in English in print in 1969 [Bra2]:

> It seems that the title of van der Waerden's paper "Beweis einer Baudetschen Vermutung" [Wae2] is not justified. Certainly [sic] van der Waerden heard about the conjecture from Baudet, a student at Goettingen.

When Alfred Brauer spoke about Baudet (I wish he did not!), he entered the area not personally known to him. Consequently, Brauer presented his hypotheses as if they were truths. Moreover, I found Brauer's hypotheses to be dramatically false: Baudet "at that time" was not "an unknown student at *Göttingen*," but instead a brilliant young Ph.D. from *Groningen*. Brauer's suggestion that Baudet "later made no mathematical discoveries" was as gratuitous as it was incorrect. In addition to publishing his doctoral thesis [Bau1] and the inaugural speech [Bau2], Baudet wrote three papers [Bau3], [Bau4] and [Bau5] that appeared in *Christiaan Huygens*—not bad for someone who left this world at the untimely age of 30. Baudet was a Full Professor at Delft University at 27—can this be said of many mathematicians?

---

[35] Translated from the German original.

Alfred Brauer's valuable testimony about Schur's creation of the conjecture, as well as his regrettable misrepresentations about Baudet, were repeated in the standard text of the field [GRS1], [GRS2] and from there were copied by a good number of publications. It is time, therefore, to set the record straight, and convey how great a man the world lost in Pierre Joseph Henry Baudet.

The following account of Baudet's life was made possible only due to the indispensable assistance of Henry Baudet II, the son of the mathematician Pierre Joseph Henry Baudet. Unless otherwise credited, the following information, slightly edited, comes from Henry Baudet II's letters to me [[BII1] – [BII13]] and my personal interviews with him:

My father was born on January 22, 1891 in Baarn (province of Utrecht, The Netherlands) in nothing less than a psychiatric clinic, where my grandfather—a neurologist—was medical superintendent. A few years later my grandparents moved to The Hague, where my grandfather started a private practice. So it was in The Hague that my father grew up, attended the elementary school and then the Gymnasium from which he graduated in 1908. He was a dedicated chess player and cellist. (In this, he followed the family tradition: we all are musicians and chess players, though not on his level).

In September 1908 my father enroled as a student of mathematics at Leiden University, where he studied under Kluyver. I know next to nothing about his study in Leiden, except the fact of his early fame as a chess player, a musician and a future mathematician. He obtained his master's degree in 1914, as far as I know just on the eve of the World War I, and became a mathematician at the same Gymnasium in The Hague where my father had been a pupil. He stayed there until his 1919 appointment as a Professor at the University of Technology at Delft (then still named the Technical High School).

As a student at Leiden he met my mother [Ernestine van Heemskerck] who studied in the Faculty of Arts, and my parents got married on April 7th of 1914... My sister (also a mathematician) was born in 1915 on the 31st of January. I myself arrived 4 years later on January 29th, 1919. So all of us are Aquarius.

How my father and Schuh[36] met, I don't know; probably in the Society of Mathematicians. They were, however, close friends since 1914 or 1915... With Schuh as supervisor, my father began to work on his thesis, but he could not take his doctor's degree with him, as Delft had no doctorate in mathematics. And [Johan A.] Barrau [a professor of mathematics at Groningen University (1874–1953)] ultimately took over Schuh's job.

In Memoriam Prof. P. J. H. Baudet" [Arr] by Dr. E. Arrias appeared on January 28, 1922. The author, who had known Baudet for 15 years, reported the astonishing talents of Pierre Joseph Henry. At 15, Baudet was "known for virtually never losing a game [in chess] and playing several simultaneous games blindfold... But all these achievements were outshone by the miraculous things he has done with the Laskagame, invented by Dr. E. Lasker [a mathematician and the world chess

---

[36] Frederick Schuh, 1875–1966, Ph. D. under Diederik Korteweg, as was L. E. J. Brouwer after him, a very versatile mathematician, with numerous publications in analysis, geometry, number theory, statistics, recreational mathematics, teaching of mathematics, etc.

champion during the incredibly long period 1894–1921]. Before Lasker had his new game published, he submitted it to Baudet for evaluation. With his characteristic tempestuous application Baudet mastered this game; it was as if he finally had found something that could fully satisfy his wits. This exceptionally intricate game with its discs in four different colours, its capricious, almost incalculable combinations, suited his mathematical brain exactly. It is, therefore, not surprising that having studied the game for half a year, he could scarcely be beaten by Lasker himself. . . Thanks to his enthusiasm a Lasca society was founded in The Hague, and even a first national tournament organized, but after everything had taken shape, he died one day before the tournament, to which he had been looking forward like an eager child (for in spite of his scientific greatness he was a child in joy). . .[37]

Pierre Joseph Henry Baudet (1891–1921). Courtesy of Henry Baudet II

---

[37] As I learned from Professor N. G. de Bruijn [Bru1], "in his *Brettspiele der Völker* (Berlin 1931) Lasker describes a game of 'Laska' he lost to Baudet at a tournament in The Hague 1920. ('Laska' was Lasker's own invention, which he tried to promote at a time he thought that eventually all serious chess games would lead to a draw.)"

As proficient as he was at board games, as high was his reputation as a musician... Being an extraordinarily sensitive cellist, he completely mastered the technique of this instrument. Many were the times that he contributed to the success of concerts by his impassioned playing. And all this without score; a feat only very few people are so privileged. With him it was not a matter of learning notes, but he absorbed the complete picture of the composition, and even when he had not seen the composition for 10 years, he was able to conjure it up clearly and to play it from memory, when only hearing the piano part... He was excellent at reading scores and he conducted already during his grammar school period. He was fully familiar with theory and counterpoint. Only recently, he could prove this when the vice-chancellor of Delft University asked him to orchestrate the Don Juan for the students' string orchestra. Next to all his excessively many occupations, this task could be added without any problem. He finished it just before death overtook him...

It was pure scientific curiosity that had made him master this as well as everything he did: learning Hebrew and the four Slavic languages simultaneously was no trouble at all, since he was learning anyway—and in fact this was far more interesting—that comparative linguistics... Stacks of work are lying in his study; constantly new ideas suggested themselves to him which he noted down only in lapidary form. He did not get around publishing much, but his confrere friends will need years of hard work to sort out and work up his sketchy notes.

On the birthday of Jesus this highly gifted man with his magnificent Christ like features parted from this earthly life, at the same age, as his greatest master. But in our thoughts he will rise again and stay alive for us as long as we keep breathing!"

Baudet defended his doctoral thesis cum laude in 1918 at Groningen University, and became a professor in Pure and Applied Mathematics and Mechanics at Delft Technical University in 1919. He was 28. Pneumonia brought his life to an untimely end on the Christmas Day of 1921. His first obituary [Schu] was written by his friend and teacher Frederik Schuh.

Pierre Joseph Henry Baudet was an extraordinary man indeed. But did he create the conjecture?

34.6 Argument for Baudet's Authorship of the Conjecture[38]

What evidence do we have to assign the credit for the conjecture to Baudet? The credit given to Baudet by Van der Waerden [Wae2] appears to be insufficient, especially since Van der Waerden wrote to me [Wae25]: "I never met Baudet" and also "I heard of 'Baudet's Conjecture in 1926" (i.e., 5 years after Baudet's death). However, by back-tracking the link from Van der Waerden to Baudet, we reach firmer grounds.

---

[38] This is an expanded version of my tribute [Soi11] to P. J. H. Baudet that was published on the occasion of his 105th birthday.

The search, in fact, was started by Henry Baudet II. Born on January 29, 1919, Henry lost his father at the age of 2, and always wanted to find out more about him. In 1962–1963, Professor of tax law and an amateur mathematician Tj. S. Visser gave a talk *Attack on Sequences of Natural Numbers* attended by Henry Baudet and his 15–16-year-old son Rémy. Surprisingly, the four-page brochure (in Dutch) of this talk survives, and thus we are granted an attendance to Visser's talk:

> My story is about the most beautiful statement of number theory, The Theorem of Baudet. The pearl of Baudet...
> Baudet is the early departed in the beginning of this century Delft's Professor of mathematics, born in Nenegouw...
> His pearl of the theory of numbers is: If one divides the natural numbers 1, 2, 3, 4... *ad infinitum* into a random number of boxes, then there is nevertheless always at least one box which contains an AP of arbitrary length...
> This proposition was formulated by Prof. Dr. P. J. H. Baudet in 1921. He died shortly after, leaving a wife and a baby. Many celebrities tried to find a proof of this theorem. The young, also Dutch mathematician succeeded. His name was B. L. van der Waerden. He published his proof in 1927 in *Het Nieuw Archief* under the title *Beweis einer Baudetschen Vermutung*.
> It takes five pages, uses no higher mathematics but is very heavy. He seems to have found it during a holiday session at Göttingen where his astuteness rightly won large admiration. Bartel van der Waerden is a son of the engineer-teacher Theo, doctor in technical sciences, a very prominent person elected to Parliament from the S.D.A.P., known as '*rooie Theo*' [Red Theo]. The young Bartel became professor at Groningen, was later oil-mathematician, is now at Zurich director of the mathematical institute, and is world-renown.
> After 1927, the statement and its proof fell asleep.

Tj. Visser then conveyed how the Russian mathematician Aleksandr Yakovlevich Khinchin brought the theorem back to life by publishing it, with a slightly different proof found by his student M. A. Lukomskaja, as one of the pearls in his book *Three Pearls of Number Theory*, which appeared in Russian, German, and English ([Khi1], [Khi2], [Khi3], [Khi4]).

As an amateur mathematician, Henry was fascinated by the conjecture. "Could I write to him [i.e., Van der Waerden]?" he asked the family friend and his father's mentor Frederick Schuh (February 7, 1875–January 6, 1966). "Of course," Schuh replied. Henry recalls:

> In the context and the fact that I proposed to Schuh to give me the address of Van der Waerden, it was clear that Schuh considered it [the conjecture] to be an important affair. He agreed that I should write.

I asked Henry (we had long interviews in his centuries-old Bourgogne stone house):

> Does it appear indirectly that Schuh was in total agreement that it was Baudet's conjecture that Van der Waerden proved?

"Absolutely, absolutely yes, absolutely," replied Henry. And so, on September 1, 1965 Henry Baudet II wrote to Van der Waerden in a style already known to the reader from Henry's first letter to me:[39]

I am the son of my father. It is always the case, but you understand what meaning this introduction has in this case. Somehow from afar I was following your publications, and thus I was able to get into my hands your work in the *Abhandlungen aus dem Mathematischen Seminar Hamburg* [Wae16]. For me this is not a completely closed book. Having at one point started in mathematics, I have become a historian in the end, and it is something entirely different.

In this letter to you, a fairly remarkable fact is taking place. It is a fact that I cannot say anything special, but nevertheless I wanted very much to establish a contact with you. Of course, I would like to ask you whether you have a reprint of your publication of 1926, in which you present a well known proof; possibly also other publications, if such exist related to my father, especially to the abovementioned work in *Hamburgsche Abhandlungen*.

Last year in Zürich I tried to find your name in the telephone book. Unfortunately I was unable to find there your name. I also tried to contact you at the University of Zürich, but also without result.

As far as I can follow number theory, I find it exciting. If I were to become a mathematician, my inclinations would have certainly led me in this direction – in the direction of numerical mathematics and number theory. In my free time I continue to deal with Fermat and Mersenne; although "in general" with the history of mathematics. I would appreciate it very much if I could hear something from you and possibly you could send me one or several copies of your works of those where you have written about my father.

On October 20, 1965 Van der Waerden replied:

It was very nice to receive a letter from you. I have not known your father and have never written anything about him. I heard about his conjecture which he had posed at *Het Wiskundig Genootschap* (Mathematical Society) in Amsterdam.

I am sending you a *overdrukje* (reprint) of my work from *Hamburger Abh.* and on loan a photocopy of my work in *Het Archief* from 1926.[40] I will further ask the publisher Birkhäuser to send you a copy of my psychological research "Einfall und Überlegung" in which the history of the solution of this problem is also considered.

Thus, Van der Waerden stated that *P. J. H. Baudet posed his conjecture at the Mathematical Society in Amsterdam.* Van der Waerden attached to his letter copies of his original proof [Wae2] and his just published reminiscences [Wae16]. Henry Baudet II discussed this correspondence with Frederick Schuh, who was a major figure in the Amsterdam mathematical circles in the 1920s. This is why the following Henry Baudet's May 27, 1996 reply to my inquiry is the crux of the matter [BII12]:

---

[39] Henry Baudet II provided me with copies of his correspondence with Vander Waerden — thank you, Henry!

[40] Actually, 1927.

When I told Schuh about my correspondence with Van der Waerden, he would have definitely told me that the conjecture was not my father's, if it had been not his.

Schuh did not correct Henry Baudet, because for him P. J. H. Baudet's authorship of the conjecture was a long known fact.

After Henry Baudet the son, the next person, who showed an active interest in the authorship of the conjecture, was N. G. de Bruijn. *Wiskundig Genootschap* (Mathematical Society) decided to publish a 2-vol. edition entitled *Two Decades of Mathematics in the Netherlands: 1920–1940: A retrospection on the occasion of the bicentennial of the Wiskundig Genootschap*. The book was to reproduce short works of the leading mathematicians of the period, such as Van der Corput, Van der Waerden, Van Danzig, each followed by a commentary. Van der Waerden was to be represented by *Beweis einer Baudetschen Vermutung* [Wae2], with a commentary by de Bruijn, who in his March 29, 1977 letter posed Van der Waerden several questions about the history of the conjecture. The latter replied on April 5, 1977 [Wae19]. I thank N. G. de Bruijn for sharing with me this important letter of Van der Waerden:

I will happily answer your questions.

1. I am quite sure that I heard about the conjecture for the first time in 1926, around the time I got my Ph.D. in Amsterdam. I probably picked it up at one of the monthly meetings of the *Wiskundig Genootschap*, where Schuh appeared regularly. I do not know if it was Schuh himself or someone else who made me aware of this.
2. Yes, the entire affair happened on a single afternoon. Only the cases k = 2, k = 3 I had already figured out before.
3. I think I only later heard of I. Schur's proposition.
4. No, I do not know anything about Baudet. I have a vague memory that he was a friend or pupil of Schuh.
5. My biography: I have studied mathematics, physics, astronomy and chemistry. Mathematics mostly with Mannoury, Hendrik de Vries and Brouwer. Astronomy with the excellent Pannekoek. In 1972 I retired in Zürich. Not "emeritus" because that does not exist in Switzerland.

Included is the Bibliography with a few corrections. Furthermore, I have nothing to add to your piece. Your praise "A thing of beauty is a joy forever,"[41] pleases me.

Thus, Van der Waerden got the conjecture directly or indirectly from Frederick Schuh, Baudet's mentor and close friend, and the authorship of Baudet came to Van der Waerden with the conjecture. Van der Waerden has even "a vague" but correct memory that Baudet was Schuh's "friend and/or pupil." Thus, we have traced the way the conjecture traveled from Baudet to Van der Waerden via Schuh. However, one question remains open: Did Baudet independently create the conjecture or received it indirectly from Schur (try not to mix up here Schur and Schuh)? This is the question I was unable to resolve until December 18, 1995, when Henry Baudet II, the son and historian, came up with what he humorously named "*A Second Conjecture of Baudet*" [BII4]:

---

[41] The text in quotation marks is in English in Van der Waerden's otherwise Dutch letter.

It seems reasonable to suppose that neither Professor F. Schuh nor my father were informed of Schur's work. Though Germany was 'next door,' the World War broke nearly all contacts, which were only slowly restored in the course of the '20s.

Henry Baudet II found convincing evidence for his conjecture. The first major mathematical event after World War I was unquestionably *Congrés International des Mathématiciens*, which took place during September 22–30, 1920, in Strasbourg, France. The whole world was represented there, with the notable exception of the German mathematicians, even though Strasbourg was located right by the French border with Germany. The wounds of the World War I were still very painful. On the French initiative, the Germans were banned from 1920 and the consequent 1924 International Congresses of Mathematicians. It was not until the congress of 1928 that they were allowed to rejoin the world of mathematics.

Both J. A. Barrau and P. J. H. Baudet were in attendance at the Strasbourg's 1920 Congress. Baudet mailed to his wife *daily accounts* of his meetings at the Congress, and these letters have survived the long years, including another war that followed. The letters report the meetings with a most impressive group of mathematicians: Denjoy, Fréchet, Valiron, Châtelet, Dickson, Eisenhart, Le Roux, Typpa, Lebesgue, Larmon, Young, De Vallée Poussin, Deruyts, Jordan, Montel, Volterra [Bau3]. The letters also captured impressions and emotions of days long gone by [Bau5]:

> I am in nearly permanent contact with the Americans here. They are after all the nicest people here. And this is not only my opinion but also Barrau's. The nicest of all is Eisenhart. [Letter of September 29, 1920]
>
> At 11 P.M. all the cafés here are closed. You understand that this is not our cup of tea. It will be much better at our next Congress. That will be in the U.S. in 1924. The Americans here are really very nice people. Dickson and Eisenhart are their principle representatives, Eisenhart brought his wife who is quite a nice person. We talked a lot in these days and she definitely expects you [Ernestine Baudet] too in the U.S. next time. You see: nothing can change it, you *must* join me next time. Barrau told Dickson about the critical review he [Barrau] had written and has modified after my severe critical comments. The consequence of the discussion was that Dickson asked me to write him about the matter, as Barrau and I had here no copy of our controversial texts. [Letter of September 23, 1920]

Thus, Baudet and Barrau met Princeton's dean and mathematics chair Luther Pfahler Eisenhart. Do not forget his name: we will meet Eisenhart again in Chapter 37, when he will invite Van der Waerden to come to Princeton.

They also met and had discussions with the famous American number theorist Leonard Dickson. The meeting with Dickson attracted my attention in particular, because Dickson's result inspired Issai Schur to come up with Schur's Theorem of 1916. However, this route only confirmed Baudet II's conjecture. Right before the Congress, in April of 1920, Dickson had completed volume 2 of his monumental *History of the Theory of Numbers* [Dic2]. He did cite (p. 774) Schur's 1916 paper [Sch]: "* J. Schur gave a simpler proof of Dickson's theorem." But in the Preface Dickson explained that "the symbol * before the authors' names" signified "that the papers were not available for review," i.e., even Leonard Dickson, the most informed number theorist of his time, had not himself seen the Schur's 1916 paper before the Congress.

Geographically speaking, Baudet and Schur had one chance to meet in August of 1921, when Henry and Ernestine Baudet with their daughter Puck visited their friend Emanuel Lasker and had a short stay in his Berlin house. Puck "still has clear recollection of their stay at the Laskers, particularly when their rowing boat on the Wannsee[42] was wrecked,"[43] because neither Lasker nor Baudet could not swim and had to be rescued. We are fortunate to have a photograph from this visit (page 345). However, Puck, does not remember visiting the University.

Seated Ernestine, Puck and P. J. H. Baudet; standing (from the left) Emanuel Lasker and a Gymnasium Rektor, Lasker's house, August 1921, Berlin, Courtesy of Henry Baudet II

The family correspondence has survived, and it does not indicate that any new acquaintances were made during this trip, which took place just a few months before the untimely passing of P. J. H. Baudet.

---

[42] The reader would recognize the name of *this* lake. Lasker–Baudet humorous episode took place at the place where on January 20, 1942 fifteen high-ranking civil servants and SS-officers decided on "The Final Solution" of the Jewish question in Europe. They agreed to deport European Jews to the East and murder them all.

[43] [BII7]

Thus, it is plausible to conclude that Baudet and Schur never met and that P. J. H. Baudet discovered the conjecture independently of Issai Schur.

My investigation into the life of Pierre Joseph Henry Baudet was possible due to help from Henry Baudet II and Nicolaas G. de Bruijn, to whom I extend my deepest gratitude.

34.7 Epilogue

The evidence presented here clearly shows that two brilliant men, Issai Schur and Pierre Joseph Henry Baudet, independently created the third conjecture of Ramsey Theory before Ramsey. From now on, let it be known as *The Baudet–Schur Conjecture*. What can be a happier conclusion to a historical research!

Obviously, without the conjecture no proof would have been possible. To conjecture such a pioneering result was surely as great a contribution as its proof by B. L. van der Waerden; it is therefore fitting to call the monochromatic arithmetic progressions theorem after all three contributors: *The Baudet–Schur–Van der Waerden Theorem*.

At the time when Alfred Brauer's work [Bra1] proving two original Schur's conjectures appeared in 1928, Frank Plumpton Ramsey was working on his pioneering work [Ram2] that he submitted for publication later that year. A few years later, in 1932, Issai Schur's student Richard Rado defended his doctorate dissertation [Rad1], which was Rado's first fundamental contribution to Ramsey Theory. During the winter of 1932–1933 Paul Erdős and George Szekeres wrote their first Ramseyan paper [ES1]. Since then Paul Erdős had inspired many mathematicians to enter the field. A new era of maturing Ramsey Theory has began.

I ought to point out amazing ways in which the lives of the players of this story are interwoven. Mentor and friend of Baudet, Frederik Schuh was instrumental in Van der Waerden getting to know the Baudet–Schur conjecture. Baudet's Ph. D. thesis *Promotor* (supervisor) was the very same Johan Antony Barrau, who in 1928, while moving to Utrecht, would offer Van der Waerden his chair at Groningen, and again in 1942 propose Van der Waerden for his chair at Utrecht. Read much more about it all in the following chapters, dedicated to vast generalizations of the Baudet–Schur–Van der Waerden Theorem (Chapter 35) and to my search for Van der Waerden (Chapters 36–39).

The brutal war separated the authors of the Baudet–Schur–Van der Waerden Theorem and their families. As we have seen here, Baudet's son Henry Baudet II and his girlfriend Senta worked in the Dutch underground saving lives of Jews. Issai Schur was thrown out of the University of Berlin, and following years of humiliation escaped to Palestine; his tired heart soon gave up. Being Dutch, Van der Waerden served as a Professor in Germany throughout the entire Nazi time.

35

Monochromatic Arithmetic Progressions: Life After Van der Waerden

35.1 Generalized Schur

> *And God said, "Let there be light"*
> – Genesis

And there was light: Issai Schur—who else—produced the first spark, a generalization of Baudet–Schur–Van der Waerden's Theorem. In fact, his result generalized both Schur's and Baudet–Schur–Van der Waerden's Theorems. With all of the search engines of today's Internet, one would be hard pressed to find it, for it did not appear in a Schur's paper: this modest man gave it to his former student Alfred Brauer to publish!

Alfred Brauer writes [Bra2] that a few days after his and Richard Brauer's 1927 visit of Schur (we peeked at this meeting in Section 34.3), he proved Schur's conjecture about quadratic residues (Conjecture 34.2) with the use of Baudet–Schur–Van der Waerden's Theorem. Schur then noticed that Brauer's method of proof can be used for obtaining a result about sequences of n-th power residues. Soon Issai Schur found a short, Olympiad-like, brilliant way to prove the following result that generalized *both* Schur's Theorem and Baudet–Schur–Van der Waerden's Theorem.

Generalized Schur's Theorem 35.1 (Schur, [Bra1], [Bra2]) For any k and l there is $S(k, l)$ such that any k-coloring of the initial set of positive integers $[S(k, l)]$ contains a monochromatic arithmetic progression of length l together with its difference.

Proof For 1 color we define $S(1, l) = l$, and the statement is true.

Assume the theorem is true for k colors. We define

$$S(k + 1, l) = W\left(k + 1, (l - 1) S(k, l) + 1\right),$$

where $W(k, l)$ is as defined in Theorem 33.1. Let the set of integers $[S(k + 1, l)]$ be colored in $k + 1$ colors. Then by Theorem 33.1 (see the right side of the equality above), there is a $(l - 1) S(k, l) + 1$ term monochromatic arithmetic progression

$$a, a + d, \ldots, a + (l - 1) S(k, l) d.$$

A. Soifer, *The Mathematical Coloring Book*,
DOI 10.1007/978-0-387-74642-5_35, © Alexander Soifer 2009

For every $x = 1, 2, \ldots, S(k, l)$, this long monochromatic arithmetic progression contains the following l-term arithmetic progression:

$$a, a + xd, \ldots, a + (l - 1)xd.$$

If for one of the values of x, the difference xd is colored the same color as the progression above, we have concluded the proof of the inductive step. Otherwise, the sequence

$$d, 2d \ldots, S(k, l)d$$

is colored in only k colors, and we can apply to it the inductive assumption to draw the required conclusion. ∎

A great proof, don't you think! It is interesting to note here that unlike Baudet–Schur–Van der Waerden's Theorem, Generalized Schur's Theorem does not have a Szemerédi-style density generalization (see more about it later in this chapter).

Schur wanted Alfred Brauer to include this theorem (as well as the one about n-th power residues) in Brauer's paper because Schur believed to have used Brauer's method in these proofs. Schur did not want to take away any credit from his student. The student had to oblige but he "always called it Schur's result"[44] and gave Schur credit everywhere it was due in his paper [Bra1], which appeared in 1928. A few weeks later Brauer also proved Schur's conjecture about quadratic non-residues (Conjecture 34.3), which appeared in the same wonderful, yet mostly overlooked paper [Bra1].[45]

Schur's ingenious contributions to Ramsey Theory before Ramsey apparently do not end here. We will come back to them later in this chapter, for I wish to speak about density results now.

35.2 Density and Arithmetic Progressions

Let us look how this flourishing field has evolved. We will start with the key definition from the Erdős–Turán 1936 paper [ET]. Denote by $r_l(N)$ the maximum number of integers less than or equal to N such that no l of them form an arithmetic progression. Paul Erdős and Paul Turán proved a number of results about $r_3(N)$ and conjectured that

$$r_3(N) = o(N).$$

This conjecture was proven in the 1953 by Klaus F. Roth [Rot]. The only conjecture about the general function $r_l(N)$ in Erdős–Turán paper was attributed to their

---

[44] [Bra2]

[45] I say 'overlooked' because the leading Ramsey Theory book [GRS2] contains almost identical result (Theorem 2, p. 70) without reference or credit to Schur.

friend George Szekeres, and was later proven false. Sixteen years had passed before in 1969 Endre Szemerédi proved [Sz1] that

$$r_4(N) = o(N).$$

In a 1973 paper Paul Erdős [E73.21, pp. 118–119] remarked: "[this] very complicated proof is a masterpiece of combinatorial reasoning." A very surprising paragraph followed [ibid.]:

> Recently, Roth [1970] obtained a more analytical proof of $r_4(n) = o(n)$. $r_5(n) = o(n)$ remains undecided. Very recently, Szemerédi proved $r_5(n) = o(n)$.

Clearly, Erdős added the last sentence at the last moment, and should have removed the next to last sentence. The latter result has never been published, probably because Endre Szemerédi was already busy trying to finish the proof of the general case. On April 4, 2007, right after his talk at the Princeton's Discrete Mathematics Seminar I asked Szemerédi whether he had that proof for 5-term APs, and what came of it. Endre replied:

> Hmm, it was so close to finding the proof of the general case, maybe two months before, that I did not check all the details for 5. It was more difficult than the general case.

Indeed, in 1974 he submitted, and in 1975 published [Sz2] a proof of the general case, i.e., for any positive integer l

$$r_l(N) = o(N).$$

This work in one stroke earned Szemerédi a reputation of a wizard of combinatorics. By then the terminology has changed, and I wish to present here the more contemporary formulation that is used in Szemerédi [Sz2]. We will make use of the notion of "proportional length," known as *density*, in the sequence of positive integers $N = \{1, 2, \ldots, n, \ldots\}$. The *density* is one way of measuring how large a subset of N is. Its role is analogous to the one played by length, in the case of the line R of reals.

Let A be a subset of N; define $A(n) = A \cap \{1, 2, \ldots, n\}$. Then *density* $d(A)$ of A is naturally defined as the following limit if one exists:

$$d(A) = \lim_{n \to \infty} \frac{|A(n)|}{n}.$$

The *upper density* $\bar{d}(A)$ of A is analogously defined as

$$\bar{d}(A) = \lim_{n \to \infty} \sup \frac{|A(n)|}{n}.$$

Now we are ready to look at a classically simple formulation of the Szemerédi's result.

Szemerédi's Theorem 35.2 Any subset of N of positive upper density contains arbitrarily long arithmetic progressions.

In various problem papers, Erdős gives the date of Szemerédi's accomplishment and Erdős's payment as 1972 (sometimes 1973, and once even 1974). The following statement appears most precise as Erdős made it very shortly after the discovery at the September 3–15, 1973 International Colloquium in Rome [E76.35] and places Szemerédi's proof around September 1972:

> About a year ago Szemerédi proved $r_k(n) = o(n)$, his paper will appear in "Acta Arithmetica"...

Erdős was delighted with Szemerédi's result, and awarded him $1000 in the late 1972–1973 [E85.33]:

> In fact denote by $r_k(n)$ the smallest integer for which every sequence $1 \leq a_1 < a_2 < \ldots < a_l \leq n$, $l = r_k(n)$ contains an arithmetic progressions of k terms. We conjectured
>
> $$(15) \qquad \lim\ r_k(n)/n = 0.$$

> I offered $1000 for (15) and late in 1972 Szemerédi found a brilliant but very difficult proof of (15). I feel that never was a 1000 dollars more deserved. In fact several colleagues remarked that my offer violated the minimum wage act.

On April 4, 2007, Szemerédi confirmed my historical deductions:

> I proved [the] general case in fall 1972, and received Erdős's prize in 1973.

I refer you to the original paper for the proof which is brilliant and hard. Partial results are proven in [GRS2] (it is remarkable that even this standard text in the field did not include Szemerédi's complete proof!).

While Szemerédi's Theorem is a very strong generalization of Baudet–Schur–Van der Waerden's Theorem, Paul Erdős and Ronald L. Graham observe in their 1980 problem book [EG, p. 19] that the analog of Szemerédi's theorem does not hold for Generalized Schur's Theorem 35.1. Can you think of a counterexample before reading one below?

Observation 35.3 (Erdős–Graham, 1980) Szemerédi-like generalization does not hold for Generalized Schur's Theorem.

Proof The set of odd integers of density $1/2$ cannot contain even a 2-term arithmetic progressions and its difference! ∎

35.3 Who and When Conjectured What Szemerédi Proved?

Throughout this book (and my life) I have given credit for a result to both the creator of the conjecture and the author of the first proof. Truly, without good conjectures we would not have many results. Moreover, pioneering conjectures, such as Baudet–

Schur, played a major role in paving the way for new mathematics. Our question here naturally is: Who and when conjectured what Szemerédi proved?

No one would expect a mystery here—just look in Szemerédi's 1975 paper, where he presents the history of advances in good detail. He starts with giving credit for conjecturing his theorem to Paul Erdős and Paul Turán in their 1936 paper [ET]. And so I am looking at this short important paper—without finding the conjecture, except for the case of 3-term arithmetic progressions. This incorrect credit is then repeated in the standard Ramsey Theory texts [GRS1] and [GRS2] in 1980 and 1990 respectively, and from there on everywhere else, until in 2002 leaders of the field Ronald L. Graham and Jaroslav Nešetřil notice the discrepancy, and explain it in the following way in two important memorial publications [GN1, p. 204]: and [GN2, p. 356]:

> Although they [Erdős and Turán] do not ask explicitly whether $r_l(N) = o(N)$ (as Erdős did many times since), this is clearly on their mind as they list consequences of a good upper bound for $r_l(N)$: long arithmetic progressions formed by primes and a better bound for the van der Waerden numbers.

Clearly, my friends Ron and Jarek, and I agree that the conjecture does not appear in the 1936 [ET]. Their argument was that the young Erdős and Turán had the conjecture "clearly on their mind" could be viewed more as an eloquent homage to the two great mathematicians rather than an historical truth. We therefore must research further.

In his 1957 first-ever open-problem paper [E57.13], Paul Erdős indicates that before him and Turán, Issai Schur (!) called on studying the longest arithmetic progressions-free opening segments of positive integers. Erdős writes:

> The problem itself seems to be much older (it seems likely that Schur gave it to Hildegard Ille, in the 1920's).

Erdős returns to Issai Schur's contribution in his 1961 second open-problem paper [E61.22], which in 1963 also appears in Russian [E63.21]:[46]

> The problem may be older but I can not definitely trace it. Schur gave it to Hildegard Ille around 1930.

Paul told me that he "met Issai Schur once in mid 1930s," more precisely in 1936 in Berlin. They shared a mutual admiration (as we have seen in Section 34.3). Undoubtedly, they discussed prime numbers, but likely not arithmetic progressions. Erdős learned about Schur's interest in arithmetic progressions and early Ramsey-like conjectures and results from Hildegard Ille (1899–1942). Now, this requires a bit of explanation, because they probably had never met!

Erich Rothe (1895–1988), *Dr. phil. Universität Berlin* 1926 under the eminent Erhard Schmidt and Richard Mises, married a fellow student Hildegard Ille, *Dr. phil. Universität Berlin* 1924 under Issai Schur. They taught at *Universität Breslau*, Germany (later and earlier Wrocław, Poland) until, as Jews, they were forced to flee the Nazi Germany in 1937, and came to the United States. Hildegard passed

---

[46] This Russian publication does not appear in any of Paul Erdős's bibliographies.

away at a young age in 1942. The accomplished mathematician Erich Rothe held a professorship at the University of Michigan, from 1941 until his retirement in 1964. His eulogy (*Notices of Amer. Math. Soc.*, 1988, 544) quotes Chair of the University of Michigan D. J. Lewis saying that "Rothe was a scholar of the old school. He was very broadly educated. He was a wise and judicious man of much wit. His companionship was very much in demand."

Erich Rothe was Paul Erdős' source of reliable information on problems and conjectures in number theory that Issai Schur had shared with Rothe's wife Hildegard (Ille) Rothe. From Rothe, Erdős learned about Schur's authorship of the arithmetic progressions conjecture, proven later by Van der Waerden (Chapter 34). Erdős learned from Rothe that Issai Schur yet again contributed to number theory and Ramsey theory when he asked his graduate student Hildegard to investigate arithmetic progressions-free arrays of positive integers. To my surprise, no one acknowledged the credit Erdős gave to Schur in his first open-problem papers [E57.13], [E61.22] and [E63.21].

However, I believe, that Erdős learned about Schur being first to investigate this subject after Erdős and Turán independently rediscovered it: their paper [ET] was published in 1936, while Erich and Hildegard Rothe came to the Unites States in 1937. Moreover, Erdős–Rothe conversations took place after Hildegard's passing in 1942. Paul was certainly correct when in both his 1957 and again 1961 open-problem papers he wrote, "The first publication on the function $r_k(n)$ is due to Turán and myself." This was an important paper, and Paul knew that. Yet, it contained the "density" conjecture for only 3-term arithmetic progressions. Graham and Nesetril are correct when they write in [GN1] and [GN2] that "Erdős did [pose the general case conjecture] many times," but the real question is: when did he pose the conjecture for the first time?

I am reading again Erdős's first 1957 open-problem paper, where Paul writes:

In [ET] we stated our conjecture that $\lim r_3(n)/n = 0$ ... Roth [Rot] proved that $r_3(n) = o(n)$ ... The true order of magnitude of $r_3(n)$ and, more generally, of $r_k(n)$, remains unknown.

Paul discusses the general function $r_k(n)$, but the conjecture of the general case is not here. If the conjecture were to exist consciously in his mind, he would have included it in this open-problem article, I am almost certain of it. Paul had not, and this, in my opinion, is a reliable indicator that the general conjecture did not yet exist in 1957.

In the second 1961 open-problem paper, Paul publishes the general conjecture explicitly for the first time:

For $k > 3$ the plausible conjecture $r_k(n) = o(n)$ is still open.

This "still open" indicates that Erdős created the problem well before he submitted this paper, which was "Received October 5, 1960." This suggests the birth of the general conjecture in 1957–1959.

During his December 23, 1991 "favorite problems" lecture at the University of Colorado at Colorado Springs, Paul indicated when he offered first the high prize of $1000 for this conjecture:

Twenty-five years ago I offered $1000 for it.

This places the $1000 offer in 1966 or so. In early January 1992, in Colorado Springs Paul confirmed that this was the highest prize he has ever paid:

> The maximum amount of money I paid was $1000 to Szemerédi in 1972. This was a conjecture of Turán and myself. If you have a sequence of positive density, then it contains arbitrary long arithmetic progressions.

Paul also told me then, "Turán and I posed this problem in the early 1930s." However, I hope, that my argument, presented here, indicates that it took time for the plot to thicken that it was a long pregnancy, and from the early seeds in the 1930s the great conjecture had grown inside Paul Erdős' head and was born in 1957–1959.

Even after Szemerédi, Erdős was not quite happy with the state of knowledge in this field. In 1979 he offered an extravagant prize for the discovery of the asymptotic behavior (published in 1981 [E81.16]):

> It would be desirable to improve [lower and upper bounds] and if possible to obtain an asymptotic formula for $r_3 (n)$ and more generally for $r_k (n)$. This problem is probably enormously difficult and I offer $10,000 for such an asymptotic formula.

Erdős's $10,000 Open Problem 35.4 Find an asymptotic formula for $r_3 (n)$ and more generally for $r_k (n)$.

We have already witnessed Erdős directing research on the chromatic number of the plane and creating a good number of related problems. Here too Erdős was in the driver's seat (well, actually, Paul did not drive), following a prophetic start by Issai Schur.

Endre Szemerédi (born August 21, 1941 in Budapest; Ph.D. 1970, Moscow State University under Israel M. Gelfand) is a professor of computer science at Rutgers University and is a researcher in combinatorics and discrete mathematics division of Alfréd Rényi Institute of Mathematics in Budapest. In 1989, he was elected to the membership in the Hungarian Academy of Sciences.

35.4 Paul Erdős's Favorite Conjecture

During our joint work on the (not yet finished) book *Problems of pgom Erdős*, between December 24, 1991 and January 9. 1992, I asked Paul which of his open problems were his favorites. Paul gave me a list of a few. He started with this problem [Soi13]:

> ... one of the most interesting problems is this: If you have a sequence [of positive integers] the sum of whose reciprocals diverges, then for every r, there are r terms that form an arithmetic progression.

On another occasion during these two working weeks, Paul told me that he offered, not surprisingly, the highest prize for the same problem:

The largest amount of money, which I offered really is: if you have a sequence of [positive] integers the sum of whose reciprocals diverges, then it contains arbitrarily long arithmetic progressions. This would imply in particular that the primes contain arbitrary long arithmetic progressions. That is $3000.

The Erdős \$3000 Conjecture 35.5 A set $A = \{a_1, a_2, \ldots, a_n, \ldots\}$ of positive integers, where $a_i < a_{i+1}$ for all i, with the divergent sum $\sum_{n \in N} \frac{1}{a_n}$, contains arbitrarily long arithmetic progressions.

What brought Paul Erdős to this conjecture? On September 15, 1979, in the problem paper [E81.16] submitted to the premier issue of *Combinatorica*, Paul writes:

In this connection I conjecture that if $\sum_{r=1}^{\infty} \frac{1}{a_r} = \infty$ then for every k there are k a_r's in arithmetic progression. Since Euler proved that the sum of the reciprocals of the primes diverges, our conjecture would settle the conjecture of primes... I offer 3000 dollars for the proof or disproof of the conjecture.

It appears that Paul Erdős *first* offered his (then) largest prize, \$3000 in his 1976 talk "To the memory of my lifelong friend and collaborator Paul Turán" at the University of Manitoba, Canada Conference [E77.28]. (In the paper [E77.26] submitted the previous year, 1975, the prize was \$2500). The highest prize and high frequency of including this conjecture in talks and papers indicate that this was one of Erdős's favorite conjectures. During his second talk at the University of Colorado at Colorado Springs on March 17, 1989, referring to this conjecture, Paul said [E89.61]:

I should leave some money for it in case I leave. "Leave" means, of course, get cured of the incurable decease of life.[47]

The prize stood at \$3000 for nearly two decades, when in one of his last problem papers [E97.18], written in 1996 and posthumously published in 1997, Paul raised the prize to \$5000:

I offer \$5000 for a proof (or disproof) of this [problem]. Neither Szemerédi nor Furstenberg's methods are able to settle this but perhaps the next century will see its resolution.

Since, as Paul believed, it may be a while before this conjecture is proven, we ought to record it with the new, highest ever Erdős (serious) prize:

The Erdős \$5000 Conjecture 35.6' A set $A = \{a_1, a_2, \ldots, a_n, \ldots\}$ of positive integers, where $a_i < a_{i+1}$ for all i, with the divergent sum $\sum_{n \in N} \frac{1}{a_n}$, contains arbitrarily long arithmetic progressions.

One question remains: when did Erdős first pose this problem? I searched for evidence in the ocean of his writings, and found three indicators. First, in a paper submitted on September 7, 1982 to *Mathematical Chronicle* (now called *New Zealand Journal of Mathematics*) that appeared the following year [E83.03], Paul writes:

---

[47] Quoted first in [Soi14]. Earlier Paul mentioned leaving some money for this conjecture in some of his papers, e.g., [E77.28].

This I conjectured more than forty years ago.

In the same year, 1982, Paul spoke at the Conference on Topics in Analytic Number Theory in Austin, Texas. I read in the proceedings (published in 1985 [E85.34], p. 60):

I conjectured more than 40 years ago that if $a_1 < a_2 < \ldots$ is a sequence of integers for which $\sum_{i=1}^{\infty} \frac{1}{a_i} = \infty$ then the a_i's contain arbitrarily long arithmetic progressions.

Thus, both of these publications indicate that the conjecture was posed before 1942. On the other hand, in the 1986 Jinan, China, Conference proceedings (published in 1989 [E89.35]), Paul writes (p. 142):

About 30 years ago I conjectured that if $\sum_{n=1}^{\infty} 1/a_n = \infty$, then the a's contain arbitrarily long arithmetic progressions.

This would date the birth of the conjecture to about 1956. This information only allows us to conclude that this important conjecture is old, and was born somewhere between the very early 1940s and mid 1950s. The conjecture is obviously hard, for in spite of all assaults, it remains open. Moreover, even its weakest $250 version has not been conquered:

Paul Erdős's $250 Conjecture 35.7 A set $A = \{a_1, a_2, \ldots, a_n, \ldots\}$ of positive integers, where $a_i < a_{i+1}$ for all i, with the divergent sum $\sum_{n \in N} \frac{1}{a_n}$, contains a 3-term arithmetic progression.

In his 1983 survey, Ronald L. Graham proposed a "related perhaps easier conjecture." This beautiful conjecture is still open today, which is a good indicator that it is not so easy as it may seem. Z^2 will denote the set of points in the plane (i, j) with integral coordinates i, j.

Graham's Conjecture 35.8 [Gra3]. If A is a subset of Z^2 and $\sum_{(i,j) \in A} \frac{1}{i^2 + j^2} = \infty$, then A contains a monochromatic square.[48]

The Erdős $5000 Conjecture 35.6' is still open. However, the existence of arbitrarily long progressions of primes has been proven by two brilliant young mathematicians, Ben Green and Terence Tao [GT] (they first submitted their proof on Aril 8, 2004; the 6th revision is dated September 23, 2007). Quite expectedly, their result is an existence proof, and does not help to construct long arithmetic progressions of primes. The longest actually constructed example consists of 24 terms. On January 18, 2007 at 3:06 AM, Jarosław Wróblewski, a mathematician from Wroclaw University, Poland informed the world of his new world record:[49]

$$468395662504823 + 205619 \times 23\# \times n,$$

---

[48] In our convention, a square is a set of its 4 vertices.

[49] http://tech.groups.yahoo.com/group/primeform/message/8240

where $n = 0, \ldots, 23$, and $p\#$, called "p primordial," stands for the product of all primes not exceeding p (in particular, $23\# = 2 \times 3 \times 5 \times 7 \times 11 \times 13 \times 17 \times 19 \times 23 = 223092870$).

35.5 Hillel Furstenberg

Two years after Szemerédi's combinatorial proof was published, which incidentally used Baudet–Schur–Van der Waerden's Theorem, in 1977 Harry Furstenberg published a totally different proof [Fur1], using tools of Ergodic Theory. In fact, in doing so Furstenberg created a new field, *Ergodic Ramsey Theory*. "Both results are beyond the scope of this book," write the authors of the standard text of the field [GRS2] about Szemerédi's and Furstenberg's proofs—the more they are beyond the scope of this book, whose goal is to introduce ideas and the excitement of mathematics of coloring, and "meet" the people behind these results.

I wish to share with you here an autobiography of Hillel Furstenberg, ever slightly edited by me. I first met this remarkable mathematician in Keszthely on Lake Balaton in the July of 1993, where we celebrated Paul Erdős's 80th birthday with a fitting conference. Hillel (Harry) looked like Moses. In my opinion, he looked infinitely more like the Prophet than Charleston Heston ever has, Hollywood make-up trickery notwithstanding. Hillel was born exactly when Adolf Hitler fired Issai Schur from his professorship, in the same city of Berlin:

> I was born in Berlin on 29/9/35. I have few recollections of Berlin of the time. I remember my sister (older than myself by 3 years) pointing out a boarded up bakery, saying this was Hitler's bakery. Apparently she (over)heard that because of Hitler this Jewish establishment had been closed off, I remember some visits to a synagogue. We actually lived next to one (33 Brunnenstrasse) which today is a perfume factory, with only a lintel giving evidence of the one time use as synagogue, because the words "This is the gate to the Lord, the righteous shall pass through" appear on it.
>
> Already before Krystallnacht (8/11/38) some of my parents' Jewish friends had received expulsion orders from the Nazis. Our own expulsion order came soon after Krystallnacht and my parents frantically searched for shelter. One of my early recollections is that of the morning after Krystallnacht when the four members of my family lined up underneath the broken windows of our basement apartment viewing the damage. I was old enough to realize the seriousness of the occasion.
>
> From letters I later found I discovered that my parents had sent a request to the Australian government for asylum, and were refused. I have no idea to how many other places we applied. Fortunately an aunt of mine was able to deposit 1,000 pounds sterling with the bank of England, thereby obtaining for us temporary asylum. We arrived in England sometime in 1939, shortly before the Blitzkrieg over London. I remember the shelters in London, the women knitting, I remember the skies at night criss-crossed with searchlights, and I remember my mother, sister and myself being sent to Norfolk, out of the London danger. My father hoped very much at that time to come to America and join my mother's brother who had recently bought a poultry farm in East Brunswick, NJ. He had a health problem (a thyroid condition), and knowing the Americans were strict, he underwent what was at that time risky surgery to rectify the problem. He did not survive the surgery and my widowed mother took her two children

to the U.S. where we arrived shortly before the outbreak of WW II. We stayed at my uncle's poultry farm for over a year, and I attended McGinnis Elementary School. Kindergarten, first and second grades were in one room. Two years in the room were enough for me, so that when after two years we moved to Manhattan, I found myself in third grade in PS 169, near 168th street where we lived. The Rabbi of the nearby synagogue that we attended convinced my mother that I should go to a Jewish Day School, and that she needn't worry that I'd become a Rabbi myself. I attended Yeshiva Rabbi Moses Soloveitchik through eighth grade, and got the rudiments of a traditional orthodox education. I graduated that institution in 1948 and, again, with some persuasion by the Rabbi, continued at a Jewish High School called Talmudical Academy, now know as Yeshiva University High School. Spending some summers in summer-school I finished high School in 1951 and continued at Yeshiva University, Since the college and highschool were located in the same building I had already in High School come under the influence of Professor Jekuthiel Ginsburg, editor of Scripta Mathematica, a journal devoted to historical and recreational aspects of Mathematics, and from whom I first heard of Paul Erdős, and believed even then that he must be very old. (Shlomo Sternberg was also a student at the high school at the time and we both had found our own proofs of the famous problem of showing that if two angle bisectors of a triangle are equal then the triangle is isosceles, and we went to share our discoveries with Professor Ginsburg. Thenceforth he would regularly give us problems to solve.) Prof. Ginsburg realized that for me to devote myself to mathematics, I would need an income, which he obtained for me by having me do editorial work for Scripta. I learned to draw diagrams that were used in the magazine, and i sharpened my mathematical German and French by translating papers sent to the journal in those languages. I don't recall now if any of those translations were ever actually used. In the early fifties, Ginsburg took advantage of his friendship with various prominent mathematicians and set up a graduate school in mathematics at Yeshiva University. The first staff members traveled to Y. U. from their home institutions: Eilenberg and Kolchin from Columbia, Jesse Douglas from City College, Gelbart from Syracuse. I graduated in 1955 receiving both a B.A. and an M.Sc. degrees.

I continued at Princeton, having made the decision not to pursue a rabbinic career at Y. U. and quickly came under the influence of Salomon Bochner who took an interest in me because of his own religious background, and I imagine he found in me someone with whom he could share ideas in a long abandoned area of his past experience. (His father was an accomplished Jewish scholar, and Bochner kept in his office a portion of his father's library which with its annotated volumes attested to his father's scholarship.)

I received my Ph.D. in 1958 and two weeks later was married to Rochelle Cohen from Chicago, whose grandparents had immigrated to the U.S. from Poland. I spent one year as instructor at Princeton, followed by two years at M.I.T. as C.E. Moore Instructor. Following a path taken by Eugenio Calabi (Bochner student - MIT and University of Minnesota) we moved to Minneapolis where we lived from '61 to '65 except for one year I spent as a visitor to Princeton ['63–'64]. During this time I was negotiating taking a position in Israel at the Hebrew University and in the summer of 1965 we made our move, spending first several months in Paris with a Sloan fellowship which provided our income during the year of our move. I also took a half-time position at Bar-Ilan University, and I'm proud particularly of Alex Lubotzky who was my Ph.D. student at Bar-Ilan, and is now my colleague at the Hebrew University. ∎

Furstenberg both created a new field of mathematics, the Ergodic Ramsey
Theory, and founded a school in this new field. This manifested itself in 1996,
when Furstenberg's scientific son and grandson joined together in generalizing the
Furstenberg's result. Vitali Bergelson (Ph.D. under Furstenberg 1984 at the Hebrew
University, born in Kiev in 1950) and Alexander Leibman (Ph.D. under Bergelson
in 1995 at the Technion, born in Moscow in 1960), both presently at Ohio State
University obtained [BL] what is often called the *Polynomial Szemerédi Theorem*.
I prefer to give credit to the authors, who in their paper give several versions of
their result. Here is one, most relevant to our theme (it is the authors' Theorem B_0
for $l = 1$):

Bergelson–Leibman's Theorem 35.9 [BL] Let $p_i(x)$, $i = 1, \ldots k$, be polynomials with rational coefficients taking on integer values on integers and with the zero
last coefficients, i.e., $p_i(0) = 0$. Then any subset of N of positive upper density
contains for any array of integers $v_1, v_2, \ldots, v_n$ a set of the form

$$\{a + p_1(x)v_1, a + p_2(x)v_2, \ldots, a + p_n(x)v_n\}$$

for some $a, x \in N$.

in particular,

Bergelson–Leibman's Theorem, Version II, 35.10 Let $p_i(x)$ be polynomials with
integer coefficients with the zero last coefficients, i. e., $p_i(0) = 0$. Then any subset
of N of positive upper density contains a set of the form

$$\{a + p_1(x), a + p_2(x), \ldots, a + p_n(x)\}$$

for some $x \in N$.

You can easily observe the validity of the following corollary that we will use in
Chapter 44:

BLT's Corollary 35.11 For any positive integers m, r, any r-coloring of the set N
of positive integers contains arbitrarily long monochromatic arithmetic progressions
whose common difference is an m-th power of a positive integer.

Presently, new exciting developments came from the pen of Vitali Bergelson. I
will share those with you briefly.

35.6 Bergelson's AG Arrays

In 2005, Vitaly Bergelson [Ber] extended the Ramseyan hunt for arithmetic progressions to *geoarithmetic progressions*. The following two easy exercises highlight the
setting better than any words can.

Proposition 35.12 Any coloring of positive integers N in finitely many colors contains arbitrarily long monochromatic *geometric* progressions.

Proof Given m-coloring C of the set N, and a positive integer k. Pick an integer t, $t > 1$. The coloring C of the whole set N, of course, assigns colors to all elements of the set $\{t^n : n \in N\}$. Now we get the new coloring C' of the set N by assigning the color of t^n to n. For the coloring C', Baudet–Schur–Van der Waerden's theorem guarantees the existence of an n-term monochromatic arithmetic progression $a, a + d, \ldots, a + (k - 1)d$. The numbers $t^a, t^{a+d}, \ldots t^{a+(k-1)d}$ form a *geometric* progression, and under the original coloring C they are assigned the same color. ∎

This proposition shows that we need to look for the existence of something more sophisticated than geometric progressions. Bergelson looked for an appropriate new term: he used *AG set*, then *geoarithmetic progression*. I propose a term *array* as more descriptive, as we really have here a square array of numbers.
Geoarithmetic array—or for short *AG array*—of rank k is a set of the form

$$\left\{ r^j \left(a + id\right) ; i, j \in \{0, 1, \ldots, k\} \right\}.$$

Observe: an AG array contains lots of arithmetic and geometric progressions, and more.

Proposition 35.13 There is a set of positive (additive) density that contains no 3-term *geometric* progressions.

Proof Just pick the set of square-free positive integers. ∎

This proposition shows that we need a different notion of density, a sort of geometric density here. In his introduction, Bergelson offers an example of what this means.
A set $A \subseteq N$ is *multiplicatively large* if for some sequence of positive integers $a_1, a_2, \ldots, a_n, \ldots$

$$\lim_{n \to \infty} \sup \frac{|A \cap a_n F_n|}{|a_n F|} > 0,$$

where $F_n = \left\{ p_1^{i_1} p_2^{i_2} \ldots p_n^{i_n} : 0 \leq i_j \leq n, 1 \leq j \leq n \right\}$ and where $\{p_i\}$ is the sequence of primes in some arbitrarily preassigned order.

We are ready now to look at a special case of Bergelson's result.

Bergelson's Theorem 35.14 Let $A \subseteq N$ be a multiplicatively large set. Then A contains AG arrays of arbitrarily large rank.

Observe: for any coloring of N in r colors, at least one of the monochromatic sets is multiplicatively large, and thus contains AG arrays of arbitrarily large rank. It is clear that Vitaly Bergelson and his coauthors are up to vast generalizations of the celebrated results of Ramsey Theory. I wish them much success.

35.7 Van der Waerden's Numbers

Through Issai Schur, Richard Rado became aware of Baudet–Schur–Van der Waerden's theorem from the beginning, and generalized it. However, in his early years he did not seem to be much interested in numerical bounds. On the other hand, already in his 1935 celebrated joint paper with George Szekeres, Paul Erdős showed interest in numerical bounds of combinatorial functions. So, when the leaders of Ramsey Theory Erdős and Rado got together in 1951, the result was the paper ([ER] (read November 15, 1951; published 1952) that poineered quantitative evaluation of Van der Waerden's numbers. Having addressed the Ramsey Theorem, Erdős and Rado created *Van der Waerden's function*, and therefore *Van der Waerden's numbers* (they do not use the word "numbers" per se, but what are Van der Waerden's numbers if not values of Van der Waerden's function?), and introduced a natural notation $W(k, l)$ for both:

> The last example of the paper is not concerned with Ramsey's theorem but with the following theorem due to van der Waerden [Wae2]. *Given positive integers k and l, there is a positive integer m such that, if the set* $\{1, 2, \ldots, m\}$ *is divided into k classes, at least one class contains* $l + 1$ *numbers which form an arithmetic progression.* The least number m possessing this property is denoted by $W(k, l)$ (*van der Waerden's function*). Our final example yields what seems to be the first non-trivial, no doubt, extremely weak, lower estimate of W, namely $W(k, l) > ck^{\frac{1}{2}}l^{\frac{1}{2}}$. An upper estimate of W, at any rate one which is easily expressible explicitly in terms of the fundamental algebraic operations, seems to be beyond the reach of methods available at present.

Erdős–Rado notation $W(k, l)$, in today's conventions, would stand for $W(k, l + 1)$. The second variable, as used today (Theorem 33.1), stands for the number of terms in the arithmetic progression. When the number of colors is $k = 2$, we simply omit the first variable: $W(l) = W(2, l)$.

Observe, the Erdős–Rado's interpretation of the notation simplifies statements of some results. For example, best lower estimate, due to Elvyn R. Berlekamp, is simpler in the Erdős–Rado notation, which he used in his paper [Berl]:

Lower Bound 35.15 (Berlekamp, 1969) $W(k) > k2^k$ if k is a prime (Erdős–Rado's understanding of the notation is used).

In today's standard notation (where the variable stands for the number of terms in arithmetic progression), the result reads as $W(k + 1) > k2^k$.

Surprisingly, Berlekamp's result remains the best known for primes after nearly four decades. In 1990 Zoltán Szabó, using Lovász' Local Lemma, found the best known lower bound for all n [Sza].

Lower Bound 35.16 (Szabó, 1990) For any $\varepsilon > 0$, $W(k) \geq \dfrac{2^n}{n^\varepsilon}$ for large enough n.

The problem of finding a "reasonable" upper bound has withstood all attempts for decades. Erdős writes in 1957 (I have just changed the notation to the one used today), [E57.13]:

> All known functions $W(k)$ increase so rapidly that they do not even satisfy the condition

$$W(k) = k^{k^{\cdot^{\cdot^{k}}}} \quad (k \text{ exponents}).$$

The problem was that all of the known proofs of Baudet–Schur–Van der Waerden's Theorem used double induction. This prompted doubts even of such mathematical optimists as Erdős, who wrote in 1979 [E81.16]:

Until recently nearly everybody was sure that $W(k)$ increases much slower than Ackermann's function. I first heard doubt expressed by Solovay which I more or less dismissed as a regrettable aberration of an otherwise great mind. After the surprising results of Paris and Harrington [PH] Solovay's opinion seems much more reasonable, and certainly should be investigated as much and as soon as possible.

Yet, Ronald L. Graham persisted with optimism and bet $1000 on it in his 1983 survey [Gra2]:

There is currently no known upper bound for $W(k)$ which is primitive recursive.[50] This is because all available proofs leading to upper bounds involve at some point a (perhaps intrinsic) *double* induction, with k as one of the variables. This leads naturally to rapidly growing functions like the Ackermann function which may help to explain the enormous gap in our knowledge here. The possibility that $W(k)$ might in fact actually have this Ackermann-like growth has been strengthened by the work of Paris and Harrington [PH], Ketonen and Solovay [KS], and more recently Friedman [Fri], who show that some natural combinatorial questions do indeed have *lower* bounds which grow this rapidly (and even much more rapidly...). In spite of this potential evidence to the contrary, I am willing to make the following [conjecture].

Graham then formulated the conjecture for first proof (or disproof) of which he had been offering $1000 since the late 1970s:

Graham's $1000 Van der Waerden's Numbers Conjecture 35.17 [Gra2].

$$W(k) < 2^{2^{\cdot^{\cdot^{2}}}}$$

for $k \geq 1$, where the number of 2's is k.

Paul Erdős asked for less, just for a primitive recursive upper bound, in the 1984 conference talk in Japan, published the following year [E85.33, p. 75]:

I give 100 dollars for a proof that f(n) is primitive recursive and 500 dollars for a proof that it is not.

Ron's and Paul's expectations were soon rewarded. Saharon Shelah proved exactly what the doctor ordered (I mean Doctor Erdős): Shelah's *Primitive recursive bounds for van der Waerden numbers* [She1] was published in 1988 "with a beautifully transparent proof," as Gowers commented later [Gow, p. 466].

---

[50] See [Soa] for definitions and comparison of rapidly growing functions.

Shelah's Theorem 35.18 [She1] Van der Waerden's numbers are primitive recursive.

Graham described this event in the December 29, 2006 e-mail to me:

I gave Shelah the check [a consolation $500 prize for Conjecture 35.16] when he was lecturing at Rutgers (as you know, he visits there for 2 months each year). It was shortly after he proved his bound, which was somewhat before it was published. Incidentally, the original title of his paper was quite different from what appeared!

Erdős too gave Shelah the highest praise in many talks. Here, for example, is a quotation from Erdős's 1988 talk at the 7th Fischland Colloquium in Wustrow, Germany [E89.27]:

This was certainly a sensational triumph.

Shelah's result inspired Paul Erdős to pose a new, most challenging conjecture. In [E94.21], first submitted on January 25, 1993 and published a year later, Paul Erdős wrote:

It was a great achievement when a few years ago Shelah gave a primitive recursive bound for $W(k)$. Probably, this bound was still much too large perhaps $W(k) < 2^{2^k}$.

We thus get Paul Erdős's conjecture, which he repeated in 1996 (posthumously published in 1997 [E97.18]):

Paul Erdős's 1993 Van der Waerden's Number Conjecture 35.19 [E97.18].

$$W(k) < 2^{2^k}.$$

In 1998 Timothy Gowers announced, and in 2001 published his incredible 124-page *A New Proof of Szemerédi's Theorem*. His upper bound for the Van der Waerden numbers appears on the next to last page as "Corollary 18.7":

Gowers' Upper Bound 35.20 [Gow] Let k be a positive integer and let $N \geq 2^{2^{2^{2^{2^{k+9}}}}}$. Then however set $\{1, 2, \ldots, N\}$ is colored with two colors, there will be a monochromatic arithmetic progression of length k.

In other words,

$$W(k) \leq 2^{2^{2^{2^{2^{k+9}}}}}.$$

In answering my inquiry, Ronald L. Graham wrote in the December 28, 2006 e-mail:

Regarding the payment to Gowers, I gave him the check during a talk I gave in Hungary (again in connection with celebrating Erdos' mathematics but I'm not sure of the exact year). I attach a photograph showing the actual presentation. I interrupted my talk and came down into the audience to give him the check!

Ronald L. Graham presenting the check to Timothy Gowers

Tim Gowers [Gow, p. 586] seemed to question whether he had completely earned the $1000 reward:

> Ron Graham has conjectured in several places (see e.g. [GRS2]) that the function $W(k)$ is bounded above by a tower of twos of height k. Corollary 18.7 [i.e., result 35.19 above] proves this conjecture for $k \geq 9$, and indeed gives a much stronger bound. It looks as though more would be needed to prove it for $k = 7$ (for example) than merely tidying up our proof. For $k \leq 5$, the exact values of $W(k)$ are known and satisfy the conjecture.

Gowers should not worry. Graham's $1500 ($500 to Shelah and $1000 to Gowers) is clearly the money best ever spent in encouragement and support of mathematical research.

As for Ronald Graham, as soon as he paid Tim Gowers, he offered another $1000 conjecture [Gra7], [Gra8]. Prefacing the New $1000 Conjecture, Graham wrote [Gra7], [Gra8]:

> In particular, this [Gow] settled a long-standing conjecture I had made on the size of $W(n)$..., and as a result, left me $1000 poorer (but much happier). Undaunted, I now propose the following:

Graham's 2007 $1000 Van der Waerden's Numbers Conjecture 35.21   [Gra6], [Gra7] For all k,

$$W(k) < 2^{k^2}.$$

Observe that for $k > 3$, we have $2^{k^2} < 2^{2^k}$, thus, Graham's new conjecture is harder—if true—than Paul Erdős's 1993 Conjecture 35.19. Which one is "better" (i.e., true and stronger)? Only time will tell—a very long time, I believe.

We have discussed here asymptotic behavior of the function $W(k)$. So little is known about its exact values for small k that in their 1980 monograph [EG] Erdős and Graham exclaimed "It would be very desirable to know the truth here." A few values were found in 1969 by Vašek Chvátal [Chv] (first three) and in 1978 by R. S. Stevens and R. Shanturam [StSh] (the last one):

$$W(2) = 3$$
$$W(3) = 9$$
$$W(4) = 35$$
$$W(5) = 178$$

With all of the dramatic improvements of computers, no further values have been computed in the past three decades. For other Van der Waerden's numbers known today (cases when more than two colors used, or non-symmetric setting), please, refer to Section 2.3 of the impressive 2004 monograph [LRo] by Bruce M. Landman and Aaron Robertson.

It is time to say a few words about our genius record holders.

The time was the late 1974; the place was Moscow. I went to Anna Petrovna Mishina's Abelian Group Seminar at the Moscow State University. She told us that the young Israeli mathematician Saharon Shelah had just published a solution of the Whitehead problem.[51] This was a sensational news, for everyone, who was somebody in Abelian Group Theory, tried to solve this problem—and failed. Better yet, the answer was not a yes or a no, as we all expected, but "it depends"—depends upon the system of axioms for set theory!

Nine years later: Roll forward to the Orwellian year 1984. As an American,[52] attending the Abelian Groups and Modules conference in Udine, Italy, dedicated to László Fuchs's 60th birthday, I was introduced to Saharon Shelah at dinner the night before the opening. I shared with him my problems and conjectures. The following day Saharon invited me to his hotel room and, to my surprise and delight, offered to collaborate on my problems. Right there he handed in to me a page with a *finite* lemma, which was the only element I was missing for settling one of my conjectures dealing with *uncountable* groups! His question "Why do people attend conferences?"— I answered, "To show their latest results, to learn about achievements of others, and to socialize." "None of this makes any sense," Saharon replied, and added "People should attend conferences in order to solve together problems they could not solve on their own." And so, I missed many talks, was not allowed

---

[51] Must an Abelian group G with $Ext(G, Z) = 0$ be free?

[52] I received my American citizenship days before leaving for Italy.

by my new coauthor to drink wine (and that is in Italy!), but in the end we solved all of the problems and proved all the conjectures—this was the subject of two fine papers in the *Journal of Algebra*. It was a special, inspirational experience to work with Saharon; it also required full concentration, for he was such a quick learner and thinker. On the conference's excursion day, I was sharing a bench on the bus with László Fuchs. "I am working with Saharon, and he is a genius," I told László. "But of course," he replied, as if it was something obvious.

Nine years later: The day before the opening of Paul Erdős's 80th birthday conference in 1993, Saharon arrived very late to Keszthely on Lake Balaton, Hungary, and invited me to join him right away for an 11 PM supper. During the meal, I told Sharon all I knew about the chromatic number of the plane problem. He was excited, and after the supper left to sleep on it. The next morning Saharon said, "I have not seen the light." He has a philosophical view on choosing his battles, which he shared once with me: "Nobody cares how many problems I cannot solve—people care only how many I can."

Nine years later: In the fall 2002, for the first time we met in the United States. Saharon invited me to his Rutgers University in New Jersey for a week of fun of the mathematical kind. This was a productive week. To our own surprise, we showed that the chromatic number of the plane may depend upon the system of axioms we choose for set theory. We also constructed a distance graph on the line whose chromatic number was 2 in the standard **ZFC** system of axioms for set theory, and uncountable in **ZFS**. I will tell you more about this meeting and its results in Chapter 46. Saharon worked in such a complete concentration that I found him wearing one blue and one brown sock. The next day the color coordination remained unchanged. On the third day (like in fairytales) it ended with the matching pair of socks—this is how I was able to conjecture that his wife Yael arrived from Israel and joined Saharon in New Jersey. We met again in the fall 2003 and extended our construction from the line to the plane.

Saharon Shelah was born in Jerusalem, Israel on July 3, 1945. He is the Abraham Robinson Professor of Mathematical Logic at the Einstein Institute of Mathematics of the Hebrew University, Jerusalem, and the Distinguished Visiting Professor at Rutgers University, Piscataway, New Jersey, where he spends every September and October. He is one of the great problem solvers of all time, who has won numerous awards, including the George Pólya Prize (1992), János Bolyai Prize (2000), and Wolf Prize (2001). The count of his papers now approaches 1000. Saharon has also authored some seven major books with two more in his pipeline. As Saharon has had some 200 coauthors, we can initiate the Shelah number not unlike the Erdős number!

William Timothy Gowers, born on November 20, 1963 in Wiltshire, received his doctorate at the University of Cambridge under the famed Hungarian combinatorialist Béla Bollobás. Following the productive years 1991–1995 at the University College London, he has been a Fellow of Trinity College and the Rouse Ball Professor of Mathematics at the University of Cambridge. In 1998 Gowers won the Fields Medal and a year later elected Fellow of the Royal Society. Having attended his talks at Princeton-Math, I can attest to the elegance and lucidity of Tim's presentations of his great combinatorial results. He is an expositor of mathematics as well, with *Mathematics: A Very Short Introduction* to his credit, and a longer introduction, *The Princeton Companion to Mathematics* in the works (to appear in 2008).

35.8 A Japanese Bagatelle

Having done the heavy lifting, I invite you to take a breather by reading a simple cute proof by the Japanese mathematicians Kakeya and Morimoto, who were among the earliest fans of Baudet–Schur–Van der Waerden's Theorem. At the end of Chapter 33, I promised the proof of their result 33.5—it is time to keep my word.

Problem 35.22 (Kakeya–Morimoto, 1930, [KM]) If $A = \{a_1, a_2, \ldots\}$ is an increasing infinite sequence of integers with $a_{k+1} - a_k$ bounded, then A contains arbitrarily long arithmetic progressions.

Proof The differences $a_{k+1} - a_k$ are bounded by, say, the constant c. This suggests $(c + 1)$-coloring of the set of positive integers in colors $0, 1, \ldots, c$ as follows: given a positive integer n, find the smallest term a in the sequence A such that $0 \le a - n$. Obviously $a - n < c$. We then color n in color $a - n$. By Baudet–Schur–Van der Waerden's Theorem, for any length l there is a monochromatic arithmetic progression $b_1, b_2, \ldots, b_l$ of color, say, i. But then by the progression $b_1 + i, b_2 + i, \ldots, b_l + i$ is both arithmetic and is entirely contained in A! ∎

36
In Search of Van der Waerden: The Early Years

*To Dorith & Theo van der Waerden whose help made
this part of my search possible this Chapter is
gratefully dedicated.*

The past is never dead. It's not even past.
  – William Faulkner *Requiem for a Nun,* 1951

*I did read your articles. Very interesting, amazing,
new for me. . .It is strange to get new information
about your family and ancestors, and realize so
many things were never talked about, and one can
never ask straight about it anymore.*
  – Dorith van der Waerden [WaD4]

*Thank you for sending me your triptych, which I read
with great interest! This history is so complex, but
you got so much information, I was astounded.
Reading was very compelling – my greatest
compliment for the study you made.*
  – Theo van der Waerden [WaT3]

36.1 Prologue: Why I Had to Undertake the Search
    for Van der Waerden

Bartel Leendert van der Waerden was a distinguished algebraist, physicist,
statistician, historian, author, and above all one of the leading algebraic geometers
of his time. He published the classic 1927 theorem on monochromatic arithmetic
progressions in finitely colored integers, which we have discussed in Chapter 33.
The proof of this magnificent theorem by Van der Waerden was made possible by the
pioneering conjecture by Pierre Joseph Henry Baudet and Issai Schur (Chapter 34),
hence I named this classic result the *Baudet–Schur–Van der Waerden Theorem.*

A. Soifer, *The Mathematical Coloring Book,*
DOI 10.1007/978-0-387-74642-5_36, © Alexander Soifer 2009

Professor Van der Waerden made major contributions to algebraic geometry, abstract algebra, quantum mechanics, and other fields. He liberally published on the history of mathematics. Among the many books, Van der Waerden wrote the two-volume *Moderne Algebra* [Wae3], one of the most influential and popular mathematical books ever written. It is therefore surprising that no monograph has been dedicated to his life and work. Why is that, I once asked Professor N. G. de Bruijn, who shared with me his theory of matters biographical [Bru8]:

> My advice to scientists who would like to have books about them after their death is (apart from obvious things like doing important work and having lots of students):
>
> 1. Stay in your country.
> 2. Stay in a single subject.
> 3. Don't get old.
>
> And, if you do happen to get old: try to write an autobiography.
>
> Van der Waerden missed the points 1, 2, 3, and was too modest to write an autobiography.

Yes, there are no books on the life of Van der Waerden: none of his homelands— the Netherlands, Germany, and Switzerland—produced any. However, there are numerous biographical articles on Van der Waerden. Some would argue that Van der Waerden's life in general and the turbulent years 1931–1951 in particular have been addressed in [Eis], [Fre1], [FTW], [Dol1], [Dol2], [Fre2], etc. While this is true, understanding his life during these years in a satisfactory way requires two indispensable components: a thorough search for the numerous key documents and a great deal of impartiality and desire to understand. So far no one has demonstrated either of these qualities. These authors apparently believed that a personal acquaintance with Professor Van der Waerden automatically made them experts on his life. Their repeating Van der Waerden's words and explanations did contribute to mathematical folklore. However, these repetitions, mixed with "cheerleading" and lacking in archival research and critical examination of facts, hardly added up to history.[53]

---

[53] My search was largely finished and my three essays waiting in *Geombinatorics'* queue when I received from a German colleague a new long Centenary article with the title nearly identical to my Chapter 37 here: "Van der Waerdens Leipziger Jahre 1931–1945" by the Leipzig University's Professor of the History of Mathematics Rüdiger Thiele (*Mitteilungen der DMV* 12-1/2004, 8–20). It has turned out that the title was about the only thing in common between our works. It would require a long article for me to correct Thiele's errors and challenge his prejudices. For example, Thiele alleges "It is natural that in particular Jewish emigrants have attacked van der Waerden for his stay in Nazi Germany." It appears as if Thiele blames the Jews for their "attacks" on Van der Waerden. Everyone—and particularly the German historian—should have exercised a better judgment and respect for the Jews who were harassed, thrown from their jobs, forced into exile, sent to death camps, killed, committed suicide. Moreover, there is no truth to Thiele's allegation: Van der Waerden's critics Otto E. Neugebauer and Oswald Veblen, for example, were not Jewish. Thiele quotes Veblen writing in December 1933 about "signs of growing anti-Semitism," as if establishing moral equivalence between the Nazi Germany and the United States. Yes, there was anti-Semitism in America, as in all places where Jews lived — but the Nazis gave a particularly bad name to anti-Semitism. There is no moral equivalence, Professor Thiele: the difference is 6,000,000 dead bodies. Thiele promotes a pre-ordained advocacy at the expense of an impartial analysis

Prof. Miles Reid's approach in his 1988 Cambridge University Press's book [Rei] did not contribute to history either:

> Rigorous foundations of algebraic geometry were laid in the 1920s and 1930s by van der Waerden, Zariski and Weil (van der Waerden's contribution is often suppressed because a number of mathematicians of the immediate postwar period, including some of the leading algebraic geometers, considered him a Nazi collaborator).

Even if "leading algebraic geometers" (presumably Zariski and Weil) had such an opinion, their fine mathematical achievements did not automatically make them custodians of the truth. It was very unfortunate that such a heavy accusation was leveled by Prof. Reid without any substantiation at all.

I will grant my predecessors one thing: it is hard to understand B. L. van der Waerden. During the 12+ years of my research, I have assembled a great wealth of material related to his life, especially the life during the trying years, 1931–1951. In some instances Prof. Van der Waerden is worthy of high praise. Other cases illustrate ever so clearly that one's response to living under tyranny can only be to leave, to die, or to compromise.

I wanted to learn about the man behind the classic 1927 result of "Ramsey Theory before Ramsey" (Chapter 33), as I named relevant results that appeared before the F. P. Ramsey's 1930 paper. The triptych of my findings, *In Search of Van der Waerden*, parts I, II, and III, first appeared on the pages of *Geombinatorics* [Soi20, 21, 24]. Part Zero [Soi26] of the series, *The Early Years*, came out later. During the time that has passed since part I appeared in 2004, I have been able to find additional important documents, and further analyze the record I have assembled. Here you will find the most complete to date version of these series of four essays.

It is important to examine Van der Waerden's early years and elucidate his relationship with his distinguished family, which included two members of the Dutch Parliament and an Amsterdam judge.

36.2 The Family

According to Theo van der Waerden, Bartel's nephew [WaT1], [WaT2],

> The Van der Waerden family originates (from what we know in the 15th century) from the *Zuidelijke Nederlanden* (the South of the country) later called *Noord-Brabant* (after the secession of Belgium in 1830), around (what is now) Eindhoven, in small villages, Catholics, agriculturists.

This family tree is difficult to construct and is not central to our purposes. Let us fly over half a millennium, to the hero of our investigation, Bartel (Bart) van der Waerden, who was born in Amsterdam on February 2, 1903 in the family house

---

of even *his* University's archival documents available to both of us. As a result, in my opinion Prof. Thiele's article contributes little to history in general, and to our understanding of Van der Waerden in particular.

at *Hondecoeterstraat* 5. He was the first child of Dr. Theodorus (Theo) van der Waerden (August 21, 1876 Eindhoven – June 12, 1940 Laren) and Dorothea van der Waerden, born Dorothea Adriana Endt ((late 1876 or 1877 Wageningen—November 14, 1942 Laren), who got married in Amsterdam on August 28, 1901. Two more sons, Coenraad (Coen) and Benno (Ben), followed on December 29, 1904 and October 2, 1909, respectively.

Bart's father, Theo van der Waerden, was third of the eight children, three girls and five boys, of Hendricus Johannes van der Waerden,[54] the owner of a large blacksmithing business, and Johanna Huberta Cornelia Goossens. Theo's granddaughter and Ben's daughter, Dorith van der Waerden provided lively details [WaD2]:

> In sequence of ages [the 8 children were] Pauline, Justine, Theo, Jan, Herman, Harry, Tjeu, and Anna. The 3 girls didn't marry. The oldest, Pauline, became a nun, the second, Justine, took care of the family and later of her parents and her brother Tjeu who was a bit retarded. The 5 boys were all sent to the Technical University of Delft where one could become an engineer or architect. They had to study quickly in order to make room (financially) for the next to study. Anna, the youngest of the family, was very intelligent and wanted to study like her brothers but was not allowed. While working, she went on studying and later became a math teacher in secondary school.

Theo and his younger brother Jan studied civil engineering at the Delft Technical University, where they both became socialists among the first student-socialists in the Netherlands [WaT1]. Upon graduation Theo taught mathematics and mechanics in Leeuwarden, Dordrecht and finally for 20 years, 1902–922, in Amsterdam. In 1911, he earned the degree of Doctor of Technical Sciences by defending the thesis *Education and Technology (Geschooldheid en Techniek)*.

A year earlier, on June 28, 1910 he was elected as a representative of *SDAP (Sociaal-Democratische Arbeiderspartij)*, to the Provincial government of North Holland, where he remained until 1919. Theo was also the editor of *The Socialist Guide (De Socialistische Gids)*, where after 1916 he started publishing articles on economic issues. From September 17, 1918, up until his passing on June 12, 1940, he was a *SDAP*'s universally admired member of the House of Representatives (*Tweede Kamer*) of the Dutch Parliament. Published on the day of his passing, Dr. Theo's moving eulogy[55] was entitled "A worker with a warm heart and a sober mind" (*"Een werker met een warm hart en een nuchtere geest"*):

> The working class loses in him one of the pioneers of the socialism in the Netherlands, who has not saved himself, a man, who always gave the best he can offer to the people.
>   We remember him in gratitude and respect.

Bart's mother, Dorothea van der Waerden, a daughter of Coenraad Endt and Maria Anna Kleij, came from a Dutch Protestant family. She was very much loved by her three sons.

---

[54] You can see his portrait hanging on the wall in the family pictures on page 371.

[55] *Het Volk*, June 12, 1940.

Dr. Theo, Bart, Dorothea, Ben and Coen van der Waerden, 1916, the collection of Dorith van der Waerden

Dr. Theo, Bart, Dorothea, Ben and Coen van der Waerden, 1925, the collection of Dorith van der Waerden

When the sons left the family's Amsterdam house at *Hondecoeterstraat* 5 in the late 1920s, Dr. Theo and Dorothea van der Waerden moved 30 kilometers out of Amsterdam to the town of Laren, well-known as the home to many famous Dutch artists and intellectuals, including the Netherland's leading mathematician Luitzen Egbertus Jan Brouwer. Theo built there a magnificent house at *Verlengde Engweg* 10.

The Netherlands was overtaken by the German invaders over the course of five short days of 1940: May 10–15. The Socialist-Democrat Dr. Theo van der Waerden would have likely been on an early list of the Dutch arrested and sent to a concentration camp. Records show that he denied the German invaders that pleasure by succumbing to cancer at 8 in the morning on June 12, 1940. He was 63-years-old. After Dr. Theo's passing, his wife Dorothea lived in the Laren house together with her sister. Unable to cope with depression caused by the German occupation of the Netherlands, Dorothea drowned herself in a nearby lake on November 14, 1942. Laren record books show that she was found at 10 in the morning. The granddaughter Dorith van der Waerden, named in honor of Dorothea, informs [WaD1]: "My father [Ben, Bart's brother] was called by her sister who lived with her after Theo died. She said Do (as she was called) was missing. My father went there and found her in a small lake." She was 60 years old.

Bart's middle brother, Coen (December 29, 1904–December 24, 1982), who must have been named after his maternal grandfather, studied at the Delft Technical University as his father and uncle Jan before him. Coen's son Theo, named in honor of his grandfather, provided me with much of the information about his father [WaT1], [WaT2]:

> After the war, in 1947 he [Coen]. . . became Secretary of the Board of the *Arbeiderspers* [The Workers Press], a few years later C.E.O. of this company. The company was the biggest publishing company in the Netherlands, editing the biggest newspaper *Het Vrye Volk* and editing an enormous quantity of books. He left as C.E.O. in 1966 because his wife (my mother) was very ill. She died in 1968 at the age of 65.

During two periods Coen was a member of the Senate (*Eerste Kamer*) of the Dutch Parliament for *PvdA* (*Partij van de Arbeid*)[56] for a total of 10 years (1957–1966 and 1970–1971)[57] and was one of the leaders of his party. Coen was also a spokesman on economic issues and a member of the union wing of *PvdA*.

Coen and Johanna Cornelia Teensma, whom he married in 1931, had three children, Carla, Theo, and Dorien, born in 1935, 1937, and 1941 respectively. Carla, a TV producer, married the well-known journalist Johannes Christiaan Jan (Han) Lammers, who was an active member of *PvdA* just like his father-in-law Coen. He served as an Alderman of the City of Amsterdam and later, in 1985, became Queen Beatrix' High

---

[56] *PvdA* was founded in 1946 as a continuation of *SDAP*, the party of Coen's father Theo, which was joined by the Liberal-Democratic Association (*Vrijzinnig-Democratische Bond*, or *VDB*) and the Christian-Democratic Union (*Christelijk-Democratische Unie*, or *CDU*).

[57] First time he left the Senate due to his wife's poor health; the second time due to his own health problems.

Commissioner (1986–1996) of the large new province Flevoland recovered from the sea. Theo studied Law at the University of Amsterdam and became Director of the Dutch Cocoa and Chocolate Association. Dorien became a painter.

I learned much about Bart's youngest brother Benno (Ben) and his heroic conduct during the Nazi time in Germany and the German occupation of the Netherlands from his daughter Dorith [WaD1]:

> My father, Benno, born 2 October 1909, died 9 of May 1987. My mother's name was Rosa Eva Louise Weijl – here comes the Jewish root – born 26 July 1909. She died 4 years ago. They met in 1939 and married 4 month later in the same year. He attended what we call a gymnasium – contrary to his father and 2 brothers he had no inclination towards mathematics. He was the youngest. He studied law [University of Amsterdam, 1927–1932] and became a lawyer. He had his own office, one room, in a canal house with other lawyers, and lived in the attic, 2 rooms. There my mother also came to live and the three children were all born there during the war. This is somewhat amazing, but I think they were too old to wait with children and hoped the war would be over soon. During the occupation, there was no work for a lawyer but after that he started again but applied for the job of judge. This was always his dream, and he became appointed in 1949 [to a judge of the City of Amsterdam]. As a judge, he was very much interested in the rehabilitation of criminals after their punishment was over. He started an organization in Holland for help to prisoners and especially help to re-socialize them afterwards and help them to find jobs and so on. He was very well-known for being a humane judge interested in personal circumstances of the people in front of him and always being polite and respectful. Politically he was a socialist like his father and brother Coen, but as a judge, he found it not right to be a member of any particular political party, so he was no longer active here. My mother was a [medical] dr. but most of her life she was a housewife.
>
> The fact that my father married a Jewish woman was no coincidence I believe. In the thirties my father was active in helping German Jews to escape from Germany to Holland. During the occupation, he made false identity cards for Jews and helped them to change identity. I do not know much more about it as this period was never spoken about in our family as in most families.
>
> My parents had 3 children: myself: Dorothee Louise, born 13 May 1941; brother Han, 14 April 1943; and sister Anneke, 8 February 1945. My brother has a shop in old vintage posters. My sister is a well-known artist, ceramics. I am a psychologist. I am the only one who is again politically active in local politics for a green leftish party *GroenLinks*.

Bartel Leendert van der Waerden was understandably proud of belonging to this distinguished family of public servants. In the difficult postwar times, he will invoke his father and brothers as high arbiters of his character and integrity.

36.3 Young Bartel

The family's collective memory preserves a funny, but telling story about the young Bartel. It was shared with me by his aunt, Ms. Annemarie van der Waerden:

When Bart was a youngster his father told him not to hang onto cars with his bicycle. Next time he was spotted hanging to a tramway. His father was angry of course. But Bart said totally innocently: but father, you said not to hang onto **cars**!?

In 1919, Bartel entered the University of Amsterdam at a very early age—he was only 16 (as was L. E. J. Brouwer before him when the latter entered the same University). Dirk van Dalen in his remarkable two-volume monograph [Dal1,2][58] on the Netherlands' greatest mathematician L. E. J. Brouwer, provides very lively and telling remarks on the student life and personality of young Bartel van der Waerden:[59]

> The study of mathematics was for him the proverbial 'piece of cake'. Reminiscing about his studies, he said: 'I heard Brouwer's lectures, together with Max Euwe and Lucas Smid.[60] The three of us listened to the lectures, which were very difficult, he treated the integration theory of Lebesgue along intuitionistic lines, and that works. It was very curious, Brouwer never paid any attention to the audience. All the time he gazed at a point on the opposite wall. He lived in Laren, rather isolated... He immediately departed after the lecture, so that it was very difficult to make contact with Brouwer.' Van der Waerden meticulously took notes in class, and usually that was enough to master all of the material. Brouwer's class was an exception. Van der Waerden recalled that 'at night he actually had to think over the material for half an hour and then he had in the end understood it.'
>
> Van der Waerden was an extremely bright student, and he was well aware of this fact. He made his presence in class known through bright and sometimes irreverent remarks. Being quick and sharp (much more so than most of his professors) he could make life miserable for the poor teachers in front of the blackboard. During the, rather mediocre, lectures of Van der Waals jr. he could suddenly, with his characteristic stutter, call out: 'Professor, what kind of nonsense are you writing down now?' He did not pull such tricks during Brouwer's lectures, but he was one of the few who dared to ask questions.

As we will see, such sarcasm towards fellow mathematicians would become quite characteristic for Van der Waerden.

When the time came for the final examination and the doctoral thesis, Van der Waerden's supervisor was not Brouwer as one could expect. Van Dalen explains [Dal2]:

> One would think that such a bright student was a man after Brouwer's heart. The truth is that Brouwer had no affinity with Van der Waerden's mathematics; furthermore, Brouwer wanted to be left alone to do his own mathematics. A clever young man who would interrupt his own contemplation with bright remarks and questions, was the last thing in the world he wished for. He certainly appreciated Van der Waerden's mathematical gifts.

---

[58] See my review of it in *Geombinatorics* XVI(2) and in *Zentralblatt für Mathematik*.

[59] [Dal2], pp. 516–519.

[60] Max Euwe, the 1935 world chess champion; Lucas Smid, an insurance mathematician.

Indeed, on October 21, 1924, Brouwer wrote a letter of introduction for Van der Waerden, addressed to Göttingen's *Privatdozent* topologist Helmut Kneser:[61]

> In some days my student (or actually Weitzenböck's) will come to Göttingen for the winter semester. His name is Van der Waerden, he is very clever and is already published (namely, on Invariants Theory). I do not know whether for a foreigner, who wants to register there are difficult formalities to fulfill; nevertheless, it would be of high importance for Van der Waerden, if he were to find some assistance and guidance. May he call perhaps once on you? Thank you in advance.

This letter of introduction must have been very important to Van der Waerden: in his ETH archive, I found both Brouwer's original and a few copies in Van der Waerden's handwriting. Brouwer, who appeared so self-centered to many of his colleagues, actually showed almost motherly care about the young Van der Waerden when he tried to get him the Rockefeller (International Education Board, or IEB for short) fellowship. On April 8, 1925, in handwritten English, Brouwer sent a letter to Dr. Augustus Trowbridge (1870–1934), Head of IEB Office in Paris (formerly Physics Professor at Princeton):[62]

> I am somewhat anxious that the blank forms filled up for Van der Waerden may not reach you before the date of April 15. I sent them to Miss Professor Noether (Van der Waerden's proper teacher in Göttingen) who has to sign them as seconder next to me as proposer, but they do not come back, so I suppose Miss Noether to be absent from Göttingen, and out of regular postal communication with her home (March and April are vacation months in Germany).
>
> On the blank forms Van der Waerden requests a stipendium for seven months (a summer semester of three and a winter semester of four months) to the amount of $100 a month.

Van der Waerden was awarded this Rockefeller fellowship at Göttingen University for 7 months (1925–1926) for studying abstract algebra under Emmy Noether. Van Dalen observes:

> Given Van der Waerden's algebraic interests, the person to take care of him was Emmy Noether. Once in Göttingen, under Emmy's wings, Van der Waerden became a leading algebraist. Emmy was very pleased with the young Dutchman, 'That Van der Waerden would give us much pleasure was correctly foreseen by you. The paper he submitted in August to the *Annalen* is most excellent (Zeros of polynomial ideals). . .,' she wrote to Brouwer [on November 14, 1925].

Van der Waerden was indeed well received at Göttingen. He impressed not only the officially under-appreciated Jewish liberal woman Emmy Noether, but also Göttingen's official leaders David Hilbert and Richard Courant. Both would write letters of recommendation for the young Dutchman in the near future.

---

[61] ETH, Hs 652 10563, 10563a, and 10563b.

[62] Rockefeller Archive Center (RAC). I thank Reinhard Siegmund-Schultze for providing me with this and a few other 1925, 1927, and 1933 documents from RAC related to Van der Waerden.

Curiously, Van der Waerden wrote his thesis in 1925 in the Netherlands, while fulfilling his military duty at the marine base in Den Helder. Van Dalen brings up an episode, which is typical of impressions Van der Waerden would leave on people throughout his life [Dal2]:

> In mathematics Van der Waerden was easily recognized as an outstanding scholar, but in the 'real world' he apparently did not make such a strong impression. When Van der Waerden spent his period of military service at the naval base in Den Helder, a town at the northern tip of North-Holland, his Ph. D. adviser [Hendrik de Vries] visited him one day. He said later that the commander was not impressed by the young man, 'he is a nice guy but not very bright.'

One question remains a mystery to me: why did Van der Waerden not defend his doctorate at his beloved Göttingen? Van Dalen seems to be equally puzzled [Dal2]:

> Notwithstanding his popularity in Göttingen, Van der Waerden came back to Amsterdam for his doctor's degree. Perhaps, he would have liked Brouwer as a Ph. D. adviser, but Brouwer systematically discouraged students from writing a dissertation under his supervision. Brouwer was not interested in the honour, pleasure and toil of the Ph. D. adviser role... It was de Vries who took the role of Ph. D. adviser of the young Bartel upon himself. The topic of Van der Waerden's dissertation was enumerative geometry, a subject that was later treated in a monograph by de Vries himself [1936]. Van der Waerden's dissertation ['*De algebraiese grondslagen der meetkunde van het aantal*' ('The algebraic foundations of the geometry of numbers'), 1926] earned him instant fame in the world of algebraic geometers for its importance as a solid basis of the subject.

Van Dalen's assessment, "Instant fame in the world of algebraic geometers," is a high bar. To verify it, I went to the Princeton University's Fine library and became the first person ever to check out this 1926 dissertation [Wae1]. This obscure 37-page brochure (plus a few-page foreword), in Dutch, without any proofs printed, I conjecture, in a tiny number of copies (who would buy it?) could not have possibly made the author famous. Van der Waerden's algebraic geometry fame was earned, of course, but later, by his long series of articles on the subject published in the most prestigious journal *Mathematische Annalen*.

In the foreword to his dissertation, Van der Waerden, of course, thanked his *Promotor* (thesis advisor) Hendrik de Vries,[63] and his professors Weitzenbök, Emmy Noether, Brouwer, and Mannoury. He also gave credit to Professor Johann Antony Barrau, whom we have already met in Chapter 34 – he was *Promotor* of the Ph.D. dissertation of P. J. H. Baudet —and will soon meet again on these pages:

> The first one [weakness in the argument] was brought to my attention by a remark of Professor Barrau, who had observed that the theorem concerning the number of inter-

---

[63] N. G. de Bruijn informs [Bru8]: "The following story might interest you. I guess I once heard it from [Arend] Heyting. At the University of Amsterdam there was a well-known geometry professor H. de Vries. The story is that H. de Vries told later that in one particular year he had three brilliant students: B. L. van der Waerden, Max Euwe, and C. Zwikker, and that [the world chess champion] Euwe was the best one of the three. Zwikker became a physics professor."

sections of a curve and a surface in the projective R^3 space that is generally credited to Bezout, had only been proved by Bezout in the very special case in which the curve is a complete intersection of two surfaces. Professor Barrau outlined for me two possible proofs in the ensuing interchange of letters, one of them, indicated by Professor Wolff, relied on the theory of Riemann Surfaces.

36.4 Van der Waerden at Hamburg

In 1975 Van der Waerden commences to tell the Story of Hamburg [Wae20]:

[In 1926] I went to Hamburg as a Rockefeller fellow to study with Hecke, Artin, and Schreier.

He confirms it to the interviewer in 1993 [Dol1]:

Yes, after one semester at Göttingen, Courant started to take notice of me. He procured for me, on the recommendation of Emmy Noether, a Rockefeller grant for 1 year. With this I studied another semester at Göttingen and one semester at Hamburg with Artin.

In his 1930 *Moderne Algebra* [Wae3], Van der Waerden enumerates his Hamburg duties when he lists the sources of this book:

A lecture by E. Artin on Algebra (Hamburg, Summer session 1926).
    A seminar on Theory of Ideals, conducted by E. Artin, W. Blaschke, O. Schreier, and the author [i.e., Van der Waerden] (Hamburg, Winter 1926/27).

I asked Hamburg University what position Van der Waerden occupied at Hamburg in 1926–1927. My inquiry was answered by *Dekan Fachbereich Mathematik* Prof. Dr. Alexander Kreuzer on January 11, 2006:

For sure he [Van der Waerden] was not a *"Wissenschaftliche Hilfskraft"* of the *Mathematische Seminar* of the University of Hamburg and therefore not an *"Assistent"* of any of the Professors. (At this time the word *"Assistent"* was not used).
    He is not mentioned in the *Vorlesungsverzeichnis* (like every official member of the University)[64] and he has not given a lecture. He was here for one Semester and we believe that he has still had a Rockefeller fellowship (or any other money not from the University Hamburg).

Hamburg University's Prof. Dr. Karin Reich of *Geschichte der Naturwissenschaften, Mathematik und Technik* and of the *Department Mathematik*, confirmed Kreuzer's words (while hinting that I would learn it all if I only read other biographers):

---

[64] Formally Dean Kreuzer is correct. However, we see in winter 1926–1927 semester's *Vorlesungsverzeichnis* (schedule) in the section *Für höhere Semester*, "561 Vortragsseminar über Algebra: Prof. **Artin**, Prof. **Blaschke**, Dr. **Schreier**. Fr[eitag] 12-2 **MathS**[eminar]," just as Van der Waerden reported in *Moderne Algebra* above, except his name is missing in *Vorlesungsverzeichnis*. I venture to conjecture that he was simply added to the leaders of this seminar too late for *Vorlesungsverzeichnis* to reflect his participation.

As far as van der Waerden is concerned, I can't give you any other information than R. Thiele or A. Kreuzer have done. There was no affiliation, van der Waerden was a Rockefeller Fellow, which is mentioned in all [sic] the biographies on van der Waerden.

Prof. Reich is right: it *is* "mentioned in all the biographies", but does it make it true? One must pause and retreat: after all, the German authors, especially Hamburg University historians, know Hamburg University history best, or do they?

Van der Waerden's Göttingen mentor Richard Courant—who would know better than my present Hamburg colleagues whether Van der Waerden was a member of the Mathematics Seminar, on November 29, [19]26 addresses his letter "*Herrn Dr. v.d. Waerden, Hamburg. Mathem. Seminar der Universität.*"[65] Furthermore, according to Reinhard Siegmund-Schultze, the author of the definitive book [Sie] on the Rockefeller mathematical charities, the Rockefeller archive contain no mention of Van der Waerden ever receiving another Rockefeller fellowship: not in 1926–1927, nor in 1933 (in 1933 the record shows that at least he applied for it).

We know for certain that Van der Waerden was at Hamburg on January 15, 1927, for the Rockefeller official Wilbur Earle Tisdale, the new assistant to Augustus Trowbridge, the head of the Paris Office of the International Education Board (IEB), wrote in his diary on January 15, 1927:[66]

> I talked for more than an hour with van der WAERDEN, who finished his fellowship this [i.e., 1926] spring. He is now Assistant [sic] to Prof. Hecke, but will go in April to Göttingen as Assistant to Prof. Courant and Prof. Emmy Noether. This is quite a boost for him and he attributes it to the opportunities afforded by his fellowship.

So the man, who would have provided the Rockefeller money to Van der Waerden, states that Van der Waerden was *not* a Rockefeller fellow at Hamburg, but rather "Assistant to Prof. Hecke"—moreover, he states that contemporaneously. Further in his notes, Tisdale records Van der Waerden describing himself in January 1927 as "*van der Waerden, assistant [sic] in algebraic geometry and algebra.*"

On the other hand, on July 23, 1928, the Curators of the University of Groningen submitted the following information to the Minister of Education, Arts and Sciences of the Netherlands:

> He [Van der Waerden] received his doctorate in Amsterdam in 1926; after that he was Assistant to Prof. [Wilhelm] Blaschke at Hamburg.[67]

This is repeated in the appendix to the Dutch mathematics magazine *Euclides*,[68] where under the June 1931 photograph of the young and handsome Bartel, we read among other:

---

[65] New York University, Archive, Richard Courant Papers.

[66] The Rockefeller Archive Center, "Tislog" (Tisdale's Log). I am most grateful to Reinhard Siegmund-Schultze for providing me with this and other Rockefeller Archive Center's documents related to Van der Waerden.

[67] *Het Nationaal Archief*, Den Haag, finding aid number 2.14.17, record number 73 (Archive of the Ministry of Education).

[68] Euclides, 7th year (i.e., 1931), No. 6. By the way, they erred in the dates, which should have been 1926–'27.

Assistant to Prof. Blaschke in Hamburg 1927–'28.

Thus, Van der Waerden was at Hamburg University in the position of an Assistant, without teaching duties, but taking part in running the seminar together with Artin, Blaschke, and Schreier. Formally, he assisted Hecke—as I view Tisdale's notes to be the most reliable document—or else Blaschke, but of course his main goal for being at Hamburg was to learn abstract algebra from Emil Artin. From Van der Waerden–Courant correspondence,[69] we do know that Van der Waerden was at Hamburg during summer and winter semesters of 1926–1927 (not one semester as reported by *Dekan* Kreuzer). We also know that this was, perhaps, the most important time of his mathematical life.

Hamburg time also allows an insight in the views and personality of Van der Waerden. During the already mentioned January 15, 1927 interview with Van der Waerden, Tisdale notes Van der Waerden's predilection for categorical opinions:

While he [van der Waerden] is young, he has very clear and definite opinions – perhaps too much so. I talked to him concerning Kloosterman[70] and, in his frank way, he told me he considered Kloosterman to be lazy, an average straight forward worker, but temperamental and requiring conditions to be just right before he can work... His feeling is that [Edmund] Landau, at Göttingen, is a man without particular vision.[71]

Still, Van der Waerden leaves a positive impression on Tisdale:

Van der Waerden appeals to me as a very intense, gifted and enthusiastic individual. He has the unfortunate defect of stammering, especially in his more intense moments, but he is so agreeable to talk to that the defect is rather minimized. I explained to him how the seriousness of such fellows as himself might be influential in justifying the appointment of future fellows, to which he reacted most enthusiastically and agreeably.

During the interview, Van der Waerden favorably evaluates his Hamburg mathematical group, as Tisdale records:

He feels that the school at Hamburg is exceptionally strong, especially considering its youth. Prof. Blaschke in differential geometry, Prof. Hecke in algebraic numbers and Prof. Artin in algebraic numbers and algebra in general form a very strong nucleus with [Otto] Schreier, private lecturer and assistant, in theory of groups; van der Waerden, assistant in algebraic geometry and algebra; [Hans] Petersson, assistant in analytical theory of numbers; [Heinrich] Behnke, assistant to Hecke, in analytical functions; with Kloosterman, I.E.B. fellow in analytical theory of numbers; Zwirner, in algebraic numbers; and Haacke (late assistant at Jena) in geometry.

[69] New York University, Archive, Richard Courant Papers.

[70] Hendrik Douwe Kloosterman (1900–1968), later a professor at the University of Leiden (1947–1968).

[71] A year later this celebrated number theorist, or according to Van der Waerden "man without particular vision," will be asked—and will write —a glowing recommendation for Van der Waerden's successful appointment to a full professorship at Groningen.

36.5 The Story of the Book

Emil Artin, a framer of abstract algebra, promised Richard Courant to write a book on abstract algebra for the Courant-edited "Yellow Series" of *Springer-Verlag*. During the summer of 1926 he was giving a course on algebra attended by Van der Waerden who was taking meticulous notes. Artin agreed to write this book, based on his lectures, jointly with the 23-year young Dutchman. However, as we all know, the book appeared a few years later under one name, that of the Student and without the Master.

What happened is a question of enormous importance, for The Book has become one of the most famous and popular books in the history of mathematics. Yet, I have found no research published on this subject. Van der Waerden told his story, his interviewers and his former Ph.D. students repeated it, and the historians and mathematicians uncritically accepted thus invented fairytale! I invite you to join me and take a look at the documents. It is most appropriate first to give the podium to Prof. Van der Waerden [Wae20], who (in 1975) tells us how enormous Artin's contribution to the book really was:

> Artin gave a course on algebra in the summer of 1926. He had promised to write a book on algebra for the "Yellow Series" of Springer. We decided that I should take lecture notes and that we should write the book together. Courant, the editor of the series, agreed. Artin's lectures were marvelous. I worked out my notes and showed Artin one chapter after another. He was perfectly satisfied and said, "Why don't you write the whole book?"
>
> The main subjects in Artin's lectures were fields and Galois theory. In the theory of fields Artin mainly followed Steinitz, and I just worked out my notes. Just so in Galois theory: the presentation given in my book is Artin's.
>
> Of course, Artin had to explain, right at the beginning of his course, fundamental notions such as group, normal divisor, factor group, ring, ideal, field, and polynomial, and to prove theorems such as the *Homomorphiesatz* and the unique factorization theorems for integers and polynomials. These things were generally known. In most cases I just reproduced Artin's proofs from my notes.
>
> I met Artin and Schreier nearly every day for two or three semesters. I had the great pleasure of seeing how they discovered the theory of "real fields," and how Artin proved his famous theorem on the representation of definite functions as sums of squares. I included all this in my book (Chapter 10). My sources were, of course, the two papers of Artin and Schreier in *Abhandlungen aus dem mathematischen Seminar Hamburg 5* (1926), p. 83 and 100.

Van der Waerden gives further credits to Artin (ibid.):

> In Chapter 5 (*Körpertheorie*) I mainly followed Artin and Steinitz...
>> Chapter 7 on Galois Theory was based on Artin's course of lectures...
>> In Chapter 10... (a) the Artin-Schreier theory of real fields and representation of positive rational functions as sums of squares... In treating subject (a) I closely followed the papers of Artin and Schreier.

Van der Waerden repeats the story in 1994, and Prof. Dold-Samplonius publishes it [Dol1]:

Artin was supposed to write a book and wanted to write it with me. Having finished the first chapter, I showed it to Artin. Then I sent him the second and asked him about the progress of his part of the book. He hadn't yet done anything. Then he gave up the idea of writing the book with me. Nevertheless, the book is based on lectures of Artin and Noether.

The idyllic picture is further enhanced by Dold-Samplonius's in her 1997 eulogy of Van der Waerden [Dol2]:

Artin gave a course on algebra that summer, and, based on van der Waerden's lecture notes, the two planned to coauthor a book on algebra for Springer-Verlag's "Yellow Series." As van der Waerden worked out his notes and showed Artin one chapter after another, Artin was so satisfied that he said "Why don't you write the whole book?"

"Artin was so satisfied," Van der Waerden and Dold-Samplonius lead us to believe. In fact, Artin was so outraged that he obviously refused to write the book together with the discourteous and ungrateful student. I read—in disbelief—Richard Courant's August 6, [192]7 letter to Van der Waerden:[72]

Dear Herr v.d. Waerden!
Herr Artin has sent me a copy of the enclosed letter about which I am somewhat astonished and concerned. Do you understand Artin's attitude? I don't. Is there any personal sensitivity behind this or are these differences of an objective nature? In any case, one cannot force Artin. But I would like to hear your opinion before I answer him.
I hope you have not angered him.

Clearly, Artin refused to write the book with Van der Waerden, and thus "astonished" Courant. He was obviously offended by Van der Waerden, but how? Let us look at the surviving shreds of evidence. The skies are cloudless on November 29, 1926, as we glance into Courant's letter to Van der Waerden:[73]

Dear Herr van der Waerden!
What about this admission of your *Habilitation*. It would be very good to get this thing moving.
How are you doing otherwise? How is the book by Artin and you coming along?

We see first clouds in Van der Waerden's December 2, 1926 reply:[74]

The Yellow Book is making progress; I have finished writing a large part; I have half-finished other parts, and the plan for the whole is becoming more precise in details through the conversations with Artin, the only thing is Artin himself writes very little.

So, Artin has given his course, Artin is making his material "more precise in details through conversations," but "Artin himself writes very little," or – as Rudyard Kipling would have put it [Kip] – Artin won't "fetch and carry like the rest of us."

---

[72] New York University, Archive, Courant Papers.

[73] Ibid.

[74] Ibid.

Two months later, on February 2, 1927, we sense more overcast from the Student dissatisfied with his Master:[75]

> My coexistence with Artin is still very fruitful. He forever digs up nice things that will also have to come into the book, and from our conversations many details emerge by which the proofs are simplified or new contexts are uncovered. Even if he does not work on the book directly, it is still coming forward.

It seems like Artin has not only provided a well thought out lecture course, ready for the note-taking, but further contributes to the joint book: "he forever digs up nice things", "many details emerge by which the proofs are simplified or new contexts are uncovered." But Artin won't "plough like the rest of us" (Kipling again), and the Student is upset and, just as in his letters to Courant, probably accuses the Master of *not writing down his fair share* of "nice things." Van der Waerden alleges that "He [Artin] hadn't yet done anything." That would explain Artin's explosion and refusal to write *his* book with Van der Waerden. Now we can better understand the quoted above 1994 interview [Dol1]. Van der Waerden, in fact, tells us the truth, but without the context behind it. Let us revisit it, now that we know the context. Van der Waerden says:

> [I] asked him [Artin] about the progress of his part of the book. He hadn't yet done anything. Then he gave up the idea of writing the book with me.

But never mind the Master, the Student has gotten everything he needs, and can now publish the book by himself, with the blessing of his mentor and "Yellow Series" Founder and Editor Richard Courant.

As a mathematician, I have coauthored a number of works with others. It never mattered to us who wrote down joint ideas and proofs. Such great mathematicians as Paul Erdős, Israel M. Gelfand and Saharon Shelah often left the writing of joint works to their coauthors. I am amazed at Van der Waerden's narrow notion of coauthorship. Producing a book requires not merely writing it down, but first of all discovering and assembling numerous ideas, theorems, proofs, giving the whole material structure and style. In all of these chores Artin's contributions were overwhelming, and to publish the book of Artin's ideas without Artin as at least a coauthor was grossly unfair, in my opinion. It could be classified as an act of "nostrification." I let Richard Courant [Cour] define the term:

> A certain duty exists, after all, for a scientist to pay attention to others and give them credit. The Göttingen group was famous for the lack of a feeling of responsibility in this respect. We used to call this process – learning something, forgetting where you learned it, then perhaps doing it better yourself, and publishing it without quoting correctly – the process of "nostrification." This was a very important concept in the Göttingen group.

On the title page of the book—what an unusual place for acknowledgements—Van der Waerden did give credit to Artin's lectures (and Noether's lectures) as being "used" in the book—but was it enough? How many theorems, proofs, ideas required—and did not get—specific credits to Artin?

---

[75] Ibid.

Van der Waerden published the two volumes in 1930 and 1931 in the Richard Courant-edited *Yellow Series*. The great book had a great success. It excited generations of mathematicians (I included), and made B. L. van der Waerden famous.

Surely, Van der Waerden deserves credit for writing down and editing the book. How much credit, depends upon how close the book was to Artin's lectures and how publishable Artin's lectures were. Those who attended Artin's summer 1926 lectures cannot testify today. However, here at Princeton University I found a good number of Artin's students from his Princeton's 1946–1958 years: Gerard Washnitzer (who took all Artin's courses 1947–1952), Harold W. Kuhn, Robert C. Gunning, Hale F. Trotter, Joseph J. Kohn and Simon B. Kochen. Independently interviewed, they were amazingly unanimous in their assessments and even in epithets they used. Tall, slender, handsome, with a cigarette in one hand and chalk in the other, without ever using any notes (well, sometimes a small piece of paper extracted for a second from a jacket pocket), Artin delivered elegant, smooth, well thought out lectures, so much so that notes, carefully taken, would be quite close to a finished book. Harold Kuhn, who took Artin's 1947 course, recalls:

> Artin's lectures were composed like a piece of music, with introduction, exposition, development, recapitulation and coda.

"So, would transcribed lectures form a book?" I asked Harold, who replied:

> Absolutely. In fact, lecture notes formed several of Artin's books, on Galois Theory, on Cauchy Theorem, etc.

Van der Waerden took such notes in his generation; Serge Lang did so in his.[76] In his book [Lan], p. vi], Lang calls Van der Waerden's book "Artin–Noether–Van der Waerden" – fair enough – but then shouldn't he have called his book "Artin-Lang," *n'est-ce pas*?

There was another way to credit and honor the teacher. Van der Waerden gave an example of it, when he had not "nostrified" somebody else's lecture notes. But of course, this was a special case of his admired mentor, *Fräulein* Emmy Noether [Wae20]:

> I took notes of the latter [Emmy Noether's] course, and these notes formed the basis of Emmy Noether's [sic] publication in *Mathematische Zeitschrift* 30 (1929) p. 641.

36.6 Theorem on Monochromatic Arithmetic Progressions

As we have discussed in Chapter 34, at the Bad Kissingen September-1927 annual meeting of the *Deutsche Mathematiker Vereinigung* (*DMV*, German Mathematical Society), Bartel L. van der Waerden announced a proof of the following result [Wae2]:

---

[76] "Since Artin taught me algebra, my indebtedness to him is all-pervasive," wrote Lang in the foreword of his *Algebra* book [Lan].

For any k, l, there is $N = N(k,l)$ such that the set of whole rational numbers 1, 2, ..., N, partitioned into k classes, contains an arithmetic progression of length l in one of the classes.

The Dutch Professor Wouter Peremans, Ph.D. 1949 under Van der Waerden, writes [Per, p. 135] that this "result...made him [Van der Waerden] at one stroke famous in the mathematical world."[77]

I truly love this classic result, and this is precisely why I have become interested in Van der Waerden's life in the first place. However, I confess, that the appearance of this result could not have possibly made Van der Waerden "at one stroke famous in the mathematical world" – indeed, it took time for this publication to be noticed and taste for such new Ramsey-type results to develop. Initially Van der Warden himself must have not thought highly of his now classic result and he did not expect others to appreciate it, for he published it in an obscure Dutch journal *Nieuw Archief voor Wiskunde*, whereas his algebraic geometry papers that he considered important, were published in the most prestigious journal *Mathematische Annalen*. Nicolaas G. de Bruijn, who knows best, confirms [Bru3, p. 116] :

> Old and respectable as the "*Wiskundig Genootschap*" may be, it has never been more than a small country's mathematical society. Accordingly, it is not surprising that the society's home journal, the "*Nieuw Archief voor Wiskunde*", has a relatively small circulation, and, as a second order effect, that the *Nieuw Archief* does not get more than a small part of the more important contributions of the Dutch to mathematics.

From Van der Waerden's captivating account of *How the Proof of Baudet's Conjecture Was Found* [Wae13, 14, 16, 18, and 26], we learned that the proof was obtained as a result of the collaboration of three mathematicians: Emil Artin, Otto Schreier and Bartel L. van der Waerden, but credited to just one, who published the result. Let me repeat just one passage from Van der Waerden's reminiscences, which we have read in full in Chapter 33:

> Finding the proof of Baudet's conjecture was a good example of team-work. Each of the three of us contributed essential ideas. After the discussion with Artin and Schreier I worked out the details of proof and published it in *Nieuw Archief voor Wiskunde* **15**, p. 212 (1927).

A thorough historian of mathematics (if such an endangered species exists) would contradict me by pointing out credit to Artin in the footnote of this 1927 publication [Wae2]. Indeed, we read:

> The conjecture that the generalization from k = 2 to arbitrary k would work by induction, comes from Herr Artin.

---

[77] Peremans also writes: "The problem circulated in German mathematical circles in the twenties and famous mathematicians like Artin and Schreier tried in vain to solve it. Van der Waerden succeeded." No substantiation of this myth is known to me. In fact, Van der Waerden contradicts it himself [Wae, 14, 16, 18, and 26].

Artin and Schreier contributed much more—Van der Waerden told us so in detail (Chapter 33)—thus, the theorem could have been published under the names of all three coauthors. Perhaps, Van der Waerden simply did not realize the significance of the result and thus gave no thought to joint authorship, for as I mentioned earlier, he published it in an obscure journal of the Dutch Mathematical Society.

As we have learned in Chapter 34, Van der Waerden in fact proved the conjecture discovered independently by Pierre Joseph Henry Baudet and Issai Schur. He never met either of his coauthors of the classic Baudet–Schur–Van der Waerden Theorem.

36.7 Göttingen and Groningen

In the waning days of February 1927, Van der Waerden successfully passed *Habilitation* at Göttingen University under the wing of Richard Courant, thus "curing" his Dutch doctorate. In April 1927 he became Courant's *Assistent*.

History possesses its own sense of humor; it also repeats itself. We will see both attributes at the junctions of the lives of Bartel L. van der Waerden and Johan Antony Barrau.

Act One of their story, according to Van der Waerden, took place during Van der Waerden's high school years. In 1994, when he conveyed this story to the interviewer [Dol1], Van der Waerden ridiculed the Groningen professor Barrau for allegedly making numerous mistakes in his book on analytical geometry. He wrote to the author about it. Barrau was impressed and—in an elegant compliment—informed Van der Waerden that he would like Van der Waerden to succeed him if he were to leave Groningen. To the contrary, we know for a fact that Van der Waerden acknowledged with gratitude—and with no disrespect—his correspondence with Barrau in the preface of his 1926 doctoral dissertation.

Act Two of the story took place in 1927, when Barrau moved from Groningen to Utrecht, when his chair was indeed offered to Van der Waerden. The following year, on May 6, 1928, Van der Waerden entered the Barrau's chair at Groningen, with the assistance of glowing recommendations by such celebrities as David Hilbert, Edmund Landau and Richard Courant. Surely, Van der Waerden could have found a lesser ranked professorship at a higher ranked German university. However, Groningen made the 25-year old young man "Ordinarius," i.e., a Full Professor.

Act Three of Barrau–Van der Waerden story will have to wait until December 1942. We will play it out it in the next section. Meanwhile, here at Groningen another important event took place.

In the midst of his Groningen years, in 1929 Van der Waerden accepted a particularly productive visiting appointment at Göttingen: in July he met there his future wife. Camilla Rellich, two years Bartel junior (born September 10, 1905) was the sister of Franz Rellich, who in the same year (1929) defended his Ph.D. dissertation under Richard Courant. Already on September 27, 1929 Bartel and Camilla united in a marriage that lasted a lifetime. Their first child, Helga, was born in Groningen on July 26, 1930. Their other two children would be born in Germany: Ilse on October 16, 1934, and Hans Erik on December 7, 1937.

Groningen seems to have been a stepping-stone for a number of fine mathematicians. Van der Corput was also there, and Van der Waerden recalled learning a lot of mathematics from him. Most importantly, at Groningen, Van der Waerden finished "The Book."

36.8 Transformations of The Book

"The Book" was the main outcome of Van der Waerden's years at Groningen. Everyone who has written a book would agree that Van der Waerden proved to be a great expositor of the new abstract view of algebra. He writes in the preface of the 1930 first edition of volume 1 that the book, started as Artin's lecture notes, has substantially changed, and by the time of its release it was difficult to find Artin's lectures in it. I know of no way to verify this statement today. Granted, Van der Waerden's contribution must have grown from 1927 to 1930. However, it is also clear that an unusually large contribution of the non-author Artin remained as we have seen above when we cited Van der Waerden's own 1975 words. The book became an instant classic, which many generations of mathematicians enjoyed. I too remember reading, during my freshman year (1966–1967), the early Russian translation (vol. 1, 1934; vol. 2, 1937) with great delight and profit.

Unlike his mentors Brouwer and Hilbert, Van der Waerden apparently did not have firm mathematical principles that he was willing to fight for, as the story of changing—and changing back—his *Moderne Algebra* book shows. It is surprising that the quick learner, Van der Waerden seemingly failed to see the importance of the battle over the foundations of mathematics that raged for decades and take a firm position on it. The leading historian of the Axiom of Choice Gregory Moore writes [Moo]:

> In 1930, van der Waerden published his Modern Algebra, detailing the exciting new applications of the axiom [of choice]... Van der Waerden's Dutch colleagues persuaded him to abandon the axiom in the second edition of 1937. He did so... [which] brought such a strong protest from his fellow algebraists that he was moved to reinstate the axiom and all its consequences in the third edition of 1950.

Indeed, in January 1937, in the preface to the second edition of volume 1, Van der Waerden himself discloses the surprising transformation of his book [Wae6]:

> I have tried to avoid as much as possible any questionable [sic] set-theoretical reasoning in algebra. Unfortunately, a completely finite presentation of algebra, avoiding all non-constructive existence proofs, is not possible without great sacrifices. Essential parts of algebra would have to be eliminated, or the theorems would have to be formulated with so many restrictions that the text would become unpalatable and certainly useless for a beginner...
>
> With the above mentioned aim in mind, I completely omitted those parts of the theory of fields which rest on the axiom of choice and the well-ordering theorem. Other reasons for this omission were the fact that, by well-ordering principle, an extraneous element [sic] is introduced into algebra and, furthermore the consideration that

in virtually all applications the special case of countable fields, in which the counting replaces the well-ordering, is wholly sufficient. The beauty of the basic ideas of Steinitz' classical treatise on the algebraic theory of fields is plainly exhibited in the countable case.

By omitting the well-ordering principle, it was possible to retain nearly the original size of the book.

Then in the preface to the third edition, in July 1950, we read about Van der Waerden's puzzling reversal [Wae11]:

> In response to many requests, I once again included sections about well-ordering and transfinite induction, which were omitted in the second edition, and on this foundation, I presented theory of fields developed by Steinitz in all its generality.

36.9 Algebraic Revolution That Produced Just One Book

Van der Waerden's book became so popular because of its high quality, but also, it seems to me, because no competition occurred. Indeed, started by Emmy Noether and Emil Artin, algebraic revolution swept mathematics during the 1920s and 1930s, yet for decades only one book on the new algebra was published. Why did this happen? Documents show that a three-volume book by one of the leading algebraists Richard Dagobert Brauer (1901–1977) was under contract with Springer and in the works, and Van der Waerden had something to do with blocking this competition book. Let us look at the facts.

Even though Richard Courant was Jewish, as a combatant in World War I, he was exempted from the April 7, 1933 Third Reich's civil service law that removed Jews from the ranks of professors. Nevertheless, on May 5, 1933 he was served a letter of dismissal. He accepted an invitation for a year's visit from Cambridge University and informed Van der Waerden accordingly:[78]

> Between the 24th and 28th of October I am supposed to give lectures for students in Holland in Amsterdam, Utrecht, Groningen and Leiden, and for that reason I want to depart from here on the 22nd. Presumably from Holland I will travel directly to England. I have an invitation to Cambridge for the next academic year.

As a result of this departure, Courant – and Ferdinand Springer[79] – wanted to find someone, who could serve as a figurehead editor, while Courant would pull all of the strings from Cambridge. The choice naturally fell on the Courant's protégé, Van der Waerden, to whom Courant offered the job on October 10, 1933:[80]

> I want to ask you therefore the following on the basis of a conversation that Neugebauer and I recently had with Springer. Because it is probable, because of my uprootedness

---

[78] Courant to Van der Waerden, letter of October 10, 1933. New York University, Archive, Courant Papers.

[79] Ferdinand Springer, Jr. (1881–1965).

[80] New York University, Archive, Courant Papers.

and my work as the editor of Yellow Series at Springer, I will be a little bit hampered, we thought about whom one can take in as a representative in case of such hindrance, and in the process we agreed, without any trouble, that you would be the obvious person for this role. I want to ask you today whether in principle you are inclined to do this.

Courant adds, "It would not be a large burden on you," as Courant expects to make all decisions himself. The protégé immediately accepts (October 13, 1933 letter to Courant[81]). However, things change, when Courant is not offered a longer stay at Cambridge, and so during Courant's Christmas visit of his family in Göttingen, he elevates Van der Waerden to a more-or-less real editor of the Yellow Series, and a member of editorial board of the *Mathematische Annalen*:[82]

> During the short visit last week I spoke explicitly with Springer about different things, among others about the case of the *Annalen*. In the meantime, as Hilbert told me, you have accepted the invitation to join the Editorial Board, and I hope that this signifies the beginning of a continual reenergizing of the *Annalen* Editorial Board. Springer feared that Hilbert has somewhat mixed up thing, which can happen easily, but still it is no longer necessary that I do anything in this case.
>
> In regard to the editing of my [sic] Yellow Series, I would like, as we have already considered this, to regard you from now on as the editor, with the thought on the back of my mind, that in case I should go to America for a longer time, you could take care of the thing possibly more than in a purely formal capacity.

Now we are ready to look at the fate of Richard Brauer's *Algebra* book. The year is 1935. Richard Courant, who by now lives in New York, offers his and Emmy Noether's (now at Bryn Mawr College in Philadelphia) defense of Brauer's book against Van der Waerden's reservations, in the July 16, 1935 letter to Van der Waerden:[83]

> I find it to be a mistake to change something in the contracts and agreements that have existed for years, for example with Richard Brauer. Brauer's book, whose new plan I will soon send to [F. K.] Schmidt,[84] has been spoken through in this past year repeatedly with Emmy Noether, and will certainly not be a superfluous publication.

Van der Waerden takes a bold move of excluding Courant from the loop and going straight to Ferdinand Springer. But he cannot simply disregard Emmy Noether's opinion. He writes about it on August 10, 1935 from Laren, Holland, where he is visiting his parents:[85]

> In regard to Brauer (R.), I proceeded on the assumption that B. [Brauer] himself, as Schmidt assumed, did not really want to get too involved with the book. If that is wrong and if even Emmy Noether is in agreement on the book, then for the time being I withdraw my reservations. However, I will be very interested in looking at the plan

[81] Ibid.

[82] Ibid.

[83] Ibid.

[84] Friedrich Karl Schmidt, Van der Waerden's co-Editor of the Yellow Series.

[85] New York University, Archive, Courant Papers.

that the author of course will send us and form an opinion on that basis. In any case, I agree with Schm[idt] and Spr[inger] that there is no hurry in view of the current state of the market for books on algebra. In other words, one should definitely not try to push it forward.

Courant is outraged with his protégé for going to Springer before a consultation with him. He starts his August 20, 1935, 5-page letter as follows:[86]

I did not find it pleasant that discussions...instead of being conducted between us first were taken to Springer without an attempt at previous agreement with me, for Springer through this would get an impression, as if in a number of cases my basic point of view is being disregarded.

Courant throws his unconditional support behind Richard Brauer:

Under no circumstances could I declare myself in agreement with any step against Richard Brauer.

Courant then offers us a rare insight into the story of Brauer's book:

Once again the prehistory. An age-old plan of the appearance of the Frobenius's algebra lectures through Schur was transformed a long time ago into the plan of the appearance of Schur's lectures. Schur then named Richard Brauer as a coauthor and in the course of time rolled the whole thing off on to him. After very careful consultations at the time, also with Emmy Noether, the contract was undertaken, in which it was clearly expressed that it would be an elementary concrete algebra and in certain sense an enlargement of your book.

When long afterwards the Nazi revolution came and Brauer went to America, we expressly discussed with Springer the issue whether under these changed circumstances, also of business circumstances, the plan should be adhered to. Springer himself desired this at the time thoroughly, and even in order to help Brauer, paid him a not an insubstantial advance of royalties. Over here Brauer worked a lot on the book, by the way, continually in close contact with Emmy Noether, with whom he was more closely connected here than anyone else was.[87] The only serious competition to Brauer's book seems to me to be Perron. In the past, Springer was continually of the position that that existence of a competition book in another press posed no problems for him. Brauer's book will be very different from Perron's book in an extraordinary number of points. Therefore it can be hoped that it can still find readers in Germany. Over here where Brauer without a doubt has a big career and where he is praised and appreciated far and wide, his book has a substantial chance (by the way, Brauer has become a Professor at Toronto).

I wrote to Schmidt of a possible modification of the plan where a division in three volumes was foreseen. First is an elementary introduction, directed at wide circle of readers, the second refinements, and the third Galois Theory – all three relatively independent. The first volume could soon be ready. At this point I have pushed Brauer con-

---

[86] Ibid.

[87] Brauer and Noether saw each other regularly. Brauer spent 1934–1935 academic year at the Institute for Advanced Study in Princeton as Herman Weyl's assistant, while Emmy Noether taught at Bryn Mawr College in Philadelphia and that year conducted a weekly seminar at Princeton University.

tinually because after everything that has happened, this seemed to me what Springer wanted. But if the principle of speed is going to be explicitly given up, one can say to Brauer, you should take time, and in all probability, one can select the English language instead of the German. One can also, if you and Schmidt are in agreement, suggest changes in the plan. I believe that in both of the named cases [second being Szegö's book], today's standpoint within Germany that non-Arian authors represent a problem, should be set aside as much as possible. But it is clear to me, that for Springer, in order to exist, and also for the reason that he wants to serve the cause, such standpoints occasionally have to play a role, and force him to be especially cautious.

Van der Waerden must have felt threatened by Courant's plan to publish Brauer's book as "an enlargement of your [i.e., Van der Waerden's] book." However, the following 2 weeks, come shocking news of Springer firing his key Jewish employees. In view of this, Courant begins to think that Springer may no longer approve of books by Jewish refugee scientists. On September 3, 1935 Courant gives up his fight for the Brauer's book:[88]

> From Neugebauer I received a very short message, according to which non-Aryan employees have been released like crazy from Springer Verlag, including Arnold Berliner[89] and *Fräulein* Strelitz. . .
>
> That our correspondence regarding Szegö, Brauer, Wintner, etc., appears in a new light because of this turn of events, is of course clear. Springer must have been under heavy pressure for a long time and have become more fearful and cautious than he showed me directly.

Finally, on September 28, 1935, Van der Waerden replies with the intention to allow at most one of three Brauer's volumes:[90]

> There is no hurry with Brauer's book, since the book does not fill in a gap in the textbook literature. Since the author has started, he should complete the elementary part at his leisure – just at his leisure. But the planned second (or third? I am still unclear about this) volume "Galois Theory" Schmidt and I would like to suppress in no uncertain terms. Galois Theory is so well represented in so many books and also so completely represented in the Yellow Series that a new textbook of this kind seems completely superfluous. I assume that even Brauer, who as we know has better things to do, will realize that.

Courant tries to write a response on October 15, 1935, but does not send it (an unusual hesitation for such a confident communicator), and finally re-writes and sends one on October 18, 1935:[91]

> In the concrete publisher affairs which we are discussing I see no other deserving resort than to terminate Brauer's contract. For the present he seems to be fairly frightened and

---

[88] New York University, Archive, Courant Papers.

[89] Arnold Berliner (1862–1942), the Editor and Founder of the journal *Naturwissenschaften* [Natural Sciences], published by Springer-Verlag, who committed suicide in 1942.

[90] New York University Archive, Courant Papers.

[91] Ibid.; both the unsent and the sent copies survive.

sad concerning this prospect. However, since Brauer has the strongest rear cover by Flexner, Veblen, and Weyl, it will be easy for him to publish his book by an American publisher or by one of the publishers being in development. Without any doubt, if his book is written rather well, it will have success over here.

In reply, Van der Waerden states that "Brauer's book... is not justified by any scholarly [sic] interest." He drafts a letter to Richard Brauer terminating the contract with him and on November 1, 1935, sends it to Courant for review and delivery:[92]

Enclosed is a letter to Brauer that I ask you to read and, if you have no heavy objections, send on to Brauer, whose current address I do not know. From it you will see that after a long conversation with Schmidt, I have still come to the position that Brauer's book would represent for us a considerable impediment that is not justified by any scholarly interest. After long reflection I decided to request from him a book on the Invariant Theory. But if your efforts to find an American publisher for his book succeed, I am very much in agreement. About the Invariant Theory we can still talk to him when this book is done.

Courant is surprised by Van der Waerden's rare, in Courant's opinion, tact and delivers the bad news to Brauer on November 16, 1935:[93]

Your letter to Brauer I found – not of course in absolute terms but relative to you – so carefully diplomatic, and also so nice and heartfelt, that I sent it on to him without any reservations. For myself I have written to him several times and now that he has overcome the shock I am hoping to receive his answer.

Courant is relieved, as on November 28, 1935 he reports Brauer's acceptance of the termination of his contract:[94]

At the same time you will have received a letter from Brauer, according to which the whole affair has been rather satisfactorily taken care of.

Unlike Van der Waerden, Richard Brauer was not a charismatic expositor. Encouraged by Schur, Springer, and Courant, Brauer went along with writing the book, and even a three-volume set. Van der Waerden's opposition, coupled with the anti-Semitic and anti-emigrant pressures on Ferdinand Springer in the Third Reich, stopped this most promising project. In the end, Van der Waerden fended off the competition, and Brauer went back to his favorite pastime, research. The world of mathematics has never gotten to see the three-volume *Algebra* by Richard Brauer. However, we did, get a huge three-volume set [BraR] of Brauer's collected papers.

---

[92] Ibid.
[93] Ibid.
[94] Ibid.

36.10 Epilogue: On to Germany

Ever since his student's years, Bartel L. van der Waerden aspired to a job in
Germany, perhaps, the place-to-be for a mathematician at the time. The leading
German colleagues had a very high opinion about him. To prove it, it suffices to
mention that Van der Waerden was ranked 3rd on the list of all-important David
Hilbert's succession at Göttingen.[95] The Dutch academics knew about it, and tried
to lure Van der Waerden to remain in the Netherlands. Van Dalen informs:[96]

> There were forces that tried to keep Van der Waerden in Holland. It was in particular
> Paul Ehrenfest[97] who made an effort to get Van der Waerden appointed in Leiden... He
> was aware that Leiden could not compete with Göttingen [no place could at the time!],
> 'The idea that in the fall you will start to work here, and that Leiden will develop
> into one of the centres of mathematics has been so much fixed in my head..., that I
> would be totally discouraged if you were snapped away in the last moment' [February
> 6, 1930].
>
>   How serious the option was, appears from the fact that Hilbert had at Ehrenfest's
> request written a recommendation for Van der Waerden.

On May 1, 1930, Van der Waerden informed Erich Hecke that he intended to
remain at Groningen for the time being; "I had refused a call to Leiden," he wrote.[98]

The attempts to keep Van der Waerden in the Netherlands failed, when on May 1,
1931 he succeeded Otto Hölder as Professor at Leipzig University. Once in Leipzig,
Van der Waerden joined the seminar conducted by the physicists Werner Heisenberg
and Friedrich Hund. Van der Waerden was an extremely quick learner. He picked
up physics from them (as he did algebra from Emmy Noether and Emil Artin)
and already the following year published a book on applications of group theory
to quantum mechanics in the Springer's *Yellow Series* [Wae4].

Hitler's ascent to power at the dawn of 1933 found Van der Waerden contemplating his second Rockefeller (IEB) fellowship.

---

[95] February 9, 1930 letter from Richard Courant to Paul Ehrenfest, cited in [Dal2], p. 688, footnote 28.

[96] [Dal2], pp. 687–688.

[97] Paul Ehrenfest (1880–1933), professor of physics at Leiden (1912–1933), a close friend of Albert
Einstein and Niels Bohr.

[98] Nachlass von Erich Hecke, Universität Hamburg.

37

In Search of Van der Waerden: The Nazi Leipzig, 1933–1945

These tragic times provide such profound lessons of human nature that we have got to learn from them as much as we possibly can. We encounter heroes and villains, but also a much more numerous group in between of these two extremes. The life of one such person "in between" is the subject of the text under review.

– Alexander Soifer, [Soi29]

It is hard to be a historian. It is difficult if you have not lived in the time you write about, and if you have, it is even worse.

– N. G. de Bruijn, June 1, 2004[99]

Good "history" is possible when historians take the initiative to undertake their own investigations of what has been accepted as "fact."

– Harriet Sepinwall, February 6, 1996
College of Saint Elizabeth
Holocaust Education Resource Center

37.1 Prologue

We have examined Bartel Leendert van der Waerden's early years and elucidated his relationship with his distinguished family, which included two members of the Dutch Parliament and an Amsterdam judge. It is now time to explore the complex, controversial and largely unexplored territory: the 20 years of Van der Waerden's life in Leipzig and Amsterdam, 1931–1951. I am partitioning my report about these controversial years into three sections to allow a deeper insight.

---

[99] [Bru8]

A. Soifer, *The Mathematical Coloring Book*,
DOI 10.1007/978-0-387-74642-5_37, © Alexander Soifer 2009

Camilla, Bartel, Theo, Coen, Dorothea and Ben van der Waerden, 30th Anniversary of Theo
& Do's marriage, Circa August 28, 1931, Freudenstadt, Southern Germany. Courtesy of Theo
van der Waerden

37.2 Before the German Occupation of Holland: 1931–1940

> *From 1933 till 1940 I considered it my most important duty to*
> *help defend the European culture, and most especially science,*
> *against the culture-destroying National Socialism.*
> – Van der Waerden, *Defense*, July 20, 1945[100]

On May 1, 1931, at 28 years of age, Bartel Leendert van der Waerden started
at *Universität* Leipzig as an *Ordinarius*.[101] Germany at the time was the center of
the mathematical world, and Leipzig, although below Göttingen and Berlin, was
a fine university. This could be viewed as a promotion from his prior *Ordinarius*
position at Groningen University, the Netherlands. Bartel was accompanied by his
Austrian wife of one year, Camilla, born Camilla Rellich, and their baby daughter
Helga.

---

[100] Rijksarchief in Noord-Holland (RANH), Papers of Hans Freudenthal (1905–1990), mathematician,
1906–1990, inv. nr. 89.

[101] The highest professorial rank in Germany, roughly equivalent to a full professor at an American
university.

Bartel L. van der Waerden, Leipzig, June 1931. Courtesy of Leipzig University

Less than 2 years later, Germany made a dramatic move that diminished its position in the mathematical world. Following Hitler's January 30, 1933 assent to power, the April 7, 1933 "Law for the Restoration of the Professional Civil Service" (*Gesetz zur Wiederherstellung des Berufsbeamtentums*) rid German universities of all Jewish (in Nazi's definition) professors, except for those who had entered civil service before 1914, fought for Germany in the World War I, or lost a father or a son in that war[102]. Van der Waerden included one short sentence about the year 1933 in "The Defense," a document he wrote for de-Nazification Boards of Utrecht and Amsterdam after the World War II: "In 1933 I traveled to Berlin and Göttingen to protest the boycott of Landau's classes by Göttingen Nazi students."[103] Unfortunately, I know of no evidence substantiating or detailing Van der Waerden's objections to these 1933 mass firings of the Jews. The firings included Van der Waerden's teacher and mentor Emmy Noether, about whom Van der Waerden wrote a beautiful eulogy in 1935 [Wae5]. But when in 1933 she was thrown out of Göttingen University, Van der Waerden was busy defending himself. Friedrich, the leader of the mathematics students' organization (*Führer der mathematischen Fachschaft*), argued that as a foreigner Van der Waerden was not fit to be the Director of the

---

[102] These exceptions were pushed through by the German President Paul von Hindenburg (1847–1934).

[103] "The Defense," *Rijksarchief* in *Noord-Holland* (RANH), Papers of Hans Freudenthal (1905–1990), mathematician, 1906–1990, inv. nr. 89.

Mathematics Institute. Van der Waerden was afraid he would lose his job, and on March 29, 1933 wrote about his worries to Richard Courant in Göttingen, who replied on April 15, 1933[104] as follows:

> I find it laughable if you believe that there is any threat to your Leipzig position because you are Dutch. Instead, I am very afraid for your Leipzig colleague L. [F. Levi, who was Jewish].

In his defense, Van der Waerden wrote the following letter to *Dekan*[105] Weickmann of Philosophical *Facultät*[106] in Leipzig on May 18, 1933:[107]

> Your Magnificence!
>
> I have just learned from you that the Ministry possesses a letter in which it is claimed that I am of a non-Arian descent. I declare that I do not know how that conclusion was reached and who could have written this to the Ministry. I am a full-blooded Arian and I can prove that if necessary, because my ancestry can be tracked for three generations.
>
> With loyal regards,
>
> Yours
> B. L. v. d. Waerden[108]

The number of Arian generations in Van der Waerden's ancestry quickly grew, for the next day, on May 19, 1933, Leipzig's *Rektor* Achelis informed Minister Hartnacke of Saxony, that the accusation that Van der Waerden was Jewish was not correct, that Van der Waerden had proof that *five* generations of his ancestors were Christians, and thus Van der Waerden should be able to retain his Directorship.[109]

At the time Van der Waerden was proving his Arianness, Princeton University decided to invite him to a visiting professorship. "A meeting of the Research Committee was held on Tuesday, May 9, 1933, in Dean [of the Faculty Luther Pfahler] Eisenhart's office, Fine Hall, at 12:00 noon. Present: Dean Eisenhart, Professors [Edwin Grant] Conklin [Biology], [Rudolph] Ladenburg [Physics], [Solomon] Lefschetz [Mathematics], [Henry Norris] Russell [Astronomy] and [Sir Hugh] Taylor [Chemistry]," I read from the yellowed pages.[110] Chapter 2 is of interest:

> 2. Dean Eisenhart reported the desire of the Department of Mathematics to secure Professor van der Waerden of Leipzig on the Mathematics funds for the first term of

---

[104] Courant, letter of October 15, 1935, slightly modified later, on October 18, 1935. New York University Archives, Courant Papers.

[105] About equivalent to a dean.

[106] Equivalent to a part of a university, such as a school or a college.

[107] Universitätsarchiv Leipzig, PA 70, p. 18.

[108] See the facsimile of this letter in this chapter (p. 397).

[109] Universitätsarchiv Leipzig, PA 70, p. 21.

[110] Archive of the Department of Mathematics, Princeton University.

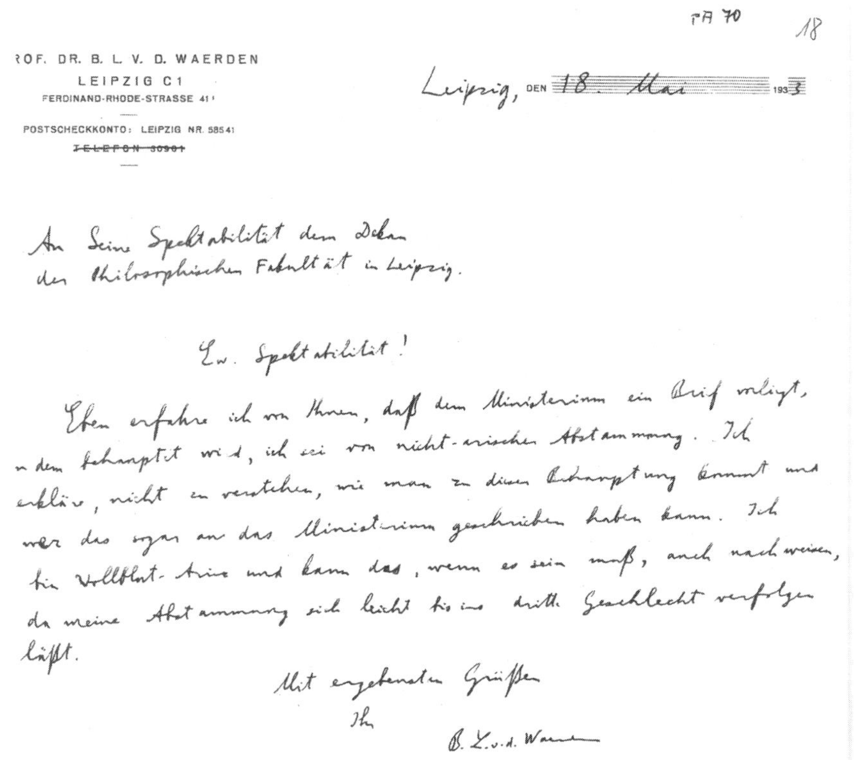

PROF. DR. B. L. V. D. WAERDEN
LEIPZIG C1
FERDINAND-RHODE-STRASSE 41
POSTSCHECKKONTO: LEIPZIG NR. 58541
~~TELEFON 30981~~

PA 70 18

Leipzig, den ~~18. Mai~~ 1933

An Seine Spektabilität den Dekan der Philosophischen Fakultät in Leipzig.

Ew. Spektabilität!

Eben erfahre ich von Ihnen, daß dem Ministerium ein Brief vorliegt, in dem behauptet wird, ich sei von nicht-arischer Abstammung. Ich erkläre, nicht zu verstehen, wie man zu dieser Behauptung kommt und wer das sogar an das Ministerium geschrieben haben kann. Ich bin Vollblut-Arier und kann das, wenn es sein muß, auch nachweisen, da meine Abstammung sich leicht bis ins dritte Geschlecht verfolgen läßt.

Mit ergebenen Grüßen
Ihr
B. L. v. d. Waerden

B. L. van der Waerden's claim of his Arianness

> 1933–34 at a salary of $3500. Dean Eisenhart reported that in this case also[111] there would be a delay on account of uncertain conditions in Germany.

Indeed, Princeton University offered Professor Van der Waerden a Visiting Professorship for the September 15, 1933–February 15, 1934 semester. Having actively sought and received approvals by *Dekan* Weickmann of the Leipzig's Philosophical *Facultät*, by von Seydewitz on behalf of the Minister of the Saxon Ministry of People's Education (*Ministerium für Volksbildung*), and having assured a replacement

---

[111] By this "also" Dean Eisenhart referred to Chapter 1 of these minutes, which is of an historical interest too, and reads as follows: "Dean Eisenhart reported the inability of Professor Heisenberg to give a definite answer to the offer of an 8 weeks engagement at a salary of $3000 at the present time owing to the uncertain conditions in Germany. Professor Heisenberg suggested that he might be able to give a definite reply at the end of the year. Dean Eisenhart has written to Heisenberg on the assumption that his letter meant the end of the academic year and suggested that decision by July would be acceptable." Chapter 7 of the minutes is relevant too: "Professor Lefschetz raised the question of alternatives to Professor Heisenberg in case it was found impossible to secure his services. After discussion, it was decided that the matter be left in abeyance until further reports were available concerning the German situation."

(*Privatdozent*[112] Dr. Friedrich Karl Schmidt from Erlangen University), Van der Waerden suddenly changed his mind and on July 29, 1933 withdrew his approved request for visiting Princeton University. This was the first major junction in the life of Van der Waerden: had he come to Princeton, as a fine and young mathematician, Van der Waerden would have most likely received further offers from Princeton University or from the recently founded Institute for Advanced Study. His life and the history of Algebraic Geometry would have been different. But Van der Waerden chose to remain in the Nazi Germany, as did his friend the Nobel physicist Werner Heisenberg, who did not accept Princeton (see a documentary proof in Footnote 13), Harvard, Yale, and Columbia job offers during the early years of the Third Reich.

Heisenberg's nationalism and devotion to Germany as reasons for staying in the Third Reich have been well established ([Wal1], [Wal2], [Pow], etc.). Van der Waerden's Princeton opportunity has never before been discussed in detail and backed by documents. His surprising rejection of the Princeton offer begs a question: why did he do it? Van der Waerden explains it in his August 12, 1933 letter to Oswald Veblen of the Institute for Advanced Study:[113]

> Like you, I am very sorry that we will not meet in Princeton in the next winter, but it was really impossible for me to leave Leipzig at the time.

As we know from the Leipzig archive, all permissions had been granted. It is therefore clear that Van der Waerden preferred courtesy to the truth in his letter to Veblen. But what was the truth?

Van der Waerden asked his mentor Richard Courant to help him receive the second Rockefeller (IEB) Fellowship, this time for work in algebraic geometry in Italy, primarily under Federigo Enriques and Francesco Severi in Rome. On March 2 1933, Courant still at Göttingen, "informally and personally" asked Dr. W. E. Tisdale, the Rockefeller Official in Paris, whether the support for Van der Waerden was possible:[114]

> Van der Waerden in spite of his considerable youth is one of the most outstanding mathematicians currently in Europe. For the occupation of the Hilbert Chair he was one of the 3 candidates of the Faculty. Now for some years van der Waerden has successfully began to deal with the problems of Algebraic Geometry and it is his serious objective to really develop this area for Germany. In fact, the geometric-algebraic tradition in Germany is almost extinct, while in Italy in the course of the past decades it has blossomed.

Tisdale received the letter on March 6, 1933, and the same day replied to Courant, asking to have Van der Waerden provide more details, which Van der Waerden did in the March 12, 1933 two-page letter (received in Paris on March 31, 1933). This

---

[112] Roughly equivalent to an associate professor, but without a guaranteed salary.

[113] Library of Congress, Veblen Papers.

[114] Rockefeller Archive Center (RAC); Collection IEB, Series 1, Sub-series 3, Box 61, Folder 1027.

letter, written in English, provides an insight into Van der Waerden's view of the state of algebraic geometry.[115]

> The algebraic geometry, originated in Germany by the work of Clebsch, Noether and others, has been continued during the last 30 years nearly exclusively by Italian mathematicians: Enriques, Castelnuovo, Severi, and others. They developed methods and theorems, which are of extremely high interest both for algebra and geometry, but which are still awaiting exact algebraic foundation: The contact between Italian geometry and German algebra is missing. I think, this is a typical case in which your foundation can help. I know the algebraic methods which can serve as a base for algebraic geometry very well, perhaps best of all German mathematicians.

Thus, Van der Waerden considered himself to be *the best* German mathematician for the job of putting algebraic geometry on the foundation of abstract algebra. Moreover, for the first time in written records found by me—Van der Waerden cast himself here as a *German mathematician*.

For his visit Van der Waerden requested the winter semester of 1933–1934:

> ... it would be desirable for me to stay half a year in Italy, and more especially in Rome with Prof. Severi and Prof. Enriques.... A winter semester should be preferable, as I can then stay half a year in full term in Italy, and need a replacement for teaching in my place only during the 4 months of a winter semester. Perhaps the replacement could be paid from your stipend, whereas I could live on my salary, if the Saxon Government is willing to consent in this... I have acquired a sufficient knowledge of the Italian language.

A successful Rockefeller (IEB) fellow first time around, Van der Waerden surely expected—and deservedly so—an easy approval of his second fellowship. So, did Van der Waerden simply choose Rome over Princeton? Indeed, I found proof of it in his own words—even before he jumped through all the Leipzig bureaucratic hoops—in a (undated, but definitely written in May or else June of 1933) letter to Richard Courant:[116]

> I still thank you many times for your efforts at Rockefeller. I only got a reply from Tisdale that now there are sufficient documents to discuss the case with his colleagues in Paris...
>
> I have an offer from Princeton University, with a stipend, to spend the coming winter semester (Sept-Jan.) there. This offer came already in the beginning of April. But it does not tempt me as much as the Rome trip; I also do not know whether the regime will allow this much of a leave of absence...

As we know, at some point—more precisely in late July 1933—Van der Waerden learned that "the regime will allow this much of a leave of absence." He may have even then, in July 1933, hoped to get the Rockefeller money for a half a year in Rome. Is this why Van der Waerden cancelled the approved by all sides visit to Princeton? Perhaps, but there could have been another important reason for not

---

[115] Ibid.

[116] New York University Archive, Courant Papers.

going to Princeton or to Rome: Van der Waerden did not really wish to leave Germany for the first winter of the Third Reich:[117]

> I cannot judge yet whether it is not cleverer to spend this winter in Leipzig.

What was so clever about staying in Germany during the winter of 1933–1934? We will never know for sure, but a plausible question is in order: Did Van der Waerden not wish to raise a suspicion of the brand new Nazi regime? Now that Van der Waerden was not going to go to Princeton anyway, it was easy for him to be generous and conscientious:[118]

> I believe I will suggest to the Americans that this time they could spend their money better than to get me out because I still have a position that I can keep.

It appears likely that the Rockefeller people, once they learned of the Princeton offer to Van der Waerden, chose to use their funds to support those mathematicians who depended solely upon the Rockefeller money, and thus decided not to fund Van der Waerden's second fellowship. According to the leading researcher of mathematics support in the Rockefeller Archive Center and the author of a monograph on the subject [Sie] Reinhardt Siegmundt-Schultze, the Rockefeller Center has no approval documents, which implies that Van der Waerden's request was not funded. In fact, Tisdale wrote in his diary on March 29, 1933:[119]

> *Van der Waerden*, past fellow now at Leipzig is excellent. As a matter of fact Princeton wants to get him in the faculty to replace shifts due to Flexner's activity [i.e., the creation of the Institute for Advanced Study]. They will probably ask him to come for a semester in which they could have a mutual exchange of view.

Yes, the Princeton position would have likely become permanent for Van der Waerden. It seems clear that Princeton mathematicians were unhappy about Van der Waerden's "clever" choice to stay in the Nazi Germany when they offered him a great opportunity to get out. As we will see later, they remembered this rejection after the war, when Van der Waerden was willing—moreover, eager—to come to Princeton from the war-devastated Netherlands.

Alas, we ought to roll back to the Nazi Germany, year 1934. As was expected of him, Van der Waerden signed and dated his oath to Hitler on November 1, 1934:[120]

> I affirm that I have taken the following oath today:
> "I swear: I will be faithful and obedient to the Führer of the German Reich and People, Adolf Hitler, I will obey the laws and fulfill my official duties conscientiously, so help me God."

May 1935 started with the Ministry dismissing five Jewish professors from the University of Leipzig – Doctor. of Medicine Bettmann, and four Philosophical *Fac-*

---

[117] Ibid.

[118] Ibid.

[119] Rockefeller Archive Center, Tisdale Log 7 (1933), p. 27.

[120] Universitätsarchiv Leipzig, PA 70, p. 33 (see facsimile on p. 401).

Ich bestätige, dass ich heute folgenden Eid geleistet habe:

„ Ich schwöre: Ich werde dem Führer des Deutschen Reiches
und Volkes, Adolf Hitler, treu und gehorsam sein, die Ge-
setze beachten und meine Amtspflichten gewissenhaft er-
füllen, so wahr mir Gott helfe.”

Leipzig, den *1. Nov. 1934*

Name... *B. L. v. d. Waerden* ...(*B.L.v.d.Waerden*)
(Bitte recht deutlich!)

Dienstbezeichnung... *Professor* ....

Fakultät bezw.Dienststelle.. *Philos. Fakultät* ...

B. L. van der Waerden's oath to Hitler

ultät professors: Wach, Landsberger, Levi, and Weigert, all veterans of the World
War I, and as such exempted from the dismissal under the April 7, 1933 law. On
Friday, May 2, 1935, Leipzig's *Rektor* Krueger discussed these firings with the
Staatssekretär Theodor Vahlen (coincidentally a mathematician himself), who was
in charge of the Third Reich's university appointments in the *Reichserziehungsmin-
isterium* and reported directly to the *Reichsminister* of Education Bernhard Rust.

The *Rektor* announced these firings on Wednesday, May 8, 1935 in the afternoon, at the Faculty meeting of the Philosophical *Facultät*. He merely wanted to
test the faculty's sentiments, and not have a full-blown discussion. But three professors questioned the legality of the firings and spoke strongly in support of the fired
colleagues: the physicists Werner Heisenberg[121] and Friedrich Hund,[122] and Bartel
L. van der Waerden.

Upon the urgent demand ("tomorrow by 1 PM") by the Saxon Ministry of People's
Education for the "precise" text, the recording secretary Hch. Junker reconstructed the
meeting's stenography on May 21, 1935 based on the notes he took during the meeting.
This surviving stenography allows us to "hear" the voices of the participants:[123]

> *Dekan*: The dismissals were done "in the interest of the Service" ("*im Interesse des
> Dienstes*"). It is not our responsibility to go further into that.

> *Heisenberg*: This action has caused dismay among many of us because they [dismissed
> professors] felt that it did not satisfy the meaning of the law. That is: combatants belong
> to the people! It is our duty to help them in every respect especially because their

---

[121] Werner Heisenberg (1901–1976); Nobel Prize "for the creation of quantum mechanics...", 1932;
Max Planck Medal, 1933.

[122] Friedrich Hund (1896–1997); Max Planck Medal, 1943.

[123] Universitätsarchiv Leipzig, PA 70, pp. 36–40.

students already stood up for them. It is necessary for the faculty to say that it is about the people who have put their life at risk for us.

Golf:[124] These are concerns that are justified. But please do not continue the discussion, and do not ask questions. The report has been sent to Dresden now. The reply will come. The Dean travels to Dresden tomorrow. Any further discussion today is therefore superfluous. We hope that we will be informed about the reply.

Hund: I believe that I cannot refrain from expressing the sentiment among a few colleagues. If these actions become a fact that would mean that a meaning of the exemption in the law, that men who have fought on the frontlines could not be expelled, would be violated. That would be a serious disappointment in the Government. Many of us, who have not been to the frontlines, including myself, would have to be ashamed before these men.

v.d.Waerden: It would be useful if a unanimous decision could be reached regarding the rights of the combatants and the meaning of the law, which has obviously been disregarded.

Dekan: I may remark that I allow this discussion *only* so that I can report in Dresden about the sentiment on the *Facultät*'s commission (*Facultätsausschuss*).

Golf: I feel satisfied with what the Rektor told us. But I want to advise Professor van der Waerden to be more cautious. He has said: a paragraph of the law has been *violated*. He obviously did not keep in mind that this amounts to saying the ReiStatth. [*Reichsstatthalter* of Saxony] has violated the law. We don't know his reasons and it is not up to us to make a judgment. So, please, be more careful, be more cautious with your comments.

v.d.Waerden: (in half-voice across the table to Golf): Thank you!

Golf: (across the table, loudly): The matter is thus closed![125]

Van der Waerden, of course, knew that Germany lived not by the law but by the latest word of the Nazi Government. However, the law as it existed since April 7, 1933, provided for an exception for veterans of W.W.I, and Van der Waerden used a violation of this law as the ground for his objection. No matter, a public protest was a brave act – especially considering his foreign (Dutch) citizenship – although his oath of loyalty to Hitler was of help, as was of help the fact that Van der Waerden was not alone but a part of a group of three protesters. The stenography of the meeting leaves an impression that Heisenberg, Hund, and Van der Waerden were "co-conspirators," who discussed politics between themselves, even though Camilla van der Waerden, in presence of Prof. Van der Waerden, denied it in 1993: "with Heisenberg and Hund

---

[124] Professor of Agriculture Arthur Golf (1877–1941), *Rektor* of Leipzig University (October 1933–March 1935, and again October 1936–March 1937), member of *NSDAP* (*Nationalsozialistische Deutsche Arbeiterpartei*, known as the Nazi Party) since 1932, the author of *Nationalsozialismus und Universität. Rektoratsrede* (Leipzig, 1933).

[125] In [Dol2] the author writes, clearly hinting at this 1935 episode, as follows: "Van der Waerden's personal file, kept in the archives of the University of Leipzig, shows, however, that he spoke out in favor of young Jewish mathematicians." This "young mathematicians" in reference to the aging by 1935 veterans of W.W.I shows that Prof. Dr. Dold-Samplonius has never read this document.

we talked about science and not about politics" [Dol1]. In fact, Mr. Thomas Powers, the author of the bestselling book *The Heisenberg's War* [Pow], has kindly provided me with the (largely unpublished) notes of the interview with Van der Waerden that his fact finder Ms. Delia Meth-Cohn conducted in Zürich on February 21, 1989:

> In Leipzig, Heisenberg, Friedrich Hund, Friedrich Carl Bonhoefer[126] and v.d.W [Van der Waerden] had formed a clique (alliance) to maintain the scientific level in mathematics and physics against the Nazis. They were all reliable anti-Nazis, met very frequently and talked a lot about political questions.

Van der Waerden even cited one such particular political discussion in his private April 28, 1948 letter to Heisenberg:

> Do you remember what I said to you when you showed me the [1937] article in *Das Schwarze Corps*? "That is a nice title, white Jew, you can be proud of that." Instead of being proud, the article annoyed you.[127]

As to Heisenberg, on July 21, 1937 he asked the *SS Reichsführer* Heinrich Himmler personally for a protection from the attack by the inventors of the notorious notion of "*Deutsche Physik*" Philipp Lenard and Johannes Stark that Van der Waerden referred to above. This protection was granted by Himmler, who on July 21, 1938 wrote to his *Gestapo* chief Reinhard Heydrich, "I believe that Heisenberg is a decent person and that we cannot afford to lose or to silence this man, who is still young and can still produce a rising generation in science."[128] The same day Himmler promised protection in a letter to Heisenberg personally. This high *SS* protection ended forever the days when Heisenberg could publicly criticize any actions of the regime, even if he were so inclined, for Heisenberg became a highly protected asset of this "gangster regime" (as Van der Waerden would call it in 1945 in a letter to Van der Corput). Heisenberg had numerous opportunities to emigrate, as he received numerous offers before the war started, from Princeton, Harvard, Yale, Columbia, etc. – but he chose to stay and to serve Germany – the Nazi Germany, as was the case.

We roll back to the Third Reich, to the year 1935. Shortly after the Leipzig faculty meeting, on September 15, 1935, the new definitions of "Jewishness" and its relation to citizenship were approved by Hitler's willing lawmakers in the so called "Nuremberg Laws." It is surprising to me that 13 days after the new law provided a blanket prohibition of Jewish civil service employment, Van der Waerden showed a certain insensitivity towards a firing by Springer-Verlag of a Jew, the founder and editor of

---

[126] Actually Karl Friedrich Bonhoeffer (1899–1957), professor of physical chemistry at the University of Leipzig from 1934 till the end of W.W.II.

[127] Private Papers of Werner Heisenberg, Max Plank Institute for Physics, Munich. Prof. Mark Walker was first to discover and use Heisenberg—Van der Waerden correspondence in his Ph.D. thesis and books. I am grateful to him for sharing it with me. I also thank Dr. Helmut Rechenberg and Werner Heisenberg Archive he directs, for permitting to quote this correspondence.

[128] [Gou], pp. 116–119.

the celebrated Springer-Verlag's journal *Naturwissenschaften* Arnold Berliner. On September 28, 1935 Van der Waerden wrote to Richard Courant as follows:[129]

> It does not seem that the Springer publishing house has been seriously attacked. The demand to dismiss the Jews who were still in service was only due to intensify. I do not understand why in foreign countries one gets so upset about the editorial change in the *Naturwissenschaften*. Berliner[130] was certainly 73 years old.

Courant had to explain to Van der Waerden what should have been obvious:[131]

> You do not understand the excitement abroad about the removal of Berliner. Of course, everything would have been in order if B. because of his age would have been retired observing the proprieties corresponding to his position and merits. In fact, however, the removal appears abroad, and it seems to be the case, as to be fired in a hurting way due to pressure coming from outside. The great reputation of B. has given in this context the reason for a heavy general criticism and for expression of doubt concerning the possibilities of Springer to pursue an objective publishing leadership. I have, partly from extremely influential people, received comments and further inquiries which I cannot describe in a letter to Germany.

A cloud of the Third Reich's suspicion hung over Van der Waerden's head ever since his May 1935 public comments at the Faculty meeting. Van der Waerden would be criticized for opposing Bieberbach, the founder of the notorious anti-Semitic notion of *Deutsche Mathematik*,[132] at the September 13, 1934 meeting of the *D.M.V.* (German Mathematical Society), even though the majority of mathematicians present opposed Bieberbach. Van der Waerden would have to defend himself from this accusation even 8 years later, in 1942 (I will present his defense later in this part). Some (but not all) of Van der Waerden's many travel requests would be denied. German representatives in the Netherlands would be asked to check on the behavior of his father, Dr. Theodorus van der Waerden, who was a member of the Second Chamber of the Dutch Parliament (1918–1940) from the Socialist Democratic Workers Party (*Sociaal-Democratische Arbeiderspartij*, or shortly *SDAP*). On May 9, 1939, the German Embassy in Haag advised the Foreign Office in Berlin as follows:[133]

> In response to the order of 11 April of this year
>
> Re: Dr. of Engineering Theodorus van der Waerden
>   Born 21 August 1876 in Eindhoven
>
> Dr. van der Waerden has been active in the Socialist Democratic movement since his days as a student. At the moment he is a Socialist Democratic representative in the

---

[129] New York University Archives, Courant Papers.

[130] Arnold Berliner (1862–1942), the Editor and Founder of the journal *Naturwissenschaften* (Natural Sciences), who committed suicide in 1942.

[131] Courant, letter from October 15, 1935, slightly modified later, on October 18, 1935. New York University Archives, Courant Papers.

[132] "German Mathematics" as opposed to "Jewish Mathematics."

[133] Universitätsarchiv Leipzig, PA 70, p. 51.

Second Chamber, and he allegedly belongs to the more moderate wing. In his attitude towards the New Germany, he probably does not differ from his Marxist comrades. He has not become apparent in this respect in public, though.

(Signed) Zech

Based on this not-too-damaging report about his father, the Saxon Ministry of People's Education advised Van de Waerden's *Dekan* at Leipzig as follows on August 23, 1939:[134]

<u>Confidential!</u>

Herr Reich's Minister of Education has sent me the attached report from the German Embassy in Haag about the father of Professor van der Waerden and remarked that a successful continuation of his teaching activity at Leipzig University requires of Professor van der Waerden, who has kept his Dutch citizenship, a loyal attitude towards the National Socialist Germany and its institutions and political restraint.
If you learn certain facts, which prove that Professor van der Waerden does not comply with this expectation, I ask for a report.

Ordered by
(Signed) Sudentkowski

Van der Waerden clearly saw the Nazi regime for what it was. On August 10, 1935[135] he wrote from his parents' house in Laren, The Netherlands, to Richard Courant, who was already in New York:

We are here in Holland for two months and rest up our souls from the constant tensions, hostilities, orders and paperwork. We do not have yet the successor to Lichtenstein; instead Ministries examine who has not yet been completely switched over [to the National Socialism], who is a friend of Jews, who has a Jewish wife, etc., as long as they themselves are not torn apart by their fight for power.

Yet, Van der Waerden chose not to remain in the Netherlands during this and several other visits (in 1933, 1935, 1938, 1939, 1940, and 1942), including those visits when he stayed in the Netherlands with his wife and children (i.e., at least in 1935 and 1939), and not to go to the United States, as we saw above. He preferred to live in the Nazi Germany. This choice appears so contradictory that it begs a question: why did Van der Waerden prefer Germany? I will attempt to answer this most important question in the course of this investigation.

Van der Waerden was no rebel: he complied with the laws of the Third Reich and with its persistent recommendations. I have copies of several official letters, which Van der Waerden closed with the recommended "Heil Hitler!" P. Peters, a student at

---

[134] Ibid., p.50.

[135] New York University Archives, Courant Papers.

Amsterdam, claimed in print[136] on February 8, 1946 that "every single day he [Van der Waerden] gave Heil Hitler salute in public at the start of his lectures to the enemy."[137]

On the other hand, his conduct as an associate editor of the major German mathematical journal *Mathematische Annalen* was commendable. Since 1934 Van der Waerden had been one of the associate editors of *Mathematische Annalen*, published by Springer. In the late 1930s, the editorial room consisted of the editor-in-chief Erich Hecke and two associate editors—Van der Waerden and Heinrich Behnke. Not only was this a trio of fine mathematicians. It was also a group that tried to be fair toward all authors, Jews included. Prof. Sanford Segal provided a wonderful description of the dynamics of this editorial room [Seg, pp. 234–244], during 1939–1941, and of the test they got from the publisher, Ferdinand Springer. Not wishing to jeopardize the journal with the Nazi authorities by publishing Jewish authors, Springer informed Hecke accordingly during their December 20, 1939 meeting. Hecke threatened to resign rather than compromise the integrity of the editorial process. To Hecke's satisfaction, Van der Waerden threatened to resign as well. "Under no circumstances do I want to form an obstacle if Springer wants to form National Socialist editorial board for *Annalen*," Van der Waerden wrote to Behnke on May 10, 1940 [Wae7]. In fact, Hecke did resign (letter of resignation was dated June 24, 1940), allowing only for his name to remain on the journal's cover (as a symbol of Hilbert's pedigree, whose student Erich Hecke was), and only under a threat by Ferdinand Springer to stop paying Blumenthal's pension.[138] Van der Waerden remained an editor, possibly convinced by Behnke who believed that their resignation would open the door to worse people in the editorial room. (Van der Waerden's editorship of the *Annalen* lasted 35 years, from 1934 until 1968.)

37.3 Years of the German Occupation of the Netherlands: 1940–1945

> *What I should explain to the Dutch people is, however, not my actions before 1940, but those after the Netherlands had been attacked by Germany. . .I have never given a class or worked on things that could be used for military purposes.*

---

[136] *Propria Cures*, a weekly of the students of the University of Amsterdam, February 8, 1946. More about this in the next Chapter.

[137] I do not know, of course, how Mr(s) P. Peters got to know this, and the degree to which we can rely on this report. But Peters replied to Prof. Van der Waerden's February 1, 1946 letter, published in *Propria Cures*, and had Van der Waerden disagreed, he could have published a rebuttal, which he did not do.

[138] Ludwig Otto Blumenthal (1876–1944), the first research student of David Hilbert, was the Editor of *Mathematicsche Annalen*. As a Jew, he had to step down from his editorship in 1938, and in 1939 emigrate from Germany to the Netherlands. His pension, paid by Springer, was the sole source of existence for his sister, who remained in Germany. In 1943 Otto Blumenthal and his wife Mali were sent to the concentration camp at Westerbork where Mali Blumenthal died. Otto Blumenthal died in Theresienstadt on November 12, 1944.

> *In 1943 the Faculty of Physics and Mathematics at Utrecht*
> *asked me whether I would accept an appointment as a*
> *Professor there. I asked them to postpone the matter*
> *if possible until after the war, because I did not want*
> *to be appointed by the Van Dam*[139] *department.*
> – Van der Waerden, *Defense*, July 20, 1945[140]

As a Ph. D. 1968 under Van der Waerden, Prof. Dr. Günther Frei writes with authority of an expert and with an understandable admiration for his teacher. The trouble is he does so without much historical research. For example, he perpetuates the myth [Fre1] that "Baudet"—and Baudet alone—authored the conjecture Van der Waerden proved in [Wae2]. This is why I read with caution his passionate words in the 1998 eulogy [Fre2]:

> Before and during the war van der Waerden lived in Leipzig as a foreigner, who had always refused to surrender his Dutch citizenship, which exposed him to many hostilities from the Nazis. This could not however affect his morally upright bearing.

Unlike Frei, who apparently believes that history is beneath mathematics and needs no proofs, I will present here documents that enable you to decide whether Van der Waerden's "morally upright bearing" was affected. Van der Waerden did retain his Dutch citizenship – and this fact could be perceived as showing his distaste for the Third Reich's citizenship. However, the following episode allows an unexpected insight and a different interpretation.

When Germany treacherously attacked the Netherlands on May 10, 1940 and conquered the country by May 15, 1940, the Dutch inside Germany were at first treated as enemies. In fact, right on May 15, 1940, Van der Waerden was suspended by the *Rektor* from teaching at Leipzig University:[141]

> I already asked you yesterday over the phone to refrain from any teaching activity until further notice. I herewith repeat this order in writing and ask you to discontinue your administrative activity as Director of Mathematical Seminars and of Mathematical Institute.
>
> Meanwhile I have asked the Ministry for a decision whether in view of your being an official and your oath to the *Führer* my order regarding your activity as a Professor and Director of the Institute should stay.

Van der Waerden was very soon reinstated. His reaction to this brief suspension allows us to better understand Van der Waerden's views of Germany and the Netherlands. He understood that the suspension was likely to be short-lived, but that he could be asked to accept the Third Reich's citizenship. The day following the suspension,

---

[139] Prof. Dr. Jan van Dam, Secretary-General of the Ministry of Education of the Netherlands (*Opvoeding, Wetenschap en Cultuurbescherming*) during the war.

[140] Rijksarchief in Noord-Holland (RANH), Papers of Hans Freudenthal (1905–1990), mathematician, 1906–1990, inv. nr. 89.

[141] Universitätsarchiv Leipzig, PA 70, p. 55.

B. L. van der Waerden, c. early 1940s. Courtesy of Leipzig University

on May 16, 1940, Van der Waerden wrote [Wae8] to a much admired mathematician and trusted friend, Erich Hecke:[142]

> For the time being I am not allowed to teach courses. But the *Rektor* has already written to Berlin and asked for authorization to allow me to carry on my office. The *Dekan* predicts that this would be smoothly approved; maybe I would be asked to become a German citizen. You will understand that I would be uncomfortable with that at this time. In principle I have no objections against German citizenship, but at this moment when Germany has occupied my homeland I really do not want to abandon my neutrality and take the German side.

How can one explain Van der Waerden's "*neutrality*" when Germany brutally invaded his homeland? Could it be that Van der Waerden believed he belonged to German culture in general, and to German science and mathematics in particular? If so, this would also explain why Van der Waerden did not accept the offer to leave Germany that he received in the middle of World War II.

In the previous section, we "attended" the May-1935 faculty meeting at Leipzig, where Van der Waerden bravely criticized the Saxon Minister for violating the law and firing four Jewish professors. Of course, the Minister wanted to retaliate, and Van der Waerden was accused of anti-Nazi conduct at the 1934 *D.M.V.* (German Mathematical Society) meeting at Bad Pyrmont that took place nearly a year earlier. Amazingly even for the massive Nazi bureaucracy, 8 years later this case was still open, and Van der

---

[142] Archive of *Mathematische Annalen*'s Editor Erich Hecke; Prof. Dr. Holger P. Petersson private collection.

Waerden still had to defend himself for his 1934 conduct. On June 13, 1942 Van der Waerden explained what had happened in 1934 in a letter "to *Dozentenschaftleiter* Prof. Dr. M. Clara, with copy to the *Rektor* and the *Dekan*." Let us listen:[143]

> In defense against an accusation directed against me I report about the events at the annual meeting of the *D.M.V.* in Pyrmont on Sep 13, 1934.
>
> The Danish mathematician Harald Bohr sharply attacked the German mathematician Ludwig Bieberbach in a newspaper article.[144] Mr. Bieberbach defended himself against this attack and has published his reply in the Annual Report of the *D.M.V.* vol 44. In this reply, Bohr was personally insulted and labeled "parasite [*schaedling*] of all international cooperation." The publication happened against the stated will of the both co-editors of the Annual Report, Hasse and Knopp. For that reason, Mr. Bieberbach was held responsible during the Annual Meeting. The publication was sharply criticized by me and many others. All of us regarded it as harmful to the reputation of the German Science [*die deutsche Wissenschaft*][145] abroad. A couple of good Germans and National Socialists sided with me, among them Mr.'s Hasse and Sperner, now treasurer of the *D.M.V.* and editor of the Annual Report. Finally, the assembly approved by a large majority a motion critical of Mr. Bieberbach's action, and Mr. Bieberbach has stepped down from his office of the Secretary of *D.M.V.*

This was an accurate description of the 1934 meeting. However, Van der Waerden continued with words that would contradict his postwar claims of being "a strong anti-Nazi":

> I firmly declare that I only had the interest and the reputation of the German Science [*die deutsche Wissenschaft*][146] in mind. By no means did I comment [*stellung nehmen*] against National Socialist principles or actions. The question of race in mathematics and Mr. Bieberbach's speech about it, which had formed the origin of Bohr's attack, had not been discussed during the meeting in question, just the form of the Bieberbach's personal counterattack and its publication in the Annual Report of *D.M.V.*

You may recall Johan A. Barrau, who arranged for Van der Waerden to succeed him at Groningen in 1928, when Barrau moved to a chair at Utrecht (Chapter 36). As his retirement at 70 was approaching, Barrau envisioned Van der Waerden as his successor again, this time at Utrecht. On December 16, 1942 he wrote to Van der Waerden about it, and asked for a clear answer:[147]

> Dear Colleague,
> At the end of the current semester, in Sept. 1943, it is my turn to resign and to be replaced. The *Faculteit* choice of the successor is dependent on knowing with certainty whether

---

[143] Universitätsarchiv Leipzig, PA 70, p. 59.

[144] During 1933–1934 the German mathematician Ludwig Bieberbach, who later founded the movement and the journal of the same name *Deutsche mathematik*, started to spread his race-based anti-Semitic view of mathematics. The Danish mathematician Harald Bohr published a stern rebuttal of Bieberbach prior to the Bad Pyrmont meeting of *D.M.V.*

[145] The term *die deutsche Wissenschaft* had race-based anti-Semitic connotation.

[146] See Footnote 145.

[147] Utrecht University, Archive of the Faculty of Mathematics, Correspondence, 1942.

you would be willing to return to the Netherlands. We are asking you politely to give us certainty. If you are not at all inclined to do that, then it is easy for you to inform me as soon as possible. However, if you want to think about it, then please tell me that too, and we will then be waiting for your decision.

On December 28, 1942, Van der Waerden replied by postcard, stamped twice on each side with "*Geprüft. Oberkommando der Wehrmacht*" [Examined. Supreme command of the Armed Forces] in a round seal:[148]

Thank you very much for your letter of December 16, 1942. With the reference to your last sentence, I want to keep this matter in mind, it is very important to me. I will write to you in early January.

Promptly on January 4, 1943, Van der Waerden elaborated, but refused to give a clear answer:[149]

I feel honored by your request. I am pleased with it. I am not rejecting the idea to return to Holland, on the contrary, I have always considered this possibility with respect to my plans for the future.

That possibility has merits. I am sorry I cannot give you the certainty that you are asking me. Whether I will accept a position or not is dependent upon circumstances, and I can only judge them when the appointment is actually there. A lot depends how the circumstances will be at that moment at the University of Leipzig, and I cannot judge that right now and I will not be able to judge that in two weeks either.

I would very much like you to keep me informed about this case in the future.

On the same day, January 4, 1943, Van der Waerden promptly met with his Leipzig University bosses and put it in writing on January 5, 1943:[150]

To the *Rektor* of the University via the *Dekan* [Heinz] of the Philosophy *Facultät*.

Magnificence!

The *Facultät* of Natural Philosophy of the Utrecht University (Holland) has asked me whether I would possibly be willing to accept the Ordinarius position in mathematics when it becomes vacant in September 1943. As I have already told you orally yesterday, this inquiry is tantamount to an offer since negotiations about offers are unusual in the Netherlands. I have informed the Utrecht *Facultät* that I cannot yet decide whether or not I will accept the appointment by the [Dutch] Ministry. I ask you to inform the Saxon Ministry of People's Education of this development.

Heil Hitler!

[signed] B.L.v.d.Waerden

Three days later, *Dekan* Heinz added his own text below Van der Waerden's:

---

[148] Ibid.

[149] Utrecht University, Archive of the Faculty of Mathematics, Correspondence, 1943.

[150] Universitätsarchiv Leipzig, PA 70, p. 66 [the document is mistakenly dated 1942 by van der Waerden].

Forwarded to His Magnifizence Herr *Rektor* of the University of *Leipzig* for his information.

Heil Hitler!

[signed] Heinz

d.Z.*Dekan*

On July 27, 1943, the Utrecht's *Faculteit* of Mathematics and Physics officially proposed to make Van der Waerden their first choice and informed the latter of this decision:[151]

Faculteit of mathematics and physics is honored to let you know that *Faculteit* is proposing to put you in the first place for the vacancy that is caused by the retirement of prof. dr. J. A. Barrau as professor in synthetic and analytical, descriptive and differential geometry. We would like to know if you are willing to accept the eventual position in Utrecht.

Chair and Secretary of the Faculteit

Even though Van der Waerden knew and informed his Leipzig bosses in early January 1943 that "this inquiry is tantamount to an offer since negotiations about offers are unusual in the Netherlands," he again, even on Sept 19, 1943, avoided giving a clear answer to Utrecht:[152]

I am very pleased that the *Faculteit* has the intention to put me first on the list for the Barrau opening. The possibility to return to my country is attractive to me, but I am sorry that in current circumstances of the war I cannot give you certainty that I will accept the appointment.

Finally, on January 18, 1944, the Secretary-General of the Ministry of Education of the Netherlands (*Opvoeding, Wetenschap en Cultuurbescherming*) Prof. dr. Jan van Dam asked Van der Waerden for a definitive answer:[153]

The presiding [*forsitzende*] *Kurator* of the University of Utrecht suggested to me to appoint you as a Professor of synthetic, analytical, descriptive and differential geometry and to fill the position that was vacated by Prof. J. A. Barrau's retirement. In fact, Professor Barrau has recently turned 70. This nomination is in accordance with the recommendation of the Faculty of mathematics and physics.

Since it is very important to me that the vacant position be filled as soon as possible, I would like to ask you to let me know whether you wish to be considered for this appointment.

The Secretary General of the Ministry for Education, Science and Administration of Culture.

[Signed] J van Dam

---

[151] Utrecht University, Archive of the Faculty of Mathematics, Correspondence, 1943.

[152] Ibid.

[153] Universitätsarchiv Leipzig, PA 70, p. 69.

At this point, Van der Waerden pleased his Leipzig bosses with his desire to remain in Germany for the duration of the war, as he did not want "to become a deserter." The *Dekan* informed the *Rektor*, who in turn reported to the Ministry:[154]

25 February 1944
Dekan of the Philosophical *Facultät* of Leipzig University
To His Magnificence Herr *Rektor* of the University
The colleague van der Waerden informed me about an offer to him from the University of Utrecht. During my discussion with him he expressed his intention to stay in Leipzig for the duration of the war. I feel satisfied with his attitude.

Heil Hitler!
[signed] Heinz

[BACK SIDE:]

Rektor   Leipzig, 1 March 1944
of Leipzig University   Beethovenstrasse 6 I. Mü

Nr. A: 73

To the *Reichsstatthalter* of Saxony, Ministry for People's Education,
Dresden – N 6
The Ordinarius of mathematics Professor Dr. van der Waerden has informed me of an offer to him from the University of Utrecht and he has expressed that he wants to stay in Leipzig during the war since he does not want to become a deserter. I welcome this decision but—without having addressed the official side of the matter—for general reasons I would deem it worth considering to enable Prof. van der Waerden to move to a different university later.

Taken into consideration Professor van der Waerden's behavior in connection with the terror attack on Dec 4, 1943, which I got to know from the *Dekan* of the Philosophical *Facultät*, Math-Scientific Division, I would be grateful if Prof. van der Waerden were invited to the Ministry to discuss the job offer which he has received.

[Signature] Wilmanns

Thus, the Third Reich's education executives were assured by Van der Waerden of his loyalty to Germany (whether he really did not wish to be a "deserter" of the Third Reich or was lying, we will never know). Only the Dutch Faculty at Utrecht was kept in limbo, until Secretary-General of Education J. van Dam wrote to the President-*Kurator* of Utrecht University on May 22, 1944:[155]

In agreement with your proposal concerning the filling of the vacancy in the synthetic, analytical, descriptive and differential geometry, I have given your proposal to the German authorities for their judgment. While at the same time I have written to Prof. dr. Van der Waerden to ask him if he would be willing to accept this position.

From the German side I received some time ago the request to "let go of this idea" [in German: "*Abstand nehmen zu wollen*"].

[154] Universitätsarchiv Leipzig, PA 70, pp. 79–80 (in fact, 3 pages).

[155] Utrecht University, Archive of the Faculty of Mathematics, Correspondence, 1944.

Prof. Van der Waerden has written to me that at this time he does not have permission from the German Ministry of Education to leave his position in Leipzig. From his letter, I would draw a conclusion that he would be willing to come to Utrecht.

After more discussion with the German authorities here in this country, one has told me that they indeed would not give permission for the departure of Prof. Van der Waerden from Leipzig. They are not against him personally.

Under these circumstances, I ask you to think about in which manner we can provide education on a temporary basis and to give me a proposal concerning this matter.

How does one interpret this document? On February 25, 1944 Van der Waerden informed his German bosses, who in turn reported to the Ministry for People's Education of Saxony that Van der Waerden wanted to remain at Leipzig. It seems logical that then, according to Van der Waerden's wishes, the Saxon Ministry informed Van Dam that they would not allow Van der Waerden to leave Leipzig. Separately, Van der Waerden answered Van Dam's January 18, 1944 letter by asserting his interest in the Utrecht job, but claiming that he did not have the German permission to leave for Utrecht. It is logical to conjecture that the latter Van der Waerden's assertion was false. Indeed, after the war, when Van der Waerden defended himself from the suspicions of his collaboration with the Germans, he never once mentioned that the German government did not allow him to accept the Utrecht job. Van der Waerden did not wish to go to Utrecht, and blaming the Germans for it appeared to be a convenient excuse for him at the time, in 1944.

Even much later, in 1993, Van der Waerden would recall the Utrecht story without mentioning this alleged German prohibition of the Dutch employment [Dol1]:

I had an offer from Utrecht. During the war they had written asking if I wished to come to Utrecht. I answered, "Not now, but after the war I shall come."

What did Van der Waerden mean here by "after the war?" If this was his reply to the original December 16, 1942 Utrecht offer, how could he have possibly known *then* how and when the war would end? The first real predictor, the German loss in the battle at Stalingrad was to come on February 2, 1943, and the D-day, June 6, 1944, was much farther away.

Written right after his arrival in Amsterdam after the war, Van der Waerden's explanation on July 20, 1945 was more detailed, but again lacked the blame of the German prohibition to leave his Leipzig's job:[156]

In 1943 the Faculty of Physics and Mathematics at Utrecht asked me whether I would accept an appointment as a Professor there. I asked them to postpone the matter if possible until after the war, because I did not want to be appointed by the [Secretary-General of the Dutch Ministry of Education Jan] Van Dam department.

Did Van der Waerden really believe that an approval by the Secretary-General of Education van Dam (who served the German occupiers) of his *faculty-initiated* Utrecht appointment would stain his reputation more than spending the entire period of the brutal German occupation of the Netherlands in the Third Reich? Is it possible

---

[156] RANH, Papers of Hans Freudenthal, inv. nr. 89.

that the real issue was "neutrality" again: going to Utrecht would have been perceived as abandoning the "German side" and violating Van der Waerden's "neutrality"?

One thing is clear: professorship at Utrecht appears to have been a fallback position for Van der Waerden – in case of Germany's defeat in the war. This conjecture is confirmed by Constantin Carathéodory's March 25, 1944 reply[157] to an apparent request for advice from Van der Waerden:

> As concerning Utrecht, I very well understand your standpoint. But it is not yet the end of all days, and it would be unfortunate if you would not be able to arrange for keeping open for some time the possibility to move there.

In his March 14, 1944 letter Prof. Van der Waerden asked *D.M.V.*[158] President Wilhelm Süss whether he should accept the Utrecht's offer. The Utrecht's offer was apparently used in this letter by Van der Waerden to obtain another position. What Van der Waerden really longed for was a professorship at Göttingen. He asked Süss to use the latter's influence to help Van der Waerden obtain the position, and in particular to contact Helmut Hasse at Göttingen, whom Van der Waerden had already written to earlier. In support of his request for a Göttingen job, Van der Waerden enumerated – more openly than ever before – the evidence of his deep attachment to Germany:[159]

> I have spent my best energies for Germany which I have applied to The German Science [*die deutsche Wissenschaft*][160]. I have written practically all my works and books in German language, I learned and also taught a major portion of my mathematics in Germany; I have a German wife and my children have been raised as pure Germans.[161]

Quite expectedly, after the war Van der Waerden's "German wife" would become "Austrian" again. After the war, Van der Waerden would never again take pride in raising his children as "pure Germans." "Morally upright bearing," Prof. Günter Frei alleges? No Sir, in tyranny one can only leave, die, or compromise.

The Utrecht's offer was apparently used by Van der Waerden to obtain a salary raise as well. On July 6, 1944 he wrote the following rather angry words to the Saxon Ministry of Education's *Ministerialdirector* Dr. Schwender:[162]

> My problem is as follows:
> As it was conveyed to me with A : 18^bSt 5 [letter reference number?] on May 12th, the Reich's Education Minister has said that the requested and again approved raise of my teaching salary by the Saxon Ministry would not be addressed. A reason was

---

[157] Carathéodory to Van der Waerden, letter in German of March 25, 1944; ETH-Bibliothek Zürich, Wissenschaftshistorische Sammlungen HS 652:10611.

[158] *Deutsche Mathematiker-Fereinigung* (German Mathematical Society).

[159] See the facsimile of this passage on page 415.

[160] The term "The German Science" (similarly to "The German Mathematics", "The German Physics") may have had a different meaning here than would, say, "Science in Germany," as it was used at the time to refer to the particular Third Reich's variety of race-based science.

[161] ETH, Hs 652:12031.

[162] ETH, Hs 652: 11835.

sehr großen Wert beilegt. Ich selbst würde
das sehr bedauern. Denn ich habe das Gefühl
immer meine besten Kräfte für Deutschland
und die deutsche Wissenschaft eingesetzt zu
haben. Ich habe praktisch alle meine Arbeiten
und Bücher in deutscher Sprache geschrieben,
ich habe den wesentlichen Teil meiner Mathematik in Deutschland gelernt und wieder
gelehrt; ich habe eine deutsche Frau ge-
heiratet und meine Kinder rein deutsch
erzogen.

B. L. van der Waerden's "Germanness," a facsimile

not given. I assume that the basis is in that the Reich's Ministry does not appreciate my work in Germany. Just a few months ago one of my colleagues by his own word received a raise of his salary by 3,000 RM [Reich's marks]. In view of the fact that my mathematical colleagues also have higher salaries, I believe that this denial [in salary raise] is a demotion. For me it is not only about the money but also about the recognition of my work.

I am still dealing with the Dutch Ministry about my call to Utrecht. I have conveyed to them that I will not come during the war, but that my final decision is dependent upon success of my dealings in Dresden and Berlin.

I would therefore request you to convey to me what the reasons are in whatever form would be appropriate for you.. . . Perhaps the reasons will reawaken the old accusations which one had against me in Berlin.

Meanwhile Van der Waerden was, apparently, active in the affairs of his Leipzig University. In the summer of 1944 when Leipzig University was filling a professorship in physics, Van der Waerden offered an inclusion in the short list of candidates to the closest personal and professional friend of Werner Heisenberg, Carl-Friedrich von Weizsäcker, a professor at the University of Strasburg, who [Wal1, p. 108] was a member of the *Nationalsozialistischer Lehrbund* (National Socialist Teacher League of Germany). Van der Waerden did not know, of course, that together with Heisenberg, Weizsäcker had been a key researcher in the "*Uranverein*" ("Uranium Club") of the "*Heereswaffenamt*" (The Army Weapons Bureau), a group that strived to create a German atomic bomb and an atomic reactor.

On July 24, 1944 Weizsäcker replied in the style reminiscent to that of Van der Waerden's letters to Utrecht, for he wanted Leipzig's professorship to be his fallback position:[163]

> The decision is not very easy for me to take. I do have the wish to have an assistant of my own; under this condition the University of Leipzig would attract me. But even then I would stay here if the conditions remain as they are in Strasburg. But this is difficult to foresee.

Before this Chapter can be concluded, I have got to mention the "Furniture and Books Affair." On December 4, 1943, Van der Waerden's house was damaged by the allied bombardments. On February 19, 1944 he asked for—and on 11 May 1944 received[164]—the *Reichsminister* for Science's approval for travel to the Netherlands:[165]

> To the Reichsminister for Science via Herr Rektor of the University of Leipzig
>   I ask for permission for a private trip to Holland in March 1944 to buy furniture and scientific books.
>   I have lost all my furniture and books during the air raid on Leipzig on Dec 4, 1943, and I have learned that in Holland there are still possibilities for replacements.
>
>   B.L.v.d. Waerden

How does one interpret this document? Is it possible that the occupied Dutch people needed food more than books and furniture, and thus "in the Netherlands there [were] still possibilities for replacements"? Did Professor Van der Waerden view the Netherlands as just a source of supplies for himself? Or, as Prof. N. G. de Bruijn has hypothesized in his June 1, 2004 e-mail to me [Bru9], did Van der Waerden, perhaps, have other motives and used books and furniture as an excuse to go to the Netherlands? But why would Van der Waerden seek the opportunity to go to the Netherlands, to defect from Germany? This was unlikely, for if he went to the Netherlands at all, he quickly returned back to Germany.

37.4 Epilogue: The War Ends

When D-Day came on June 6, 1944, the outcome of the war became clear. The end of the war found the Van der Waerden family—Bartel, Camilla, and their three children Helga, Ilse, and Hans—in the Austrian countryside at Tauplitz, near Graz, in the house of his mother-in-law [Dol1]. Van der Waerden did not wish to return to Leipzig. He and his family allowed the American liberators to transport them from Austria to the

---

[163] ETH, unlabeled letter.

[164] ETH, Hs 652:12289.

[165] Universitätsarchiv Leipzig, PA 70, pp. 70–71, 76–78, 81–82.

Netherlands, where Van der Waerden thought he still had that job offer, at Utrecht University—after all, in the two and a half years of Utrecht's courting him, he never said "no" to them.

In the next Chapter, we will follow Professor B. L. van der Waerden to the Netherlands.

38

In Search of Van der Waerden: The Postwar Amsterdam, 1945[166]

> *It was not at all fitting for a Dutchman to make mathematics in Germany flourish in those years when Germany was preparing for war and was throwing out the Jews everywhere.*
> – J. G. van der Corput, August 20, 1945[167]

> *It is my sincere desire to keep you for the Fatherland and for higher education.*
> – J. G. van der Corput, August 28, 1945[168]

38.1 *Breidablik*

Dr. Theo van der Waerden[169] was a Member of the Second Chamber of the Dutch Parliament from *SDAP* (*Sociaal-Democratische Arbeiderspartij*) and a universally beloved politician. When in the mid-1920s his and Dorothea's three sons—Bart, Coen, and Ben—left their Amsterdam house to live on their own, Theo built a house in the Amsterdam's suburb of Laren at *Verlengde Engweg* 10. The magnificent house even had a name, proudly displayed right below the large bay window of the second floor: *Breidablik*. Ben's daughter Dorith explains [WaD3] that *Breidablik* "means wide view and comes from the old Norwegian saga about the gods *Wodan* and *Donar*." Coen's son Theo adds [WaT1]:

> *Breidablik* means "with a wide view" (the view was beautiful) and figuratively: "people with a broad view".

---

[166] This part II of the triptych first appeared in April 2004 [Soi21]; it covers only the second half of 1945, but this is a very important time, well worth the effort. Part I (Chapter 37) was first published in July, 2004 [Soi20]; part III (Chapter 39) first appeared in the January-2005 issue of *Geombinatorics* [Soi24].

[167] Van der Corput to Van der Waerden; ETH-Bibliothek Zürich, Wissenschaftshistorische Sammlungen, ETH, Hs 652: 12161.

[168] Van der Corput to van der Waerden; ETH, Hs 652: 12162.

[169] See more about the Van der Waerden family in Chapter 36.

A. Soifer, *The Mathematical Coloring Book,*
DOI 10.1007/978-0-387-74642-5_38, © Alexander Soifer 2009

Breidablik. Courtesy of Dorith van der Waerden

Dr. Theo van der Waerden succumbed to cancer on June 12, 1940 in *Breidablik.* His wife Dorothea, depressed by the German occupation of the Netherlands, took her own life on November 14, 1942. The magnificent house stood empty—or so it appeared. In fact, *Breidablik* was used to save lives during the German occupation [WaT1]:

> When grandmamma died in 1942, the house was rented to people. They have hidden people sought by the Nazis.

Now that the war was over, occupation ended, and *Breidablik* stood empty indeed, ready for its new tenants. Following the war's last "three months, distant from all culture and barbarism"[170] in the Austrian Alps, the Van der Waerdens were liberated by the American armed forces. Van der Waerden was not thrilled about the hardships of his liberation, and he described it on July 1, 1945 in a letter to Otto Neugebauer[171] from the camp for displaced persons at the town of Sittard in the southern most Dutch province of Limburg:[172]

---

[170] Van der Waerden, July 1, 1945 letter to Otto Neugebauer; Library of Congress, Manuscript Division; a copy sent to me without identification of its location within this vast archive – likely from the Veblen Papers.

[171] Otto E. Neugebauer (1899–1990), a historian of mathematics, an anti-Nazi, the founder of *Zentralblatt für Mathematik* (1931) and of*Mathematical Reviews* (1940).

[172] Library of Congress, Manuscript Division, Washington, D.C.; no holding location has been provided to me.

When the Americans have liberated us, we were like cows pushed together in cattle wagons and transported to Holland, my wife, 3 children and I. The transport lasted 16 days, it was horrible. The children were of course sick but then recovered here in a camp.

Months later, in November 1945, Van der Waerden was still angry at the Americans, whose "friendly" offer turned into a distasteful experience, as he wrote to Richard Courant of New York:[173]

~~When the Americans came, and we were given a friendly offer to get a direct flight to Holland, the misery began. Three weeks we spent in hard freight wagons [*Güterwagen*] and in dirty unsanitary camps with poorly prepared and hard to digest food~~[174]

However, Van der Waerden knew, that by comparison with many other survivors, he did all right, or, perhaps, he did not wish to appear as a whiner to his friend Courant, so he crossed out the above description and replaced it with the softer words:

The repatriation was less than attractive. Three weeks in freight wagons and camps, but of course one can survive that.

On July 1, 1945 Van der Waerden was a free man. He expected to get a ride from the camp to Laren very soon, for in writing from the Sittard camp to his American colleagues Lefschetz, Veblen, and Neugebauer on that day he gave the *Breidablik* return address. Indeed *Breidablik* was ready to provide the roof over the heads of Bartel and Camilla van der Waerden and their children Helga, Ilse, and Hans. In a few days the Van der Waerdens made it to this magnificent house. Now they needed to find bread for their table.

The Van der Waerdens had had it easier in Germany during the war than in the occupied Dutch. After the 5 years of occupation and a devastating last winter, life in the Netherlands immediately after the war was no bed of roses. Van der Waerden assessed it so on July 1, 1945: "Holland is freed from oppression, but it is—like Germany and Austria—in a desolate state. Food supply is sufficient, but all other necessities of life are lacking. . ."[175] The postwar life in the Netherlands must have been even harsher on the Van der Waerdens, who arrived in the Netherlands with practically nothing. Even half a year later they were so short on bare necessities that Dr. Van der Waerden had to step on his (considerable) pride and ask Richard Courant in New York for help:[176]

---

[173] November 11, 1945 Van der Waerden's letter to Richard Courant; ETH, Hs 652:10649 (unfinished and unsent, 2 pages survive). The complete 3-page letter was sent on November 20, 1945. It is located in New York University Archives, Courant Papers.

[174] Here and throughout the book, strikethrough text represents words carefully crossed out and readable in the original manuscript.

[175] Van der Waerden, July 1, 1945 letter to Solomon Lefschetz; ETH, Hs 652:11346.

[176] Van der Waerden, December 29, 1945 letter in English to Richard Courant; New York University Archives, Courant Papers.

I thank you very much for sending me the two volumes of Courant-Hilbert. Your kindness gives me courage to utter another wish. We are so short of underwear and warm cloths for the children. Helga is 15, Ilse 11, Hans 8 years old. My father's house is extremely cold. Perhaps your wife has got some wool or things the children don't wear any more? They can be as old and ugly as they may: my wife can change nearly anything into anything. And further: Would it be possible to send a sheet (of bed)? We have only 4 sheets for 5 beds, and it is quite impossible to get any here.

I hope that you and your wife will not be angry with me for asking so much. If it is difficult for you, or if your people need the things more than I, please don't send anything.

38.2 New World or Old?

> *I do not mind his remaining a German Professor*
> *until the end – I do mind his remaining a German*
> *Professor at the beginning!*
> – Otto Neugebauer, August 15, 1945[177]

The reader may recall the 1933 invitation by Princeton University that Professor Van der Waerden received in April of 1933, shortly after Hitler's ascent to power (Chapter 37). Upon receiving all of the approvals, Van der Waerden suddenly decided against going to Princeton at the last moment in late July 1933. After just one week in the Netherlands, on July 1, 1945, he wrote from the Sittard camp to the Princeton's new mathematics chair Solomon Lefschetz.[178] After the war, Van der Waerden would have been given a hero welcome at Leipzig University – why did he not return there? This question occupied me for many years, until I found the answer in this letter, addressed to Lefschetz. Even Lefschetz never learned the answer, for it was contained *only* in the copy Van der Waerden kept for himself in which the answer was written and then crossed out! We read here—and nowhere else—that Van der Waerden particularly did not wish to go back to Leipzig because Leipzig was now in the Russian zone of occupation, and he had no desire to live under the Russian rule. More generally, Van der Waerden did not wish to stay in the Netherlands, Austria or Germany due to their "desolate state." He believed he could get a position in the Netherlands, likely referring to his old Utrecht offer, but preferred to come to America:[179]

Peace at last, thank God! By the help of our mighty allies, Holland is freed from oppression, but it is – like Germany and Austria – in a desolate state. Food supply

---

[177] Letter to Heinz Hopf; Hopf Nachlass, ETH, Hs 621:1041.

[178] Solomon Lefschetz (Moscow, 1884-Princeton, 1972), professor (1924–1953) and chair of mathematics department (1945–1953) at Princeton University. In 1945 he replaced the long term chair (1929–1945) Luther Pfahler Eisenhart.

[179] Letter written in English; ETH, Hs 652:11346.

is sufficient, but all other necessities of life are lacking: not even railways are going. Scientific work and international contact are practically impossible.

In March, my home in Leipzig having been destroyed by bombs, I could escape with my family from the bomb hell to Austria. From there we have just been repatriated to Holland. ~~Returning to Leipzig, which belongs now to the Russian zone of occupation, seems impossible and, even if possible, not advisable.~~ I can get a position in Holland ~~probably~~ but Holland is in a heavy political and economical crisis, as I said before. For all these reasons I should like to go ~~temporarily or definitively~~ to America.

In particular, Van der Waerden wanted to be invited to Princeton again:

Several years ago, you encouraged me to write to you if I wanted to be invited to America. In the year 1939[180] I was invited to come to Princeton as a guest for half a year. Do you think that this invitation could be repeated? I should enjoy very much getting into contact with the American mathematicians again, especially with those of Princeton. I shall accept with joy any invitation of this kind... With best greetings to Veblen,[181] Neumann[182] and the other Princetonians.

That same day, July 1, 1945, Van der Waerden wrote a nearly identical letter to Oswald Veblen at the Institute for Advanced Study in Princeton.[183] The only difference was in the justification for wanting to come to America: to "a desolate state" of Holland, Germany and Austria, Van der Waerden added a high praise of mathematics in the U.S.:

I have been cut off from international mathematics, whose heart pulses in America, for five years, and I want to regain contact as soon as possible.

The third July 1, 1945 letter Van der Waerden sent to Otto Neugebauer.[184] On August 20, 1945 Solomon Lefschetz replied:[185]

I was very sorry to hear about your losing your home in Leipzig and can well understand your desire to come to the United States (who does not feel the same way in Europe just now?). However, we are in a complete state of flux here and the time does not seem very propitious for bringing in scientists from the outside, especially professors in former German universities. I have transmitted copies of your letter to some mathematicians that know you, in particular to the members of the Institute for Advanced Study, for the pre-war invitation that you mention can only have come from them. They have informed me that there is nothing available at the present time. One of them did express the hope that you would accept the position at Utrecht since, no doubt, you are very badly needed there. I confess that I agree a little bit with him.

---

[180] True, but he was invited six years earlier, in 1933, see Chapter 37.

[181] Oswald Veblen (1880–1960), professor at Princeton (1905–1932), the first professor at the Institute for Advanced Study (1932–1950), instrumental in saving European scientists from Hitler and bringing them to the U.S.

[182] John von Neumann (Budapest, 1903-Washingron D.C., 1957), mathematician and physicist, one of the great scientists of the twentieth century, professor at the Institute for Advance Study.

[183] Letter written in English; ETH, Hs 652:12193.

[184] Library of Congress, Manuscript Division, Washington, D.C.; no holding location has been provided to me.

[185] ETH, Hs 652:11347.

Van der Waerden could not have found Lefschetz's letter particularly encouraging. No doubt he sensed a thinly concealed irony behind Lefschetz's rhetorical question: "who does not feel the same way in Europe just now?" Lefschetz was even more blunt when he acknowledged that the time was not "very propitious for bringing in scientists from the outside, especially professors in former German universities." Lefschetz seemed to be implying that Van der Waerden made a wrong choice by staying in Germany, and that now had to pay the price for being on the wrong side of the divide during the war. In Lefschetz's "defense" one should note that he treated harshly and sarcastically the vast majority of humans around him. Even in Lefschetz's 1973 eulogy [Hod] Sir William Hodge quoted Princeton students' song:

> Here's to Lefschetz (Solomon L.)
> Who's as argumentative as hell,
> When he's at last beneath the sod,
> Then he'll start to heckle God.

One must add that in his reply Prof. Lefschetz was factually wrong: not only did the 1933 invitation come from Princeton University and not from the Institute, but Lefschetz himself attended the meeting of the Princeton's Research Committee that decided to invite Prof. Van der Waerden (Chapter 37).

Moreover, Princeton did need an algebraist. Lefschetz simply had someone else in mind, and was willing to charm him in. On Wednesday, October 17, 1945, he wrote to the algebraist of his choice:[186]

> Dear Artin,
>   Owing to recent losses in our department,[187] to which now must be added Wedderburn's[188] retirement (soon to be official), I feel very strongly that we should add a major scientist to our staff. You are the first person of whom I thought in this connection and, if possible, I would just as soon not go further in my search. Your achievements as a mathematician, together with your well-known sympathetic influence on the younger men, do indeed make you the man of the hour.

Artin happily responded on October 21, 1945:[189]

---

[186] Personnel File of Emil Artin, Princeton University.

[187] Lefschetz likely refers here to the July 1, 1945 retirement of the long term professor, chair, and dean of graduate school Luther Pfahler Eisenhart, and September 1945 departure of (associate) professor Henri Frederic Bohnenblust for Indiana.

[188] Joseph Henry Maclagan Wedderburn (February 2, 1882, Scotland – October 9, 1948, Princeton), a Scottish born algebraist and Princeton professor. On the occasion of his retirement, on October 29, 1945 all members of Mathematics Department of Princeton signed the following resolution, drawn by A. W. Tucker and A. Church: "RESOLVED that the Department of Mathematics record its appreciation of the long and distinguished service of Professor J. H. M. Wedderburn as a member of the faculty of Princeton University and its appreciation of the signal contribution he has made to the reputation of the Department by his outstanding mathematical research and his unstinted efforts as editor of the Annals of Mathematics. It is the hope of his colleagues that retirement will not bring these contributions to an end but that he will continue to add to scientific life of the Department for many years to come."

[189] Personnel File of Emil Artin, Princeton University.

Dear Lefschetz:

It is with very great joy that I received your letter and I feel deeply honored that you are thinking of me. I would not be a mathematician if I would not feel greatly interested and attracted by a chance to go to Princeton. Princeton is now after all the center of all mathematics.[190]

As if especially for my narration, Artin then asked:

How did the case of van der Waerden go on after his letter?

This Artin's question shows that Lefschetz circulated Van der Waerden's July 1, 1945 letter asking for a Princeton job, likely together with Lefschetz's negative reply. On October 27, 1945, Lefschetz informed Artin that Van der Waerden had not been invited to Princeton:[191]

Nothing has been done regarding van der Waerden—nothing, at least from this side.

Before replies from America could arrive, Dr. Van der Waerden wrote two letters to his friend Heinz Hopf, a German mathematician, now a Swiss citizen, working at the ETH in Zürich.[192] I have been unable to locate these letters, but according to Hopf's August 3, 1945 reply,[193] they were written on July 19 and 21, 1945. Hopf praised Switzerland and its neutrality:[194]

Here is Switzerland one is naturally less fanatical, in my opinion, as a particularly important and happy consequence of our neutrality...It goes well for my wife and me...we are happy that we are Swiss.

It is plausible that this praise of the Swiss neutrality and Hopf's happiness with the Swiss citizenship may have planted in Van der Waerden an interest in living in Switzerland. Hopf was unhappy that the Swiss considered—as they should have, in my opinion—"Hitlerism" to be a part of the German culture:

I ask you not to misunderstand the above comment about neutrality, the open opinions here are completely unified against Germany, the bitterness about the Nazis is gigantic, but the boundaries between Hitlerism and the German culture are not always observed here either.[195]

---

[190] The Lefschetz—Artin correspondence was kept "entirely confidential," as Lefschetz put it in his October 17, 1945 letter. The Mathematics Department of Princeton was briefed on Artin's acceptance only at the March 22, 1946 faculty meeting, two days after Artin's formal acceptance telegram.

[191] Personnel File of Emil Artin, Princeton University.

[192] Heinz Hopf (1894–1971), one of leading topologists, professor at ETH since 1931, from a Jewish German family.

[193] Letter to Van der Waerden, August 3, 1945, ETH, Hs 652:11129.

[194] In view of revelations of recent years, the Swiss neutrality during the war has been questioned.

[195] In my opinion, Nazism (to a great regret of so many) was a product and part of the German culture, every bit as Marxism or music of Bach and Beethoven were. Of course, there is high culture and low culture, but both of them are parts of culture in a broader sense of the word, and who can—or should—split them apart?

Nevertheless, Hopf was optimistic that these uncertain "boundaries" would not affect the Swiss mathematicians' perception of Van der Waerden:

> The question is "How would Swiss mathematicians today personally view you?" – I believe I can answer this way: Certainly almost all, perhaps, absolutely all, would not only see in you an important intellectual, but those with real interest in you, and who have known about you over the last few years across the borders, would be clear that you have been no Nazi, and indeed that the Nazis could not stand you. The Carathéodory's situation over the last few years has been the same as yours, and for his 70th birthday many Swiss mathematicians dedicated their works to him.

Hopf understood the liability of Van der Waerden's spending the entire Nazi time in Germany, including the 5 years of the Nazi occupation of the Netherlands, and offered Van der Waerden a line of defense:

> ... one would probably have to argue this way: he worked as a professor in Germany even during a period of abuse of his homeland by Germany because he believed that he could somehow accomplish to save the culture in Europe this way, and consequently he tried to save Germany in some way. I believe it would be very difficult to argue against this..."

It is unclear whether Prof. Hopf sincerely believed that one could save the German culture by serving and thus empowering the Nazi state. Van der Waerden would indeed use this line of defense in the Netherlands, as we will see later, but not altogether successfully. Hopf meanwhile admitted poor prospects of finding a job in Switzerland:

> ... your finding a job here is a more problematic question...the possibilities of a position in Switzerland at the moment are very minor, almost impossible.

And so Hopf suggested Van der Waerden to consider a job in Germany, that advice Van der Waerden probably did not appreciate:

> I believe that for someone who believes himself to be youthful, has a strong ability to work and has energy, it could really be satisfying to work right now in Germany in pure science. Perhaps, because the situation in Germany is now so miserable and possibly without hope that the younger powers could more intensively work on pure intellectual and cultural ideas, which they have not been able to do before, or even anywhere else...

Hopf also advised exploring employment opportunities in the U.S.:

> I would in this situation also write to America, perhaps to Weyl.[196] (By the way, I wrote to Neugebauer, a few days ago, right after Kloosterman's visit, I wrote to him briefly about you.).

Finally, Hopf scolded the Dutch for not immediately jumping on the opportunity to hire Van der Waerden:

---

[196] Hermann Klaus Hugo Weyl (1885–1955), professor at ETH (1913–1930), Göttingen (1930–1933) and the Institute for Advanced Study (1933–1952).

When the Dutch, whom you can approach with clean conscience and offer them your services, do not want you, then in my opinion they hurt themselves, and that is their business. I consider it certain that in a few years, when the waves have calmed down a bit, somewhere in the world you will work again in the profession – assuming naturally that you with your family can economically survive until then, which I am not sure about.

Van der Waerden would quote these lines to the Dutch almost immediately, within 2 weeks. Four and a half months later, in his next letter of Dec 18, 1945, Heinz Hopf explained his long silence by his inability to invite Van der Waerden even for a short visit. What was the reason? The Swiss treasured their neutrality more than Van der Waerden's expertise:

All my attempts to invite you here for a few presentations ended up without success. It was very strictly suggested to avoid right from the beginning any kind of conflicts with friendly governments. I am not the only one here that regrets this.[197]

So, after all, there was a price for the Swiss neutrality: in 1945 Switzerland did not allow even a brief visit to the former German Professor Van der Waerden. As we will soon see (Chapter 39), the Swiss would drop this caution the very next year.

Some time in July–August, 1945 Hopf wrote about Van der Waerden's plight to the German historian of mathematics Otto Neugebauer, who now lived in the U.S. and edited *Mathematical Reviews* that he created in 1940 after Springer-Verlag put pressure on Neugebauer to Nazify *Zentralblatt für Mathematik*. On August 15, 1945 Neugebauer replied to Heinz Hopf in English as follows:

... I have heard directly from van der Waerden. I do not mind his remaining a German Professor until the end – I do mind his remaining a German Professor at the beginning! However, I feel very differently than the Lord and [thus] I do not intend to do anything positive or negative.[198]

Meanwhile, Van der Waerden had heard neither from Hopf (since the early August) nor from Neugebauer. Thus, on November 11, 1945, Van der Waerden wrote to his early mentor and friend Richard Courant[199] in New York about the bombings of the late months of the war, his tough repatriation, and his new job at the "Royal Dutch Oil."[200] On December 13, 1945 Courant, a refugee from the Nazi Germany himself, sent a guarded reply. Before deciding whether to renew their old friendship, Courant wanted to know why Van der Waerden chose to stay in Germany:

I wish very much that there were an opportunity of talking to you personally and for that matter to other old friends who have been in Germany during the war. Of course, so

[197] ETH, Hs 652:11130.

[198] Hopf Nachlass, ETH, Hs 621:1041.

[199] Richard Courant (1888–1972), a Jewish German Émigré to the U.S., active in helping refugees from the Nazis find jobs in the U.S., the founder of the *Mathematics Institute* at Göttingen and the *Courant Institute of Mathematical Sciences* at the New York University.

[200] ETH, Hs 652:10649.

much has happened in the meantime that in many cases much will have to be explained before one can resume where one left off. Your friends in America, for example, could not understand why you as a Dutchman chose to stay with the Nazis.[201]

Moreover, Courant made his request for an explanation public: at the top of the letter, I see a handwritten inscription:

cc. sent to: Reinhold Baer[202], U. of Ill. Urbana Herman Weyl – Inst. for Advanced Study Princeton Veblen

(Courant's papers include both, Van der Waerden's November 20, 1945 hand-written letter and its typewritten copy, which suggests that Courant had it typed and copies sent to the same addressees as his reply.) As Lefschetz before him, Courant too apparently believed that Van der Waerden made the wrong choice. On December 20, 2004, I had an opportunity to ask Ernest Courant,[203] the elder son of Richard Courant and a prominent nuclear physicist himself, a natural question: "What did your father think about Van der Waerden?" He replied as follows:

He [Richard Courant] considered him [Van der Waerden] a great mathematician, and was a bit critical of him for being perhaps too comfortable in the Nazi Germany.

Thus, America and Switzerland had to wait. Beggars could not be choosers, and so dr. Van der Waerden was now—finally—willing to seriously entertain a professorship in his "desolate" (his word) homeland.

38.3 Defense

> Some of the stories are difficult to believe. Part of all this is the way people always talk about their past. The reasons they give for their behaviour in the past may be just inventions, colored by how history took its course.
>
> – N. G. de Bruijn, June 1, 2004 [Bru9]

Van der Waerden expected that the Utrecht chair, first offered to him in December of 1942, was still waiting for him. He also did not mind a chair at Amsterdam. But fol-lowing the liberation, the *Militair Gezag* (Military Authority) installed *Commissie van Herstel* at each of the five Dutch universities, which gradually became known as *College van Herstel* (Recovery Board, or Restoration Board), formed to advise

---

[201] New York University Archives, Courant Papers.

[202] Reinhold Baer (1902, Berlin-1979, Zurich), a famous group theorist, who was a professor at University of Illinois (1938–1956) and then at Frankfurt.

[203] Ernest David Courant, born in 1920 in Germany, came to the U.S. in 1934 with his family; a nuclear physicist, member of the National Academy of Sciences, distinguished scientist emeritus of Brookhaven National Laboratory.

the Military Authority on how to act against collaborators and other pro-German professors and staff members, and when the university could be re-opened. It was expected that all suspect staff would be removed in a few months time. However, the removal took much longer.[204]

The Utrecht University's *College van Herstel en Zuivering* (Board of Recovery and Purification), as it was called there, was installed on 18 June 1945,[205] while the University of Amsterdam's *College van Herstel* (Board of Recovery) came into being on June 8th, 1945.[206] At the time the University of Amsterdam belonged to the City. Yet *B. & W.*, the executive, consisting of the *Burgemeester en Wethouders* (mayor and aldermen), could not appoint professors; only the city council that numbered 45 could appoint them. However, an appointment of a professor needed a Royal assent. The Queen could never give her assent if the government did not submit to her a request for assent. On the other hand, the government would not submit a request for assent if there was even a slight chance that the Queen would refuse it, as she had a few times during these postwar years.

Originally Dutch, Professor of the History of Mathematics at the Massachusetts Institute of Technology Dirk J. Struik maintained close ties with the leading Dutch colleagues, and based the following statement to me [Str] on a letter he had received from Jan Schouten[207] in 1945–1946:

> Though he [Van der Waerden] stayed at the University of Leipzig during the Hitler days, he was able to protect Jewish and left wing students.[208] This was brought out after the war when his behavior in Leipzig was scrutinized by a commission of his peers in the Netherlands. He was entirely exonerated.

---

[204] I am grateful to Dr. Peter J. Knegtmans, the University Historian at the University of Amsterdam, for the information on *Colleges van Herstel* and the workings of the City of Amsterdam, contained in his e-mails [Kne4] and [Kne5] to me. See more in his book [Kne2]. The Dutch postwar educational and governmental systems were a "jungle", and it was invaluable to have such a uniquely qualified guide!

[205] It consisted of Jonkheer Mr. Dr L.H.N. Bosch ridder van Rosenthal, president (and also former president, 1930–1940, until he was dismissed during the war by the German authorities); Dr. H.W. Stenvers; Dr. A.J. Boekelman; and Miss Marie-Anne Tellegen as an extra member, who must have combined this appointment with her job as director of the Queen's Cabinet. The Utrecht *College van Herstel en Zuivering* was converted into the (normal) *College van Curatoren* in June 1946.

[206] It consisted of the neurologist Prof. C. T. van Valkenburg, who during the German occupation initiated the resistance of general doctors and medical specialists; the architect Wieger Bruin who had been an active member of the resistance movement among artists; and Gijs van Hall, a fundraiser and banker for the resistance, who later became mayor of Amsterdam. It was to investigate staff against whom suspicion had risen, but in fact it did so only in cases of doubt and then very superficially due to its acting at the same time as the *College van Curatoren*. It was converted and extended into the *College van Curatoren* on May 19, 1947.

[207] Jan Arnoldus Schouten (1883–1971), professor of mathematics and mechanics at Delft (1914–1943), and extraordinary professor (similar to associate professor) at Amsterdam (1948–1953).

[208] As we have seen in Section 37, Van der Waerden spoke against firing of Leipzig's Jewish professors in 1935, and published papers of Jewish authors in the *Annalen*. I found no evidence of him "protect Jewish and left wing students."

On April 12, 1995, I quoted this statement in my letter to Professor Van der Waerden and asked him to describe for me in detail this "commission of his peers." On April 24, 1995, Van der Waerden mailed his reply [Wae25] (see the facsimile of this letter on p. 429):

> Before your letter came, I did not know that a commission was formed to investigate my behaviour during the Nazi times.

However, I have established that the University of Amsterdam's *College van Herstel (CvH)* did investigate Van der Waerden, for the City executive, *B. & W.*, wrote about Van der Waerden to a de-Nazification board, such as the *CvH*.[209] Moreover, Waerden knew about the investigations: on July 20, 1945, just a few weeks after he had returned to the Netherlands, he wrote in his own hand his "Defense" and forwarded it to the Amsterdam's *College van Herstel*, which was also submitted to the Utrecht's *College van Herstel en Zuivering*,.[210] This was Van der Waerden's defense of his reasons for staying in the Nazi Germany, and his activities in the Third

Van der Waerden, April 24, 1995 letter to Alexander Soifer

---

[209] Dr Knegtmans [Kne2] refers to the April 17, 1946 letter from *B. & W.* of Amsterdam to *CvH, archief Curatoren* nr 369, which says that "the [Van der Waerden's] appointment did not go through also because the Minister had told the City Council beforehand that he would not ratify it."

[210] On August 14, 1945, the Faculty of Mathematics and Physics of Utrecht University forwarded this document to *de Commissie tot Herstel en Zuivering*, when they recommended van der Waerden as their

Reich. This is a very important testimony, that has never before discussed by historians. I am compelled to include the translation of this Dutch handwritten document in its entirety (with my commentaries; see also its facsimile in this chapter):[211]

Defense

Since 1931 I have been a Professor at the University of Leipzig. The following serves as an explanation as to why I stayed there until 1945:

1) From 1933 till 1940 I considered it to be my most important duty to help defend the European culture, and most especially science, against the culture-destroying National Socialism. That is why in 1933 I traveled to Berlin and Göttingen to protest the boycott of Landau's classes by Göttingen Nazi students. In 1934[212] Heisenberg and I strongly protested against the dismissal of 4 Jews in a faculty meeting at Leipzig. Because of that I got a reprimand by the Saxon Government (*Untschmann*[213]) and an admonition that as a foreigner I should not interfere in German politics. What my wife and I have personally done to help Jewish friends with their emigration is not relevant here, but what is, is that as [an] editor of the *Math. Annalen* I accepted until 1942 articles of Jews and "*jüdische Mischlinge*"[214], furthermore that in the *Gelbe Sammlung* [Yellow Series] of Springer which I was partially responsible for, an important work by a Jewish author appeared in 1937 (Courant-Hilbert, *Methoden der Mathematischen Physik II*), and that in 1941 a non-Arian was promoted by me. In 1936,[215] when my esteemed teacher Emmy Noether died, I pointed out the great merits of this Jewish woman.

I could not have known in advance that all this would be like "punching a brickwall" [*vechten tegen de bierkaai*] and that the Nazis would drag the entire German culture with them into their destruction. I still hoped that the German people would finally see reason and would put an end to the gangster-regime. Meanwhile my work was not altogether for nothing because my students, such as [Herbert] Seifert, Hans Richter, Wei-Liang Chow, Li En-Po, Wintgen, etc., whose dissertations were accepted in the *Math. Annalen*, have done an excellent work at Leipzig. If I had not been in Germany, these [students] would likely not have encountered the problems that I have given them.

Van der Waerden was meticulous in adhering to the facts of his activities under the Third Reich. His record of noble and courageous behavior toward the Jews during the Nazi years can withstand the most prejudiced scrutiny. However, Van der Waerden was not telling the whole truth. Thus, he did not mention his statement about being a "full-blooded Arian" (1933), his oath of allegiance to Hitler (1934), and his use of the recommended "*Heil Hitler*" salute in lectures and letters (Chapter 37).

---

first choice for J. A. Barrau's position. Utrecht University, Archive of the Faculty of Mathematics, Correspondence, 1945.

[211] Utrecht University, Archive of the Faculty of Mathematics, Correspondence, 1945. This was an important document for Dr. Van der Waerden: even half a year later, on January 22, 1946, Van der Waerden included a copy of "The Defense" in a letter to his friend Hans Freudenthal. Subsequently, another copy of this document is held at RANH, Papers of Hans Freudenthal (1905–1990), mathematician, 1906–1990, inv. nr. 89.

[212] True, but it took place in 1935, see Chapter 37.

[213] This must be the last name of an official in the Saxon Government.

[214] In this Dutch document, this Nazi term for people of Jewish and Arian mixed blood, appears in German in quotation marks. The rough English translation is "Jewish miscegenants".

[215] True, but it took place in 1935.

As Heinz Hopf advised, Van der Waerden justified his staying in the Nazi Germany by stating that it was his "most important duty to help defend the European culture, and most especially science, against the culture destroying National Socialism." However, as is evident from Hopf's description of the public opinion in Switzerland, many of Van der Waerden's contemporaries found it difficult to separate the "German culture" from "Hitlerism." Given Van der Waerden's scruples regarding the "gangster-regime," his fellow scientists—then and now—considered his willingness to serve that regime naïve at best and hypocritical at worst. Van der Waerden continued his "Defense" with part 2, dedicated to the 5 years of the German occupation of the Netherlands:

> This all may serve for closer understanding of my attitude towards the Nazis. What I should explain to the Dutch people is, however, not my actions before 1940, but those after the Netherlands had been attacked by Germany.
>
> 2) From 1940 to 1945. After the breakout of the war with the Netherlands, I was first locked up and then released on the condition that I do not leave Germany.[216] So I was practically in the same position as those who were forced laborers in Germany.
>
> If I had given up my position, then I would have probably been forced to work in an ammunitions factory.

To say that a university full professor was "in the same position as those who were forced laborers in Germany," was to make a dramatic exaggeration, and it likely appeared as such to the Dutch who read the "Defense."

> I have never worked for *Wehrmacht* [the German Army], I have never given a class or worked on things that could be used for military purposes.

This could be true, but Van der Waerden taught students, many of whom may have served in *Wehrmacht* and some definitely "worked on things that could be used for military purposes."[217] Besides, by working in the Nazi Germany's Civil Service, Van der Waerden contributed to what, after the war he called "the gangster regime," and lent his credibility and his acclaim as a distinguished scientist to that of the Third Reich.

> In 1943[218] the Faculty of Physics and Mathematics at Utrecht asked me whether I would accept an appointment as a Professor there. I asked them to postpone the matter if possible until after the war, because I did not want to be appointed by the Van Dam[219] department.

---

[216] Leipzig professor Hans-Georg Gadamer, in a letter to Y. Dold-Samplonius [Dol1], claimed for himself "a little act of heroism" for getting an immediate help of police chief in Van der Waerden's release. Van der Waerden, however, seems to credit Nobel Physicist Werner Heisenberg with it. "I am still in your debt: in the past when I was arrested, you helped me to something much greater, and that is freedom," he wrote to Heisenberg on December 22, 1947 (see more on this correspondence in Chapter 39).

[217] For example, J J O'Connor and E F Robertson write, "When war broke out [Van der Waerden's Ph D student] Seifert volunteered for war work with the *Institut für Gasdynamik* which was a research centre attached to the German Air Force" (http://www-history.mcs.st-andrews.ac.uk/Mathematicians/Seifert.html).

[218] Actually, in December 1942.

[219] Prof. Dr. Jan van Dam, Secretary-General of the Ministry of Education (*Opvoeding, Wetenschap en Cultuurbescherming*)

Verdediging.

Sinds 1931 ben ik professor aan de Universiteit Leipzig. Ter verklaring, waarom ik aldaar tot 1945 gebleven ben, diene het volgende:

1) van 1933 tot 1940. Ik beschouwde het als mijn belangrijkste taak, de Europeese cultuur, en in het bijzonder de wetenschap tegen het cultuurvernietigende nationaalsocialisme te helpen verdedigen. Daarvoor ben ik in 1933 naar Berlijn en Göttingen gereisd om tegen de boycot van de colleges van Landau door Göttinger nazi-studenten te ageren. In 1934 hebben Heisenberg en ik in een faculteitsvergadering in Leipzig met klem tegen het ontslag van 4 joden geprotesteerd. Daarvoor heb ik van de Saksische regering (Mutschmann) een berisping gekregen en een vermaning, dat ik mij als buitenlander niet in de Duitse politiek mocht mengen. Wat mijn vrouw en ik persoonlijk hebben gedaan om Joodse vrienden bij hun emigratie te helpen, doet hier niet ter zake, maar wel, dat ik als redacteur van de Math. Annalen tot 1942 nog artikelen van joden en „jüdische Mischlinge" heb aangenomen, verder dat in de Gelbe Sammlung bij Springer, waar ik mede verantwoordelijk voor ben, in 1937 nog een belangrijk werk van een joods auteur (Courant-Hilbert, Methoden der mathematischen Physik II) is verschenen en dat er in 1941 nog een niet-ariër bij mij is gepromoveerd. In 1936 heb ik bij de dood van mijn vereerde leermeesteres Emmy Noether op de grote verdiensten van deze jodin gewezen.

Dat dit alles „vechten tegen de bierkaai" zou zijn en dat de Nazis de hele Duitse cultuur in hun ondergang mee zouden slepen, kon ik niet van tevoren weten. Ik hoopte nog steeds, dat het Duitse volk eindelijk rede aan zou nemen en een einde aan het Gangster-regime zou maken. Intussen is mijn werk toch niet helemaal tevergeefs geweest, want mijn leerlingen, zoals Seifert, Hans Richter, Wei-Liang Chow, Li En-Po, Wintgen e.a., wier dissertaties in de Math. Annalen zijn opgenomen, hebben in Leipzig voortreffelijk werk verricht.

Als ik niet in Duitsland was geweest, waren deze niet licht met de problemen, die ik ze heb voorgelegd, in aanraking gekomen.

Dit alles moge tot nader begrip van mijn gezindte tegenover de Nazis dienen. Waarover ik tegenover het Nederlandse volk rekenschap verschuldigd ben, zijn echter niet mijn handelingen vóór 1940, maar die nadat Nederland door Duitsland aangevallen werd.

2) Van 1940 tot 1945. Na het uitbreken van de oorlog met Nederland werd ik eerst opgesloten, daarna vrijgelaten met het verbod, Duitsland te verlaten. Ik was dus praktisch in dezelfde positie als degenen, die onder dwang in Duitsland werkten. Had ik mijn betrekking neergelegd dan was ik waarschijnlijk in een munitiefabriek te werk gesteld. Ik heb nooit voor de weermacht gewerkt, nooit college gegeven over of gewerkt aan dingen, die voor militaire doeleinden gebruikt zouden kunnen worden.

In 1943 vroeg de faculteit voor wis-en-natuurkunde in Utrecht mij, of ik een benoeming tot hoogleraar aldaar aan zou willen nemen. Ik heb verzocht, de zaak zo mogelijk tot na de oorlog uit te stellen, omdat ik niet door het departement-Van Dam benoemd wenste te worden.

Ik hoef hier niet aan toe te voegen, dat ik nooit lid van enige NS organisatie ben geweest of daarmee gesympathiseerd heb, want dat spreekt voor een fatsoenlijk denkend mens vanzelf. Het was in Duitsland algemeen bekend, dat ik geen Nazi was, vandaar dat de regering mij wantrouwde en mij geen toestemming gaf om in 1939 naar het Voltacongres in Rome te gaan, om voordrachten in Hongarije of voor Franse krijgsgevangenen te houden of in Rome aan een bijeenkomst van wiskundigen deel te nemen. De faculteit in München had mij als opvolger van Carathéodory voorgedragen, maar de partij-instanties verklaarden mij „untragbar"; en de benoeming kwam niet tot stand.

Ook mijn vrouw, die een Oostenrijkse is, is van 't begin af aan fel tegen het Nazi-regime gekant geweest.
Laren N-H, 20 Juli 1945   B.L.v.d.Waerden

Van der Waerden, "Defense"

For the discussion of the Utrecht offer, I refer you to the previous chapter. It suffices to say here that coming home to Utrecht on the request of the Utrecht Faculty, even with the approval by the Nazi-collaborating Minister Jan van Dam, would have been much better for Van der Waerden's reputation in his homeland and the rest of the postwar world than continuing to serve the Third Reich to the end.

> I do not need to add to this that I have never been a member of any *NS* [National Socialist] organization or have sympathized with them, because that is self-evident for a decent thinking human being. It was commonly known in Germany that I was not a Nazi and because of that the government distrusted me and did not give me permission to go to the Volta Congress in Rome in 1939, and to have lectures in Hungary or to French prisoners of war, or to partake in the convention of mathematicians in Rome.

However, the Nazi government did allow Professor Van der Waerden to travel in and out of Germany: for example, to the Netherlands in 1933, 1935, 1938, 1939, 1940, 1942, and 1944.

> The Faculty at Munich suggested me as a successor to Carathéodory, but the party authorities declared me *"untragbar"* [intolerable], and the appointment did not happen.
> Also my wife, who is Austrian, has been strongly opposed the Nazi regime from the very beginning.
> Laren, N-H [North-Holland], 20 July 1945   B.L.v.d.Waerden

This is true that in the Munich deliberations Van der Waerden was perceived as a philo-Semite, not subscribing to the Nazi ideology of anti-Semitism [Lit], and this may have cost him the Munich job. As to Mrs. Camilla van der Waerden, who is now Austrian again (she was "German" in the March 14, 1944 letter – see Chapter 37.3) we will meet her soon and gain some insight into her views.

With the "Defense" submitted, Van der Waerden hoped to get a professorship at Utrecht or Amsterdam. Van der Corput was the key man to this end.

38.4 Van der Waerden and Van der Corput: Dialog in Letters

Johannes Gualtherus van der Corput (1890–1975) was a professor of mathematics at Groningen (1923–1946) and Amsterdam (1946–1954). During the war and the German occupation of the Netherlands, he took an active part in the Dutch underground, and in 1945 spent a week in the Nazi jail for hiding people from the occupiers in his house. According to Dr. Knegtmans (June 10, 2004 e-mail to me, [Kne7]), "Van der Corput belonged to a small group of Groningen professors that had developed some ideas about the postwar university in the sense that it had to become a moral community that would be able to withstand any authoritarian threat or defiance. Van der Leeuw, the first postwar minister of Education, had belonged to the same group." Right after the war Van der Corput was the organizational leader of the Dutch mathematicians and in 1946 became one of the founders and the first director of the *Mathematisch Centrum* (Mathematics Center) in Amsterdam.[220]

Van der Corput knew Van der Waerden from their 1928–1931 years together at Groningen, where young Bartel learned much mathematics from him [Dol1]. They met in 1939 when the whole Van der Waerden family visited the Netherlands, and Bartel gave a talk at Van der Corput's Groningen University. The colleagues corresponded even during the war. A thick file of their 1945 correspondence, lying in front of me as I am writing these lines, is an invaluable resource for understanding their views on moral standards during the Nazi time and the occupation of the Netherlands, and more generally, the moral dilemmas raised by the war and its aftermath. I will let the correspondents do most of the talking. A number of handwritten versions of some of these letters exists. Some copies were sent to third parties, such as Van der Waerden's close friend mathematician Hans Freudenthal (1905–1990). All of this indicates that Van der Waerden took this exchange extremely seriously. The closeness of the two correspondents, who with no exception address each other with *amice* (friend), allows for a rare insight.

On July 29, 1945 Van der Corput sends Van der Waerden a letter in which he informs Van der Waerden of his (Van der Corput's) new critical role in the mathematical higher education of the Netherlands:[221]

> ... I have been appointed chairman of a commission to reorganize higher education in mathematics in the Netherlands, which will have as its primary duty to offer advice for the filling of vacancies in mathematics.[222]

Van der Corput realizes that his new authority to advise the minister of education calls for a new responsibility, and so he continues with probing questions:

> Your letter made me do a lot of thinking. I never understood why you stayed in Germany between 1933 and 1940,[223] and also why after 10 May 1940[224] you did not return to the Netherlands as so many succeeded in doing, if need be to go into hiding here. Rumors went around about you that you were not on our side any more, at least not entirely. That could have been slander. I would find it important if you could explain to me the situation completely and in all honesty.

---

[220] The Center still functions today, but under a new name *CWI, Centrum voor Wiskunde en Informatica* (Center for Mathematics and Computer Science).

[221] ETH, Hs 652: 12159.

[222] Prof. Dr. Gerardus J. van der Leeuw, Minister of Education, Arts and Sciences (*Onderwijs, Kunsten en Wetenschappen*) appointed J. G. van der Corput to be the chair of the Committee for the Coordination and Reorganization of Higher Education in Mathematics in The Netherlands (*De Commissie tot Coördinatie van het Hooger Onderwijs in de Wiskunde in Nederland*). Members of the committee were J. G. van der Corput, D. Van Dantzig, J. A. Schouten, J. F. Koksma, H. A. Kramers and M. G. J. Minnaert. The Committee became known as the "Van der Corput Committee."

[223] Indeed, even some Germans went into exile: "Between 1933 and 1941, an estimated thirty-five thousand non-Jewish Germans, not all of them Socialists, went into exile" [Scho, xiii].

[224] The day of the Nazi Germany's unprovoked attack on the Netherlands.

Van der Corput concludes by relating his own resistance activities:

People were in hiding in my house throughout the entire war, 23 in total, of which 5 were Jews; I was a representative at Groningen of the Professors Resistance Group. When I was arrested in February 1945, they found two people in hiding in my house, of which one was Jewish. I was suffering from angina and was released from prison after a week. My house and all my furniture were impounded [by the authorities] but we moved back on the day of liberation... I was on the Committee of Free Netherlands[225] and was arrested for disseminating illegal literature.

Van der Waerden replies on July 31, 1945.[226] He expresses a delight with his friend being in charge of appointments in mathematics, almost too much of a delight – but then, understandably, crosses most of the delight out:

I am very happy to be able to direct my defense to the right address against the things that have been blamed on me completely unexpectedly from all sides. ~~So you are chairman of the commission who will decide on the future occupation of the professorships of mathematics, perfect! An illegal work of the highest order and what is more, beneficent towards me. Delightful!~~

From the following lines we discover how the writing of the "Defense" has come about. We also learn that Van der Waerden attaches a copy of the "Defense" to this letter:

I have heard from Pannekoek[227] and Clay[228] that people were thinking about suggesting me for the Weitzenböck[229] vacancy at Amsterdam. When I spoke with Freudenthal about it and told him that I was looking forward to possible collaboration with him, he firstly pointed out the difficulties, especially from students' circles, that could be expected, and for the aspersions that would be cast upon me because of my stay in Germany since 1933. He advised me to write down my defense, which I had presented to him verbally. I have done it, and after conversations with others, I have added a few more things... In this situation you now come forward and ask for my justification. Voila! I hereby include a copy of the piece.[230]

Van der Waerden then explains why he did not return to Holland when Germany waged the war against his homeland:

I truly did not come to the idea to return to the Netherlands after 1940 and to go into hiding here. At the end of 1942 I came to Holland and have spoken with all sorts of

---

[225] *Vrij Nederland*, an underground newspaper.

[226] ETH, Hs 652: 12160.

[227] Antonie (Anton) Pannekoek (1873–1960), professor of astronomy at the University of Amsterdam and a well-known Marxist theorist.

[228] Born Jacob Claij (1882–1955), a major supporter of Van der Waerden's appointment, professor of physics at the University of Amsterdam, 1929–1953, who played a major role in the reconstruction of applied scientific research in the Netherlands after W.W.II.

[229] Roland W. Weitzenböck (1885–1955), a professor of mathematics at the University of Amsterdam, whose pro-German views cost him his job after the war.

[230] Actually, ETH Archive, the holder of this letter, does not have a copy of the "Defense". Fortunately for us, the copy sent to Freudenthal has survived. See the text and the analysis of the "Defense" earlier in this chapter.

people (honestly no *NSB*-ers[231] because those do not belong to my circle of friends) but there was nobody who gave me [such] advise; the concept of going into hiding, furthermore, did not exist at that time.

Van der Waerden is not being entirely open when he claims that he "truly did not come to the idea to return to the Netherlands." In fact, as we have already established that in December 1942 he received a job offer from Utrecht, which he discusses with Carathéodory and likely other colleagues. Van der Waerden then spells out the *real* reason why he did not wish to come home during the war:

Why would I go to Holland where oppression became so intolerable and where every fruitful scientific research was impossible?

Therefore, it appears that Van der Waerden never seriously considered going to the Netherlands during the war. In a statement that Van der Corput must have found particularly disingenuous, Van der Waerden claims that his "struggle" for the German culture and science was at least as noble as his correspondent's underground activities in the Netherlands:

For your struggle of which I have heard with great delay and only in part, I had great admiration and undivided sympathy, but I could not partake in it from that distance, because I did not have enough contact with you. Since 1933 I waged another struggle, together with other reasonable people such as Hecke,[232] Cara,[233] and Perron against the Nazis and for the defense of culture and sciences. That I was on the good side of that struggle was, as I thought, universally known. I did not expect that people here in Holland would have so little understanding of it.

In the next letter, dated August 20, 1945, Van der Corput makes his displeasure known to Van der Waerden, stating that he was not completely satisfied with his friend's explanations:[234]

Your letter has not completely satisfied me. You complain that we here in Holland have no sufficient understanding of your troubles, but after reading your letter I wonder whether you have a sufficient understanding of troubles with which we had to deal with here and of what was to be expected of a Dutchman in these years. It is not clear to me from your letter whether you consider your attitude in the past as faultless or whether you plead mitigating circumstances.

Van der Corput refuses to condone Van der Waerden's actions during the war, comparing them unfavorably to his own unambiguous rejection of the Nazism:

Concerning me personally, in January 1939 I refused Hecke's invitation, given to me by [Harald] Bohr, to give one or more lectures, because I refused to come to Germany

---

[231] *Nationaal Socialistische Beweging* (National Socialist Movement, a Nazi party in the Netherlands).

[232] Erich Hecke (1887–1947), one of the best students of David Hilbert, a famous number theorist, professor at the University of Hamburg (1919–1947), a man of highest integrity, who allegedly never used Hitler's salute, and who in 1940 resigned as the Editor-in-Chief of the leading journal *Mathematische Annalen* in protest of the publisher Ferdinand Springer's demand not to publish Jewish authors.

[233] Constantin Carathéodory (1873–1950), a German mathematician of Greek ancestry, professor of mathematics at Göttingen (1913–1918), and Berlin (1918–1920), and Munich (1924–1938).

[234] ETH, Hs 652: 12161.

as long as Hitler was in power. Consequently I have not gone to Germany after 1932. In connection with this position of mine that was shared by very many, I do not understand how you can so easily gloss over those years between 1933 and 1939. Indeed it was not at all fitting for a Dutchman to make mathematics in Germany flourish in those years when Germany was preparing for war and was throwing out the Jews everywhere.

Van der Corput contrasts his and Van der Waerden's positions with regard to the German and the American mathematical reviewing journals:

Speaking of Jews, when Levi-Cività was thrown out of *Zentralblatt*, I withdrew as an associate (while giving my reasons) and suggested all Dutch associates to do the same and to become associates for the *Mathematical Reviews*. Contrary to that, you suggested to a couple of associates to stay on and, if I am not mistaken, you invited new associates.

Van der Corput cites the 1939 incident that, apparently, still bothers him, and directly asks whether Van der Waerden and his wife were Nazi sympathizers:

Furthermore I remember that after a lecture at Groningen in the *Doelenkelder*,[235] you spoke appreciatively about the regime in Germany, more especially about Göring,[236] upon which I advised you to stop because this was not well received by the students of Groningen. I have to add to that that I do not know whether or not you were being serious at that time, but it made a strange impression on us, who considered Hitler a grave danger for the humanity. From different sides I was furthermore told that you wife was pro-Hitler, and that when she was supposed to come stay in Holland, she even gave as a condition that no bad could be spoken about Adolf. I say this, because you write that your wife was always against the regime. It is better that these things are discussed in the open, because then you can defend yourself.

Nevertheless, Van der Corput clearly wants to help Van der Waerden, and through him, Dutch mathematics:

I for myself think that the Netherlands has to be very careful not to lose [*zuinig*] its intellect and especially the one like yours. I have always regretted that you went to Germany and I will look forward to it if you can be won back completely for the Netherlands. . .

I would want nothing better than for everything to be all right. Because there is no Dutch mathematician with whom I like working more than with you. I would find it fantastic if we could work on mathematics at the same university again. I think that we could found a mathematical center then.

Van der Corput then explains why he needs to get information from Van der Waerden:

I hope that you will not just excuse me for these questions but that you would also understand them. Before the government can appoint someone it will conduct a very

---

[235] The steakhouse *De Doelenkelder* still exists in Groningen: call 050-3189586 for reservations:-).

[236] Hermann Göring, Commander-in-Chief of *Luftwaffe* (German Air Force), President of the Reichstag, Prime Minister of Prussia and Hitler's designated successor.

detailed investigation, and it is to be expected that it will also ask for my advice. It is therefore necessary for me to be well informed.

Perhaps to Van der Corput's surprise, Van der Waerden remains nonchalant. He asserts his complete innocence, and quotes the letter (see earlier in this chapter) he received from Heinz Hopf just about 2 weeks prior:[237]

> You ask whether I want to plead mitigating circumstances. Absolutely not! I demand a complete exoneration because I do not think that I can be blamed for anything. And I am also convinced that when my case ~~now or after a few years when the understandable commotion and confusion caused by the German terror has calmed down~~ is looked at objectively, that this exoneration will be given me. This conviction I shared with Hopf in Zürich who (following a conversation with Kloosterman about me) writes: "When the Dutch, whom you can approach with clean conscience and offer them your services, do not want you, then in my opinion they hurt themselves, and that is their business. ~~I consider it certain that in a few years, when the waves have calmed down a bit, somewhere in the world you will work again in the profession.~~"[238]

Van der Waerden continues by presenting, again, his (and Hopf's) opinion that one must differentiate between "the Hitler regime" and "the German culture":

> Your most important accusation, I assume, is the words "it was not at all acceptable for a Dutchman to make mathematics flourish in Germany in the years when Germany was preparing for the war and was throwing the Jews everywhere."
> In this sentence two things are identified with each other that I see as the strongest opposites: the Hitler regime and the German culture.[239] What was preparing for the war and was throwing out the Jews was the Hitler regime; what I was trying to make flourish or rather to protect against annihilation was the German culture. I considered and still consider this culture to be a thing of value, something that must be protected against destruction as much as possible, and Hitler to be the worst enemy of that culture. Science is international, but there are such things as nerve cells and cell nuclei in science from which impulses are emitted, that cannot be cut out without damage to the whole. And I mean that this standpoint is principally defensible even for a Dutchman, and I should not be in the least ashamed for having taken this position.
> Of course, it is understandable that people today here in Holland do not want to know, to make a difference between the Nazis and Germany or the German culture. Germany attacked the Netherlands and shamefully abused it, and the whole German people are also responsible for that. For the duration of the war this position is completely true, but one must not use this as measure to measure things that happened before the war.
> By the way, nobody at the time thought to condemn my actions. The Dutch Government itself allowed officially in 1934 or 1935 that I could continue my activities in

---

[237] ETH, Hs 652: 12153. The 4-page letter is undated; I am certain, however, that it was written between August 21 and August 27, 1945.

[238] The text in quotation marks is in German, see the discussion of this H. Hopf's letter earlier in chapter.

[239] Cf. H. Hopf's letter to Van der Waerden of August 3, 1945 earlier in this chapter, from which this idea may have come from. Could not have these two brilliant minds seen that the "German culture" gave birth to the "Hitler regime"?

Leipzig. The student organization invited me in 1938 for a series of talks, among other places at Groningen, a certain Van der Corput asked me in 1943 to write a book for his "*Weten Sch. Reeks*" [Scientific Series], and I could name a whole other series of things like that.

~~Also the English and the Americans, and above all the Russians, make a distinction between the Nazis, whom they want to destroy, and the German culture, which they want to help resuscitate. Should we not try to make this objective way of judgment acceptable also in the Netherlands again?~~[240]

Van der Waerden then explains his statement about Göring made during his 1939 visit to the Netherlands:

> You seem to remember that I spoke appreciatively in the *Doelenkelder* about the regime in Germany and more specifically about Göring. You must therefore consider me as somebody without an elementary sense of right and wrong; because Göring is, as everybody knows, a clever crook, whose henchmen burned the Reichstag and who used that to abolish socialist parties. An unprecedented deception of the people that was used to destroy the democracy and the parties to which I, because of tradition, friendship, and because of my own father, was connected. And I would have defended that criminal? And the Hitler regime moreover? And now I would twist around like a weathervane and contend that I was always against Hitler? In other words, that makes me a deceiver, a cunning liar! Nevertheless you always willingly offer me your mediation, not only with words but also with deeds, with Noordhoff, give my defense to Minnaert, and write that you do not like to work more with anybody than me. I do not understand that attitude. Or rather I can only give one explanation to it, namely that deep in your innermost a voice tells you: no, I know that man from before as decent and truth-loving, let me give him an opportunity to defend himself.
>
> Well, I can guarantee you, that what you write about the *Doelenkelder* must be a misunderstanding. I have never uttered a word of defense of the Nazi regime to anybody. The question which we spoke about in the *Doelenkelder* was, if I am not mistaken, not whether this regime was defensible, but how can people cope in Germany in spite of this regime. How is science under these circumstances possible? Then I may have mentioned a few facts from which it was apparent that at Leipzig especially and more importantly in mathematics, the pressure from above was not as oppressive as people imagined it here. I may have mentioned in connection with something or other that Göring was not an anti-Semite and even appointed Jews in his ministry, or I have told how popular he was with the people and with his subordinates or something like that. But to defend Hitler or Göring? Impossible!

Thus, Van der Waerden asserts that he did nothing wrong and "only" said the Nazi pressure at Leipzig was not too bad, and that the second man of the Nazi Germany, Göring, was not an anti-Semite! He then goes on to explain his wife Camilla's statements:

> Now about my wife supposedly being a Nazi. Would you believe that this is the third time that I hear this spiteful slander? I can not figure out where this slander is coming from. We, my wife and I, have avoided any contact with the Nazis in Leipzig like the black plague. Our acquaintances were only people who shared our horror for the

---

[240] This paragraph is thinly crossed out in this version, but was not crossed out in another, unfinished version in my possession.

Nazi regime. And then, when she stayed in Holland, she asked that nothing bad be said about Adolf? Do you honestly believe that my father, when we stayed with him in 1939, would have accepted such a condition, or whether my brothers would have kept themselves to it? The truth is that my wife could not tolerate it when bad was spoken about the Germans. Indeed, German is her mother tongue, and she knew so many kind people in Germany. If you do not want to believe all of this on my word, then please write a letter to Frau Lotte Schoenheim, Hotel Stadt Elberfeld, Amsterdam. From 1932 up until her emigration to the Netherlands in 1938, she has been frequently in conversation with my wife and me, and after that in Holland has stayed in contact with my family. She knows our opinion not only from words but also from deeds.

Thus, according to Van der Waerden, Camilla "could not tolerate it when bad was spoken about the Germans." Were all of the Germans in the Third Reich above criticism in 1939? Did not Van der Waerden himself write above in this very letter that "Germany attacked the Netherlands and shamefully abused it, and the whole German people are also responsible for that"?

This handwritten letter is particularly important to Van der Waerden: he encloses a large handwritten part of it, entitled "From a letter to Prof. J. G. van der Corput", in his January 22, 1946 letter to Freudenthal[241] together with "The Defense," which was earlier submitted to the Amsterdam's *College van Herstel* and Utrecht's *College van Herstel en Zuivering*.

In his immediate, Aug 28, 1945 reply, Van der Corput soft pedals on his probing questions and assures Van der Waerden of his support:[242]

> Am I mistaken if I have an impression that you wrote your letter in somewhat irritated state? I believe that I have consistently acted in your interest; also during a conversation with the minister I pointed out that the Netherlands should be very careful not to lose [*zuinig*] a man like you. I even said that the Netherlands should rejoice if we get you back for good. But there are general rules and it needs to be determined how much those apply to you.
>
> I have always considered it impossible that you are a "weathervane, a hypocrite, and a cunning liar," and I still consider it impossible. With my remark I wanted to show that you in my opinion did not sufficiently realize how we thought of the Hitler regime even then. It was all joking, and I never attached much significance to it, but when afterwards remarks were made indicating doubt, I thought it was important for you to mention this in my letter. I would be very sorry if I hurt you with it but it is still better to bring these things out in the open and to give you an opportunity to rebut them. To my great pleasure I found out today that it was said at the Mathematical Congress in Oslo that you were known as a strong anti-National Socialist.
>
> Immediately after receiving your letter I made sure that this week Friday night or Saturday morning there will be meeting between me and the minister of education about this matter. The minister has already told me in the first conversation that the cabinet has spoken about general rules concerning the persons who were in German service during the war. Those rules were not finalized then. Whether or not this happened since then I will find out this week.

---

[241] RANH, Papers of Hans Freudenthal, inv. nr. 89.

[242] ETH, Hs 652: 12162.

Van der Corput leaves the last two points of Van der Waerden's letter (Göring and the Nazi sympathies of Mrs. Van der Waerden?) to a confidential in-person conversation, and thus, to my regret, out of reach for this study:

> About the different other points of your letter, I would like to speak with you orally next week. Tuesday September 4 I hope to get to Laren for this before 9 o'clock in the morning.

But not to worry anyway:

> Be convinced that it is my sincere desire to keep you for the Fatherland and for higher education.

Van der Corput communicates the first hopeful signs on Sept 11, 1945:[243]

> ...I have discussed your case with Oranje[244] and Borst,[245] leaders of the Professors' Resistance. After my explanation neither one of them saw any problem with your appointment with one of the Dutch universities. They of course cannot decide anything, but as is evident to me, it is much easier for the minister and his department if they know that there is no opposition from that particular side. I have the impression that things will be all right and that after a few months we will be able to collaborate more...
>
> PS:... During my absence Van der Leeuw[246] has called to tell me that both parts of my most recent letter were "good". One of the parts concerned my statement that we do not need to fear any opposition from Borst and Oranje... It will all work out, that is my opinion.

Five days later, Van der Corput is ready to celebrate "mission accomplished":[247]

> I have just received a written confirmation from Van der Leeuw... He writes: "We should now figure how good the Van der Waerden's matter is."
>
> This means that he is ready to nominate you.

On September 22, 1945, Van der Waerden describes the state of his employment affairs to his friend and confidant Hans Freudenthal as follows:[248]

> Minister Van der Leeuw told Van der Corput that now that Van der Corput and Borst and Oranje of the Professors Resistance Group consider me as sufficiently "pure", he also considers the affair "OK". My appointment at Utrecht is therefore very close.

---

[243] ETH, Hs652: 12163.

[244] Prof. J. Oranje, professor of law, Free University (*Vrije Universiteit*, a Calvinist university). During the occupation Prof. Oranje was chair of *Hooglerarencontact*. According to Dr. Knegtmans, the illegal during the German occupation *Hooglerarencontact* (Contact Group of Professors) tried to persuade professors and university boards to close their universities in 1944.

[245] Prof. Dr. J. G. G. Borst, professor of medicine, University of Amsterdam, one of the leaders of *Hooglerarencontact*.

[246] Prof. dr. Gerardus van der Leeuw, the Minister of Education, Arts and Sciences ("Onderwijs, Kunsten en Wetenschappen"), 1945–1946.

[247] September 16, 1945 letter from Van der Corput to Van der Waerden; ETH, Hs652: 12164.

[248] RANH, Papers of Hans Freudenthal, inv. nr. 89.

Then on September 29, 1945, Van der Corput informs Van der Waerden by a telegram that *College van Herstel en Zuivering* of the University of Utrecht got on Van der Waerden's board as well:[249]

> Minnaert[250] signals *College van Herstel* considers Van der Waerden sufficiently polit-
> ically reliable and desires appointment at Utrecht
>   Van der Corput

However, about a month later, unexpectedly, new problems surface. Van der Corput informs Van der Waerden about them in his October 24, 1945 letter:[251]

> Indeed there now come again difficulties concerning your appointments. As there is
> someone in higher education, who works against you and among other things main-
> tains that you had to use – and did regularly use – the Hitler salute at the inception of
> your classes in Germany. Be so kind to give very clear answer to this question, so that
> I can contradict it if this slander comes about again.
>   ... This week I received an invitation from Faculty of Natural Sciences in Ams-
> terdam to become Weitzenböck's replacement. This shows that the opposition against
> your nomination in Amsterdam is too strong. I do not know what I am going to do.
> Personally, I like Utrecht better, but maybe I can do more for mathematics in Amster-
> dam....
>   I am not happy about the turn that the mathematical problems have taken. I would
> be particularly sorry if certain illegal circles [*illegale kringen* – he probably means
> former underground circles] will successfully delay your appointment at a Dutch uni-
> versity.

Van der Waerden answers right away, on October 26, 1945. He does not give
a "very clear answer to this question" of the Hitler salute, or any answer for that
matter. He shares Van der Corput's pessimism about his academic prospects in the
immediate future, and blames the students and Minister of Education Van der Leeuw
for it:[252]

> After what I have read in the *Vrij Katholiek*[253] about the radical demands of the stu-
> dents and the willingness of Van der Leeuw to listen [to them], I think it will take some
> time before I can get a position at Utrecht. I have something else now, as of October 1,
> 1945 I am working for *Bataafsche*.[254]

---

[249] ETH, Hs 652: 12158.

[250] Marcel Gilles Jozef Minnaert (1893–1970), a member of the "Van der Corput Committee," see footnote 54 for more information on the Committee. Documents in the archive of the University of Utrecht
show that Minnaert—in a sense—represented Van der Waerden to the Utrecht's *College van Herstel en
Zuivering*, which most likely had never met with Van der Waerden in person. This was a very beneficial
representation for Van der Waerden, because as an outspoken critic of Nazism Minnaert spent nearly 2
years in a Nazi prison, from May 1942 until April 1944 [Min].

[251] ETH, Hs652: 12166.

[252] ETH, Hs652: 12167

[253] *De Vrij Katholiek* (The Free Catholic) monthly of the Free Catholic Church in the Netherlands, was
published 1926–1992.

[254] *Bataafsche Petroleum Maatschappij (B.P.M.)*, today known as the Royal Dutch Shell.

Van der Corput's reply comes a full month later, on November 26, 1945. He opens his letter with good news:[255]

> I very much want you to have a position in higher education. The Committee for Mathematics [*Wiskundecommissie*] intends to create Center for Pure and Applied Mathematics, most likely in *A'dam* [Amsterdam], and if the Center comes into being, I want you to work there.

Then there come the bad news:

> But there are problems, and I hate time after time to ask you these questions and ask you for clarifications but I have to do this. In order to support you I need the answers to these questions.
>
> It now centers around three clearly indicated points:
>
> The first. Your father and your uncle repeatedly and with a lot of emphasis have insisted before the war that you should leave Germany. They felt it was your duty to leave but you refused, and they considered it as neglect of your duties.
>
> The second. some people are certain that your wife is an anti-Semite, others believe that this is too strong a statement, but she did not want to have anything to do with Jews.
>
> The third. During the war there was an opportunity for you to go to America, but you refused, for you [argued that you] needed to stay because you could do a good work for your students, some of whom were Jews. If this is true that even during the war, when you had a chance to go to the United States, you still did not want to leave, this will create definite difficulties for you.

Apparently, without receiving a reply, Van der Corput writes again on December 8, 1945, this time quite apologetically:[256]

> I am not asking you these things for myself... I want to collaborate [with you] as much as I can...It would be very unpleasant if these questions would somehow cause the deterioration of my relations with you or your wife. Please, understand I only need it for the government.

Van der Waerden immediately, replies on December 10, 1945. He first reassures Van der Corput of his friendship:[257]

> This correspondence will not have negative consequences on our relationship and friendship; there is no danger for that.
>
> While in my letters there is sometimes a tone of annoyance, it is against people who disseminate certain gossip against me and definitely not against you because I know that you have worked tirelessly in my interest and in the interest of the Dutch science.

Then there come the words that spell out Van der Waerden's fundamental principles:

> On the other hand, I also cannot imagine that you are irritated by my democratic anti-Fascist point of view that I have expressed in my letter. My point of view is that when appointments are concerned, only capacities of the appointee should be taken into

---

[255] ETH, Hs652: 12168.

[256] ETH, Hs652: 12169.

[257] ETH, Hs652: 12170.

account, and not – as it is usual in the Fascists regimes – the person's character, his past, and his political trustworthiness.

I had been raised under the influence of my father, who was a principled democrat, then I had been under the influence of Hitler, and that counter-point of view led to terrible consequences. You too actively fought against the Nazis, and fought for democracy and freedom of our people. So I cannot imagine that you would have a problem with my point of view even if you do not share all aspects of it.

This dialog in letters is so alive that I feel compelled to enter it and say: Bartel, you invoke your principled democrat father as your great influence. But according to Van der Corput (12/26/1945 letter above), "Your father and your uncle repeatedly and with a lot of emphasis have insisted before the war that you should leave Germany. They felt it was your duty to leave but you refused, and they considered it as neglect of your duties." You did not contradict, and so I believe this to be truth. I agree with your father Dr. Theo and his brother Jan: you should not have served the "gangster-regime" (your words), occupying and terrorizing your people. Mathematicians do not live in a vacuum, and thus their "character," their "past", and their morality matter, especially in Holland right after five years of brutal occupation.

Meanwhile, Van der Waerden ends the letter with the major good news, promising an Amsterdam professorship to him very soon:

Revesz[258] has told me yesterday that the Amsterdam Faculty recommended me for appointment to ordinary [professor] to *B. en W.* Thus things have started to happen now!

On December 22, 1945, Van der Waerden writes again.[259] This three-and-a-half-page letter is full of technical negotiations. One alderman prefers Van der Waerden's appointment to be at an extraordinary (*bezonder*) professor level (rather than at the level of a full professor). Van der Waerden does not mind that, but then he wants to keep his oil industry position as well. Jacob Clay thinks that this solution is very good. Van der Waerden further asks for a lectureship at Amsterdam for his friend Hans Freudenthal. And he wants a clear definition of the boundaries between duties of Van der Corput, Freudenthal, Brouwer, and himself. Van der Waerden discusses these details, because in his mind, his appointment at Amsterdam is a done deal.

In the end, Van der Corput is not completely satisfied with the positions of Van der Waerden the man. But Van der Corput has a great respect for Van der Waerden the mathematician, and he believes that if he were to help Van der Waerden get a fine position at Amsterdam, then Van der Waerden would spend his career there, and thus would greatly benefit their homeland.

It is worthwhile to note here that Van der Waerden is much more open and harsh in his criticism of the Dutch in his November 20, 1945 letter (in German) to Richard Courant of New York University than in all of his correspondence with Van der Corput:[260]

---

[258] Hungarian born (fled in 1920) Geza Révèsz (1878–1955) was the first and founding professor in psychology at the University of Amsterdam; a close friend of L. E. J. Brouwer.

[259] ETH, Hs652: 12171.

[260] New York University Archives, Courant Papers.

The Dutch are completely crazy. They have no concept in their heads except "cleansing" ("*Sauberung*"): they punish all those who had worked together with the Germans. There are managers, bosses who would not employ any workers who were forced to work in Germany.[261] There are more political prisoners in Holland than in all of France, even though the Dutch showed much more character in the war than the French did. So is my appointment to Utrecht, which ran into great difficulties, even though it was a done deal with the faculty for years. I am very happy that I currently have a pleasant job in the industry and can await the return to normal circumstances.

And while Van der Waerden demands "a complete exoneration" from Van der Corput, Van der Waerden sounds more conciliatory in his December 29th, 1945 letter (in English) to Courant:[262]

I am much pleased that you have the intention to resume the old friendship with me and other old friends as far as possible, and that old Göttingen will keep a warm place in a corner of your heart. And just for that reason, I am convinced that you at least will understand a little bit what my other friends in America could not grasp, namely "why I as a Dutchman chose to stay with the Nazis". Look here, I considered myself in some sense as your representative in Germany. You had brought me into the *redaction* [editorial board] of the *Yellow Series* and the *Math Annalen*, I thought, in order to watch that these publications were not nazified and that they might maintain their international character and *niveau* [standard] as far as possible. This I considered as my task, and together with Hecke and Cara [Carathéodory] I have done my best to fulfill it, which I could do only by staying in Germany. [It] is not that plain and easy to understand, apart from other sentimental and familiar [familial] links attaching me to Germany. I have made some mistakes perhaps, but I have never pacified with the Nazis.

Under "other sentimental and familiar [familial] links" to Germany, Van der Waerden no doubt refers to his "German wife," to raising his three children "pure German,"[263] and to his sense of belonging to the German culture in general, and the German mathematics in particular. For the first—and to the best of my knowledge the only—time Van der Waerden admits making "some mistakes."

The *Dialog in Letters*, presented here will undoubtedly force the reader to define his or her positions on many fundamental issues, such as the place of a scientist in a tyranny.

38.5 A Rebellion in Brouwer's Amsterdam

For decades mathematics at the University of Amsterdam had been run by the most famous Dutch mathematician of the twentieth century, Luitzen Egbertus Jan Brouwer (1881–1966), an ordinarius at Amsterdam ever since 1913. Brouwer had

---

[261] Van der Waerden refers here to *Arbeitseinsatz*, the Nazi forced labor program.

[262] Ibid.

[263] Van der Waerden, Letter to Wilhelm Süss, March 14, 1944, ETH, Hs 652: 12031.

a famous feud with the leading German mathematician David Hilbert. Likely due to Van der Waerden's closeness to Hilbert, Brouwer did not wish Van der Waerden to get a chair at Amsterdam. But following the liberation, Brouwer was suspended from office for a few months while the Amsterdam's *College van Herstel* investigated his behavior during the occupation. This suspension and Brouwer's advanced age allowed younger charismatic mathematicians to wage a power struggle with him. J. G. van der Corput was the leader of this new generation. He and Jacob Clay, professor of physics at the University of Amsterdam, undertook what one might call "The Battle of Van der Waerden."

Professor Dirk van Dalen has kindly shared with me relevant chapters of his then not yet submitted manuscript of his brilliant, comprehensive biography of L. E. J. Brouwer [Dal2]. Van Dalen believes that Van der Waerden did not get a university job in 1945–1946 because of Brouwer's opposition. I respectfully disagree and believe that Brouwer's opposition to Van der Waerden's chair at Amsterdam only strengthened Clay's and Van der Corput's resolve, and thus increased Van der Waerden's chances. Professor Nicolaas G. de Bruijn, who succeeded Van der Waerden at the University of Amsterdam in 1952, seems to agree with my vision of this complicated affair. Following are my questions and his answers [Bru10]:

A.S.: Was Brouwer against hiring Van der Waerden at Amsterdam in 1945–46? If "yes" why was Brouwer against? How influential was Brouwer in such matters in 1945–46...?

N.G.B: Brouwer did not have much influence. He had a fight with the rest of the world, in particular with his Amsterdam colleagues and with the Amsterdam mathematical centre...

A.S.: As I understand, Van der Waerden's strongest supporters were Clay and Van der Corput, am I right?

N.G.B.: You may be right. Along with Schouten they were the older people, and in those days the older people dominated the networks. But the support of the younger generation, like Koksma, Van Dantzig and Freudenthal, must have been very essential. In particular the fact that Van Dantzig and Freudenthal were Jewish may have impressed the authorities.

In fact, on September 22, 1945 Van der Waerden assured Van der Corput of being ready to join in the war against Brouwer if necessary:[264]

Dear Colleague!
I would of course have preferred if the whole *Faculteit*, including Brouwer, approved my appointment. If you are prepared together with me to make something good of mathematics in Amsterdam even against Brouwer, if that is necessary, I will be collaborating in that effort.

On the same day Van der Waerden summarized for his close friend Hans Freudenthal the state of the Brouwer's Amsterdam:[265]

---

[264] ETH; Hs652: 12165.

[265] RANH, Papers of Hans Freudenthal, inv. nr. 89.

Van Dantzig saw the future of math in Amsterdam as rather bleak. Unless a counterweight to the influence of Brouwer could be formed by the filling of the second professorship by somebody who can stand up to Brouwer, he feared that Brouwer would want to rule with 4 lectors dependent on him.

Clay told me that Brouwer had answered evasively his question whether he supports my candidacy, and he obviously does not want to work with me (something I have already known). Clay, however, wanted to nominate me against Brouwer's will if I can guarantee him that I would accept the appointment. I answered him today:

"I would have of course preferred if the entire faculty including Brouwer were to approve my appointment. But if you are prepared to literally go to war with me and to try to make something good out of mathematics in Amsterdam, even against Brouwer if that is necessary, then I would like to offer my help. However, if the appointment at Utrecht comes first, then I would take it as you understand. I desire to take my part in the reconstruction of the Dutch science as soon as possible, be it at Utrecht or at Amsterdam."

Clay did not seem to want to conclude the matter soon, so I think nothing will come of it. If something were to come of it, I would also try to find a beneficent solution to the conflict between Freudenthal and Bruins.[266] Because my coming to Amsterdam only makes sense when you and I can set the tone there, and not when you stay in a subservient position and both of us have to fight Brouwer and creatures all the time.

Three months later Van der Corput was able to talk Brouwer out of opposing Van der Waerden's appointment at Amsterdam. On December 30, 1945 Van der Corput reported this development to Van der Waerden:[267]

With Brouwer I have come to an agreement that he will only cover the courses about intuitionism, that he will give exams only to the students that have an interest in that particular discipline. And he liked my willingness in this. He agrees with your appointment to extra-ordinary also with an appointment of Freudenthal as a lecturer...

He [Brouwer] has $5\frac{1}{2}$ years left before his retirement, while I have about 15 years left. I can count on his help, and we can work together...

Apparently, Brouwer, even agreed to pass on to Van der Corput and Freudenthal his "baby," the journal *Compositio Mathematica* that he founded in 1934—in spite of his falling out with Freudenthal:[268]

He [Brouwer] asked if I was willing to take over the Secretariat of the *Compositio*, together with Freudenthal. I wrote about it to Fr. [Freudenthal]. If he is willing then I would be too. The result of this is that the Br.'s [Brouwer's] name would remain but that Freudenthal and I would publish *Compositio*, while Fr. [Freudenthal] and Br. [Brouwer] would not have anything to do with each other.

With Minister Van der Leeuw's support and Brouwer's blessing, Van der Waerden was on course to a professorship, when his ship ran into an "explosion" in the sea of public discourse. To be continued in the next chapter.

---

[266] Evert Marie Bruins (1909–1990), a mathematics faculty at the University of Amsterdam.

[267] ETH, Hs652: 12172.

[268] Ibid.

39
In Search of Van der Waerden: The Unsettling Years, 1946–1951

39.1 The *Het Parool* Affair

> *When in May 1940 the Germans conquered our*
> *country Mr. Van der Waerden was still standing*
> *behind his podium at Leipzig.*
> – *Het Parool*, January 16, 1946

I find it surprising that the early media records have been completely overlooked and never mentioned by any of the many prior biographers of Van der Waerden. Did they view the news reports to be too much off the calf and not carrying lasting truths? Yes, the shelf life of a newspaper is one day, but it captures—and preserves—the *zeitgeist*, the spirit of the day, better than anything else available to a historian. Moreover, in our *Drama of Van der Waerden*, a newspaper was an important player. I therefore will use newspapers here liberally and unapologetically.

If after the war Bartel L. van der Waerden were to go back to Leipzig—or any other place in Germany—he would have been received with open arms. After the war both East and West Germanies were quite soft even on Nazi collaborators, which Van der Waerden certainly had not been. In addition, Van der Waerden's loyalty to Germany and German mathematics had been unquestionably great.[269] The Netherlands was another matter. Its standards of "good behavior" during the Nazi time and occupation were much higher, especially when judged by the editors of a publication like *Het Parool*, a newspaper that had been heroically published underground ever since July 1940,[270] and had paid for it by lives and freedom of many of its workers. After the war and the occupation, at the circulation of 50,000 to 100,000 copies in Amsterdam alone, and local editions appearing in more than ten cities in the country [Kei], *Het Parool* had an enormous moral power.

---

[269] On June 12, 1985 Leipzig University awarded Prof. Dr. Van der Waerden the honorary doctorate.

[270] It started under the title *Nieuwsbrief van Pieter't Hoen* on July 25, 1940 and became *Het Parool* on February 10, 1941 [Kei].

In early January 1946 everything was in place for Dr. Van der Waerden's professorship at the University of Amsterdam. The City Council's meeting with his
appointment on the agenda was about to begin in the afternoon of January 16, 1946,
when just hours earlier a "bomb" exploded on page 3 of *Het Parool* [Het1]:[271]

Him??
No, not him!

The proposal to appoint Dr B C [sic] van der Waerden as professor in the faculty
of mathematics and physics at the University of Amsterdam should surprise all those
who know that Mr. Van der Waerden served the enemy throughout the entire war. His
"collaboration" is not today's or yesterday's news. When the war broke out in September 1939, and the Netherlands, fearing invasion, mobilized, Mr. Van der Waerden was
standing behind his podium at the University of Leipzig. He had stood there for years.
And he *continued* to stand there. He saw the storm coming as well, but he did not think
about coming back to his fatherland. When in May 1940 the Germans conquered our
country Mr. Van der Waerden was still standing behind his podium at...Leipzig. And
he continued to stand there. For five years the Netherlands fought Germany and for all
those five years Mr. Van der Waerden kept the light of science shine in...Leipzig. He
raised Hitler-followers. His total ability – a very great one – and all his talent – a very
great one – were at the service of the enemy. Not because Mr. Van der Waerden had
been gang-pressed (*geronseld*) to the forced *Arbeitseinsatz* [labor service], not because
it was impossible for Mr. Van der Waerden to go into hiding; no, Mr. Van der Waerden
served the enemy, because he liked it at Leipzig; he was completely voluntary a helper
of the enemy, which – and this could not have remained unknown to Mr. Van der
Waerden – made all of higher education plus all results of all scientific work serve
enemy's "*totale Krieg*" [total war].

When asked, Mr. Van der Waerden cannot answer what an average German answers
when he hears of the boundless horrors done in the country: "*Ich habe es nicht
gewusst*" [I did not know]. In the middle of the war years Mr. Van der Waerden came
back to the forgotten land of his birth and he heard and saw how disgracefully his
patrons (*broodheeren*) were acting here. Did he not care at all? (*Liet het hem Siberisch
koud?*) A few weeks later Mr. Van der Waerden was standing behind his podium
at...Leipzig again. In the Netherlands firing squads shot hundreds. In the concentration
camps, erected as signs of *Kultur* (culture) by the Germans in Mr. Van der Waerden's
second fatherland, many of the best of us died; as did a few Dutch colleagues of Mr.
Van der Waerden. Did that do anything to him? The story is becoming monotonous:
Mr. Van der Waerden raised the German youth from behind his podium at...Leipzig.

However that is where the house of cards collapsed. Germany, including Leipzig,
surrendered. The Third Reich, which Mr. Van der Waerden had hoped would last, if not
a thousand years, then at least for the duration of his life, became one great ruin. And at
that very moment Mr. Van der Waerden remembered that there existed something like
the Netherlands and that he had a personal connection to it. He looked at his passport:
yes, it was a Dutch passport. He packed his bags. He traveled to "the fatherland." Now
Leipzig was not that nice anymore. All those ruins and all those occupying forces –

---

[271] In search for greater expressiveness, the authors included in this Dutch article some passages in
German. I am leaving them in German, and add translation in brackets. I also include in parentheses
some Dutch expressions that are particularly hard to adequately translate into English.

yuk (*bah*). After five years of diligent service to the mortal enemy of his people, Mr. Van der Waerden was now prepared for the other camp.

There are more like him. But what is worse, the University of Amsterdam seems willing to give this Mr. Van der Waerden another podium immediately. Mathematics has no fatherland, you say? Yes, sir (*tot uw dienst*), but in the Netherlands in the year 1946 it should be desired of a professor of mathematics that he does have one, and that he remembers it more timely than on the day on which his podium in the land of the enemy became too hot under his shoes.

This passionate article, which circulated throughout the whole country, with "*Mr. Van der Waerden was standing behind his podium at Leipzig*" being repeated over and over like a refrain in a song, must have made the Amsterdam City Council concerned, if not embarrassed. While Nazi collaborators or even volunteers of the German labor service (*Arbeitseinsatz*) among the faculty, staff, and students were removed from the University, the City Council was planning to approve the appointment of a professor who voluntarily served Germany the entire Nazi time, including the five terrible years of the German occupation of the Netherlands. The approval of Van der Waerden's appointment was postponed. The following day, on January 17, 1946, *Het Parool* reported the outcome [Het2]:

Prof. VAN DER WAERDEN NOT YET APPOINTED
Appointment halted

After the Amsterdam City Council convened yesterday afternoon in the Committee General (*Comité Generaal*), Mayor de Boer announced that, as a result, the nomination to appoint Professor Dr. B. L. van der Waerden, Professor of Mathematics at Leipzig, as Extra-Ordinarius (*Buitengewoon Hoogleeraar*) at the University of Amsterdam has been put on hold.

Because of the publication in *Het Parool* about Professor Van der Waerden, the Council suggested that there should not be a rush action. Further information was demanded.

On behalf of *B. en W.*,[272] City Alderman (*Wethouder*) Mr. De Roos responded that Professor Van der Waerden had good papers. Leipzig was a mathematical center. Beforehand many authorities were asked for information; among others also the Commission of Learned People (*Gestudeerden*) in Germany. The *College van Herstel* (College for Restoration)[273] of the University and also the faculty supported the appointment. For now, however, the appointment has been halted; *B. en W.* will consult later with the *College van Herstel*.

Van der Waerden was outraged not only by the City Council's refusal to approve his appointment, but also by such heavy and public accusations by the newspaper

---

[272] "*B en W*" stands for *Burgemeester en Wethouders*, i.e., Major and Aldermen.

[273] Once again, we have here a definitive proof that Professor van der Waerden was not correct when he wrote to me on April 24, 1995 [Wae25] "Before your letter came, I did not know that a commission was formed to investigate my behavior during the Nazi times." *College van Herstel* was precisely such a de-Nazification commission; see the previous chapter for more on this subject.

that was read and respected practically by everyone in the postwar Netherlands. On January 22, 1946 he wrote the following lines to his friend Hans Freudenthal:[274]

> Amice,
>
> Thank you for your kind letter. It did us a lot of good to have at least one loyal friend in the midst of this enemy world.
>
> I have sent the enclosed rebuttal to *Het Parool* and to *Propria Cures*. Already before that I supplied Clay with the necessary data for the Alderman's[275] defense of [Van der Waerden]. I have the impression from the report of the council meeting in *Het Parool* that the Alderman is fighting for me like a lion.
>
> The attitude of the students gives me a great joy. As soon as I am there I will win them for me completely. I am convinced of that.

I am not certain as to why Dr. Van der Waerden got "a great joy" from the students' attitude. As we will soon see, students presented a vocal opposition to his appointment. Also, note Van der Waerden's line "I supplied Clay with the necessary data for the Alderman's defense": we will soon learn about the content of this data from a *Het Parool*'s article.

In this letter to Freudenthal, Van der Waerden enclosed two documents—the "Defense" and "From a letter to Prof. J. G. van der Corput"—both discussed in great detail in the previous chapter, as well as the following handwritten letter to the editor,[276] which he sent to both papers, *Het Parool*[277] and *Propria Cures*, even though the latter had not run any commentary on Dr. Van der Waerden's impending appointment:

Correction [*Rechtzetting*]

> In the "*Het Parool*" dated Jan 16, my person was sharply attacked. I do not wish to go into this at great length. The question of whether or not I acted wrongly is being carefully researched by the concerned services.[278] But I have to correct two untruths. It is said that I hoped that the Third Reich would last for as long as I would. This is slander. I was known in Germany and outside as a strong opponent of the Nazi regime; I can prove this with witnesses.
>
> It furthermore says that I returned because my podium became too hot under my feet. This is also not true. I returned because the *Faculteit* of Math and Physics of the State University of Utrecht asked me to take up a professorship in mathematics.
>
> B. L. van der Waerden

---

[274] *Rijksarchief* in *Noord-Holland* (RANH), Papers of Hans Freudenthal (1905–1990), mathematician, 1906–1990, inv. nr. 89.

[275] Here Van der Waerden clearly refers to one particular Alderman (there were six): Mr. Albertus de Roos (1900–1978), the alderman (1945–1962) for Education and Arts.

[276] RANH, Papers of Hans Freudenthal, inv. nr. 89.

[277] Van der Waerden's letter to *Het Parool* was dated January 21, 1945, as seen from *Het Parool*'s January 23, 1945 acknowledgement sent to Van der Waerden and signed by Secretary Hoofdredactie: see ETH, Hs 652: 11631.

[278] Cf. Footnote 273.

There existed words—about patriotism, love of the fatherland, contributions of the Van der Waerden family to the Netherlands, desire to return home—which could have touched the readers' hearts and made a strong case for Van der Waerden's acceptance. Van der Waerden's dry and proud text about returning because of a job offer could not have possibly made things better for him. The self-description as being a "strong" anti-Nazi could not be accepted by the editors who risked their lives daily during the occupation. Both *Het Parool* ("Prof. Van der Waerden defends himself," [Wae9]) and *Propria Cures*[279] ("Correction," [Wae10]) published the complete text of the above Van der Waerden's "Correction" on February 1, 1946. *Het Parool* added the following editorial response: [Het4]

> We are pleased to give Mr. Van der Waerden the opportunity to defend himself. Has he made his case stronger with this? No, not quite. Unless there are Dutchmen who truly believe that the Germans from 1940 to 1945 allowed "strong" (!) opponents to occupy professorships. Which acts show this strong anti-Nazism of Mr. Van der Waerden? And the timing of his return to the fatherland in 1945 is then one of those rare coincidences that one should believe as such...or not. Mr. Van der Waerden – and this is the heart of the matter – from the first until the last day of the war served science in the land of the enemy and this was compensated by the enemy's money. He who has voluntarily served the enemy from May '40 to May '45 is a bad Dutchman. Those who unleash him afterwards on the Dutch youth do not understand the demands of this time. And if the appointment of Van der Waerden is approved, then one should immediately stop objecting to workers and students who volunteered for the labor service [*De Arbeitseinsatz*],[280] etc., for the labor service [*De Arbeitseinsatz*] of Van der Waerden was more complete than that of any other Dutchman. "Rewarding" ("*Belooning*") that with a professorship would mean that all the others who worked for the enemy voluntarily deserve a feather and a bonus.
>
> – *Red* (Editors) *Het Parool*

Earlier, on January 25, 1946, *Het Parool* had already reported the postponement of the approval of Van der Waerden's appointment [Het3]:

Prof. dr. B. L. van der Waerden

---

[279] University of Amsterdam students' weekly.

[280] Under the *Arbeitseinsatz* program, the Dutch (and other) people were sent to work in Germany (or "Greater" Germany). Those who went were punished after the war. In a 2004 e-mail to me, Dr. Knegtmans comments as follows [Kne8]: "As far as I know, only very few people actually volunteered for the *Arbeitseinsatz*. Most (several hundreds of thousands) did so under pressure and among them were three thousand students of all Dutch universities and a few staff members. After the war, however, there was some criticism of these men. Could they not have evaded conscription, some asked publicly. I think they could not, because their names and addresses were known and most needed the income for their families. This was of course not the case with the students, but in fact most students fled from the *Arbeitseinsatz* in Germany back to Holland, while others did not return to Germany from their holidays. I think that none of the students, staff members or professors of the University of Amsterdam was punished for voluntarily joining the *Arbeitseinsatz*. Probably no one did join voluntarily. But some of the Nazis among the students and staff joined the German army (or the Dutch Volunteer Corps) or paramilitary German organizations. The staff members among them were removed from the university, the students simply did not return to the universities."

The nomination of *B. en W.* to appoint Prof. dr. B. L. van der Waerden, which was put on hold at the previous session of the city council, because of the article in *"Het Parool"*, does not appear on the agenda for January 30th. It was put there initially, but it has been scrapped off by *B. en W.*, from which it can be deduced that further consultation has not yet ended.

On February 13, 1946, *Het Parool* published its last commentary on the Van der Waerden affair [Het5]. From it we can understand which data Van der Waerden supplied to Prof. Clay for the Alderman Albertus de Roos (recall Van der Waerden mentioning this data in his January 22, 1946 letter to Prof. Freudenthal):

Concerning Van der Waerden

The city council has circulated a little piece of advertising for the benefit of Prof. Van der Waerden, of which the main points are that he protested against the firing of the Jews in 1934 (even though he himself continued teaching classes) and that during the war, with the exception of a family visit in November 1942, he was not allowed to leave Leipzig, while, the little piece says, at that moment "going into hiding was out of the question", so that it could not be expected of Van der Waerden to "go under", even less so because he would have had to leave [his] wife and children in Germany.

This writing makes us slightly nauseous. November 1942! Pieter 't Hoen[281] has been in prison for eleven months, Wiardi Beckman[282] is in prison, Koos Vorrink[283] is in hiding, indeed all *Parool* people are in hiding; the *O.D.*[284] trial is over [resulting in] 70 people shot. The entire *O.D.* leadership is in hiding. All *Vrij-Nederland* people and those of *De Geus*, and *Je Maintiendrai*, and *Trouw*, and *De Waarheid* are in hiding.[285] In hiding, leaving behind wives and children! No, the little piece of advertising says "going into hiding was out of the question." And then the explosion comes: "...and there was also no clear resistance [to the Germans] yet"!!! See above, reader! November 1942. Hundreds have been shot for the resistance. Thousands are in camps. Other thousands have gone under. The illegal press flourishes (Parool 15,000 copies!). "No, there was no clear resistance yet," the writer of the little piece of advertising says.

---

[281] Pieter 't Hoen was the pseudonym of the Amsterdam journalist Frans Johannes Goedhart (1904–1990), the founder of *Het Parool*, who was arrested in January 1942. [Kei] reports that "Goedhart was one of the twenty-three suspects to be brought to trial before the German magistrate in the first *Parool* trial in December 1942. Seventeen death sentences were pronounced and thirteen *Parool* workers were executed by firing squad in February 1943. Goedhart managed to obtain a reprieve. He escaped in September 1943 and resumed his position on the editorial board."

[282] Herman Bernard Wiardi Beckman, (1904-Dachau, March 15, 1945), a member of the Editorial Board of *Het Parool*, one of the intellectuals of the *SDAP* (*De Sociaal-Democratische Arbeiders Partij*), arrested in January 1942, he ended his life in the Nazi concentration camp Dachau.

[283] Jacobus Jan (Koos) Vorrink (1891–1955), a member of the Editorial Board of *Het Parool*, chairman of *SDAP* (*De Sociaal-Democratische Arbeiders Partij*) and later of *PvdA* (*De Partij van de Arbeid*, labor party), was arrested on April 1, 1943, and later sent to the Nazi Concentration Camp Sachsenhausen, from which he was liberated by the Soviet Army in 1945.

[284] *"O.D."* stands for *Orde Dienst*, a national resistance organization.

[285] *Vrij-Nederland, De Geus, Je Maintiendrai, Trouw*, and *De Waarheid* were Dutch underground publications of the occupation period.

There was such a clear resistance that Van der Waerden was advised by his immediate environs not to return [to Germany]. He went anyway. For three more years he
taught in the enemy's country for the enemy's money. Who could stomach to suspend
an art student from the university for a few years while at the same time make Van der
Waerden a professor?

Clearly, the use of the expression "strong" anti-Nazi in Van der Waerden's reply
to *Het Parool* was treated as an exaggeration and, understandably, it backfired. Now
that Van der Waerden initiated a discussion on the pages of the student's weekly
Propria Cures, he received a reply from P. Peters, apparently a student, in the next
February 8, 1946 issue of the weekly [Pete]:

To Mr. Editor

During the last weeks there has been repeated mention in the press of the appointment of Prof. B. L. van der Waerden to a professor in the group theory of algebra
at our University. Still cloaked in the clouds of dust blown up by the return of other
professors one should be surprised by the fact that no attention has been devoted by *P.
C.* (*Propria Cures*) to the discussion of Prof. Van der Waerden.

Prof. Van der Waerden, as is well-known, taught during the entire war at the University of Leipzig.

In "*Het Parool*" he recently declared having been anti-Nazi. Be it as it may, it is
not entirely clear how to square this with his collaborative attitude, most tellingly illustrated by the fact that after the defeat of the Netherlands, as he had grown used to doing
before that time, every single day he gave *Heil Hitler* salute (*Heil Hitlergroet*) in public
at the start of his lectures to the enemy. Given the circumstances, it is hard to accept that
he continued to fulfill his function in Germany under duress; even more so because,
as was said, he was offered a professorship in the Netherlands. Subsequently, in his
defense (in "*Het Parool*") he does not discuss the voluntariness of his collaboration.

How tedious the subject of the purification might have become, let there be no
double standard.

Would it therefore be more tactful if the [City] Council, which is still contemplating
his appointment, avoids the provocation here, and that Prof. Van der Waerden remains
content with his present job (with *B.P.M.*) for now?

P. Peters

This was not an opinion of just one student: Dr. P. J. Knegtmans in his monograph
[Kne2] reported about the protest of a major student organization:

The *ASVA*[286] protested heavily against the coming of the mathematician Professor Van
der Waerden to the University of Amsterdam because he had taught throughout the
entire war at a German university.

---

[286] According to Dr. Knegtmans [Kne3], *ASVA* stands for *Algemene Studenten Vereniging Amsterdam*, a
new general student union that had emerged from the circles in the Amsterdam student resistance. During
the first postwar years it was very keen on matters involving the behavior of old and new professors during
the war.

More precisely, Knegtmans reports in an e-mail to me [Kne3] that on February 5, 1946 *ASVA* wrote a letter to *B. & W.*[287] According to Dr. Knegtmans's notes (translated by him for me), the letter said:

> Word has reached the *ASVA* that *Burgemeester & Wethouders* have proposed prof. Dr. B. L. van der Waerden as professor at the University of Amsterdam. This proposal has surprised the *ASVA*, considering the fact that during the war prof. Van der Waerden has been professor at a German university.
>
> The *ASVA* is under the impression that the *College van Herstel* also had had some doubts, before it eventually advised *Burgemeester & Wethouders* to go ahead with this proposal. However, the facts that have surfaced about Van der Waerden's behaviour during the war are so serious, that his assignment would be unacceptable for the students, as long as the results of the investigations by the *College van Herstel* remain unknown.
>
> Therefore, the *ASVA* requests to reveal the grounds on which *Burgemeester & Wethouders* think Van der Waerden is qualified for a position of professor at a Dutch university.[288]

On April 17, 1946 the *Burgemeester & Wethouders* replied[289] not to *ASVA*, but to a "de-Nazification committee," the *College van Herstel* of Amsterdam. Dr. Knegtmans's notes (translated by him for me) convey the following [Kne3]:

> *Burgemeester & Wethouders* inform the *College van Herstel* that they felt obliged to withdraw the nomination to appoint dr. B. L. van der Waerden as extra-ordinary professor in group theory and algebra that they submitted to the city council on 4-January-1946, as it turned out that the government would withhold its assent in the event of an appointment of Dr. Van der Waerden.

This is an important document. It shows that:

1. Prof. Dr. Gerardus van der Leeuw, the minister of Education, Arts and Sciences (*Onderwijs, Kunsten en Wetenschappen*), who initially did not object to Dr. Van der Waerden's appointment, changed his mind,[290] likely due to *Het Parool's* and students' inputs, and informed *B. & W.* accordingly.

---

[287] *B. & W.* stands for *Burgemeester & Wethouders*, or Mayor and (at the time six) Aldermen, or the Executive Committee of the City of Amsterdam.

[288] Archives of the *ASVA* in the International Institute for Social History in Amsterdam.

[289] The Archives of the *College van Curatoren* in the Municipal Archives of Amsterdam (*Gemeentearchief Amsterdam*).

[290] In fact, Minister Van der Leeuw telephoned the Mayor of Amsterdam de Boer and asked for information about Van der Waerden. On February 15, 1946, Mayor de Boer sent the Minister a two-page glowing report, prepared by Van der Corput and signed by the Mayor. It included a mention of *Samuel Goudsmit*, whom we will soon meet in this chapter: "Professor Goudsmit who as chair of American bureau in Paris had a task of investigating political activities of professors in Germany has told Professor Clay and Professor Michels that his investigation did not show anything against Prof. vdW. And a telegram was received by Clay from Goudsmith 'Preliminary Informations favorable." (*Het Nationaal Archief*, Den Haag, finding aid number 2.14.17, record number 73, Archive of the Ministry of Education.)

2. Contrary to Dr. Van der Waerden's statement to me [Wae25], the "de-Nazification" committee, *College van Herstel* (*CvH*) of the University of Amsterdam did investigate Van der Waerden, since the *B. & W.* letter about Van der Waerden was addressed to *CvH*. Moreover, I have now received documents of *College van Herstel en Zuivering* of Utrecht, which specifically deal with Dr. Van der Waerden's case among other matters.

3. Royal assent was required for a professorial appointment at *any* Dutch university, including the municipal University of Amsterdam. Prof. van der Waerden (as well as, apparently, his colleagues Van der Corput and Clay) had never understood this last point, for even in 1993 he told his interviewer Prof. Dold-Samplonius [Dol1] that "Amsterdam is a city university, and there the queen was unable to interfere." In fact, The University Historian of the University of Amsterdam Dr. Knegtmans advises me as follows [Kne5]:

> If Clay and Van der Corput really thought that an appointment as professor at the University of Amsterdam by the city council did not need approval by the queen, they were mistaken. It did so by law of 1876 and this procedure was not changed until sometime around 1980. However, approval by the queen did and does in fact mean approval of the minister (of Education, in this case). The queen was and is not supposed to have an opinion of her own. This [is] the minister's responsibility. It is the minister who advises the queen what to do: to give or not to give her approval. In Van der Waerden's case this meant that the then minister of Education, professor Gerardus van der Leeuw, professor of theology [as well as religions and Egyptology] at the Groningen University, who was minister in the first postwar year, withheld his approval of Van der Waerden's appointment as professor in Utrecht as well as in Amsterdam. Van der Waerden was probably not appointed in Utrecht at all, because it was Van der Leeuw who had to appoint him. He was probably only proposed as professor by the *College van Herstel* in Utrecht.

In the end, the media and students held feet of the academics and the governments to such a hot fire that the latter, convinced or not of the validity of the arguments, were so scared to err in the public eye on the serious issues raised by the press and students, that they gave up trying to place Dr. Van der Waerden in a Dutch university. Moreover, on March 13 1946, this was formalized in a letter from Dr. Gerardus J. van der Leeuw, Minister of Education, Arts and Sciences (*Onderwijs, Kunsten en Wetenschappen*) to *College van Herstel en Zuivering* of the Utrecht University:[291]

> I notify you that the Council of Ministers has decided that persons, who during the occupation years have continuously worked in Germany out of their free will, cannot now be considered for government appointments.

---

[291] This letter is a part of the documents provided to me by the Utrecht University Archives. These documents show that the Utrecht's *College van Herstel en Zuivering* was impressed by Van der Waerden retaining his Dutch citizenship while in Germany during the years 1931–1945, and thus favored Dr. Van der Waerden for the Utrecht job until this letter arrived.

The reason for the decision was the discussion of a possible appointment of Dr. B. L. van der Waerden to professor in Amsterdam.

It will be clear to you that the appointment of dr. Van der Waerden either in Amsterdam or in Utrecht cannot take place.

The Minister of Education, Arts and Sciences

Signed for the Minister by Secretary-General H. J. Reinink

Amazingly, Van der Waerden's individual case prompted the government of the Netherlands to pass a new order, banning *all* "persons, who during the occupation years have continuously worked in Germany" from *all* of the government jobs!

I was unable to find Van der Corput's opinion about the *Het Parool* Affair, but I have just found the second best thing: the opinion of the second major supporter of Van der Waerden at Amsterdam, Prof. Jacob Clay. Clay wrote to Van der Waerden as follows just 6 days after the Minister's decision, on March 19, 1946:[292]

Dear v d Waerden,

To my great regret our plan has not materialized at the last moment. The City government had already been convinced that the appointment was appropriate when the decision from the Minister came that nobody who has worked in Germany during the war, without any exceptions, for the time being would receive an appointment in public service. The response that I had prepared was not looked at, and in retrospect I am sorry that I have allowed the Alderman[293] to keep me from responding to *Het Parool*. When so much time has passed, it seems better not to bring these things up again. I now hope very strongly that we will receive a better collaboration for the Mathematical Centre and that in time this matter will still work out OK, and I do not doubt that this is going to happen in time.

39.2 Job History 1945–1947

Upon his return to the Netherlands in late June 1945, Dr. Van der Waerden needed a job as soon as possible. His friend, Hans Freudenthal, came through. He introduced Van der Waerden to *Bataafsche Petroleum Maatschappij* (*B.P.M.*), today known as Royal Dutch Shell, and on October 1, 1945 Van der Waerden got his first post-W.W.II job as an analyst for *B.P.M.*. In 1994 Van der Waerden recalled [Dol1]:

One day Freudenthal called me and wanted me to come to Amsterdam to talk. I went to Amsterdam, and Freudenthal told me that he was able to find a position for me at Shell. "Would you accept it?" Yes, of course; I accepted it most willingly.

Yet, Mrs. Van der Waerden was clearly bitter. We see it even half a century later in this 1994 interview [Dol1] which continues with her words:

So we were saved. I have always said that they can take everything away from us but our intellect.

---

[292] ETH Hs652: 10646.

[293] Clay here clearly refers to Albertus de Roos, the Alderman for Education and Arts.

Who are "they"? Who was taking "everything away" from the Van der Waerdens, the Dutch people and the Queen, who refused to sign off on a university professorship?

In 1946 a group of mathematicians, lead by Professor. Van der Corput, founded the *Mathematisch Centrum, MC* for short (Mathematics Center) in Amsterdam. As *MC's* first director, Van der Corput hired Dr. Van der Waerden to a part-time (one day a week) position as the applied mathematics director of the *MC*.

At this point Zürich enters the stage in our narration. The life-long ETH-Zürich[294] Professor Beno Eckmann recalls [Eck1]:

In 1944 Speiser[295] left Zurich for Basel. Finsler[296] was promoted and became his successor in Zürich; Finsler had been associate professor of applied mathematics. So in 1944 the chair of applied mathematics became vacant. Lars Ahlfors[297] was appointed in 1945, but he left after 3 semesters. . .

Olli Lehto explains [Leh]: "Ahlfors did not stay long in Zürich; later he confessed that he did not have a good time there." Consequently Ahlfors gladly accepted an offer to return to Harvard University (where he worked 1935–1938), and he remained there for decades (1946–1977, afterward as an active Professor Emeritus). The University of Zürich upgraded Ahlfors' position (who was an extraordinary professor) to a full ordinarius and started the search for his replacement.

In a fateful coincidence, the search started on March 13, 1946, the very same day the Dutch Minister van der Leeuw announced to Utrecht the prohibition of all governmental appointments for persons with backgrounds similar to Van der Waerden's.

Dr. Heinzpeter Stucki, *Universitätsarchivar*, found only one document related to this search, which, actually, proved to be of great significance: the six-page July 15, 1946 report by *Dekan* H. Steiner to Executive authority (*Regierungsrat*) Dr. R. Briner of the Education Directorate (*Erziehungedirection*) of the Zürich Canton.[298] Steiner chose two foreign mathematicians:

Prominent mathematicians are available today for a short time, and the two world-famous mathematicians in question are: *Rolf Nevanlinna*[299] (Finland) and Prof. *van der Waerden* (Holland).

---

[294] ETH, short for the *Eidgenössische Technische Hochschule* Zürich, often called Swiss Federal Institute of Technology is one of world's premier universities and research centers.

[295] Andreas Speiser (1885–1970), a professor of mathematics at the University of Zürich (1917–1944) and then at the University of Basel.

[296] Paul Finsler (1894–1970), a professor of mathematics at the University of Zürich (1927–1959) and Honorary Professor thereafter.

[297] Lars Valerian Ahlfors (Finland, 1907–USA, 1996), a professor of mathematics at Harvard University (1946–1977), one of two first Fields Medal winners (1936).

[298] Universität Zürich, Universitätsarchiv, ALF Mathematik 1944–1946.

[299] Rolf Herman Nevanlinna (1895–1980), a professor of mathematics (1926–1946) and *Rektor* (1941–1944) at Helsinki University; professor of applied mathematics at the University of Zürich (1946–1963, Honorary Professor starting in 1949).

Dekan assessed the candidacy of Prof. Nevanlinna first. After praising his mathematical achievements, *Dekan* addressed the personality of the candidate:

> He was born on October 22, 1895 in Joensuu (Finland) and for many years he was *Rektor* of the University of Helsinki. He had to leave this position in the consequence of the political circumstances after the end of the war. Consequently, as he briefly communicated, he is ready for an appointment at Zurich. . .

This was a rather short assessment: born-rektored-forced to resign. Looking at this 15-page summary [Ste] of the 317-page biography of Rolf Nevanlinna, written by his student (Ph.D., 1949) and advocate Olli Lehto, one is compelled to quote at least some information, which should have been relevant to the neutral Switzerland just 1 year after World War II:

> In 1933 Hitler became the German *Reichskanzler*. Up to the year 1943 Nevanlinna was of the opinion that Hitler in German history can be compared to Friedrich the Great and Bismarck. . . He and other members of his family regarded the cause of the Nazi Germany as their own cause. Germany was Nevanlinna's motherland (his mother was German). . . This contributed to. . . his Nazi-friendly convictions in particular, which he expressed in a series of speeches and publications. Nevanlinna, however, has never been a member of a National Socialist party and did not held anti-Semitic positions. . .
>
> When in Finland as well as in Germany the thought arose to establish a Finnish Volunteers Battalion, Nevanlinna welcomed this idea and agreed to the deployment of volunteers unreservedly. On the demand of [*Reichsführer SS*] Himmler there was developed the [Finnish] *SS* Battalion, and in the summer of 1942 Nevanlinna became the Chairman of the *SS* Volunteers Committee of this Battalion!

Prof. Nevanlinna was the first choice of the University of Zürich. *Dekan* Steiner then moved on to his second choice, Dr. Van der Waerden. Steiner admitted that

> since he [Van der Waerden] became politically strongly disputed in Holland, the real state of affairs had to be clarified.

Dekan then quoted a clarification supplied by the Dutch mathematician Jan A. Schouten,[300] who at that time lived in seclusion in Epe, The Netherlands:

> Herr van der Waerden. . . remained during the war in Germany, to which he did not have any military obligation, and he always behaved there as an enemy of Nazism and in particular did much good for the Jews. The State Commission for Coordination of Higher Education, which has been established here after the war, and of which I have the honor to be a member, would have liked to have Herr van der Waerden in Amsterdam or Utrecht. After putting him through the test, the "Cleansing Commission" found him pure, and the Minister of Education was ready to appoint him. Then a Jewish brother-in-law of Herr v. d. Waerden, who disliked him and particularly his (German) wife for already a long time, unleashed a terribly dirty (*hundsgemeine*) agitation in the

---

[300] Jan Arnoldus Schouten (1883–1971), from a well-known wealthy family of shipbuilders, a professor of mathematics and mechanics at the Delft Technical University (1914–1943), extraordinary professor (without teaching) of mathematics at Amsterdam University (1948–1953). Schouten was President of the 1954 International Congress of Mathematicians at Amsterdam.

press. The Minister, who is no strong personality and who already had heavy unpleas-
antness with others such agitations, has thereupon allowed to intimidate himself. They
[the Dutch] cannot imagine themselves at all, what unhealthy conditions prevail here,
dirty malicious agitation with self-interest and political purposes, stemming from the
agenda many times born from craving for revenge. . .

   Our principal purpose was to keep Herr v. d. Waerden provisionally for Holland, so
that as soon as the wave of hate and suspicion passes, to give him Ordinarius Professor,
as to a great mathematician.

These harsh words, directed at the recently liberated Netherlands, were intended
to make Dr. Van der Waerden appear as a victim of extremism. It must be said
that Dr. Schouten peddled gossip to the Swiss: Van der Waerden had no sisters and
thus had no brother-in-law, Jewish or otherwise. Regardless, so many Jews recently
had been killed, that it was in poor taste to blame a Jew for Van der Waerden's
employment difficulties. But to claim that one ordinary person, Jewish or not, was
able to "unleash a terribly dirty agitation in the press" meant to take Zürich Faculty
for fools. Unbelievably, *Dekan* had accepted Dr. Schouten's words as the truth, and
concluded Van der Waerden's political evaluation with

> No reason is thus present to refrain from a possible appointment of Herr v. d. Waerden
> in Zurich.

Thus, two top choices, two world-class mathematicians, two individuals whose
political and moral choices were questioned during the immediate post-W.W.II time,
ended up at the top of the Swiss wish list. Nevanlinna was chosen for the position.
Prof. Beno Eckmann has summed up this succession as follows [Eck3]:

> If I may make a remark as I see it today [in 2004]: Politically Nevanlinna and vdW
> [Van der Waerden] were not easy cases for Switzerland one year after the war. But Uni-
> versities tried to forget the past and look into the future. The decision for Nevanlinna
> must have been mathematical: he was absolutely world famous and at that time many
> mathematicians still considered analysis to be the most important part of mathemat-
> ics – this has changed soon, algebra and topology became more and more important.

Indeed, this affair showed that the Swiss neutrality was a pragmatic rather than a
moral choice, façade rather than substance. Later *Dekan* Boesch would write about
this search as follows:[301]

> It is explicit from the Faculty proposal for filling a new position of Professor of Applied
> Mathematics dated July 15, 1946, that Prof. Van der Waerden was thoroughly considered.

Indeed, Dr. Van der Waerden was thoroughly considered, and the interest in
hiring Van der Waerden was high. In a few years this 1946 consideration would
bear fruit. Meanwhile, Van der Waerden continued his full-time work at *Bataafsche
Petroleum Maatschappij* (*B.P.M.*), and part-time work at the *Mathematisch Centrum*
(Mathematics Center).

---

[301] *Dekan* Hans Boesch to Education Directorate [*Erziehungedirection*] of the Canton of Zürich, June 9,
1950; *Universität Zürich, Universitätsarchiv, Lehrstuhlakten Mathematik.*

39.3 "America! America!"[302]

Van der Waerden surely wanted a university professorship—he held one since the tender age of 25. As we know from his letters to Lefschetz, Veblen, Neugebauer, and Courant, his first choice was an academic job in the United States. In early 1947, Dr. Van der Waerden received a letter from Baltimore, Maryland that offered him both a university professorship and an opportunity to live in America. Frank Murnaghan,[303] the Johns Hopkins University's chair of mathematics, offered Van der Waerden the position of a Visiting Professor. In his May 5, 1947 letter, Van der Waerden informed Johns Hopkins' President Isaiah Bowman of his acceptance "with much pleasure."[304] Coincidentally, on the same day, May 5, 1947, the Board of Trustees of Johns Hopkins approved the appointment. From their minutes we learn that the appointment was effective July 1, 1947 to June 30, 1948.[305] On May 13, 1947 the Provost Stewart Macaulay specified Prof. Van der Waerden's salary as $6,500 for the year.[306] The Van der Waerdens—Bartel, Camilla, Helga, Ilse, and Hans Erik—boarded the ship called *Veendam*, which arrived in Port New York on September 29 or 30, 1947.[307]

At Johns Hopkins, Professor Van der Waerden was well respected, and was offered a permanent professorship. This offer was made suddenly, and appeared to have been the result of an unspecified "emergency," as it was called in a number of documents,[308] which happened at Johns Hopkins University in the early February 1948.[309] On February 6, 1948 President Bowman swiftly formed a special committee and wrote to its members the following letter:

> An emergency has arisen in the Department of Mathematics that calls for early action on an appointment recommended by both Dr. Murnaghan and Dr. Wintner[310]. The candidate is Dr. van der Waerden...You have received telephone notice of an Academic Council meeting at 8:30 a.m. on Monday, February 9, in Room 315 Gilman Hall. You will want to study the enclosed material on Professor van der Waerden before the meeting.

---

[302] From *America the Beautiful*, a song by Katharine Lee Bates.

[303] Francis Dominic Murnaghan (1893–1976), mathematics chair at Johns Hopkins University (1928–1948).

[304] Johns Hopkins University (JHU), The Milton S. Eisenhower Library, Record Group 01.001 Board of Trustees, Series 2, Minutes, May 5, 1947.

[305] Ibid.

[306] Ibid.

[307] Ibid.

[308] JHU, Record Group 01.001 Board of Trustees, Series 2, Minutes, 2/9/1948.

[309] J. J. O'Connor and E. F. Robertson write as follows in The MacTutor History of Mathematics archive: "He [Murnaghan] held this post until 1948 when he retired after a disagreement with the President of Johns Hopkins University, and went to Sao Paulo, Brazil" (http://www-history.mcs.st-andrews.ac.uk/Mathematicians/Murnaghan.html). Did this disagreement take place? If so, was this chair's departure (chair did depart) the "emergency" that prompted such a rush in making Prof. van der Waerden this offer? I was unable to confirm it.

[310] Aurel Friedrich Wintner (Budapest, 1903- Baltimore, 1958), one of the leading mathematics professors at Johns Hopkins University (1930–1958).

This was a short notice indeed. The next day, on Saturday (!), on February 7, 1948, the special committee, chaired by the chemist Alsoph H. Corwin, unanimously approved the mathematics department's recommendation without the usual external letters of reference. On Monday, February 9, 1948, the Academic Council, also chaired by Prof. Corwin, at its special 20-minute meeting (8:30 A.M. to 8:50 A.M.) "voted to suspend its hold-over rule and unanimously recommend the appointment of Dr. van der Waerden" to the President. That same day (!), the Board of Trustees approved the appointment of Prof. van der Waerden to a Full Professorship that paid "$8,000 first year; $9,000 second year; and $10,000 third year."[311] Amazingly, Van der Waerden turned down this offer and chose to return to the Netherlands. Instead of himself he recommended his former Leipzig Ph. D. student (Ph. D., 1936) and coauthor Wei-Liang Chow for the position. Chow would indeed be hired the following year, and would serve as Professor at Johns Hopkins for nearly three decades (1949–1977), including over ten years as the chair.

In 1945, Van der Waerden wanted badly to come and live in America. He got such an opportunity in 1947. Why then in 1948 did he decide to reject a prestigious, well-paid professorship at Johns Hopkins and leave America? He returned to Amsterdam, where, rightly or wrongly, he was not treated particularly warmly during 1945–1947. Was his treatment in the United States worse? I tried—and failed—to find answers in the Archives of Johns Hopkins University. The investigative thread seemed to have run into the dead end.

The time had passed. One day in my office I glanced at the many books on the shelves, and picked one to read at home. It happened to be *Heisenberg's War: The Secret History of the German Bomb* by the Pulitzer Prize winner Thomas Powers. It was a great read; moreover, Van der Waerden made a cameo appearance on the pages of the book. So far no surprises: as we know from chapter 37, Van der Waerden was Heisenberg's friend at Leipzig and attended Heisenberg and Hund's seminar on quantum mechanics. However, here Van der Waerden appeared as Heisenberg's American pen pal in 1947–1948. The letters were quoted from the 1987 Princeton-History Ph. D. thesis of Mark Walker, defended under the supervision of my dear Princeton friend and the founder of the history of science program Charles Gillispie. I was intrigued—and telephoned Tom Powers at his Vermont country home. Powers led me to Walker—Walker sent me copies of the Heisenberg–Van der Waerden correspondence. The answers to my questions were hidden in these letters!

Yes, the surprising answers were hidden in the Werner Heisenberg Archive in Munich, in the unpublished December 22, 1947 letter from Van der Waerden to the 1932 Nobel Laureate and his friend Werner Heisenberg. I read in excitement and disbelief:[312]

---

[311] JHU, Record Group 01.001 Board of Trustees, Series 2, Minutes, February 9, 1948.

[312] Van der Waerden's letter to Heisenberg, December 22, 1947, Private Papers of Werner Heisenberg, Max Plank Institute for Physics, Munich. I am most grateful to Prof. Mark Walker for sharing with me the 1947–1948 correspondence between Van der Waerden and Heisenberg, and Van der Waerden and Goudsmit, as well as Heisenberg's unpublished work *Die aktive und die passive Opposition im*

Dear Herr Heisenberg,

On the 9th of October I have sent you a care-package, write to me please if it has arrived and how you are doing with groceries. I would be very glad to send you more next year. I am still in your debt: in the past when I was arrested, you helped me to something much greater, and that is freedom.

I need your advice: you are a reasonable man and at the beginning of this war, you predicted who in the end would be the victor. I think I will receive an offer to be a professor in Baltimore, and then I must decide either in favor of Baltimore Johns Hopkins or Holland. In Holland, I would do for the most part applied mathematics and I would train applied mathematicians at the newly founded Math. Centrum and at my oil company. I like this work very well and my work at Johns Hopkins I like too, so this [aspect] is equal. The people here are unbelievably nice and helpful: you know that. Nevertheless, I would rather stay in Europe: I love Old Europe and so does my wife.

Thus, Van der Waerden liked his job at Johns Hopkins and considered American people to be "unbelievably nice and helpful." Yet, Bartel and Camilla van der Waerden preferred "Old Europe." Fair enough, one can relate to that. However, his main concern about living in Baltimore popped up in the next paragraph:

Now my question: how do you judge the prospects for war, and how do you judge the question whether one could better secure one's family in America or Holland if the insanity would break out? The people here and in Europe are telling us that it is crazy, that it is insanity, and that if you have a possibility to stay in America, it is insanity to go back to Holland. Personally I do not believe there will be war, but if it nonetheless should come, then an American big city does not seem to me to be the most secure place in the world, but in the past I have been very mistaken in similar cases and do not want to have a responsibility on my shoulders for leading my wife and children to ruin. You understand more about nuclear physics than I do; what do you think about this?

Here I have spoken with different people, and gotten a definite impression that America would never start a war on its own, which has set me to rest.

Van der Waerden was afraid that in a large American city—Baltimore—his wife and children would be in a real danger of a Russian atomic bomb attack! This may sound irrational to us looking from today to the year 1947. However, I, recall similar fears experienced by Van der Waerden's successor at the University of Amsterdam, Prof. N.G. de Bruijn, who wrote to me about it in his June 1, 2004 e-mail [Bru9]:

...in 1952 I got a professorship in Amsterdam and...I preferred not to live in town but in a village 20 kilometers to the east of it. Nobody would believe now that one reason I had at that time was that in a Russian atomic attack my family would be pretty safe at that distance. A few years later atomic bombs would be hundred times as strong as the Hiroshima type, so the whole argument became utterly silly.

Dritten Reich used in this chapter. Walker was *first* to discover and use them in his research, dissertation [Wal1] and the book [Wal2]. His main interest was the physicists Heisenberg and Goudsmit and their debate; mine is Van der Waerden, hence I am quoting somewhat different passages from these important materials, and offer my own analysis of them. I also thank Dr. Helmut Rechenberg, Heisenberg's last Ph. D. student and Director of the Werner Heisenberg Archive for the permission to reproduce these materials.

Van der Waerden concluded his December 22, 1947 letter to Heisenberg with the hope that Germany would be rebuilt and they would once again work *there* together:

> They [Americans] even see in all seriousness a desire to support the reconstruction of Germany, which I am very happy about. [Richard] Courant thinks that because of the Marshall plan, in some years Germany would once again reach the heights. Maybe we will get together again!

In the March 16, 1948 letter, Van der Waerden informed Heisenberg that "In principle, I have accepted the job offer from [the University of] Amsterdam." Before we follow Professor Van der Waerden to Amsterdam, we will briefly visit a "letteral triangle," which could be a subject of a wholly separate book.

39.4 Van der Waerden, Goudsmit and Heisenberg: A 'Letteral Triangle'

Over the last 2 years of the World War II, Dr. Samuel A. Goudsmit,[313] an American nuclear physicist born in the Netherlands, served as the Chief of Scientific Intelligence of the U.S. War Department's *Alsos Missions*, dedicated to gathering information about the German nuclear program, capturing its materials, equipment and records, and capturing and interrogating its leading scientists. In his 1947 book, entitled *Alsos* [Gou] Goudsmit attributed the German fiasco in building the atomic bomb to the treatment of science in the totalitarian Nazi state and scientific blunders of Werner Heisenberg and other scientists, rather than to Heisenberg's concerns for the fate of the humanity. The book prompted public and private debate between the two old friends, Goudsmit and Heisenberg. Much more about the *Alsos Mission* and the debate can be found in [Pow] and [Wal1] respectively. Of course, we have rich eyewitness accounts written by Alsos's major players in [Gou], [Pash], and [Grö].

Upon reading *Alsos*, on March 17, 1948, Van der Waerden wrote a letter to Samuel A. Goudsmit that opened with high praise:

> With great interest I have read *Alsos*. It has kept me in tension during half of the night. Your picture of characters is excellent: by a few strokes men like Bothe, Weiszäcker, Mentzel, Osenberg are drawn down to their feet. Also the main actor has been well approached: the somewhat mysterious character of W. H. [Werner Heisenberg] has now become clearer to me in several respects. What you write about the causes of the German failure [to produce atomic bomb], about self-overestimation and clique-mentality is well motivated and certainly correct.[314]

He then posed a number of important questions:

---

[313] Samuel Abraham Goudsmit (1902, Den Haag-1978, USA); Max Planck Medal, 1964.

[314] I left unedited the Dutch into English translations of this letter and the following Goudsmit's reply, because these translations were made by Van der Waerden himself for Dr. Mark Walker, who has kindly shared them with me.

You write, "The bomb is what they were after." How do you know that, or rather what do you mean exactly? Do you mean that these people, knowing who Hitler is, planned the horrible crime to give into his hands an atomic bomb? If this is what you mean, what proofs do you have for this horrible accusation? As far as I can see it is only the document on page 178[315]...

Goudsmit replied in March (no date), 1948:[316]

I don't agree with you that it is a crime that they worked on it. It's a thing you cannot stop. It is a kind of scientific triumph, of which you realize the consequences only when it is too late. If they really had succeeded, I am firmly convinced that von Laue would have done his very best to prevent its use. But it would have been in vain. The same thing happened here [in the US]. Before the bomb was used, several colleagues have issued a petition not to use the bomb.

Goudsmit then sited a number of documents in his book *Alsos* and outside of it that showed that the German scientists worked in the direction that led to an atomic bomb, and advised the Nazi authorities accordingly. The main reasons for failure, in his opinion, were scientific errors of the German scientists:

... they did not understand that it was possible to make a small bomb.

In his March 16, 1948 letter to Heisenberg, Van der Waerden copied the words he wrote to Goudsmit, and assured his friend of the Leipzig years Heisenberg of his support:[317]

Since coming here I have tried in a cautious way to defend you.

Following a long conversation with Goudsmit, Van der Waerden wrote two more letters to Heisenberg. In studying Van der Waerden, I clearly see that he had always valued the character of a person (himself included) more than the person's deeds. Likewise, he advised Heisenberg in the first April 19, 1948 letter:[318]

Questions like the one about the *complacency* [English word used] of the German physicists and about things you and your friends failed to see – questions like these lose importance in my eyes, compared to the much more important ones, whether your character [sic] is to be criticized, and whether one can and should work with you together.

The following day, Van der Waerden, like a good defense attorney, decided to teach his friend Heisenberg how to defend himself by asking him a series of leading questions that contained answers desired by Van der Waerden:[319]

---

[315] A secret Gestapo summary, dated May 1943, enumerating two applications of uranium fission: the Uranium Engine, and the Uranium Bomb.

[316] Private Papers of Werner Heisenberg, Max Plank Institute for Physics, Munich.

[317] Ibid.

[318] Ibid.

[319] Ibid.

G. [Goudsmit] thinks that if you and your group found plutonium, you would have decided to make the bomb. Afterwards many of you would have tried to prevent the use but it would have been in vain. I have held on to the possibility that you would have stopped this thing. For support, I have pointed to your sentence: "We always thought to keep this thing in our hands." We have then got to the legal question of not to condemning *in dubio pro reo* [Latin: the presumption of innocence].

But personally I would like to know your answer to this question. Surely you have considered it. When you wrote to higher authorities about possible [atomic] explosives, was all that only a pretense in order to get money for physics? Did you firmly decide to never in any circumstances to let it go that far? Then everything would have been in order; in regards to these people every deceit would be permitted. Or?

You understand what I mean. As your attorney, I have enough facts to defend you. But as your friend I would so terribly like to believe that under all circumstances your decency would have been stronger than your nationalism plus ambition. Can you give this belief a support? Have you had any conversations with trusted persons that could give me a place to begin? And what does Hahn think about this question?

Indeed, was Heisenberg's decency, under all circumstances including the Nazi regime, stronger than Heisenberg's enormous nationalism and ambition? In his heart of hearts, Van der Waerden probably knew the answer but did not wish to believe it. He was determined to continue his work of defending Heisenberg "in cautious way." However, some doubt can be seen in Van der Waerden's next, April 28, 1948 letter to Heisenberg (even though he always publicly defended Heisenberg). Apparently, in the non-surviving April 1, 1948 letter, Heisenberg approved Van der Waerden as his (unofficial, of course) defense attorney, and advised Van der Waerden to start his defense by the invocation of the Nazi "atmosphere." Van der Waerden replied:[320]

Unfortunately, I cannot begin with "atmosphere." It is so inconceivable, everyone feels the atmosphere differently! What I need is concrete statements, decisions, conversations, and so on, which you have had.

Also even the mere denial "This statement has naturally never been made in this way," would be useful for me. You are supposed to have said "How nice it would have been if we had won." That is allegedly the literal statement. Can you remember what you said, if not this? Or, did you mean something different by that?

Of course, you are right, that in the questions of "German" Physics you have achieved a real success and of course it is *irrelevant* to hold it against you. Nevertheless, the reaction of the others is not inconceivable. It is not logical I admit. However, emotionally it is conceivable. Do you still remember what I said to you when you gave me to read an article in the *Schwarzen Corps*? That is a nice title: White Jew, you can be proud of that. Instead of being proud, you were angry about the article. Of course, you were right that in the interests of physics you have acted as you did in the connection with [Niels] Bohr. However, on the other side, could you have contact with these people, exercise influence over them without compromising yourself? I assume yes, but I can understand if others do not believe it.

---

[320] Ibid.

Only one Heisenberg's reply, apparently, survives that of April 28, 1948. Heisenberg warns Van der Waerden that:

> ... because every letter from Germany is read by the censor, and therefore particularly when it has to do with the matter of atomic bomb, and is finally somehow made public, I must write to you more briefly and "more officially" than I would like to.[321]

Heisenberg then repeats his, now well-known and well-contested, explanations of the German failure to produce an atomic bomb. He even goes as far as to insinuate his belief in the moral superiority of the German physicists over the Allied:

> You want to know basically my human position to this question. At the beginning of the war when I was drawn in the work on uranium, at first I found out...what was possible in this area. When I (end of 1941) knew that the uranium pile would work and that one probably would be able to make atomic bombs (...I thought the effort would be still larger than it in fact was), I was deeply horrified by the possibility that one could give such weapons to any person in power (not only Hitler).
>
> When in the fall of 1941 I spoke with Niels Bohr in Copenhagen, I directed this question to him whether physicist had the moral right to work on atomic problems during war. Bohr asked back whether I believed that the military application of atomic energy was possible, and I answered: yes, I knew it. I then repeated my question and Bohr answered to my surprise that the military involvement of physicists is inevitable in all countries, and therefore it was also justified...
>
> When at the beginning of 1942 in Germany the official discussions about the uranium problems began, I was very happy about it that the decision had been taken from us. The Führer's orders prevented large efforts for atomic bombs. Besides irrespective of that, it was clear that atomic bombs in Germany would never be completed during the war. I would have regarded it in any case a crime to make atomic bombs for Hitler. But I do not find it good that the atom bomb was given others in authority, in power, and was used by them. On the other hand, I have also learned something from the past years that my friends in the West do not really want to see, that in times like these almost no one could avoid committing crimes or supporting crimes by inaction, whether it is the Germans, the Russians or the side of the Anglo-Saxons...
>
> 1) P.S., reading this letter I see that the last sentence could be misunderstood in two ways. First, one could think that I wanted to designate Oppenheimer or Fermi as criminals or one can assume that under certain circumstances I would have been ready to commit various crimes "for Hitler." I hope you know me well enough to know that both of these are not intended.

Heisenberg seems more sincere in the following passage of his *New York Times* interview (in English):[322]

> German sciences sank to a low ebb. I think I am safe in saying that, because of their sense of decency most leading scientists [in the Nazi Germany] disliked the totalitarian system. Yet as patriots who loved their country they could not refuse to work for the Government when called upon.

---

[321] Ibid.

[322] Kaempffert, W., "Nazis Spurned Idea of an Atomic Bomb," *New York Times*, Dec. 28, 1948, pg. 10.

These words explain the rationale for Heisenberg's choices. He subscribes to the widely shared, but false notion of patriotism, according to which in times of war a true patriot has to rally behind his government, even if the government is engaging in ostensibly criminal activities.

Moreover, Heisenberg apparently considers himself to be in "active opposition" to the Nazi regime. Prof. Mark Walker found and first discussed [Wal1, pp. 335–338] an amazing unpublished November 12, 1947 Heisenberg's 4-page paper *Die aktive und die passive Opposition im Dritten Reich*.[323] I agree with Walker's analysis that while on the surface Heisenberg refers to the "active opposition" of the second highest ranked diplomat of the Third Reich Ernst von Weizsäcker (the father of Heisen-berg's friend and collaborator Kare Friedrich von Weiszäcker), he counts himself among the active oppositionists. Werner Heisenberg must have used all of his vast ingenuity to present a Nazi collaboration as "the active resistance" to the Nazies! Those who were thrown out of the Nazi Germany, he labels as being in "passive [read: worthless] opposition" (he even insinuates that they *chose* the exile). Those who actively fought the regime he believes "did not understand the stability of a modern dictatorship, tried the path of open, immediate resistance during the first years and ended up in a concentration camp [read: worthless]." Heisenberg then describes the worthy and morally noble behavior in the Third Reich:

> For the others who recognized the hopelessness of this way, there remained another way, the attainment of a certain degree of influence, i.e., the attitude that had to appear on the outside like collaboration [with the Nazi regime]. It is important to be clear that this was in fact the only way really to change anything. This attitude that alone had contained the prospect of replacing National Socialism with something better but without enormous sacrifices, I would like to designate as the attitude of active opposition [sic].
>
> On the outside the position of these people was much more difficult than that of the others. Remember, the active opposition had to repeatedly make concessions to the system on unimportant points in order to possess the influence to improve things on important points. In a certain sense he had to play a double game. The unavoidable difficult moral problem that was put before the member of the active opposition one can understand by means of the following constructed case, to which the reality may well have come close some times.

Heisenberg then illustrates his moral position with a hypothetical example. He finds it acceptable—moreover, highly moral—to prove loyalty to the Nazis by signing a death sentence for an innocent person, if this allows saving other lives:

> Let us assume that a man wishing to save human life comes into a position where he really can decide about life and death of other people. And further let us assume, and this in a really evil system as National Socialism is thoroughly thinkable that he can only prevent the execution of 10 innocent people by means of signing a death sentence for another innocent person. He knows that the 10 others will be executed

[323] I thank Prof. Walker for sharing with me the text of this document. I also thank Dr. Helmut Rechenberg and Werner Heisenberg Archive he directs, for permitting to quote this document and Heisenberg's correspondence with Van der Waerden and Goudsmit.

through the action of someone who will be put in his place if he does not sign the death sentence. The fate of the one is in any case sealed, no matter whether he signs or not, nothing is changed. So how should he act? Personally I believe that after a conscientious reflection, that in such a case the signing a death sentence is demanded of us, which entails of course that we are prepared to bear the consequences of that personally. To measure this by the ultimate moral standards, it seems to me that a person who acts and thinks in this manner stands higher than the one who simply says, I do not want anything to do with all of this.

It seems as if the theoretical physicist Heisenberg justifies collaboration with the Nazi regime and murder of an innocent person by a simple arithmetic calculation $10 - 1 = 9$. Human life, in my opinion, carries infinite value, and if Heisenberg were to understand that, his arithmetic would have given an uncertain result: $10 \times \infty - 1 \times \infty$. Heisenberg's allegory is a masterpiece of hypocrisy, in which he elevates Nazi collaborators to heroes of active resistance, and denigrates into dummies the real heroes, who fought the regime and paid the high price for it. The reader may recall (Chapter 37) that Heisenberg sought and received protection personally from the *SS Reichsführer* Himmler. Having attracted the high personal attention and patronage of Himmler, Heisenberg could hardly allow himself as much as a whisper of an opposition to the Nazi regime—but here, after the war, he insinuates that he was a hero of "active resistance"!

I have got to quote here a passionate letter that the co-discoverer of nuclear fission Lise Meitner wrote in late June 1945 to her coauthor Nobel Laureate Otto Hahn. She addresses Hahn, Heisenberg, and other scientists who collaborated with the Third Reich, and without even reading Heisenberg's manuscript (which Heisenberg wrote 2 years later), she powerfully rebuts Heisenberg's pretense of any resistance, even a passive one [LS, p. 310]:

> You all worked for Nazi Germany and you did not even try passive [sic] resistance. Granted, to absolve your consciences you helped some oppressed person here and there, but millions of innocent people were murdered and there was no protest. I must write this to you, as so much depends upon your understanding of what you have permitted to take place. Here in neutral Sweden, long before the end of the war, there was discussion of what should be done with German scholars when the war was over. What then must the English and the Americans be thinking? I and many others are of the opinion that one path for you would be to deliver an open statement that you are aware that through your passivity you share responsibility for what has happened, and that you have the need to work for whatever can be done to make amends. But many think it is too late for that. These people say that you first betrayed your friends, then your men and your children in that you let them give their lives in a criminal war, and finally you betrayed Germany itself, because even when the war was completely hopeless, you never once spoke out against the meaningless destruction of Germany. That sounds pitiless, but nevertheless I believe that the reason that I write this to you is true friendship. You really cannot expect that the rest of the world feels sympathy for Germany. In the last few days one has heard of the unbelievably gruesome things in the concentration camps; it overwhelms everything one previously feared. When I heard on English radio a very detailed report by the English and Americans about Belsen and Buchenwald, I began to cry out loud and lay awake all night. And if you

had seen all those people who were brought here from the camps. One should take a man like Heisenberg and millions like him, and force them to look at these camps and the martyred people.

As to Heisenberg's concept of moral superiority of the German physicists over the Allied scientists, it was best refuted by Prof. Philip Morrison of Cornell University in his December 1947 review [Morr] of Goudsmit's *Alsos*:

> The documents cited in *Alsos* prove amply that, no different from their Allied counterparts, the German scientists worked for the military as best their circumstances allowed. But the difference, which it will never be possible to forgive, it that they worked for the cause of Himmler and Auschwitz, for the burners of books, and the takers of hostages. The community of science will be long delayed in welcoming the armorers of the Nazis, even if their work was not successful.[324]

Regretfully, Morrison's latter prediction had not materialized. Very soon, in 1950—and again in 1954—Heisenberg was invited for V.I.P.[325] lecture tours to the United States. On May 14, 1958, he was made a Foreign Honorary Member of the American Academy of Arts and Sciences. After the war, Heisenberg could have even been offered a job in the U.S., as were many of the Third Reich's scientists and engineers. America was acquiring ammunition for the Cold War and was paying a high moral price for it.

Decades later, in 1989, Delia Meth-Cohn showed the pages of Mark Walker's dissertation [Wal1] with the Van der Waerden—Heisenberg correspondence to Dr. and Mrs. Van der Waerden during the interview in their Zürich home. She recorded their reaction as follows:

> [Van der Waerden] was quite shocked to see the pages from his [Walker's] dissertation with the letters. He had no idea that these letters still existed – and his wife copied the pages, almost in tears at how wonderful the letters were that her husband had written to Heisenberg.[326]

---

[324] I wish to note here that deplorably, the high moral authority of the Nazi years' Germany, the Nobel Laureate Max von Laue, added his insult to the Nazi injury of Goudsmit when he wrote, "We do know that Goudsmit lost not only father and mother, but many near relatives as well, in Auschwitz and other concentration camps. We realize fully what unutterable pain the mere word Auschwitz must always evoke in him. But for that very reason one can recognize neither him, nor his reviewer Morrison, as capable of an unbiased judgment of the particular circumstances of the present case" (*Bulletin of the Atomic Scientists* 4(4), 1948, p. 103). Morrison was absolutely right in his reply: "I am of the opinion that it is not Professor Goudsmit who cannot be unbiased, not he, who most surely should feel an unutterable pain when the word Auschwitz is mentioned, but many a famous German physicist in Göttingen today [Heisenberg], many a man of insight and responsibility, who could live for a decade in the Third Reich, and never once risk his position of comfort and authority in real opposition to the men who could build that infamous place of death" (*Bulletin of the Atomic Scientists* 4(4), 1948, p. 104).

[325] Abbreviation for "Very Important Person."

[326] Meth-Cohn, D., Manuscript of Feb. 21, 1989, Zurich interview with B. L. van der Waerden, courtesy of Mr. Thomas Powers, the author of the best-selling book [Pow], for whom Ms. Meth-Cohn conducted this, mostly unpublished interview.

39.5 Professorship at Amsterdam

By 1948, the de-Nazification of the Netherlands was over, and *Colleges van Herstel* were gone. In addition, the American acceptance improved Dr. Van der Waerden's standing in Europe. Yet, L. E. J. Brouwer eloquently objected to Van der Waerden's appointment in his April 15, 1948 letter to the Minister of Education. Moreover, Dirk van Dalen believes that "the feelings expressed in this passage perfectly reflected the general opinion of the Dutch, and in particular the students, in the matter":[327]

> From a researcher like Professor Van der Waerden, who is only theoretically, but not experimentally active, the scientific influence is almost independent of personal presence. Thus, as soon as a materially and scientifically favourable position has been secured, the question of his presence here in the country loses all scientific and national importance, and it becomes almost exclusively a matter of national prestige. From a viewpoint of national prestige the motivation of his appointment here in the country seems however extremely weak to the undersigned. For if it is claimed that by the presence of Professor Van der Waerden in Amsterdam the strength of our nation is enhanced, the reply is forced upon us that in that case the national strength of the German empire has been enhanced during the whole period of the Hitler regime by the presence of Professor Van der Waerden in Leipzig. And if it is argued that if Professor Van der Waerden is not offered a suitable position in the Netherlands, this will be done by America, the reply is forced upon us that if at the moment there are positions open to Professor Van der Waerden in America, this should not have been less the case between 1933 and 1940, when many prominent and right-minded German scholars and artists were welcomed with open arms in America, and that therefore one has to assume that Professor Van der Waerden had not felt the desire to turn his back on the Hitler regime.

The late Prof. Herman Johan Arie Duparc (1918–2002) wrote down the following recollections for me during our September 1996 meetings in his apartment in Delft [Dup]:

> Van der Corput and others feared again difficulties. He said to me: "Tomorrow vd Waerden gives his first lecture; interesting; let us go there." So we went there. There were no difficulties.
>
> So times seemed to have changed and they could make him professor of mathematics (analysis) in 1950.
>
> Then Van der Corput and vd Waerden had a common room in Amsterdam University. When vd Corput went to the US in 1950, I had to take over his work in 1950 and met vd Waerden regularly there.

According to Duparc, in 1948 Van der Waerden was appointed as a *bijzonder* (special) professor of applied mathematics at the University of Amsterdam. This part-time ("one day a week", according to Duparc) position was paid by the Foundation, which "was just a derivative of the Mathematical Centre, with Clay and

---

[327] [Dal2], 829–830.

Van der Corput in the driver's seat,"[328] and thus did not require an approval by the Queen. This was a far cry from a tenured full professorship at Johns Hopkins University that Van der Waerden turned down, but this was a start. Plus, this time Van der Corput hired Van der Waerden as a full-time director of applied mathematics at the Amsterdam's *Mathematisch Centrum*, where Van der Waerden worked part-time in 1946–1947. Then there came the prestigious membership in the Royal Dutch Academy of Arts and Sciences, which had to be—and was—approved by the Queen. However, this, was not the same Queen Wilhelmina, who in 1946 rejected Van der Waerden and others who voluntarily worked for the German occupiers. Her daughter Queen Juliana, who took over in 1948, presided over less principled and emotional times.

After the Minister of Education indicated that he—and Queen Juliana—would have no objections, on April 19, 1950 Van der Waerden was finally appointed to an Ordinarius chair (a full professor) by the City Council of Amsterdam, effective October 1, 1950. It appeared that the relationship between Holland and her prodigal son Bartel had been restored and was likely to grow closer in time. Van der Waerden had a fine job, and was among talented and supportive colleagues. Yet he chose to accept a chair at the University of Zürich. Van der Waerden de-facto included his notice of resignation in his inaugural [sic] speech "Concerning the Space" [Wae12], given on Monday, December 4, 1950, at 4 o'clock in the afternoon at the University Auditorium:

> Eminent Clay and Van der Corput,
>   With undaunted energy you both have organizationally prepared my appointment to a Professor regardless of all difficulties and you have finally reached your goal. I appreciate this very much and will remain grateful to you forever for it. Even though now I will soon be going to Zürich, I trust that another one would take over my job on this faculty, which was organized by your ideas.

On March 21, 1951 Prof. Van der Waerden formally asked for his resignation from the University of Amsterdam, which was granted effective May 1, 1951.

Van der Corput was proven wrong: he did all he could to support Van der Waerden in academia and government; he closed his eyes on his disagreements with some of Van der Waerden's moral positions and life's choices (Chapter 38). Yet, in the end he did not win Van der Waerden for Holland for the rest of the latter's career. Nicolaas Govert de Bruijn, who in 1952 became "another one [to] take over [Van der Waerden's] job on this faculty", wrote to me about the understandable disappointment of Van der Waerden's colleagues at the University of Amsterdam [Bru8]:

> I had regular contact with some mathematicians who knew him [Van der Waerden] better than I did, like Kloosterman, Koksma, Van Dantzig, Freudenthal, Van der Corput, who were disappointed by his leave after they had gone into so much trouble to help him with jobs in the Netherlands.

And more [Bru9]:

---

[328] [Dal2], p. 827.

Actually I do not remember anything from my own experience. I only remember that people like Koksma, Van Dantzig, Schouten[329] confidentially complained that Van der Waerden disappointed them after all the trouble they had taken. I suppose they had to fight unwilling authorities in order to let them forget the objections from the past. Step by step they got him a position with the Shell company, a part-time professorship at the University of Amsterdam, the membership of the Royal Dutch Academy of Arts and Sciences (which had to be signed by the queen) and finally the full professorship. The people who all went through this trouble of course felt they lost their face with respect to all those authorities when Van der Waerden unexpectedly left them in the lurch...

As a part-time professor Van der Waerden taught applied mathematics, maybe mainly from a pure mathematician's perspective. As a full professor he had not even started; around that time he decided to leave for Zurich. So there was hardly a Van der Waerden tradition of courses in Amsterdam.

Amsterdam appears to have been used by Professor Van der Waerden merely as a stepping stone.

39.6 Escape to Neutrality

> *Mathematics has no fatherland, you say?*
> *– Het Parool* [Het1]

Prof. Beno Eckmann remembers the Zürich 1950–1951 succession as follows [Eck1]:

In 1950 Fueter[330] retired. Shortly before I was offered that position (and to be "director"). Then the position was offered to vdW [Van der Waerden] who accepted but his appointment was finalized only in 1951 (I vaguely remember that there were discussions among Zurich authorities whether it would be appropriate to appoint a man who had remained in Nazi Germany during the war).

In fact, Eckmann was the early first choice [Eck2]:

I was asked either in 1949 or early in 1950 whether I would accept (I really cannot remember when this happened – Rolf Nevanlinna talked to me personally, had I said yes I would have received that position).

The voluminous file of Rudolf Fueter's succession opens with *Dekan* of Philosophical *Facultät* II Hans Boesch's May 5, 1950 letter[331] calling the meeting of the Mathematics Commission for Monday, May 8, 1950 at 1400 hours in *Dekanat*

---

[329] In view of Prof. Schouten's 1946 (Section 39.2) and 1950 (Section 39.6) letters of reference to the Swiss on Prof. Van der Waerden's behalf, it is hard to understand *his* disappointment.

[330] Karl Rudolf Fueter (June 30, 1880–August 9, 1950), a professor of mathematics (1916–1950) and *Rektor* (1920–1922) of the University of Zürich.

[331] *Universität Zürich, Universitätsarchiv, Lehrstuhlakten Mathematik.*

room 13. A mysterious handwritten page, the stenography[332] of this meeting, would delight any professional or amateur paleographer. The Commission nominates young Swiss mathematicians, such as Nef, Häfeli, etc., but only three candidates are numbered:

1. Van der Waerden (03), Ord. Leipzig, Hollander
2. Pólya (62),[333] Stanford University
3. Eckmann (17), ETH

References, who would be asked to evaluate the above candidates, are also listed on this page:

Fueter, Speiser, Hopf, Ahlfors, ~~Erhard Schmidt~~, Schouten

At the bottom of the page the final list appears again, without the stricken Erhard Schmidt of Germany. Schouten's name is separated by a line from the other four names, for he is to be asked only about the current political opinion about Van der Waerden in the Netherlands.

The following day *Dekan* Boesch sends identical letters[334] to Van der Waerden and Pólya, inquiring whether they would like to be considered for the position of professor and director of mathematics institute in succession to the retiring Prof. Fueter. The file does not contain a similar letter to Prof. Eckmann of ETH: he has already turned down this position, for he has been quite happy at ETH, where he would later become the founder of *Forschungsinstitut für Mathematik*.

On the same day Boesch also sends letters[335] to the four official references. Van der Waerden's old correspondent on algebraic geometry (at least since 1936), Prof. Paul Finsler of mathematics department, writes to the fifth, unofficial reference, Jan A. Schouten of the Netherlands.

Shortly after letters of reference pour in. ETH Prof. Heinz Hopf recommends considering only the top three candidates:[336]

> G. Pólya is without a doubt one of the most interesting personalities among the living mathematicians.... Professors and students at a university where Pólya works, work with him, receive his instruction, and just by dealing with his personality get education, intelligence, humor and goodness in such an unusual amount. We, colleagues at the ETH, where he has been working for so long, miss him very much....
>
> B. L. van der Waerden is one of those mathematicians who in the last 25 years has been instrumental in creating a significant change in the appearance of mathematics.

---

[332] Ibid.

[333] The apparent date of birth in parentheses should have been (87), for 1887. George Pólya, a professor of mathematics at ETH (1920–1940) and Stanford University (1942–1978, including active Emeritus Professor since 1953), a brilliant mathematician and pedagogue.

[334] *Universität Zürich, Universitätsarchiv, Lehrstuhlakten Mathematik.*

[335] Ibid.

[336] Heinz Hopf's 5-page long letter to Hans Boesch of May 14, 1950; *Universität Zürich, Universitätsarchiv, Lehrstuhlakten Mathematik.*

His "modernization" is in the first line in the area of algebra, in which very clear "conceptual", "qualitative" thinking is placed in the foreground as opposed to "numerical", "quantitative" operations.... Certainly there would be nobody better than van der Waerden to found a new algebraic school at the University of Zürich....

B. Eckmann – about 30 years younger than Pólya and 15 (or something less) years younger than van der Waerden – cannot of course have as many successes and cannot yet be called in the same sense a famous mathematician as the other two I have named. But I believe that he is on his best way to secure his place under the leading mathematicians.... Many colleagues at the ETH are happy to have Eckmann amongst us, especially I personally am very happy with the fact that he was my direct student.... If he did get a call from another university, naturally we will attempt, with great energy, to keep him with us. And I believe also that he himself does not see any enjoyment in leaving the ETH.

Lars V. Ahlfors, Chairman of Mathematics Department at Harvard University, expresses an opinion similar to Hopf's:[337]

Among the Swiss mathematicians there remains certainly Prof. Eckmann the single one whom I would consider seriously....

From the foreign mathematicians I agree with you that certainly van der Waerden and Pólya should be named in the first list. One must thank van der Waerden for having strengthened algebraic geometry, even though I know that his work has been surpassed by other people. Nevertheless, van der Waerden is a first class mathematician, but it would be important to find out whether he is still on top in his knowledge. Prof. Pólya in his own individual way stands in the zenith of knowledge. He has depth and originality. He is, and I believe most mathematicians would agree with me, not a leading mathematician but instead an extraordinarily skillful (*Geschickter*) one.

Prof. R. Fueter, shockingly, has nothing positive to say about George Pólya:[338]

Prof. Dr. Pólya, during his first years in a Zürich position [at ETH] attempted to work together with us, but then in many situations worked against Speiser and myself and fought with our students. In this situation I would also like to point out some of Prof. Speiser's views regarding this.

Fueter much prefers Van der Waerden or else one of his own many former doctoral students, such as W. Nef, H. Häfeli, Erwin Bareiss, or Kriszten:

I do not need to say anything new about Herr. Prof. Van der Waerden because in the large materials regarding the call of Herr. Prof. Nevanlinna [1946] I have spoken about him at length and all of that is still in effect today. Naturally he is much weightier than the above mentioned young Swiss. But I would still like to mention how extraordinarily desirable a Swiss would be as my successor because for so many years there was no position open for young Swiss mathematicians.

---

[337] Lars V. Ahlfors's letter to Hans Boesch of May 21, 1950; *Universität Zürich, Universitätsarchiv, Lehrstuhlakten Mathematik.*

[338] Rudolf Fueter's letter to Hans Boesch of June 1, 1950; *Universität Zürich, Universitätsarchiv, Lehrstuhlakten Mathematik.*

Andreas Speiser praises Van der Waerden and the young Swiss candidates, while putting down Pólya as a mathematician:[339]

> Of the foreigners Pólya does not even come into view. He has dealt with an enormous amount of small problems but has never seriously worked in a serious area and would rapidly sink the level of mathematics at the University. Opposite to this, Van der Waerden is an apt (*trefflicher*) mathematician, whom one would have to recommend.

Evaluating Pólya unfairly is not the only deplorable aspect of the Speiser's letter. Following praise for the (Jewish) mathematician Richard Brauer, Speiser uses—in the year 1950—the *Nazi Deutsch* to describe Brauer as "not Arian (*nicht arisch*)." Truly, old habits die hard!

Summing up, Prof. Van der Waerden is the unanimous choice of the four references. Only one question remains: has Van der Waerden been sufficiently "purified"? This is to be answered by Prof. Schouten. The latter sends his handwritten reply to Prof. Finsler on May 12, 1950. Schouten's letter deals exclusively with Van der Waerden a person, and not at all with his mathematical work. The following is its complete text:[340]

> Dear Herr Colleague!
>
> I have received your friendly letter of May 9. A few weeks ago Herr van der Waerden has been named an *Ordinarius* in Amsterdam. Political reservations do not apply here [in the Netherlands] against him. I should actually say that they do not apply any more, because certain circles had earlier tried *completely without justification* to raise their voice against him. But that has all now passed and he is also now a Member of the Royal Amsterdam Academy.
>
> Even though I hope that you will not snap this man away from us, I must absolutely tell you my opinion that he is completely politically harmless (*unbedenklich*).
>
> With friendly greeting to the entire Zürich circle,
>> Yours most respectfully
>> J. Schouten

Thus, Prof. Van der Waerden is cleared for Swiss employment again. The Mathematics Commission meets on June 3, 1950 and ends up with the same slate and order of the three candidates they had started with.[341] On June 9, 1950 *Dekan* Boesch reports the faculty findings to the Education Directorate (*Erziehungedirection*) of the Canton of Zürich in a 5-page letter.[342] He lists, with compliments, a large number of young Swiss mathematicians (no doubt to impress the government), but reserves the highest compliments for

---

[339] Andreas Speiser's letter to Hans Boesch of May 10, 1950; *Universität Zürich, Universitätsarchiv, Lehrstuhlakten Mathematik.*

[340] Jan Schouten's letter to Hans Boesch of May12, 1950; *Universität Zürich, Universitätsarchiv, Lehrstuhlakten Mathematik.*

[341] *Universität Zürich, Universitätsarchiv, Lehrstuhlakten Mathematik.*

[342] Ibid.

Mr.'s Van der Waerden, Pólya and Eckmann [who] would be the candidates for this Mathematics Professor position, whereby Herr van der Waerden would be in first place [, Herr Pólya in second place].[343]

Compliments for Prof. Pólya are outweighed by the following considerations:

One cannot ignore his advanced age, especially since Herr Pólya let us know that in the case of the call he would have to give up his pension. In addition, there is an advantage [to Pólya's age of 63] that in the foreseeable future there would possibly develop again a position for a Swiss mathematician [i.e., Pólya would die soon or else retire at the mandatory age of 70].[344] One has to also mention the rejecting positions of Mr.'s Fueter and Speiser against Herr Pólya.

Prof. Van der Waerden, on the other hand, gets a clean bill of political health from *Dekan* Boesch:

Certain problems found in Herr Van der Waerden's working at the University of Leipzig during the war, which were focused on by Holland, are no longer applicable according to the communication that Prof. Schouten has forwarded. On the contrary, it is explicit from the Faculty proposal for filling a new position of Professor of Applied Mathematics dated July 15, 1946, that Prof. Van der Waerden was thoroughly considered.

As is mentioned above, Prof. Eckmann has turned down the offer before the search begins, Prof. Pólya is rejected by Fueter and Speiser, who certainly know in advance that they do not want Pólya back in Zürich. From day one of the search, Prof. Van der Waerden is listed as the number one candidate. Thus, the elaborate search seems to have been done to satisfy the rules of decorum, but has had only one goal from the beginning—to hire Van der Waerden. He is offered the job on September 20, and accepts it with "heartfelt gratitude" on September 24, 1950.[345]

One document in the Fueter succession deserves another look: the letter from *Dekan* Boesch to the Education Directorate of the Canton of Zürich of July 14, 1950,[346] in which Boesch asks the government to not only swiftly approve Van der Waerden's appointment, but also "to find out from Herr Van der Waerden if it would be possible to begin his work in Zürich already in the forthcoming winter semester 1950/1951." Thus, Van der Waerden has an opportunity to realize his Swiss dream right away, without spending another year at Amsterdam. He apparently does not agree to an early Zürich start. I can venture a conjecture to explain this refusal: perhaps, Van der Waerden desires a vindication for the *Het Parool*ean humiliation, and the Amsterdam full professorship with its Inaugural Lecture ceremonies in December 1950 provides such an opportunity. Van der Waerden wants to leave his Home-

---

[343] Text in brackets added in pencil.

[344] Time proved Herr Boesch to be wrong: George Pólya would live to the age of 98, and give inspiring lectures at a very advanced age.

[345] *Universität Zürich, Universitätsarchiv, Lehrstuhlakten Mathematik.*

[346] Ibid.

land as a winner, by willingly giving up Holland's highest academic credentials he has finally earned.

For a decade I have been absorbed with the following question: why did Professor Van der Waerden leave the Netherlands for good in 1951? Was the University of Zürich (which, in my opinion, was no match to its famed neighbor, ETH) a better place than the University of Amsterdam? This was not at all obvious to me, so I asked Prof. de Bruijn, who replied as follows [Bru9]:

> We were looking at the US and Switzerland as a kind of paradise. Whether in the long run Zürich would be much better than Amsterdam, may be open to discussion. In 1950 Amsterdam had lost the glory of Brouwer's days of the 1920's...
>
> By the way, I really do not know the order of the events. The offer from Zürich may have come at a time where the procedures for getting him the full professorship at Amsterdam had hardly started. He may have kept the Zürich offer secret for a time, in order to keep both possibilities open. If it had happened to me, I would have felt a moral pressure against letting Amsterdam down.

H. J. A. Duparc recalled [Dup] in 1996:

> Van der Waerden's wife, Rellich,[347] was German[348] and had many difficulties in normal life in Holland because of her speaking German language (Holland was occupied 5 years by the Germans)."[349]

N. G. de Bruijn [Bru10] added:

> Justified or not justified, those anti-German feelings were very strong indeed. I can understand that Camilla was treated as an outcast, and that she therefore disliked living in Holland.[350]

Ms. Annemarie van der Waerden, Prof. van der Waerden's first cousin, recalled opinions of her parents and other family members:[351]

> Camilla was a very proud woman: 'like Carmen, from the opera'. She was furious about all the things he had to go through and the accusations. Camilla had a lot to do with the decision to move to Switzerland, a neutral country, a real centre of science. For sure, Camilla is the one that broke the connections with his family in Holland.[352]

While the role of Mrs. Van der Waerden in the decision to leave the Netherlands must have been significant, such an important decision had to be ultimately made

---

[347] Mrs. Van der Waerden's maiden name.

[348] Actually Austrian. However, as have seen, Prof. Van der Waerden also calls her some times Austrian and other times German.

[349] Mrs. Van der Waerden learned and spoke Dutch, but apparently with a German accent.

[350] Children, on the other hand, seemed to enjoy their life in Laren. Their first cousin Theo van der Waerden recalls [WaT2]: "In 1949 we moved to Amsterdam... We met the family more and more, we went to Laren, where Bart and his family lived (1945–1951). I had the impression that they loved the house, the children were happy there with the schools, the nature, etc."

[351] Communicated by Mrs. Dorith van der Waerden [WaD1].

[352] A correspondence shows that a certain degree of connections continued.

by Professor Van der Waerden. Van der Waerden wanted at all times to be at the best place for doing mathematics, which now moved to Switzerland and the United States. Which one should he claim?

He aspired to belong to the German culture; it was important—perhaps, too important—to him. The decision to move to Switzerland was the last critical decision of Van der Waerden's career. He decided to leave the Fatherland of Suffering for the Land of Neutrality, the Land of German language but not Germany.

Two years later Germany would invite Prof. Van der Waerden—to a chair at München—and that offer would be rejected, not because "children did not want to move any more," as Mrs. Camilla Van der Waerden would lead us to believe in 1993 [Dol1],[353] but no doubt because of the desire to remain in a prosperous German-speaking non-Germany.

In 1972 Van der Waerden retired from his professorship at Zürich. Germany did not forget Van der Waerden's loyalty. In 1973 he was awarded the old German *Orden Pour le Mérite für Wissenschaften und Künste* (Order of Merit for Sciences and Arts), limited to 40 German and 40 foreign living recipients. On June 12, 1985, Leipzig University awarded Prof. Van der Waerden its honorary doctoral degree.

39.7 Epilogue: The Drama of Van der Waerden

> *One's response to living under tyranny can only be*
> *to leave, to die, or to compromise.*
> — Alexander Soifer, [Soi29]

Most authors tend to concentrate on exceptional personages. This may create timeless lessons, but misses out on capturing times and places. The key times for us, 1931–1951, encompass the Nazi time and de-Nazification of Europe. These tragic times provide such profound lessons of human nature that we have got to learn from them as much as we possibly can. We encounter heroes and villains, but also a much more numerous group in between these two extremes. The life of one such person "in between" has been the subject of this research.[354] Van der Waerden was neither a villain nor a hero (as he was portrayed by all previous biographers). Studying his life allows us to pose important questions about the role of a scientist in a tyranny, and about some of the moral issues surrounding the World War II and its aftermath.

How does one understand and reconcile the contradictions in the record I have presented on these pages? We witness instances of courage and compromise with the Nazi authorities; signs of high integrity and instances of moral insensitivity;

---

[353] The statement "Children did not want to move" surely implies children's knowledge of the München's offer. However, one "child" at 23 was long married and gone from the parents' house, while another did not know about this offer until I shared this information in 2004.

[354] The last three sentences come from my review of Carathéodory's biography [Soi29], and appeared as an epigraph for this triptych (Chapter 37). There are great similarities between the life of Van der Waerden and the life of Constantin Carathéodory.

declared desire to save the German culture and little effort to contribute to the culture of the Netherlands, his fatherland that has been served with such a high distinction by the rest of the Van der Waerden family.

People, who have known and liked Van der Waerden, prove to be surprisingly unanimous in their explanations of these contradictions. Beno Eckmann, Van der Waerden's friend from 1951 to 1996, in his e-mail to me on Dec 7, 2004, writes [Eck0]:

> As a person and friend vdW [Van der Waerden] was very kind but seemed to be quite naïve.

Ms. Annemarie van der Waerden, Prof. Van der Waerden's first cousin, recalls the impressions of her parents and other family members about B. L. van der Waerden:[355]

> Childish, not in the world, no interest in politics. Never thinking about what people would think of him and not understanding worldly matters. Totally impractical.

As recently as the year 2005 [Soi24], I too believed that Van der Waerden was a naïve, stereotypical abstract mathematician, who, as I put it, "built his morality on the foundation of laws of the lands he lived in, by rules of formal logic. He seems to have been quintessentially a mathematician—and not only by profession—but by his moral fabric."

Van der Waerden might have been naïve. However, I now find this explanation inadequate. He was not a prisoner of the "Ivory Tower": he was aware of life around him. He clearly saw the Nazi regime for what it was. On August 10, 1935, from his vacation in Holland he wrote:[356] "We are here in Holland for two months and rest up our souls from the constant tensions, hostilities, orders and paperwork... Ministries examine who has not yet been completely switched over [to the National Socialism], who is a friend of Jews, who has a Jewish wife, etc., as long as they themselves are not torn apart by their fight for power." In the April 6, 1943 letter from Leipzig, Van der Waerden described to a trusted friend, Erich Hecke, the tragedy of the occupied Holland and the Holocaust:

> Maybe he [Blumenthal] is in hiding like thousands of others. Maybe he is already in Poland like ten thousand Jews from Holland.[357]

No, Van der Waerden was not naïve, or not naïve enough to thus explain his life's decisions. He knew the truth about the Nazi regime, and consciously and opportunistically chose to tolerate it.

Van der Waerden's record is complex. The evidence shows that on one hand, he strived to be a highly moral individual, a fitting member of his great family. While in Germany, he occasionally reached the moral heights he sought: in 1935 he publicly objected to the firing of four Jewish professors and tried to hold Germany to its

---

[355] Communicated by Mrs. Dorith van der Waerden [WaD1].

[356] New York University Archives, Courant Papers.

[357] Nachlass von Erich Hecke, Universität Hamburg.

own April 7, 1933 law; in 1935 he wrote a eulogy for his Jewish mentor Emmy
Noether; through 1940 he was determined to publish Jewish authors in *Mathematische Annalen*.

On the other hand, as a brilliant mathematician, he desired—perhaps felt
entitled—at all times to be in the best place for doing mathematics, even if the
time and place was the Nazi Germany, even if there was a price of moral compromise attached to it. He should have been ashamed of his 1933 invocation of his
"full-blooded Arianness" while Jews were thrown out of their university positions;
his 1934 oath to Hitler; his Hitler salutes in letters and lectures—he knew those
were not the rituals his father and brothers would have approved. Worse yet, on
May 16, 1940, the day after Holland fell to the German invaders, Van der Waerden
failed to take the side of his oppressed homeland—he chose "neutrality" between
his homeland and her oppressor.

Van der Waerden knew the moral price he had paid for the comfort of doing
mathematics in the Nazi Germany, and he admitted it once, in his December 29th,
1945 letter to Richard Courant:[358] "I have made some mistakes perhaps, but I have
never pacified with the Nazis."

Despite having no illusions on the nature of "the gangster regime" (his words),
despite his father's and uncle Jan's insistence that it was Van der Waerden's "duty"
to leave the Nazi Germany even before its occupation of Holland, Van der Waerden
chose to live there. "Why would I go to Holland where oppression became so intolerable and where every fruitful scientific research was impossible?" he wrote to Van
der Corput[359] without even realizing that the intolerable oppression of his homeland
was caused by the very country he served!

For doing his "labor of love," Van der Waerden chose the Third Reich, even
though the price was his silence, which was a form of condoning the Nazi regime.
This brings to mind the 1953 book *The Captive Mind* [Mil], by the Polish poet
and 1980 Nobel Laureate Czesław Miłosz. In this book, Miłosz coins the term
"Professional Ketman." Under such a Ketman (unwritten contract between a scientist and a totalitarian State), the scientist reasons in the following manner:

> I pursue my research according to scientific methods, and in that alone lies the aim
> of my life... Discoveries made in the name of a disinterested search for truth are
> lasting, whereas the shrieks of politicians pass. I must do all they demand, they may
> use my name as they wish, as long as I have access to my laboratory and money for
> the purchase of scientific instruments.[360]

What does the State gain?

> The State, in its turn, takes advantage of this Ketman because it needs chemists, engi-
> neers and doctors.[361]

---

[358] New York University Archives, Courant Papers.

[359] Van der Waerden, letter to van der Corput; July 31, 1945; ETH, Hs 652: 12160. Van der Waerden
refers to Holland, occupied by Germany, 1940–1945.

[360] [Mil], p. 69.

[361] Ibid, p. 70.

Likewise, Van der Waerden served for the Nazi Germany's Civil Service, and lent his credibility and his acclaim as a distinguished scientist to that of the Third Reich.

In 1994 Herbert Mehrtens [Meh] aptly coined a term "irresponsible purity" for the scientists who pursued their pure academic fields in the Third Reich and yet assumed no responsibility for thus serving and strengthening the criminal state. Van der Waerden's words and deeds serve as a clear example of this phenomenon.

The great anthropologist and my dear friend James W. Fernandez, upon reading these chapters, summarized my findings concisely during our "Fang Summit"[362] in early August 2007: "Frailty of Brilliance!"

Van der Waerden must have felt the weight of his Nazi time conflict and compromise for the rest of his life. Prof. Beno Eckmann, his friend for nearly half a century, told me that Van der Waerden always avoided any mention of his time in the Third Reich [Eck0]:

> We never really talked about his time in Leipzig, in any case not about politics. He and his wife seemed to avoid these themes.

What troubled the editors of *Het Parool*, Van der Corput and others the most was not seeing a man who aspired high moral ground and fell victim of a compromise with the Nazi tyranny. They detested the hypocrisy of denying the compromise, the invocation of high moral ground, the ground they thought had eroded.

One's response to living under tyranny can only be to leave, to die, or to compromise. Van der Waerden chose the compromise between high moral aspirations and doing mathematics in the Nazi Germany. The struggle between these two contradictory goals was the drama—perhaps, the tragedy—of the life of Bartel Leendert van der Waerden, one of the great mathematicians of the twentieth century, the century marked by the brutal war he had spent in the enemy's camp.

This has been my report on research, *In Search of Van der Waerden*. In it, I have faithfully followed the approach used by Professor Van der Waerden himself [Wae15]:

> I have tried to consider the great mathematicians as human beings living in their own environment and to reproduce the impression which they made on their contemporaries.

In fact, this work is forever in progress, in search of the hero. While I have found answers to most of the questions I posed to myself, I prefer to consider these four sections as a report on research in progress, *In Search of Van der Waerden*. A complete insight is impossible, we can only aspire to come as close as we can to it!

---

[362] Our annual meeting devoted to the art and culture of the Fang people of Gabon, Africa, extensively studied by Fernandez, and to other topics of mutual interest, such as a role of a scientist in tyranny.

VIII
Colored Polygons: Euclidean Ramsey Theory

There is a running discussion between Dieudonné and Branko Grünbaum. Dieudonné sort of says that geometry is dead and of course Branko Grünbaum disagrees with him. I think I am on the side of Branko Grünbaum and I hope that I will convince you that at least combinatorial geometry is not dead.

– Paul Erdős[1]

---

[1] [E83.03].

40
Monochromatic Polygons in a 2-Colored Plane

We have already met briefly a 2-colored plane in Problem 2.1, which can be restated as follows:

Problem 40.1 For any positive d, any 2-colored plane contains a monochromatic segment of length d.

Solve the following problem on your own.

Problem 40.2 For any positive d, any 2-colored plane contains a non-monochromatic segment of length d; if at all each of the two colors is present in the plane.

Let me remind you that in our discussions a *triangle* stands simply for a 3-element set. When these three points are on a line, we will call the triangle *degenerate*. Accordingly, a set of n points in the plane will be called an *n-gon*. An n-gon with all n vertices in points of the same color is called *monochromatic*.

You may wonder why after discussing a multi-colored plane should we now talk about a mere 2-colored plane? Would it not be more logical to put this chapter earlier in this book? Yes, it would. But this logical approach creates, as Cecil Rousseau puts it ([Soi1], introduction), "books written in a relentless Theorem–Proof style." This logical approach ignores a higher logic of mathematical discovery.

For me, personally, a fascination with the chromatic number of the plane problem came first. Then I looked into a 2-colored plane. Why? If we can prove the existence of monochromatic configurations in any 2-colored plane, we will have tools to study a 3-colored plane. And configurations present in any 3-colored plane may provide tools to attack a 4-colored plane. And it is a 4-colored plane where we "only" need to find out whether a monochromatic segment of length 1 is necessarily present![2]

With this rationale in mind, in 1989–1990 I proved some results, formulated conjectures, and thus rediscovered *Euclidean Ramsey Theory*. I published a problem essay [Soi2] about it in the first issue of volume I of the newly founded research quarterly *Geombinatorics*. On July 5, 1991, Ron Graham sent me a copy of the series of three papers by six authors, which broke the news to me: I was 15–17

---

[2] Of course, others probably had different reasons for looking into 2-colored planes. Erdős et al. in their trilogy [EGMRSS] were pursuing expansion of Ramsey theory to Euclidean Ramsey Theory.

A. Soifer, *The Mathematical Coloring Book*,
DOI 10.1007/978-0-387-74642-5_40, © Alexander Soifer 2009

years too late: Paul Erdős et al., were first to discover what they named *Euclidean Ramsey Theory*! Fortunately, some of my results remained new as you will see them in this chapter (Problems 40.7, 40.14, 40.19), which is chiefly dedicated to Erdős et al., series of papers [EGMRSS]. Paul Erdős referred to the authors as "us," or "the six." All the distinguished six authors deserve to be listed here. They are Paul Erdős, Ronald L. Graham, P. Montgomery, Bruce L. Rothschild, Joel H. Spencer, and Ernst G. Strauss.

When three of the six coauthors wrote *Ramsey Theory* monograph (1st edition [GRS1] in 1980; 2nd edition [GRS2] in 1990), they did not include many of the trilogy [EGMRSS] results in their book. Perhaps, they viewed these results as being too "elementary" for their dense monograph. On the other hand, they realized how difficult these "elementary" problems can be, for Paul Erdős and Ron Graham included open problems of Euclidean Ramsey Theory in many of their (open) problem talks and papers. It seems that most of these results of "the six" and other results of Euclidean Ramsey are appearing here for the first time in book form.

Problem 40.3 (Erdős et al., [EGMRSS]) 2-color the plane to forbid a monochromatic equilateral triangle of side d.

Solution: Divide the plane into parallel stripes, each $\frac{\sqrt{3}}{2}d$ wide ($\frac{\sqrt{3}}{2}d$ is the altitude of the equilateral triangle of side d), then color them alternatively red and blue (Fig. 40.1). Include in each stripe region its left border line, and do not include its right border line and we are done. ∎

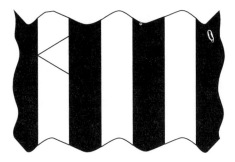

Fig. 40.1

Problem 40.4 (Erdős et al., [EGMRSS]) Find a 2-coloring of the plane different from the one in the solution of Problem 40.3 that does not contain a monochromatic equilateral triangle of side d.

Solution: Start with the coloring described in the solution of Problem 40.3 (Fig. 40.1). Draw a line making, say, a $\frac{\pi}{3}$ angle with the border lines of the stripes (Fig. 40.2), and change the colors of the points of their intersections. It is easy to verify that as before, the plane does not contain a monochromatic equilateral triangle of side 1. ∎

If you solved Problem 40.4 on your own, you have probably noticed that your and my solutions did not differ much from each other and from the solution for Problem 40.3. In fact, Paul Erdős et al., thought that the solutions cannot differ much!

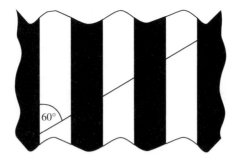

Fig. 40.2

Conjecture 40.5 ([EGMRSS], Conjecture 1 of Part III) The only 2-colorings of the plane for which there are no monochromatic equilateral triangles of side 1 are the colorings in alternate strips of width $\frac{\sqrt{3}}{2}$, as in the solution of Problem 40.3, except for some freedom in coloring the boundaries between the strips.

Decades have passed; Ronald L. Graham and Paul Erdős repeated problems and conjectures of the Euclidean Ramsey Theory, including 40.5, in their talks and papers (example [E83.03]), but no proof was found to these easy-looking, hard-to-settle triangular conjectures. However, in March 2006, a group of four young Czech mathematicians from Charles University on Malostranské plaza (I visited Jarek Nešetril at this historic place in 1996) Vít Jelínek, Jan Kyncl, Rudolf Stolar, and Tomás Valla [JKSV] disproved this 33-year old conjecture!

Counterexample 40.6 ([JKSV, Theorem 3.19]) Every zebra-like 2-coloring of the plane has a twin 2-coloring that forbids monochromatic unit equilateral triangles.

For definitions of "zebra-like" 2-coloring of the plane and of "twin" coloring, I refer you to the original work, which, while not published by Combinatorica for nearly a year (since submission in March 2006), is now made available by the authors at arXiv. Here I would like to show an example of a zebra-coloring provided to me by one of the authors, Jan Kyncl (Fig. 40.3).

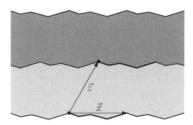

Fig. 40.3

I wish to congratulate the authors for introducing into the field a brand new rich class of 2-colorings of the plane, and for solving, in the negative, an old-standing conjecture by Erdős et al.

Any equilateral triangle can be excluded from appearing monochromatically by choosing an appropriate 2-coloring of the plane. However, some triangles, exist monochromatically in any 2-colored plane. The first such example was found by Paul Erdős et al. [EGMRSS].

Problem 40.7 ([EGMRSS]) Any 2-colored plane contains a monochromatic triangle with the small side 1 and angles in the ratio 1:2:3.

My Solution [Soi9]: Pick a monochromatic segment AB of length 2 (Problem 40.1) and construct a regular hexagon H on AB as on the diameter (Fig. 40.4). If at least one more vertex of H is of the same color as A and B, we are done. If not, we are done too! ∎

Fig. 40.4

Problems 40.7 and 40.8 were offered to high school students during the Colorado Mathematical Olympiad in 1990. The first was solved by several participants. Nobody solved the second one.

Problem 40.8 ([Soi2]) Any 2-colored plane contains a monochromatic triangle with the small side 1 and angles in the ratio 1:2:4.

Solution [Soi9] Assume that such a triangle does not exist in a 2-colored red and blue plane. Toss a regular 7-gon of side length 1 on the plane (Fig. 40.5). Since 7 is odd, two of its consecutive vertices will be of the same color. Say, A and B are blue. Then D and F must be red. Therefore, C and G are blue. We got a blue triangle CAG in contradiction to our assumption. ∎

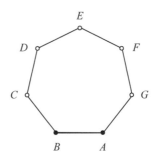

Fig. 40.5

Problem 40.9 ([Soi2]) For any positive integer n, any 2-colored plane contains a monochromatic triangle with the small side 1 and angles in the ratio:

a. $n : (n + 1) : (2n + 1)$;

b. $1 : 2n : (2n + 1)$.

Proof Assume that a 2-colored plane P (red and blue) does not contain a monochromatic triangle with the small side 1 and angles in the ratio $1 : 2n : (2n + 1)$. Let the diagonal of a regular $(4n + 2)$-gon of side 1 have length d. Due to Problem 40.1, we can find in the plane P a monochromatic (say, red) segment S of length d. We construct on S as on a diameter, a regular $(4n + 2)$-gon K. Now we number the vertices of K starting with an endpoint of red diameter S (Fig. 40.6).

Now we start a rotation. The points 1 and $2n + 2$ are red, therefore the points 2 and $2n + 3$ are blue. Thus, the points 3 and $2n + 4$ are red, etc. Finally, the points $2n + 2$ and 1 are blue, which is a contradiction.

The existence of a monochromatic triangle with angles in the ratio $n : (n + 1) : (2n + 1)$ can be proved by a similar rotation. (Instead of adding 1 to the endpoints of the diameter, we just add $n + 1$.) ∎

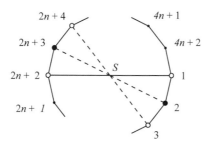

Fig. 40.6

Leslie Shader from the University of Wyoming proved an important result.

Problem 40.10 (L. Shader, [Sha]) For any right triangle T, any 2-colored plane contains a monochromatic triangle congruent to T.

As you can see, we have many examples of triangles that exist monochromatically in *any* 2-colored plane, and one example of a triangle (equilateral) that may not. Having realized this, I posed the following $25 problem to my university and high school students in 1989 (published in [Soi2]).

Open $25 Problem 40.11 [Soi2]) Find all triangles T such that any 2-colored plane contains a monochromatic triangle congruent to T.

Paul Erdős et al., tried to solve this very problem some 14 years before me. Moreover, they posed the following conjecture in 1973.

Conjecture 40.12 ([EGMRSS], Conjecture 3 of Part III) For any non-equilateral triangle T, any 2-colored plane contains a monochromatic triangle congruent to T.

This problem appears surprisingly difficult, and in 1979, Paul Erdős set the prize for it [E79.04]:

> Many special cases have been proved by us (i.e., the authors of [EGMRSS]) and others but the general case is still open and I offer 100 dollars for the proof or disproof.

In 1985 Erdős increased the pay off [E85.01]:

> Is it true that every non-equilateral triangle is 2-Ramsey in the plane (i.e., Conjecture 40.12)? I offer \$250 for a proof or disproof.

Let me formally attach Erdős's \$250 tag to the above conjecture:

Paul Erdős's \$250 Conjecture 40.13 Is it true that any 2-colored plane contains any non-equilateral triangle monochromatically?

Paul Erdős et al., also conjectured that any 2-coloring of the plane may not contain monochromatically at most an equilateral triangle of one size.

Conjecture 40.14 ([EGMRSS]) If a 2-colored plane P does not contain a monochromatic equilateral triangle of side d, then P contains a monochromatic equilateral triangle of side d' for *any* $d' \neq d$.

In 2003 Graham [Gra5] offered \$100 for the proof:

Ronald L. Graham's \$100 Conjecture 40.15 Every 2-coloring of the plane contains a monochromatic copy of every triangle, except possibly for a single equilateral triangle.

Thus, you can win three prizes by solving essentially one problem!
My intuition regarding the above conjecture agrees with the authors of [EGMRSS], except that I am not sure about degenerate triangles.

Open Problem 40.16 Is it true that any 2-colored plane contains a degenerate isosceles triangle of small side 1 (Fig. 40.7)?

Fig. 40.7

In order to solve the above open problems, you need tools. Here are two for you.

Let T be a triangle. Then T_m will stand for *the triangle whose sides are twice as long as the corresponding medians of T* (the medians of any triangle are themselves the sides of a triangle—prove this nice elementary fact on your own).

Tool 40.17 ([Soi2]) For any triangle T, any 2-colored plane contains a monochromatic triangle congruent to T or T_m.

Proof If one color is not present in the 2-colored plane, we are done; assume now that both colors are present. Let the side lengths of T be a, b, and c, and P be a plane colored red and blue. By Problem 40.2, P contains a segment AE of length $2a$ with blue A and red E. The midpoint C of AE has the same color as A or E, let it be blue as A.

We pick B and D such that $ABCD$ is a parallelogram with side lengths b and c (Fig. 40.8). If at least one of the points B, D is blue, we get an all blue triangle ABC or ADC. Otherwise BED is an all red triangle with side lengths twice as long as the corresponding medians of T (prove this nice geometric fact on your own). ∎

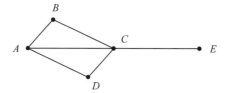

Fig. 40.8

Prove the following corollary of Tool 40.17.

Problem 40.18 Any 2-colored plane contains a monochromatic equilateral triangle of side 1 or $\sqrt{3}$.

Out of the many nice tools contained in [EGMRSS], I would like to share here with you my favorite. Erdős et al., prove it in a true Olympiad style, so I am not changing a thing in it. However, I am adding an additional diagram showing that the statement is true for the case when the triangle K is *degenerate* as well.

Tool 40.19 ([EGMRSS] Theorem 1 of Part III) Let K be a triangle with sides a, b, and c, and let K_a, K_b, and K_c be equilateral triangles with sides a, b, and c respectively. Then a 2-colored plane contains a monochromatic triangle congruent to K if and only if it contains a monochromatic triangle congruent to at least one of the triangles K_a, K_b, K_c.

Proof Consider the configuration in Fig. 40.9. The six triangles HBC, ABD, CDE, EFH, DFG, AHG all have sides a, b, and c. The triangles ABH, DFE, BCD, FGH, HEC, ADG are equilateral with sides a, a, b, b, c, c, respectively. We see that if one of the second six triangles is monochromatic, one of the first six must be monochromatic too. The converse is true by a symmetric argument.

If the triangle K is degenerate (this is one case the authors of [EGMRSS] did not explicitly address), look at the configuration in Fig. 40.10 that I added for you. No changes in the text of the proof are necessary while using Fig. 40.9 to prove a degenerate case! ∎

The following problem is a good test of your skills: try it on your own before reading the solution.

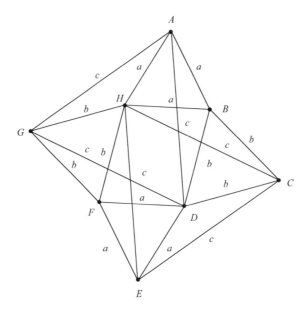

Fig. 40.9

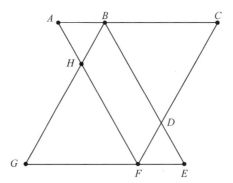

Fig. 40.10

Problem 40.20 (László Lovász, [Lov2]) Prove that any 2-colored plane contains a monochromatic triangle with side lengths $\sqrt{2}$, $\sqrt{6}$, and π.

Proof Given a 2-colored plane P. Due to Tool 40.18, P contains a monochromatic equilateral triangle of side $\sqrt{2}$ or $\sqrt{6}$ (just use as T an equilateral triangle of side $\sqrt{2}$; the sides of T_m will be equal to $\sqrt{6}$). In either case, due to Tool 40.19, the plane P contains a monochromatic triangle with sides $\sqrt{2}$, $\sqrt{6}$, and π. ∎

You may think that we are only concerned with triangles: we aren't. The following problem is a new form (and solution) of a problem that the famous American problem solver and coach of the American team for the International Mathematics Olympiad, Cecil Rousseau, once created for the 1976 USA Mathematics Olympiad (USAMO).

Problem 40.21 (Cecil Rousseau; USAMO, 1976) Any 2-colored plane contains an $m \times n$ monochromatic rectangle such that $m = 1$ or 2, and n is a positive integer not greater than 6.

Proof Toss on a 2-colored plane (red and blue) a 2×6 square lattice (Fig. 40.11).

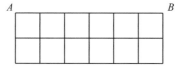

A   *B*

Fig. 40.11

By the Pigeonhole Principle, out of the seven vertices (i.e., intersections of the lattice lines) in the top row AB, at least four must be of the same color, say, blue. We keep the corresponding four columns and throw away the rest (Fig. 40.12).

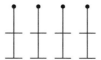

Fig. 40.12

If the second *or* third row in Fig. 40.12 contains more than one blue vertex, we get a monochromatic blue rectangle, and the problem is solved.

If the second *and* third rows contain at most one blue vertex each, then we throw away the columns corresponding these blue vertices. We are left with a monochromatic red rectangle located in the second and third rows. ∎

I was able to strengthen this result in 1990, and offered it at the 1991 Colorado Mathematical Olympiad (CMO). Try to solve it on your own first.

Problem 40.22 (CMO 1990, [Soi9]) Prove the statement of Problem 40.21 with n not exceeding 5.

Proof Given a 2-colored plane. If one color, in fact, is not present at all we are done. Otherwise, due to Problem 40.2, there are two points A and B of opposite colors distance 6 apart. Construct on AB a 2×6 square lattice like in Fig. 40.11 and repeat *word by word* the solution of Problem 40.21. ∎

This train of thought naturally runs into the following open problems.

Open Problem 40.23 [Soi9] Is the statement of Problem 40.21 true with n not exceeding 4?

Open Problem 40.24 Find the lowest upper bound for n, such that the statement of Problem 40.21 is true.

It is easy to prove the statement of Open Problem 40.23 conditionally.

Problem 40.25 If a 2-colored plane P contains a monochromatic degenerate isosceles triangle of side 1 (Fig. 40.7), then P contains an $m \times n$ monochromatic rectangle such that $m = 1$ or 2, and n is a positive integer not exceeding 4.

Proof Let a 2-colored plane P (red and blue) contain a monochromatic, say blue, degenerate isosceles triangle T of side 1. We construct on T a 2×4 square lattice (Fig. 40.13).

The first from (from the left) column is all blue. If any of the other four columns contain at least two blue vertices, then we get all-blue rectangle. Otherwise, each of these four columns has at least two red vertices. But there are only $\binom{3}{2} = 3$ distinct ways to have two red vertices in a column. Therefore, at least two of the four columns have two red vertices in the same rows, i.e., we obtain an all-red rectangle. ∎

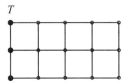

Fig. 40.13

The image of a figure F under translation is naturally called a *translate* of F. Erdős et al., found a cute use of the Mosers Spindle.

Problem 40.26 ([EGMRSS] Theorem 3 of Part II) Given a 2-colored plane P (red and blue) and a triangle T in it. Then P contains a pair of red points distance d apart for every d, or a blue monochromatic translate of T.

Proof Let A, B, C be the vertices of T. Assume that for a positive d there is no pair of red points d apart. We toss the Mosers Spindle S (Fig. 2.2) of side d on the plane, and denote by $S_1 = t_1(S)$ and $S_2 = t_2(S)$ the images of S under translations through $\vec{AB}$ and $\vec{AC}$ respectively.

Due to observation after Problem 2.2, any three vertices of the spindle S contain a pair of vertices distance d apart. Therefore, each seven-point set S, S_1, S_2 contains at most two red points. Thus, there is a vertex, say A, of S such that all three vertices A, $t_1(A)$ and $t_2(A)$ are blue. They form a translate of T! ∎

Problem 40.27 ([EGMRSS], Theorem 1′ of Part II) Any 2-colored plane (red and blue) contains a red pair of points distance 1 apart, or 4 blue points on a line 1 distance apart.

Proof Assume a 2-colored plane P does not have either a red pair of points distance 1 apart or a four blue points on a line distance 1 apart. Then P must have a red point p. The circle C_1 of radius 1 and center p must be entirely blue.

Now we add a concentric circle C_2 of radius $\sqrt{3}$ and an equilateral triangle a, b, c inscribed in C_2 (Fig. 40.14). Denote the points of intersection of C_2 and ab by d and e.

It is easy to confirm (please do) that

$$|ad| = |de| = |eb| = 1$$

Since both d and e are blue (they are on C_1), not both a and b are blue. This is similarly true for a and c, and for b and c. Therefore, at most one of a, b, c is blue. Suppose a and b are red;

Now we rotate ab about the center p, to its new position fg, such that $|af| = 1$. Then, of course, $|bg| = 1$. Therefore, f and g are both blue. So are h and i (they are on C_1). Thus, we get a blue quartet f, h, i, *and* g distance 1 apart, in contradiction to our initial assumption. ∎

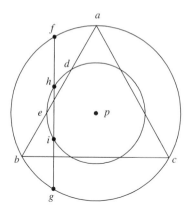

Fig. 40.14

Having proved 40.26, Erdős et al., [EGMRSS, Part II, p. 535] formulated but could not decide the following question:

Is it true that any 2-colored plane (red and blue) contains a red unit length segment or a blue unit square?

On March 25, 1977, the Hungarian mathematician Rozália Juhász submitted (and in 1979 published) an impressive paper [Juh], where in one stroke she proved a powerful generalization of Problem 40.26, and more than answered the above question by Erdős et al., in the positive:

Problem 40.28 ([Juh], Theorem 1) For any 4-gon Q, any 2-colored plane (red and blue) contains a pair of red points distance 1 apart, or a monochromatic blue 4-gon congruent to Q.

In the same paper Juhász showed that the result of Problem 40.28 is not true for an n-gon where $n \geq 12$.

Counterexample 40.29 ([Juh], Theorem 2) There is a 12-gon K and a 2-colored plane P (red and blue) such that P does not have either a monochromatic unit-distant red segment or a blue monochromatic 12-gon congruent to K.

Construction: First let us describe the 2-coloring of the plane that does the job. We start with a regular triangular lattice with distance 2 between nearest vertices (Fig. 40.15).

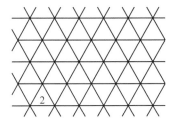

Fig. 40.15

We make every vertex of the lattice to be the center of a red circular disk of radius $1/2$. With every disk we also color red half of its boundary under its horizontal diameter and the left point of that diameter. The rest of the plane we color blue (Fig. 40.16). You can easily verify that our 2-colored plane P has no red monochromatic segment of length 1. You can also show (do!) that any closed disk (i.e., disk including its boundary circle) of radius $\frac{2}{\sqrt{3}} + \frac{1}{2}$ (shown in Fig. 40.16) in P must contain at least one of the red disks (together with its boundary).

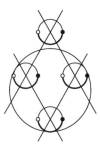

Fig. 40.16

Let us now define our 12-gon K. We draw a regular triangular lattice just like the one in Fig. 40.15, but with side $\frac{\sqrt{3}}{2}$, and a circle C of radius $\frac{2}{\sqrt{3}} + \frac{1}{2}$ with its center in the center of one of the triangles (Fig. 40.17). Inside C we have exactly 12 vertices of the lattice, they form our 12-gon K.

All there is left to show is that P does not contain a blue monochromatic congruent copy of K.

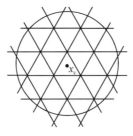

Fig. 40.17

Place a congruent copy K_1 of K anywhere in the plane P, together with a surrounding circle C_1 congruent to C. As we noticed above, C_1 will contain completely at least one of the red disks C_2. C_2 (with its red half-boundary) in turn will contain at least one of the vertices of K_1. Thus, at least one of the vertices of K_1 will always be red! ∎

About 15 years later, Rozália Juhász's 12-point counterexample was improved by the two other Hungarian mathematicians, György Csizmadia and Géza Tóth. On January 15, 1991 they submitted, and in 1994 published [CT] an 8-point counterexample, thus "almost" closing the gap.

Counterexample 40.30 (Csizmadia and Tóth) There is an 8-point set K in the plane (namely, a regular 7-gon and its center) and a 2-colored plane P (red and blue) such that P does not have either a monochromatic unit-distant red segment or a blue monochromatic set congruent to K.

Problems 40.28 and 40.30 deliver the state of the art in this direction. Can we guarantee a monochromatic blue pentagon of at least one given shape and size in a 2-colored plane without red monochromatic segment of unit length? Nobody knows! (So far pentagons have been slow to enter the Euclidean Ramsey Theory.) A small 3-number gap remains:

Open Problem 40.31 For which n in the interval $5 \le n \le 7$ is the following statement true:

For any n-gon K, any 2-colored plane (red and blue) contains a pair of red points distance 1 apart, or a monochromatic blue n-gon congruent to K?

41
3-Colored Plane, 2-Colored Space, and Ramsey Sets

Having created Problem 40.8 in 1989–1990, I tried the same construction in a 3-colored plane, and it worked! It is certainly not much but we know so little about a 3-colored plane that every little bit helps.

In 1991 [Soi3] I named an n-gon K in an n-colored plane *representative* if all n colors are represented among its vertices. (Paul Erdős and Ron Graham preferred the term *rainbow*.)

Problem 41.1 [Soi3] Any 3-colored plane contains a monochromatic or representative triangle T with the small side 1 and angles in the ratio 1 : 2 : 4.

Proof Assume that a 3-colored plane P (red, white, and blue) does not contain a monochromatic congruent copy of T. Toss a regular heptagon H of side 1 on the plane P.

H can have at most 3 vertices of the same color, because *any* 4 vertices of H contain a triangle congruent to T (prove it on your own). On the other hand, by the Pigeonhole Principle H must contain at least 3 vertices of the same color. Hence, 3 it is: H contains, say, three red vertices.

There are only three ways (up to rotations and reflection) to have 3 red vertices on H without red monochromatic copy of T (Fig. 41.1). Numbers of white and blue vertices must be 3–1 or 2–2 respectively. It is now easy to verify (do) that every completion of three colorings in Fig. 41.1, subject to the above constraints, contains a representative copy of T. ■

Fig. 41.1

We probably cannot expect a guaranteed monochromatic copy of *any* triangle in a 3-colored plane. I would like to know which ones we can guarantee:

A. Soifer, *The Mathematical Coloring Book*,
DOI 10.1007/978-0-387-74642-5_41, © Alexander Soifer 2009

Open Problem 41.2 Find all triangles T, such that any 3-colored plane in which all three colors are present, contains a monochromatic or representative triangle congruent to T.

Ronald L. Graham believes that we can exclude any triangle with an appropriate 3-coloring. He formulated the following conjecture during our July 10, 1991 phone conversation (it appeared in 1991 in [Soi3]). Now [Gra7], [Gra8] Graham is offering $25 for it.

Graham's $25 Conjecture 41.3 (R. L. Graham) For any triangle T there exists a 3-colored plane that does not contain a monochromatic triangle congruent to T.

And now, as promised in the title of this chapter, let us peek at 2-colorings of the space E^3. Unlike the case in the plane E^2, we do get a unit monochromatic equilateral triangle in *any* 2-coloring of E^3.

Problem 41.4 ([EGMRSS], Theorem 6 of Part I) Any 2-colored space E^3 contains a unit monochromatic equilateral triangle.

Proof Let the space E^3 be 2-colored, red and blue. We pick two points A and B of the same color, say red, distance 1 apart (we can pick such A and B in *any* plane of the space E^3). If there is a third red point C at distance 1 from both A and B, we are done. Otherwise, we get a whole circle S_1 of blue points that lies in the plane perpendicular to AB through the midpoint O_1 of AB (Fig. 41.2).

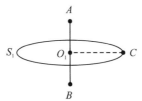

Fig. 41.2

The radius of this circle S_1 is $\frac{\sqrt{3}}{2}$. Now we pick a chord MN of S_1 of length 1. If there is a third blue point K at distance 1 from both M and N, we are done. Otherwise there is a whole circle S_2 of red points in the plane perpendicular to the plane of S_1 (Fig. 41.3). The radius of S_2 is, of course, the same as the radius of S_1 (because we really used the same construction for both circles).

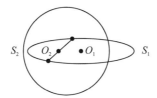

Fig. 41.3

If this second alternative holds as we rotate the chord MN about O_1, the white circle S_2 will rotate about O_1 accordingly and will create for us a degenerate torus T (a torus without a hole in the middle due to self intersection). Thus, we get a whole red torus T!

The largest horizontal circle (equator) S_3 on the torus T has diameter $d = \frac{\sqrt{2}+\sqrt{3}}{2}$ (verify that). We can inscribe in S_3 an equilateral triangle K of side $\frac{\sqrt{3}}{2}d = \frac{\sqrt{6}+3}{4} > 1$. Moving symmetrically the vertices of K along the surface of the torus T towards the middle of T (so that the plane determined by K remains horizontal), we will get a new equilateral triangle K_1 of side 1 on T. Since K_1 is on T and the whole torus T is red, K_1 is the desired monochromatic triangle. ∎

Paul Erdős et al., used a clever method similar to their solution of Problem 41.4 to prove the following stronger result. Try to prove it on your own.

Problem 41.5 ([EGMRSS]), Theorem 24 of Part PII) For any 2-colored space E^3 there is one color such that equilateral triangles of all sizes occur in that color.

This result, of course, makes one wonder whether a similar success can be guaranteed on the plane. However, this, is an open question:

Open Problem 41.6 [EGMRSS, Part III, p. 579] Is it true that for any 2-colored plane E^2 there is one color such that all triangles which occur monochromatically occur in that color?

Now we can prove for the space what is still an open problem for the plane.

Problem 41.7 For any triangle T, any 2-colored space E^3 contains a monochromatic triangle T_1 congruent to T.

Proof Let T be a triangle with sides a, b, and c, and the space be 2-colored. By Problem 41.5, the space contains a monochromatic equilateral triangle K_a of side a. Since the plane P that contains K_a is 2-colored, due to Tool 40.19 we have in P a monochromatic triangle T_1 with sides a, b, and c, which is congruent to T. ∎

For right triangles this result can be proved even for a 3-colored space, as Miklós Bóna and Géza Tóth showed in 1996 [BT]:

Problem 41.8 (M. Bóna and G. Tóth) For any right triangle T, any 3-colored space E^3 contains a monochromatic triangle T_1 congruent to T.[3]

In conclusion I would like to present here, without proofs, the two main results and the main open problem of the Erdős et al., trilogy [EGMRSS], and related results by P. Frankl and V. Rödl, and I. Křiž.

Generalizing the line R^1, the plane R^2 and the space R^3, we define the n-dimensional space R^n for any positive integer n as the set of all n-tuples $(x_1, x_2, \ldots, x_n)$, where $x_1, x_2, \ldots, x_n$ are real numbers. When the distance between two points $(x_1, x_2, \ldots, x_n)$ and $(y_1, y_2, \ldots, y_n)$ of R^n is defined by the equality

---

[3] Compare this result to Shader's Problem 40.10.

$$d = \sqrt{(x_1 - y_1)^2 + (x_2 - y_2)^2 + \ldots + (x_n - y_n)^2} \quad (*)$$

we get the *Euclidean n*-dimensional space E^n (in other words, E^n is just R^n together with the distance d defined by (*)).

Many notions are generalized from E^2 and E^3 straight forwardly to E^n. The *sphere* of radius r and center O in E^n is the set of all points of distance r from O. A 4-point set P in the plane is called a $d_1 \times d_2$ *rectangle* if it is congruent to the set $\{(0,0), (d_1,0), (0,d_2), (d_1,d_2)\}$. Similarly, a 2^n-point set in E^n is called a $d_1 \times d_2 \times \ldots \times d_n$ *rectangular parallelepiped* if it is congruent to the set

$$\{(x_1, x_2, \ldots, x_n) | x_1 = 0 \text{ or } d_1; x_2 = 0 \text{ or } d_2; \ldots; x_n = 0 \text{ or } d_n\}.$$

A finite subset C of E^n is called *r-Ramsey for* E^n if for any r-coloring of E^n there is a monochromatic subset C_1 congruent to C. If for every r there is n such that C is r-Ramsey for R^n, then the set C is called *Ramsey*.

Now you are ready for two main results by Paul Erdős et al.

Necessary Condition 41.9 ([EGMRSS], Theorem 13 of Part I) If a set C is Ramsey, then C must lie on an n-dimensional sphere for some positive integer n.

Sufficient Condition 41.10 ([EGMRSS], Corollary 22 of Part I) Any subset of a rectangular parallelepiped is Ramsey.

There is obviously a gap between the necessary and sufficient conditions for a finite set to be Ramsey. In fact, in 1986 Peter Frankl from France and Vojtech Rödl from Czechoslovakia proved that the Sufficient Condition 41.10 is not necessary by showing that even obtuse triangles (which cannot be embedded as subsets in a rectangular parallelepiped) are Ramsey:

Problem 41.11 (P. Frankl and V. Rödl, [FR1]) All non-degenerate triangles are Ramsey.

In their consequent paper they generalized this result to n-dimensional Euclidean spaces.

Problem 41.12 (P. Frankl and V. Rödl [FR3]) Any non-degenerate *simplex* (i.e., $n + 1$ points generating the whole n-dimensional Euclidean space) is Ramsey.

In 1991 Igor Křiž [Kri1], then from the University of Chicago (and presently from the University of Michigan), published powerful results that imply the following:

Problem 41.13 (I. Křiž) Any regular polygon is Ramsey.

Thus we finally get the first Ramsey pentagon: the regular one. Křiž's results also imply a similar result in three dimensions:

Problem 41.14 (I. Křiž) Any regular polyhedron is Ramsey.

This result has just (2007) been generalized by Kristal Cantwell [Can2] to all regular n-dimensional polytopes.

Problem 41.15 (K. Cantwell) All regular polytopes are Ramsey.

In his next 1992 paper [Kri2] Igor improved Frankl–Rödl result 41.11:

Problem 41.16 (I. Křiž) Any trapezoid is Ramsey.

As no criterion for a set to be Ramsey appeared, Paul Erdős attempted to speed up the process in 1985 [E85.01]:

> We (i.e., the authors of [EGMRSS]) do not know which (if any) of these alternatives characterize Ramsey sets, and I offer $500 for an answer to this question.

Paul Erdős's $500 Problem 41.17 Find a criterion for a set to be Ramsey.

Ever since 1993, if not before, up to the present [Gra3], [Gra7], [Gra8] Ronald L. Graham expressed his $1000-belief that the necessary condition 41.9 is also sufficient:

Ronald L. Graham's $1000 Problem 41.18 Prove that all spherical sets are Ramsey.

He also offered a consolation prize for a partial result [Gra7], [Gra8]:

Ronald L. Graham's $100 Problem 41.19 Prove that any 4-point subset of a circle is Ramsey.

42
Gallai's Theorem

42.1 Tibor Gallai and His Theorem

Gallai's theorem is one of my favorite results in all of mathematics. Surprisingly, it is not widely known even among mathematicians. Its creator was Tibor Gallai, born Tibor Grünwald, a member of the Hungarian Academy of Sciences, who passed away on January 2, 1992 at the age of 79. His lifelong close friend and coauthor Paul Erdős was visiting me in Colorado Springs[4] when Professor Vera T. Sós called from Budapest to give Paul the sad news of Gallai's passing. I asked Paul to write about Gallai for *Geombinatorics*. Here is Paul's *Obituary of My Friend and Coauthor Tibor Gallai* [E92.14] in its entirety, including the sketch for Sylvester–Gallai that he drew on the margin of his manuscript.

"I met Tibor Gallai in 1929 when we were both in high school. We knew of each other's existence since we both worked at the *Középiskolai Matematikai Lapok*, a journal for high school students which appeared every month and published problems and their solutions by students. This periodical had an immense influence on Hungarian mathematics; many children before the age of 15 realized that they wanted to be mathematicians, and many of the well-known mathematicians as young people worked in this journal. Gallai and I worked together on mathematics since 1930 and had many joint papers (for details, see my forthcoming obituary of Gallai in *Combinatorica* and also the article of Lovász and myself in *Combinatorica*, Vol. 2, 1982 written for Gallai's 70th birthday).

Here I just want to state some of the elementary results of Gallai which can easily be understood by beginners. In 1933 I conjectured that if $x_1, x_2, \ldots, x_n$ are n points in the plane, not all on a line, then there is always a line which goes through precisely two of our points. I thought that I will prove this in a few minutes but, in fact, I could not prove it. I told my conjecture to Gallai who found a very nice proof of it which goes as follows: Project one of the points to infinity and join it to all the other points. If my conjecture would be wrong, we would get a set of parallel lines each of which contains at least two finite points. Consider now the oblique lines, each of them contains at least three points. Take the line which has the smallest angle.

---

[4] We were working on our join project, a book of Paul's open problems: *Problems of pgom Erdős*, which I hope to finish by 2010.

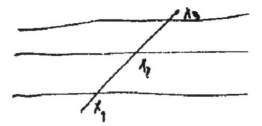

On the line in the middle there must be another point besides x_2, say y and one
of the lines $x_1 y$ or $y x_3$, clearly gives a smaller angle. This contradiction proves the
conjecture. A few years later L. M. Kelly found that my conjecture is not really
mine. It was conjectured in 1893 by Sylvester but as far as we know, Gallai was
the first who proved it. The simplest proof is due to L. M. Kelly. By the way,
I observed that Gallai's theorem implies that n points not all on a line determine at
least n distinct lines. There is a very nice related conjecture of Gabriel Dirac. Let
$x_1, x_2, \ldots, x_n$ be n points not all on a line. Join every two of them. Then at least
one of the points has $\frac{n}{2} - c$ lines incident to it. Beck, Szemerédi, and Trotter proved
that there is a point with at least $c_1 n$ lines incident to it; their c_1 is positive, but it is
very small.

Gallai was very modest—I would almost say abnormally so. Many of his beautiful
results were published with great delay. Often he did not publish them at all and they
were later discovered by others. He felt sorry for this only once. Dilworth in 1950 in
the *Annals of Mathematics* proved the following classical theorem: Let ϑ be a partially
ordered set. Assume that the maximal number of non-comparable elements is d, then
ϑ is the union of d chains. In fact, Gallai and Milgram had a complete proof of this
beautiful theorem in 1942. Milgram was a topologist who did not realize the importance of this result. Gallai wrote their joint paper in German. Milgram wanted to have
it published in English and promised to rewrite it but delayed it until it was too late. I
promised Gallai never to mention this in his lifetime since the theorem should clearly
be known as Dilworth Theorem.

Hilbert, in his beautiful obituary of Minkowski it *Math Annalen* 1909 wrote 'I can only
be grateful that I had a friend and co-worker for such a long time.' This is what I have
to say about Gallai and "May his theorems live forever."

Paul Erdős added [E92.15]:

A few years before his death he [Gallai] finally accepted the degree of Doctor of the
Academy and two years ago, much against his will, he was even granted the membership in the Academy.

Indeed, Gallai discovered a number of fabulous results, some of which were
named after other mathematicians: he preferred not to publish even his greatest
results. Why did he not publish them? On July 20, 1993 in Kesztely, Hungary during
a dinner my (then) wife Maya, our baby Isabelle and I shared with George Szekeres
and Esther Klein, the legendary couple from the legendary circle of young Jewish
mathematicians in the early 1930s Budapest, I was able to ask them about the friend
of their youth.

Tibor Gallai, 1935–1936. Courtesy of Alice Bogdán

"Gallai was so terribly modest," explained George Szekeres. "He did not want to publish because it would show the world that he was clever, and he would be restless because of it."

"But he was very clever indeed," added Esther Klein-Szekeres. Esther continued: "Once I came to him and found him in bed. He said that he could not decide which foot to put down first."

"Gallai was Paul Erdős's best, closest friend," continued George. "I was very close with Turán. It was later that Paul Erdős and I became friends."

I always thought, as probably everyone, that hypergraphs were invented by the great French graph theorist Claude Berge. Amazingly, Gallai was first here too: at the age of 18–19 (Gallai was born on July 15, 1912), he introduced hypergraphs. Paul Erdős mentioned it in passing in his 1991 talk at Visegrád (Hungary) Conference, and published 3 years later [E94.22]:

> As far as I know, the subject of hypergraphs was first mentioned by T. Gallai in conversation with me in 1931, he remarked that hypergraphs should be studied as a generalization of graphs. The subject really came to life only with the work of Berge.

Paul Erdős told me that Tibor Gallai discovered the theorem of our prime interest in the late 1930s. He did not publish it either. It *first* appeared in the paper [Rad2] by Richard Rado (with a credit to *"Dr. G. Grünwald,"* which was Gallai's last name then; the initial "G" should have been "T" and must be a typo). Rado submitted this paper on September 16, 1939; it is listed in bibliographies as a 1943 publication, but in fact came out only in 1945—the World War II affected all facets of life, and made no exception for the great Gallai result. I hope you will enjoy it as much as I have, and try your wit and creativity in proving this beautiful and extremely general, classic result. Readers, unfamiliar with m-dimensional Euclidean space, can look up the definition in chapter 41 or assume $m = 2$: plenty of fun is to be found in the plane!

Gallai's Theorem 42.1 ([Rad2]) Let m, n, k be arbitrary positive integers. If the lattice points Z^n (i.e., the points with integer coordinates) of the Euclidean space E^n are colored in k colors, and A is an m-element subset of Z^n, then there is a monochromatic subset A' of Z^n that is homothetic (i.e., similar and parallel) to A.

In fact, with out too much effort the Gallai Theorem can be strengthened as follows:

Gallai's Theorem 42.2 ([GRS2]) Let m, n, k be arbitrary positive integers. If the Euclidean space E^n is colored in k colors and A is an m-element subset of E^n, then there is a monochromatic subset A' of E^n that is homothetic to A.

In 1947, the well-known Russian mathematician Aleksandr Yakovlevich Khinchin (1894–1959) published a book *Three Pearls of Number Theory* [Khi1]. The booklet was an instant success, and second edition came out in 1948 [Khi2]. It included a new, "much simpler and transparent," in the opinion of Khinchin, exposition of Van der Waerden's proof (which we discussed in Chapter 33), proposed by the Russian mathematician M. A. Lukomsakja from Minsk. In 1951 this second edition of the book was translated into German and in 1952 into English. Each of these translations proved instrumental in bringing into existence two more independent proofs of Gallai's theorem. The 1951 German translation [Khi3] inspired Ernst Witt to discover his proof in 1951 ([Wit], submitted on September 21, 1951; published in 1952), while the 1952 English translation [Khi4] stimulated Adriano Garsia in finding his proof in 1958. Khinchin writes [Khi3]:

> It is not out of the question that Van der Waerden's theorem allows an even simpler proof, and all efforts in this direction can only be applauded.

Witt [Wit] quotes this Khinchin's call to arms in his paper, and happily reports:

> This was the occasion to strive for a new order of proof that then led directly to a more general grasp of the problem.

How does one attribute credit for this classic result? Graham, Rothschild, and Spencer call it "Gallai's Theorem" ([GRS2], while Hans Jürgen Prömel with his coauthors Vojtech Rödl and Bernd Voigt call it "Gallai–Witt's Theorem" [PR], [PV]. It is not a deciding factor for me that Gallai did not publish his proof— he shared it with Rado in 1930s, who gave Gallai credit very early, in 1939. It is not a deciding factor that Garsia did not publish his proof— he provided me with an old blue-line, faded from age, copy of his 1958 proof. Witt and Garsia appear to have discovered their proofs independently from Gallai, and their proofs constitute contributions to the field. However, the significant time that had lapsed between the 1930s and the 1950s prompts me to give credit for the discovery of the theorem to one person, and call it accordingly *Gallai's Theorem*.

Let me briefly introduce you to double induction, which will be used in the proof of Gallai's Theorem.

42.2 Double Induction

There is a curiosity related to induction in mathematical education. We all learn induction, some in high school, others in college. It is presented in numerous ways in numerous textbooks. However, when we start reading research articles, we encounter a more powerful version, called a *double induction*. So as we read, we have to understand what it is and be confident that it is a valid method. To the best of my (and Paul Erdős's) knowledge, the double induction is not presented in any textbook. To aid you, I will describe it for you here before using it.

Given lattice points on a line (Fig. 42.1):

○ ○ ○ ○ ○ ○ ○ ○ ○ ○ ○ ○ ○ . . . ○ . . .

Fig. 42.1

If we can visit the first point from the left, and we possess the translation ⇒ that takes us from any point to its neighbor to the right, then we can visit each lattice point on the line.

This is a (new) formulation of the *Mathematical Induction Principle*. It is one of the axioms of positive integers (and cannot be proven without assuming the truth of an equivalent statement).

The *Double Induction* does the same thing but on a two-dimensional lattice (Fig. 42.2). If we can visit the origin O and we possess two translations, ⇒ and ⇑ that take us from any point to its neighbor to the right or above respectively, then we can visit each lattice point of the plane.

Now you can easily envision a triple induction, a quadruple induction, etc. (even though I have not seen them actually used).

The double induction was used in the proofs of Baudet–Schur–Van der Waerden's theorem, Ramsey's theorem, and Gallai's Theorem.

42.3 Proof of Gallai's Theorem by Witt

Now we are ready to look at Ernst Witt's proof of the Gallai's theorem. In fact, it took an effort to read Witt's page and a half exposition before I completely understood his dense and beautiful proof (published in German, of course). I will try to preserve the beauty of it here, but will "unzip" it, make it more accessible and less

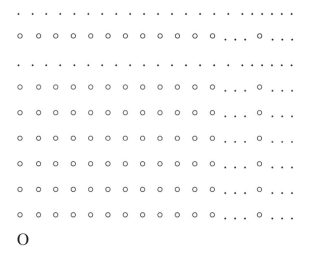

Fig. 42.2

dense. Witt proves Gallai's theorem for the plane. An n-dimensional generalization is straight forward.

Theorem 42.3 (Gallai's Theorem for the plane) For any arbitrary positive integer k and a finite n-element set S_n of points in the plane, there is a finite set $\Phi = \Phi(n, k)$ of points in the plane, such that if Φ is colored in k colors, there is a monochromatic subset S'_n of Φ that is homothetic to S_n.

Proof by Ernst Witt, "unzipped": We can think of points of the plane as of complex numbers. (Those who prefer to visualize points as ordered pairs of real numbers can do so. By addition of points in the plane we would mean the component-wise addition; multiplication of a point by a real number would be performed component-wise as well.)

Let the given figure S_n consist of n distinct complex numbers $e_1, e_2, \ldots, e_n$. Without loss of generality we can assume that $e_1 = 0$. We say that a figure S'_n is *homothetic* to the figure S_n if

$$S'_n = \lambda S_n + a,$$

where λ is a positive integer, and a is an arbitrary real number.

Given a figure $S_n = \{e_1, e_2, \ldots, e_n\}$, we would say that the two points e^1_1 and e^1_j from the set $\Phi = \Phi(n, k)$ are *connected* by S^1_n, if there is a homothetic image $S^1_n = \{e^1_1, e^1_2, \ldots, e^1_n\}$ of S_n such that the subfigure $S^1_{n-1} = \{e^1_1, e^1_2, \ldots, e^1_{n-1}\}$ of S^1_n is monochromatic.

We will prove the following Tool 42.4 and Theorem 42.3 in the same time:

Tool 42.4 (*E. Witt,* [Wit]) There is a finite set $\Phi = \Phi(n, k, m)$ of points in the plane such that for any coloring $f(x)$ of Φ in k colors, Φ contains a sequence of pairwise connected points $a_1, a_2, \ldots, a_m$.

In Tool 42.4 every pair of points a_i, a_j, $(i < j)$ is connected by a homothetic image $S_n^{ij} = a_i + \lambda_{ij} S_n$ which is monochromatic with a possible exception of its last point $a_j = a_i + \lambda_{ij} e_n$.

Proof by double induction: We will conduct a double induction in n and m with arbitrary k. For $n = m = 1$ the tool is obviously true.

In order to prove the step $m \Rightarrow m + 1$, we assume that the following sets have already been constructed,

$$\Phi_m = \Phi(n, k, m) \text{ and } \Phi_m^2 = \Phi(n, K, 2)$$

for a fixed n and m, and all k and K. Here K stands for the number of distinct ways to color the set Φ_m in k colors.

Now we can define the set Φ_{m+1} as follows:

$$\Phi_{m+1} = \Phi_{m+1}(S_n, k, m+1) = \Phi_m + \Delta_m$$
$$= \{x + y \,|\, x \in \Phi_m; \, y \in \Delta_m\}$$

i.e., Φ_{m+1} consists of all sums $x + y$ where x and y belong to the sets Φ_m and Δ_m respectively.

Assume that the set Φ_{m+1} is colored in k colors by a coloring f. Let y be an arbitrary point of Δ_m. The coloring of Φ_{m+1} uniquely determines the coloring of the subset

$$\Phi_m + y = \{x + y \,|\, x \in \Phi_m\},$$

and we can assign this coloring to the point y itself. Observe: since $\Phi_m + y$ is a translation of the set Φ_m in the plane, what we assigned to the point y is a coloring of Φ_m, one of K possible colorings of Φ_m (see the definition of K above). Thus we got the set Δ_m colored by the induced coloring, call it f^*, in K colors!

Now we can apply the inductive assumption to the set $\Delta_m = \Phi(S_n, K, 2)$: we conclude that Δ_m contains two connected by S_n points a and b, i.e., there is μ such that

$$f^*(a) = f^*(a + \mu e_i) \text{ for any } 0 \leq i < n.$$

In terms of the original coloring f of the set Φ_{m+1}, the above equality means the following:

$$f(x + a) = f(x + a + \mu e_i) \text{ for any } x \in \Phi_m \text{ and } 0 \leq i < n.$$

The set $\Phi_m + a$ can be viewed as a copy of the set $\Phi_m = \Phi(S_n, k, m)$ with the coloring in k colors that is induced by the coloring of Φ_{m+1}. Therefore, by applying the inductive assumption to $\Phi_m + a$, we conclude the existence of the sequence of m pairwise connected points $a_1, a_2, \ldots, a_m$.

Denote $a_m = a_i + \lambda_{jm}\, e_n$ for any $1 \le i \le n$ (of course, $\lambda_{mm} = 0$). We get:

$$f(a_i) = f(a_i + \lambda_{jm}\, e_j) = f(a_i + \lambda_{jm}\, e_j + \mu e_j) \text{ for any } 0 \le j < n$$

Taking into account $a_i + \lambda_{jm} e_n + \mu e_n = a_m + \mu e_n$, we obtain the extended sequence of pairwise connected points:

$$a_1, a_2, \dots, a_m, a_m + \mu e_n.$$

Starting with $\Phi_2 = \Phi(S_n, k, 2)$, we can construct $\Phi_m = \Phi(S_n, k, m)$ for any m. In particular, when $m = k + 1$, according to the Pigeonhole Principle, the sequence of pairwise connected points $a_1, a_2, \dots, a_k, a_{k+1}$ contains two points, say a_i and a_j, of the same color. Since a_i and a_j are connected, there is a set S_n^1 homothetic to S_n that contains a_i and a_j, and monochromatic with a possible exception for the color of a_j. But a_i and a_j are of the same color, therefore the set S_n^1 is monochromatic.

We are now ready to prove the second inductive step of our double induction: $n \Rightarrow n + 1$. Assume that the sets $\Phi(S_n, k, 2)$ and $\Phi(S_n, k, k + 1)$ are constructed. Let $S_{n+1} = \{e_1, e_2, \dots, e_n, e_{n+1}\}$ be a given $(n + 1)$-element set. We define the set $\Phi(S_{n+1}, k, 2)$ as the union of all homothetic images S_{n+1}^1 of S_{n+1} such that at least one of its n-element subsets lies in $\Phi(S_n, k, k + 1)$. Assume now that the set $\Phi(S_{n+1}, k, 2)$ is colored in k colors. According to the paragraph above, there is a *monochromatic* homothetic image $S_n^1 = \{e_1^1, e_2^1, \dots, e_n^1\}$ of the set $S_n = S_{n+1} \backslash \{e_{n+1}\} = \{e_1, e_2, \dots, e_n\}$ that lies in the set $\Phi(S_n, k, k + 1)$. Just add the point e_{n+1}^1 to S_n^1 that makes the set $S_{n+1}^1 = S_n^1 \cup \{e_{n+1}^1\}$ to be homothetic to S_{n+1}. We are done: the points e_1^1 and e_{n+1}^1 are connected by the homothetic image S_{n+1}^1 of the given set S_{n+1}, such that all the points of S_{n+1}^1 except possibly for e_{n+1}^1 are colored in the same color.

Ernst Witt concluded his paper with a noteworthy footnote, showing his way of visualizing the above proof:

In order to better conceptualize, one can imagine in some courtyard Φ taut clotheslines, on which from clothespins a_i hang similarly shaped laundry pieces S_n^1, which are monochromatic except for small mistake on the right end e_n^1. Every clothespin carries $n - 1$ pieces of laundry. ∎

Ernst Witt was born on the island of Alsen in 1911. Alsen together with the rest of North Schleswig became part of Germany in 1864. The island was returned to Denmark in 1920. Two-year old Witt went to China with his missionary parents. At 9 he was sent back to Germany to live with his uncle. Witt studied at the universities of Freiburg and Göttingen. His doctoral work at Göttingen was supervised by Emmy Noether. That was the dawn of Hitler's rein. Witt's former student at Hamburg (and presently a Bielefeld professor) Ina Kersten writes in his biography [Ker] that on May 1, 1933 Witt joined the Nazi party and the storm troopers SA—observe, he did it days after his teacher Noether was fired by the Nazi regime. After the war, as a proof of how little the Nazi and SA memberships meant to him, Witt claimed that his family did not know about his belonging to the SA and the Nazi party [Ker]. He was not as considered towards his fired from Göttingen Jewish mentor Emmy Noether [Bert]:

> Storm trooper Ernst Witt, resplendent in the Brownshirt uniform of Hitler's paramilitary, knocked on a Jew's apartment door in 1934. A short, rotund woman opened the door. Emmy Noether smiled, welcomed the young Nazi into her home, and started her underground math class. The Brownshirt was one of her favorite pupils.

Indeed, Witt must have been Noether's second most favorite student behind only Van der Waerden. Since Emmy Noether was forced out as Jewish and liberal, Witt defended his doctorate under Gustav Herglotz in July 1933, and joined Helmut Hasse's seminar, who entered Göttingen in 1934 as the Director of the mathematics institute and professor. Hasse's seminar was also attended by Oswald Teichmüller, who joined the Nazi party and the storm troopers even before Hitler came to power, and who was the boycott leader of Edmund Landau's classes.

Gian-Carlo Rota writes [Rota]: "There is no reason why a great mathematician should not also be a great bigot." These words are fully applicable to Witt's seminar leader Helmut Hasse. Hasse actively supported the gagster Nazi regime and its complete disregard for the most basic human rights, and expressed the most hateful attitudes towards people of other races and ethnicity. For instance, during Hasse's talk in Pisa after the start of World War II and before Italy's collapse, L. Tonelli asked Hasse about the fate of Polish mathematicians (and in particular of Schauder). Hasse replied:

> Poles should not do mathematics. They should work in coal mines and agricultural labor.[5]

It is amazing that even long years after the end of the war, Hasse did not change his racist rhetoric. Sanford Segal, who presents much material on Hasse [Seg], describes how in the 1960s at Ohio State University, Hasse claimed that "slavery in America had been good institution for blacks." Right after the war Hasse was welcomed to research positions and shortly after to professorships and other high honors by both East Germany and West Germany—*Mathematik über alles*! But let us return to our "hero," Ernst Witt.

Kersten [Ker] informs that in 1934 Witt became Hasse's Assistant at Göttingen. In 1937 Emil Artin left Hamburg for the United States. In 1939 Witt was appointed to the downgraded to an Associate Professor position of Artin and worked there until his dismissal by the British Military Authority in the fall of 1945. However, the Brits could not keep long grudges against the Nazis in Germany, and in 1947, Witt was reinstated in his position, in 1957 promoted to an Ordinarius, and remained on the job until his retirement in 1979.

Kersten describes Witt's 1960–1961 visit to the Institute for Advanced Study at Princeton, and his "astonishment" at the negative reaction of the Institute's members when Witt disclosed his Nazi past:

> On day during a discussion about a member of the National Socialist party, he [Witt] felt obligated to declare that he had also been a member of that party. To behave oth-

---

[5] The source: Jacopo Barsottti, who attended this Hasse's lecture.

erwise would have seemed insincere to him. He found, to his utter astonishment, that his contacts with his colleagues were suddenly severed.

Could Witt not comprehend that people at the Institute, some of whom escaped from the Nazis, were "sincerely" shocked to find a former Nazi in their midst? I am utterly astonished that Witt was "utterly astonished."

In 1978, Ernst Witt was given honors of membership in the Göttingen Academy of Sciences. He died in Hamburg on July 3, 1991 of natural causes.

42.4 Adriano Garsia

Let us now turn to Professor Adriano Garsia. The story of his discovery, told in the February 28, 1995 e-mail to me, is almost as intriguing as the story Van der Waerden told us in Chapter 33.

"I discovered the result in the fall of 1958. I was then a Moore instructor at MIT. We used to have fun at the time tossing each other problems at the common room. A student had asked the following question:

If we color the points of the plane in two colors can we always find a square with vertices all of the same color?

This problem frustrated everybody... including me... Until Paul Cohen[6] solved it. I didn't want to know the solution since I wanted to solve it myself... After a few days of unsuccessful attempts I finally asked somebody who knew Paul's solution how he did it!

I learned that he had used Van der Waerden's theorem on arithmetic progressions. I did not know of Van der Waerden's result at the time so I was at disadvantage on this one. So I got hold of Khinchin's book *Three Pearls of Number Theory* [Khi4] and studied Van der Waerden's proof very carefully.

I noticed then that the theorem could be generalized to higher dimensions to show that we could find any finite set of lattice points (up to scaling) with all elements of the same color.

I wrote up the proof and sent it to Van der Waerden who liked it and offered to publish it in the *Mathematische Annalen*. However a few weeks later I got another letter from Van der Waerden who had been doing some search on the literature on the subject and discovered that precisely the same generalization had already [been] published by T. Rado[7]... Under those circumstances he felt that although my proof was much neater... he didn't think it was worth publishing.

---

[6] In 1963 the American mathematician Paul Joseph Cohen (April 2, 1934–March 23, 2007) invented a technique called *forcing* and used it to prove that neither the continuum hypothesis nor the axiom of choice can be proved from the standard Zermelo–Fraenkel system of axioms (**ZF**) for set theory. In 1966 he won Fields Medal for this great achievement at the International Congress of Mathematicians in Moscow. He will appear again in Part X.

[7] Actually R. Rado, [Rad2].

In the mean time I had asked myself "what about a regular pentagon?. . ." In fact, what if we are given any geometric figure consisting of a finite set of points, can we find a stretch and translate of the figure with all elements of the same color?. . . Now it developed that my proof could be used under this more general situation as well. In fact, contrarily to P. Cohen or Rado who derived their result by applying Van der Waerden's theorem I had obtained mine by extending Van der Waerden's mechanism of proof.

Basically I showed that a sufficiently high "power" of the figure had to contain a monochromatic stretch and translate of the figure. (Power here means that we construct a figure of the form $F + a_1$, $F + a_2$, . . . $F + a_n$; with $A + B$ representing the vector sum of every point of A with every point of B.)

Although the version I had sent to Van der Waerden did not specifically address itself to this more general situation very little needed to be added to include this. Nevertheless after Van der Waerden's second letter I gave up on the idea of publishing the result. I have still some duplicates of seminar notes in which the more general result is presented. In fact the summer of 1959 I did give a lecture at Bell Labs on it. I believe G. Rodemich who is now at JPL [Jet Propulsion Laboratory], perhaps Henry Pollack was also at that lecture. . . I don't quite remember others. Jurgen Moser was at MIT at that time and I remember discussing my result with him in great detail.

This is the story. I am presently visiting UQAM and University of Montreal and it is difficult from here to locate those notes. I will get back to San Diego at the end of March. Send me your address and I will mail you a copy.

The idea of the proof is noticing that the same pigeon-hole argument of the original proof of Van der Waerden can be used in this more general situation. Inductively, we consider "colored" powers of the figure as "colors" assigned to say the center of the power. Then having proved the result for any number of colors and all figures with $n - 1$ points, we construct in a sufficiently high power of that power a monochromatic configuration of centers that is similar to the given figure minus a point. However, monochromatic centers now means that the corresponding powers centered at those points are all colored the same way!. . .

At this point we then use the Van der Waerden idea. . . which is well explained in Khinchin's book. Incidentally Khinchin states that he is presenting a simpler proof but Van der Waerden himself assured me that his proof was identical. . . I never did see Van der Waerden's original proof.

That is the story as I can remember it . . .

Best wishes on your book,

Garsia

PS: I am surprised that you call this Gallai's theorem. . . I was under the impression that a formal language version of the result which could be easily translated into mine (by sending letters into vectors) was due to Graham and Rothschild and a 3rd author I can't remember [Spencer – A.S.]."

To complement this fabulous story, Adriano Garsia sent to me the *original*, faded with age, 10 mimeographed blue-lettered pages of his notes, as he wrote on April 20, 1995 in another e-mail:

> I finally found the notes from which the paper I sent to Van der Waerden was written.
> I don't seem to have any copy of that paper. The notes are a bit faded but still readable.
> I am mailing them today. Best of luck in deciphering them.
>   – Garsia

Adriano Garsia was born in Tunisia on August 20, 1928. He received his secondary and college education in Rome, and Ph.D. from Stanford University in 1957. He was a professor at the California Institute of Technology (1964–1966), and since 1966 has been a professor at the University of California San Diego.

42.5 An Application of Gallai

A beautiful application of the Gallai Theorem was found by Alexej Kanel-Belov (listed as just Belov in this article) and S. V. Okhitin in 1992 [BO].

Theorem 42.5 ([BO]) Each cell of an [infinite] square grid contains an integer. For any given non-zero integer n there is a square with sides parallel to the lines of the grid, such that the sum of all integers inside it is divisible by n.

Proof Affix x and y axes along the lines of the grid. Now we "color" each unit cell (x, y) of the grid in one of n colors by assigning to it the remainder $S(x, y)$ upon division by n of the sum of numbers located in all cells with coordinates (a, b) such that $0 \leq a \leq x; 0 \leq b \leq y$.

The first quadrant of the grid is colored in n colors. By Gallai's Theorem, there is a monochromatic square, whose vertices have coordinates, say, $(x, y), (x + k, y), (x, y + k), (x + k, y + k)$. But this is all we need to prove the result, for it is easy to notice that the sum of all numbers inside this square is

$$S(x, y) - S(x + k, y) - S(x, y + k) + S(x + k, y + k),$$

and this sum is congruent to zero modulo n! ∎

The authors generalize this theorem on two counts at once as follows.

Theorem 42.6 ([BO]) Each cell of an [infinite] k-dimensional square grid contains an integer. For any given non-zero integer n and a positive integer m there is a positive integer $L = L(k, m, n)$ such that the grid contains a k-dimensional cube of side Lm with all edges parallel to the lines of the grid, which is partitioned into m^k "little" cubes of side L, such that the sum of all integers inside each "little" cube is divisible by n.

Hint: Instead of claiming a monochromatic square, as we did in the proof of 42.5, we can now use the Gallai Theorem to claim the existence of a monochromatic subgrid homothetic to the k-dimensional square grid of side L (which consists, of course, of m^k cells of the same color). ∎

Theorem 42.7 ([BO]) Each cell of an [infinite] k-dimensional square grid contains a real number. For any given positive integer m and a (small) positive ε there is a positive integer $L = L(k, m, \varepsilon)$ such that the grid contains a k-dimensional cube of side Lm with all edges parallel to the lines of the grid, which is partitioned into m^k "little" cubes of side L, such that the sum of all numbers inside each "little" cube differs from an integer less than by ε.

Hint: The proof repeats the proof of the previous result with the more delicate interpretation of coloring. We partition a unit segment $[0,1]$ into $N > 2^k m^k \varepsilon$ equal "little" segments – they are our "colors"–and determine the color of a cell of the grid with coordinates $(x_1, \ldots, x_k)$ by the "little" segment into which the fractional part falls of the sum of numbers in the grid's cells with coordinates $(a_1, \ldots, a_k)$, where $0 \leq a_i \leq x_i$ for $1 \leq i \leq k$. ∎

Of course, the Gallai Theorem allows us to generalize Theorems 42.5, 42.6, and 42.7 further and use k-dimensional parallelepipeds of the given in advance ratio of sides. I leave this development to you.

42.6 Hales-Jewett's Tic-Tac-Toe

Surely you played Tic-Tac-Toe in your tender years (Fig. 42.1). The goal is to mark a line of cells with your sign. In the "normal" Tic-Tac-Toe, the line can be horizontal, vertical, and diagonal (there are two diagonals). In fact, we can represent the cells by nodes, and replace X's and O's by two colors. The game then asks two players to color the nodes in turn. The winner is the one who creates a monochromatic line in his color. We will accept all the usual lines except one of the diagonals, going from the upper left to the lower right corner (Fig. 42.2).

Fig. 42.1

Fig. 42.2

In 1963 the 25-year old Alfred W. Hales and 26-year old Robert I. Jewett published the result that raised the Tic-Tac-Toe game to the level of a mathematical result of Ramsey Theory, the result of great importance. Informally speaking, they proved that the n-dimensional, r-player generalization of Tic-Tac-Toe cannot end in a draw, no matter how large n is, and no matter how many people r play so long as the playing board has a sufficiently high dimension. (In fact, the first player has a winning strategy due to the strategy-stealing argument.) As is often the case in mathematics, this is an existence result: no algorithm is known for the winning strategy.

In order to present the theorem formally, we need to define an n-dimensional cube and a *combinatorial line*, or simply a *line* in it. Given a fixed finite set, often called *alphabet*, $A = \{a_1, a_2, \ldots, a_m\}$, the *n-dimensional cube* on the alphabet A is, expectedly, the set $A^n = \{(x_1, x_2, \ldots, x_n) : x_i \in A\}$. Given a set of coordinates S, $\emptyset \neq S \subseteq \{1, 2, \ldots, n\}$, a line L is a set of the form

$$L = \left\{(x_1, x_2, \ldots, x_n) : x_i = x_j \text{ for } i, j \in S; \text{ and } x_l = a_l \in A \text{ for } l \notin S\right\}$$

We are ready to formulate Hales—Jewett's Theorem.

Hales—Jewett's Theorem 42.8 [HJ] For any finite set A and positive integer k there exists an integer $N(A, k)$ such that for $n \geq N(A, k)$, any k-coloring of A^n contains a monochromatic line.

A very clear "sketch of proof" can be found in [Gra1].

This result—as is often the case in mathematics—was obtained by the young mathematicians: Alfred W. Hales was 23, and Robert I. Jewell 24. Alfred e-mailed to me, on January 3, 2007, and recalled how it all came about:

> Bob and I were undergraduates at Caltech[8] together – he was a year ahead of me. We had common interests in both math and volleyball. We also both worked in Sol Golomb's[9] coding theory group at the Jet Propulsion Laboratory (JPL, affiliated with Caltech) during summers, and we continued doing this when we were in graduate school – he at the University of Oregon and I at Caltech.
>
> The strong connection between error correcting codes and combinatorics led Sol to steer us in various combinatorial directions and this led (eventually) to our joint paper written at JPL in 1961.

In the December 17, 2007, e-mail, Alfred added:

> I did ask Sol [Golomb] about this – You recall that he was our supervisor in the Jet Propulsion Laboratory's coding theory group. He seems to remember that a problem in Martin Gardner's column suggested to him the possibility of generalizing van der Waerden's theorem in some way, with applications to games and to coding in mind. He thinks he discussed this with us, and we proceeded to formulate and prove the eventual result.

In 1971 the Hales–Jewett Theorem earned the authors the George Polya Prize, which they shared with Ronald L. Graham, Klaus Leeb, and Bruce L. Rothschild, the authors of the Affine Ramsey Theorem.

There is a noteworthy connection between the two celebrated results (see proof in [GRS2, pp. 40–41]):

Connection 42.9 Hales–Jewett's Theorem implies Gallai's Theorem.

---

[8] California Institute of Technology.

[9] We have already met Solomon Golomb in Chapter 2 of the book.

IX

Colored Integers in Service of Chromatic Number of the Plane: How O'Donnell Unified Ramsey Theory and No One Noticed

An interesting recent result of O'Donnell [Odo4, 5], perhaps giving a small amount of evidence that
$$\chi\left(E^2\right) > 4$$

– Ronald L. Graham [Gra6]

Give a man a fish and you feed him for a day. Teach him how to fish and you feed him for a lifetime.

—老子

Indeed, I agree, Paul O'Donnell proved sensational results, showing that there are unit distance 4-chromatic graphs of girth 9, girth 12, and even an arbitrarily high girth. These results do give some evidence that, perhaps, the search for a 5-chromatic unit distance graph may celebrate its victory one day—this is the result O'Donnell was ultimately after, but has not succeeded: no one has.

The epigraph shows, of course, that Ron Graham appreciated the result, as did our other colleagues. However, what no one noticed, was how great Paul O'Donnell's proofs were. The dilemma of results vs. methods of proofs reminds me the proverb of the ancient Chinese sage, the father of Taoism 老子 (Lao Zi): "Give a man a fish; you have fed him for today. Teach a man to fish; and you have fed him for a lifetime."

Just imagine you created a huge 4-chromatic graph without cycles of order say, 100. Now you need to embed it in the plane so that every edge is a unit segment. Wouldn't you feel that this is extremely hard and messy, and you would likely waste much time and end up with nothing? Paul showed bravery and imagination when he plunged into unit distance embeddings, which we studied in Chapter 14.

He has also set world records of embedding smallest known unit distance graphs without small cycles, jointly with his friend and one time roommate Rob Hochberg—we have seen those in Chapter 15.

I, however, appreciate the most his constructions in this section. Paul uses the powerhouse of classic results of integer coloring, such as Baudet–Schur–Van der Waerden's Theorem, great results related to the search for the proof of Fermat's Last Theorem from Number Theory and Ergodic Ramsey Theory, such as Mordell–Faltings' Theorem and Bergelson–Leibman's Theorem. He applies this powerhouse of integer coloring and Number Theory sophistication to the problem of coloring the plane, the chromatic number of the plane problem. And by doing so, he is unifying Ramsey Theory in general, and this book in particular—as nothing else would.

I have followed Paul's research ever since my March 1992 talk at Florida Atlantic University, where my talk inspired him to write his thesis on these problems. He then visited me in Colorado Springs, and later claimed in his thesis that "it all came to me as I drifted off to sleep on your couch." In the end, I was a member of Paul O'Donnell's May-1999 Ph. D. defense committee at Rutgers University, together with János Komlos, Michael Saks, and Endre Szemerédi. Yet, even though I knew this dissertation well and followed through its many revisions, it took me time, well after Paul's defense, to fully appreciate how great Paul's results and moreover methods were. Enjoy!

43

Application of Baudet–Schur–Van der Waerden

At the end of Chapter 14, I left you with the embedding of the 352,735-vertex Blanche Descartes graph in the plane by Paul O'Donnell. You may ask, would attaching longer k-cycles ($k > 7$) to the foundation vertices increase the graph's girth while keeping the chromatic number at 4? The answer is no, not if k-cycles were attached to *all* k-element subsets of the foundation set, because some k-cycles would have two or more vertex intersection which could cut down the girth of the graph. We would get a chance to succeed at this construction if we were to dramatically limit the number of attached k-cycles, by, say, allowing at most a single point intersection for the k-subsets of the foundation to which k-cycles are allowed to be attached. This is exactly what O'Donnell has implemented.

We met hypergraphs at the end of Section 26.1; let us get acquainted with a special type of them here. A *k-uniform hypergraph H* is a family of k-element subsets of an *n-element set S*. The *vertices* of H are the elements of S. The *edges* (or *hyperedges*) of H are the k-element subsets. A *cycle* of length $k > 2$ in H is a sequence of distinct vertices and edges of H,

$$v_1, E_1, v_2, E_2, \ldots, v_k, E_k, \tag{43.1}$$

such that $v_{i+1} \in E_i \cap E_{i+1}$ for $1 \leq i \leq k$ (where the addition in the indices is done modulo k). The *girth* of a hypergraph is the length of its shortest cycle. The *chromatic number* of a hypergraph is the minimum number of colors needed to color the vertices so that no edge is *monochromatic*, i.e., consists of vertices which are all colored the same color.

Let n be a positive integer, H a graph on k vertices ($k \leq n$), and $S \subseteq \binom{|n|}{k}$ a k-uniform hypergraph.[1] Then $G_{n,H,S}$ would denote the Blanche Descartes graph[2]

---

[1] Here the symbol $\binom{|n|}{k}$ stands for the set of all k-element subsets of the $|n|$-element set.

[2] Defined in construction 12.10; see also examples of use 12.8 and 12.9.

A. Soifer, *The Mathematical Coloring Book*,
DOI 10.1007/978-0-387-74642-5_43, © Alexander Soifer 2009

built on the foundation vertex set $F = \{u_1^*, u_2^*, \ldots, u_n^*\}$ by attaching[3] copies of H to those subsets of F that are in S.

In this notation, the 112-vertex graph constructed in Problem 12.8 can be recorded as $G_{7,3-cycle,\binom{|7|}{3}}$. The 6448-vertex graph, first embedded by Wormald (Section 12.3), can be recorded as $G_{13,5-cycle,\binom{|13|}{5}}$. The girth 6, 352,735-vertex Blanche Descartes graph embedded in the plane by Paul O'Donnell (see the end of Chapter 14), is encoded as $G_{19,7-cycle,\binom{|19|}{7}}$.

O'Donnell came up with a brilliant idea of attaching cycles only to certain *arithmetic progressions* (for short, APs) of the foundation set and restricting APs in the following two ways:

1. the set D of allowable common differences is chosen so that APs with distinct common differences overlap by at most one element (overlaps by two or more vertices may create small cycles).
2. given D, the set S is constructed so that APs with the same common difference do not overlap.

The distance between any two points in a k-term AP is ad, where $a < k$ and d is the common difference. To prevent two APs from intersecting at two points, it suffices to ensure that $ad_1 \neq bd_2$ for all a, b less than k and distinct common differences d_1, d_2 from D. Formally, let D_j denote the set of allowable common differences less than or equal to j. We define D_j recursively:

$$D_j = \begin{cases} D_{j-1} \cup \{j\}, & \text{if for all } d \in D_{j-1} \text{ and positive integers} \\ & a, b \in [k-1], ad \neq bj; \\ D_{j-1} & \text{otherwise.} \end{cases} \tag{43.2}$$

Then the allowable set of common differences is:

$$D = \bigcup_{j=1}^{\infty} D_j \tag{43.3}$$

How dense is D? If too many numbers are in D, then the graph will have short cycles. If too few numbers are in D, then the graph will not be 4-chromatic. So, we need to perform a balancing act! The following tool gives an idea of the density of D.

Tool 43.1 For all d, at least one of $\{d, 2d, 3d, \ldots, k!d\}$ is in D.

Proof If $k!d \in D$ then we are done. If not, then there exist positive integers $a, b \in [k-1]$ and $d_1 \in D$ with $d_1 < k!d$ such that $ad_1 = bk!d$. Solving for

---

[3] Defined in Section 14.1.

d_1, we get $d_1 = \frac{bk!d}{a}$. Since $a < k$, a divides $k!$, and thus d_1 is a multiple of d, as desired. ∎

Once we get D, we can construct the set S of APs. Let us formally define S:

$$S = S(n, k, D) = \{\{a, a + d, \ldots, a + (k-1)d\} : d \in D, \tag{43.4}$$
$$a \equiv 1, 2, \ldots, d \pmod{kd}, a + (k-1)d \le n\}. \tag{43.5}$$

For example, if $D = \{1, 3, 4, 5, \ldots\}$ then $S(17, 3, D)$ is:

{1,2,3}	{1,4,7}	{1,5,9}	{1, 6,11}
{4,5,6}	{2,5,8}	{2, 6,10}	{2, 7,12}
{7,8, 9}	{3,6,9}	{3,7,11}	{3, 8,13}
{10,11,12}	{10,13,16}	{4,8,12}	{4,9,14}
{13,14,15}	{11,14,17}		{5,10,15}

Now we need to check the chromatic number and the girth of the graph $G_{n,k\text{-}cycle,S}$ for appropriate k and n, and verify that $G_{n,k\text{-}cycle,S}$ is a unit distance graph.

It is delightful to see how Paul O'Donnell uses the Baudet–Schur–Van der Waerden's Theorem (Theorem 33.1) to show that for some n, $G_{n,k\text{-}cycle,S}$ is 4-chromatic!

Theorem 43.2 There exists n such that $\chi(G_{n,k\text{-}cycle,S}) = 4$.

Proof By Baudet–Schur–Van der Waerden's Theorem, there exists n such that any 3-coloring of the integers from 1 to n contains a monochromatic AP of length $(2k - 1)k!$.

Let d be the common difference of this AP. By Tool 43.1 there exists $d' \in D$, such that d' is a multiple of d such that $d' \le k!d$. Hence there is a $(2k - 1)$-term monochromatic AP of foundation vertices with $d' \in D$

$$u_a, u_{a+d'}, \ldots, u_{a+(2k-2)d'}. \tag{43.6}$$

One of the first k of these indices is congruent to some element in $\{1, 2, \ldots, k\}$ (mod kd). The vertex with this index and the $k - 1$ vertices after it (in the AP with common difference d') form a set in S. This set has a k-cycle attached. But if all of these foundation vertices are of the same color, there are only two colors remaining to color the odd cycle. Two colors are not enough. Thus at least four colors are necessary to color $G_{n,k\text{-}cycle,S}$. ∎

Theorem 43.3 For odd $k \ge 9$, girth($G_{n,k\text{-}cycle,S}$) ≥ 9.

Proof A cycle containing no foundation vertices is a k-cycle. All other cycles consist of the foundation vertices separated by at least two vertices of an attached cycle. It is therefore impossible to have a cycle with only one foundation vertex.

A cycle has only two foundation vertices if the APs of the two attached cycles intersect in two places. However, our choices for D and S prevent this.

A cycle with at least three foundation vertices has at least nine vertices. Therefore, the girth of our graph is at least min $\{9,\ k\}$. ∎

Observation: Just like the Blanche Descartes construction, this method generalizes to arbitrary chromatic number. By attaching girth 9, $(l-1)$-chromatic graphs to appropriate APs of the foundation vertices; we obtain girth 9, l-chromatic graphs. However, we need to embed our graphs in the plane as unit distance graphs, and the 4-chromatic graphs seem to be the only reasonable candidates for it.

Theorem 43.4 There exists a girth 9, 4-chromatic unit distance graph.

Proof As we have established above, for appropriate choices of k and n, the graph $G_{n,k\text{-}cycle,S}$ is 4-chromatic of girth at least 9. Given odd $k \geq 9$, let n_0 be the smallest such n. We show that $G_{n,k\text{-}cycle,S}$ is a unit distance graph using an embedding procedure similar to that used for the Blanche Descartes graphs in Chapter 14.

By the choice of n_0, there is a 3-coloring of the foundation vertices labeled from 1 to $n_0 - 1$ such that no monochromatic set has an odd cycle attached. We place all the foundation vertices with color i in the δ-ball around C_i for $1 \leq i \leq 3$. We place vertex n_0 in the δ-ball around C_4. Since the vertices with a k-cycle attached are always in at least 2 δ-balls, the embedding tools of Chapter 14 allow the attachments of all cycles and removal of any coincidences. (Technically if the girth is more than 9, we add a 9-cycle to get a girth 9 graph.) ∎

44

Application of Bergelson–Leibman's and Mordell–Faltings' Theorems

To achieve a girth 12 unit distance graph, Paul O'Donnell alters the set D of allowable common differences. This changes which sets are in S (i.e., which sets of the foundation vertices get odd cycles attached). It's no longer enough for the sets in S to have intersection of size at most one, as we required in Chapter 43. In addition, O'Donnell requires that no three sets in S intersect pairwise. How does one achieve this?

Unexpectedly, O'Donnell uses sophisticated results of Number Theory and Ergodic Ramsey Theory. He attaches k-cycles only to specified APs whose common difference is an m-th power, for he wants to make use of BLT's Corollary 35.10 of Bergelson–Leibman'Theorem! (Chapter 35).

As was done in Chapter 43, we will again use the Blanche Descartes $G_{n,k\text{-}cycle,S}$ construction. We will then establish that the constructed graph is indeed 4-chromatic girth at least 12 unit distance graph.

However, before we dive into "O'Donnellia," we need to take a tour of Number Theory related to Fermat's Last Theorem. As is customary in this book, we will include at least a brief history of this field in our excursion.

In 1922 Louis Joel Mordell (Philadelphia, 1888-Cambridge, 1972) conjectured [Mor], and in 1983 the 29-year-old German mathematician Gerd Faltings published (and in 1986 was awarded the Fields Medal primarily for his proof) this very important result (in a more contemporary formulation than Mordell could have had). This result, among other consequences, was, of course, a major step in the ascent of Fermat's Last Theorem. In consistently following my view that creating conjecture is important (every theorem is preceded by a conjecture, and sometimes the conjecture is brought up by someone other than the one who proved it), I will call it Mordell–Faltings' Theorem. We need here precisely the consequences of this theorem that are relevant to Fermat's Last Theorem when we construct the set of allowable common differences. It deals with (integer) solutions of Diophantine equations of the form

$$ax^m + by^m + cz^m = 0. \quad (*)$$

A. Soifer, *The Mathematical Coloring Book*,
DOI 10.1007/978-0-387-74642-5_44, © Alexander Soifer 2009

Before we state the theorem, we need to introduce some preliminaries. A solution $(x_0, \ y_0, \ z_0)$ of (*) is called *primitive* if $gcd\{x_0, \ y_0, \ z_0\} = 1$; and *trivial* if $x_0, \ y_0, \ z_0 \in \{-1, \ 0, \ 1\}$. Notice that if $(x_0, \ y_0, \ z_0)$ is a solution, then any integer multiple of this triple is also a solution. Thus, if an equation has one solution it has infinitely many. However, for an appropriate choice of m, it has only a finite number of primitive solutions. For a better choice of m all primitive solutions are also trivial. For the final choice of m, all equations (*) with $a, b, c \in \{-k, \ldots, k\}$, not all zero, have no nontrivial primitive solutions. This allows us to construct the set of allowable common differences and the set of APs to which odd cycles are attached.

Mordell–Faltings' Theorem 44.1 A nonsingular projective curve of genus at least two over a number field has at most finitely many points with coordinates in the number field.

I refer you to contemporary Number Theory texts for definitions of terms used in 44.1. What we need here is the following Corollary, obviously relevant to Fermat's Last Theorem:

Mordell–Faltings' Corollary 44.2 Given $a, \ b, \ c \in Z$ not all zero, for $m \geq 4$ the equation $ax^m + by^m + cz^m = 0$ has at most finitely many primitive solutions.

Tool 44.3 Given $a, \ b, \ c \in Z$, not all zero, there exists m such that the equation $ax^m + by^m + cz^m = 0$ has no nontrivial primitive solutions.

Proof Mordell–Faltings' Corollary 44.2 states that for $m \geq 4$, $ax^m + by^m + cz^m = 0$ has finitely many primitive solutions. Given a, b, c, let w be the integer of the largest absolute value in any primitive solution of $ax^4 + by^4 + cz^4 = 0$. Choose $l = l(a, b, c)$ such that $2^l > w$. We need the following claim to complete the proof:

Claim 44.4 The equation $ax^{4l} + by^{4l} + cz^{4l} = 0$ has no primitive solutions except possibly trivial ones, in which $x, \ y, \ z \in \{-1, \ 0, \ 1\}$.

Proof of 44.4 Assume $ax_0^{4l} + by_0^{4l} + cz_0^{4l} = 0$ with $gcd\{x_0, \ y_0, \ z_0\} = 1$. Then $a\left(x_0^l\right)^4 + b\left(y_0^l\right)^4 + c\left(z_0^l\right)^4 = 0$ shows that $x_0^l, \ y_0^l, \ z_0^l$ is a primitive solution of $ax^4 + by^4 + cz^4 = 0$. By the definition of w,

$$\max\left(\left|x_0^l\right|, \left|y_0^l\right|, \left|z_0^l\right|\right) \leq |w| < 2^l,$$

therefore, $x_0, \ y_0, \ z_0 \in \{-1, \ 0, \ 1\}$. ∎

All there is left to complete the proof of Tool 44.3 is to choose $m(a, \ b, \ c) = 4l$, which in view of 44.4 satisfies the statement of Tool 44.3. ∎

Corollary 44.5 Given a positive integer k, there exists a positive integer m' such that none of the equations $ax^{m'} + by^{m'} + cz^{m'} = 0$ with $a, b, c \in \{-k, \ldots, k\}$ not all zero, has a nontrivial primitive solutions.

Proof Given a, b, c, by Tool 44.3 there exists $m = m(a, b, c)$ such that $ax^m + by^m + cz^m = 0$ has no nontrivial primitive solutions. The same holds with for any exponent which is a multiple of m. Hence

$$m' = \prod_{\{(a,b,c):a,b,c\in\{ik,...,k\},\text{ not all }0\}} m\,(a,b,c)$$

suffices. ∎

Everything is now ready for our construction. Given $m' = m'(k)$, we define $D = \{x^{m'} : x \in N\}$. This is the set of allowable common differences needed to construct the set S of APs. Each AP in S corresponds to a set of foundation vertices with an attached cycle.

Theorem 44.6 For odd $k \geq 13$, $girth(G_{n,k\text{-}cycle,S}) \geq 12$.

Proof A few cases need to be addressed depending upon the number of the foundation vertices in a k-cycle.

A cycle containing no foundation vertices is a k-cycle. All other cycles consist of foundation vertices separated by at least two vertices of an attached cycle. So a cycle with at least four foundation vertices has at least 12 vertices.

A cycle has three foundation vertices if the APs of the three attached cycles intersect pairwise. Let a_i be the starting point and d_i be the common difference, $1 \leq i \leq 3$, for the three APs. The pairwise intersections of the APs imply the existence of constants $c_1, c_2, \ldots, c_6$ between 0 and $k - 1$ such that

$$a_1 + c_1 d_1 = a_2 + c_2 d_2$$
$$a_2 + c_3 d_2 = a_3 + c_4 d_3$$
$$a_3 + c_5 d_3 = a_1 + c_6 d_1$$

Thus,

$$a_1 + a_2 + a_3 + c_1 d_1 + c_3 d_2 + c_5 d_3 = a_1 + a_2 + a_3 + c_6 d_1 + c_2 d_2 + c_4 d_3$$

or,

$$(c_1 - c_6)\,d_1 + (c_3 - c_2)\,d_2 + (c_5 - c_4)\,d_3 = 0.$$

Since the common differences are all m'-th powers and the three foundation vertices are distinct, this is an equation of the form $ax^{m'} + by^{m'} + cz^{m'} = 0$ with integral coefficients $a, b, c \in \{-k, \ldots, k\}$ not all zero. By Corollary 44.5, it has only trivial primitive solutions. Thus, any solution has all the d_i equal, yet in the construction of S, APs with the same common difference do not intersect.

A cycle has only two foundation vertices if the APs of the two attached cycles intersect in two places. Let a_i be the starting point and d_i be the common difference, $1 \leq i \leq 2$ for the two APs. The intersection of the APs implies the existence of constants c_1, c_2, c_3, c_4 between 0 and $k - 1$ such that

$$a_1 + c_1 d_1 = a_2 + c_2 d_2$$
$$a_1 + c_3 d_1 = a_2 + c_4 d_2$$

By adding up the respective side of these equalities, we get

$$(c_1 - c_3) d_1 + (c_4 - c_2) d_2 = 0.$$

Since the common differences are all m'-th powers and the two foundation vertices are distinct, this is an equation of the form $ax^{m'} + by^{m'} = 0$ with non-zero integer coefficients between $-k$ and k. As in the previous case, there are no nontrivial primitive solutions. The d_i must be equal, yet in the construction of S, APs with the same common differences do not intersect.

A cycle with only one foundation vertex is not possible. Therefore the girth is at least $\min\{12, \ k\}$. ∎

Observation: Just like Blanche Descartes's girth 9 construction of Chapter 43, this method generalizes to arbitrary chromatic number. By attaching girth 12, $(l - 1)$-chromatic graphs to appropriate APs of foundation vertices, we get girth 12, l-chromatic graphs. Again, the only reasonable candidate for embedding in the plane as unit distance graphs seem to be the 4-colorable graphs.

Theorem 44.7 There exists n such that $\chi(G_{n,k\text{-}cycle,S}) = 4$.

Proof By BLT's Corollary 35.10 of Bergelson–Leibman's Theorem 35.9 (Chapter 35), there exists n such that any 3-coloring of the integers from 1 to n contains a $(2k - 1)$-term monochromatic AP of foundation vertices

$$u_a, u_{a+d}, \ldots, u_{a+(2k-2)d}$$

where d is a m-th power. One of the first k of these indices is congruent to some element in $\{1, \ 2, \ \ldots, \ k\} \pmod{kd}$. The vertex with this index and the $k - 1$ vertices that follow it, form a set in S. This set has a k-cycle attached. But if all of these foundation vertices are of the same color, there are only two colors remaining to color the attached odd cycle. This is not enough. Thus at least four colors are necessary to color $G_{n,k\text{-}cycle,S}$. ∎

We are ready for the embedding.

Theorem 44.8 There exists a girth 12, 4-chromatic unit distance graph.

Proof From the preceding theorems we know that for appropriate choices of k and n, the graph $G_{n,k\text{-}cycle,S}$ is a 4-chromatic graph of girth at least 12. Given odd $k \geq 13$, let n' be the smallest such n. We will show that $G_{n',k\text{-}cycle,S}$ is a unit distance graph using an embedding procedure similar to that used in the previous Chapter. By the choice of n', there is a 3-coloring of the foundation vertices labeled from 1 to $n' - 1$ such that no monochromatic set has an odd cycle attached. We place all of the foundation vertices of color i in the δ-ball around C_i, for $1 \leq i \leq 3$. We place vertex n' in the δ-ball around C_4. Since the vertices with a k-cycle attached are always in at least 2 δ-balls, the embedding tools of Chapter 14 allow the attachments of all cycles and removal of any coincidences. (Technically, if the girth is more than 12, we add a 12-cycle to get a girth 12 graph.) ∎

45
Solution of an Erdős Problem: O'Donnell's Theorem

45.1 O'Donnell's Theorem

In a surprising twist, the complete solution of Paul Erdős's old July-1975 problem about unit distance 4-chromatic graphs of arbitrary girth comes out to be simpler than all partial solutions we have discussed in the previous two Chapters. In another surprise, Paul O'Donnell uses in his solution the 1966 result obtained jointly by Paul Erdős and Andras Hajnal, the result that has been known all alone, but no one has noticed any connection! You may wish to revisit definitions of uniform hypergraphs in the beginning of Chapter 43.

Theorem 45.1 (Erdős–Hajnal 1966, [EH1]) For all integers $k \geq 2$, $g \geq 2$ and $l \geq 2$ there exist k-uniform, girth g, l-chromatic hypergraphs.

This theorem gives the desired generalization of the girth 9 and girth 12 constructions. Instead of attaching cycles to APs, we attach cycles to the edges (hyperedges) of a hypergraph. Given k and g, let H be a k-uniform, girth g, 4-chromatic hypergraph. Let $n = |V(H)|$. Then $G_{n,k\text{-}cycle,H}$ is the desired graph (re-read its definition in Chapter 43 if need be).

Theorem 45.2 (O'Donnell) $\chi(G_{n,k\text{-}cycle,H}) = 4$.

Proof Since H is 4-chromatic, any 3-coloring of the foundation vertices contains a monochromatic hyperedge. In other words, any 3-coloring of the foundation vertices has a monochromatic set with an odd cycle attached. That odd cycle cannot be colored with the remaining two colors, so $\chi(G_{n,k\text{-}cycle,H}) \geq 4$. With four colors, one can be used for the foundation vertices leaving three for the odd cycles. Thus $\chi(G_{n,k\text{-}cycle,H}) = 4$. ∎

Theorem 45.3 (O'Donnell) $girth(G_{n,k\text{-}cycle,H}) = k$.

Proof The approach is to show that $girth(G_{n,k\text{-}cycle,H}) \geq \min\{k, 3g\}$ and choose $g \geq k/3$. The only cycles containing no foundation vertices are the attached k-cycles. All other cycles consist of foundation vertices separated by at least two vertices of attached cycles. Since any two consecutive foundation vertices are in the shadow of an attached cycle in G (i.e., appear in a hyperedge of H), the consecutive foundation vertices form a cycle (i.e., hypercycle) in H. So if the girth of H is g, the length of the cycle in $G_{n,k\text{-}cycle,H}$ is at least $3g$. Thus, all cycles of $G_{n,k\text{-}cycle,H}$ are either k-cycles or l-cycles for $l \geq 3g$. ∎

O'Donnell's Theorem 45.4 [Odo3]; [Odo4, 5] For any $k \geq 3$, there exists a girth k, 4-chromatic unit distance graph.

Proof Assume k is odd. Let H be a k-uniform, 4-chromatic hypergraph with girth $\geq k/3$ having the fewest vertices. Let $n' = |V(H)|$, then as we know from the previous theorems, $G_{n',k\text{-}cycle,H}$ is a girth k, 4-chromatic graph. As in the previous two Chapters, we use the embedding tool chest of Chapter 14. By the choice of n', there is a 3-coloring of the foundation vertices labeled from 1 to $n' - 1$ such that no hyperedge is monochromatic, in other words, no monochromatic set has an odd cycle attached. We place all the foundation vertices with color i in the δ-ball around C_i for $1 \leq i \leq 3$, and place vertex n' in the δ-ball around C_4. Since the vertices with a k-cycle attached are always in at least 2 δ-balls, the embedding tools allow the attachments of all cycles and removal of any coincidences. For an even k, a k-cycle is added to the 4-chromatic unit distance graph of girth $> k$. ∎

Would you like to *see* the embedded O'Donnell graph? Paul offers an illustration (Fig. 45.1):

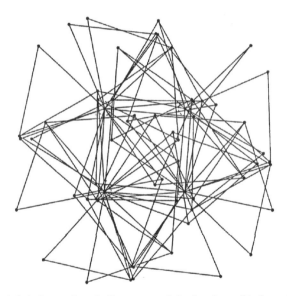

Fig. 45.1 A girth k 4-chromatic unit distance graph in the plane. (Notice, what looks like a vertex is many vertices; what looks like an edge is many almost parallel edges.)

45.2 Paul O'Donnell

My old request for a "self-portrait" has been granted by Paul O'Donnell in his March 31, 2007 e-mail:

I was born in New York City on April 18, 1968. I was adopted in October 1968 and grew up in Jackson, New Jersey. I received my undergraduate degree in mathematics and computer science from Drew University in 1989 and my Ph.D. in applied mathematics from Rutgers University in 1999. My doctoral thesis was on arbitrary girth 4-chromatic unit distance graphs in the plane, from a problem posed by Paul Erdős.

My interest in unit distance graphs sprang originally from a Putnam exam problem about them, and from undergraduate courses taught by Linda Lesniak. This interest was reawakened after attending Alexander Soifer's 1992 presentation on the interesting history of this problem at a conference at Florida Atlantic University in Boca Raton. This marked the start of our friendship.

The main idea for the arbitrary girth unit distance graph work came a few years later while dozing off to sleep on Alexander Soifer's couch in Colorado Springs after watching the Derek Jarman movie "Wittgenstein" with him.

I have taught at Rutgers University and Drew University. Currently, I am working in the Research & Development Equity department of Bloomberg L.P. In my free time I play ultimate frisbee and am a theatre/movie buff (credits include work on the L.A. and off-Broadway productions of the musical Reefer Madness, and an appearance as an extra in the movie Army of Darkness). My wife Carmelita and I also teach ballroom/latin dance and are the proud parents of daughter Kimberly.

Paul O'Donnell

X
Predicting the Future

I never think of the future – it comes soon enough.
– Albert Einstein

Prediction is very difficult, especially about the future.
– Niels Bohr

46
What If We Had No Choice?

46.1 Prologue

A prudent question is one-half of wisdom
– Francis Bacon[1]

On the pages of this book we have seen a variety of approaches used in attempts to settle the chromatic number of the plane problem (CNP). Tools from graph theory (Chapter 17), topology (Chapters 8, 24), measure theory (Chapter 9), abstract algebra (Chapter 11), discrete and combinatorial geometry (Chapters 4, 6, 7) have been tried and yet no improvement has been attained in the general case. The range for CNP remains as wide open as ever: $\chi = 4, 5, 6,$ or 7.

I felt—and wrote a number of years ago—that such a wide range was an embarrassment for mathematicians. The 4-Color Map-Coloring Problem, for example, from its beginning in 1852 had a conjecture: 4 colors suffice. Since 1890, thanks to Alfred Bray Kempe and Percy John Heawood [Hea], we knew that the answer was 4 or else 5. The CNP problem is an entirely different matter. After 58 years of very active work on the problem, we have not even been able to confidently conjecture the answer. Have mathematicians been so bad, or has the problem been so good? Have we been missing something in our assault on the CNP?

These were the questions that occupied me as I was flying across the country from Colorado Springs to the Rutgers University of New Jersey in October 2002, for a week of joint research with Saharon Shelah, in my opinion a genius of problem solving and a very quick learner (I knew that, for we produced a couple of joint papers on Abelian group theory before, in 1984, when we met in Udine, Italy).[2] On Saharon's request, I compiled a list of problems we could be interested in working

---

[1] Quoted from [Pet], p. 494.

[2] Ronald L. Graham and Joel H. Spencer [GS] agree with me: "Shelah is widely regarded as one of the most powerful problem solvers in modern mathematics."

A. Soifer, *The Mathematical Coloring Book*,
DOI 10.1007/978-0-387-74642-5_46, © Alexander Soifer 2009

Saharon Shelah (*left*) and Alexander Soifer, Paul Erdős's 80th Birthday Conference, Keszthely, Hungary, July 18, 1993. Photograph by Maya Soifer

on together, and numbered them according to set-theorists' taste, from 0 to 12. Problem 0 read as follows:

What if we had no choice?

This was a natural question for someone who grew up in the Soviet Union with very few choices: we voted for one candidate per office, ate whatever food was sold at the moment, and lived wherever we were allowed to live. But of course, I meant here something else that made mathematical sense. So, let me explain.

Nicolaas G. de Bruijn and Paul Erdős reduced CNP to finite sets on the plane, as we have seen in Chapters 5 and 26. Their famous theorem, obtained in fact shortly before Ed Nelson even posed CNP, required the Axiom of Choice (**AC**). This is the choice I referred to in my problem 0 for Saharon and me to ponder on:

What if we had no Axiom of Choice?

In the absence of the **AC**, we would not have the De Bruijn–Erdős Theorem, and so CNP would not necessarily be reduced to finite plane sets. In particular, I was interested in the following questions:

*What can and should we use in place of the **AC**?*
What results can we prove in this alternative Set Theory?
How would "choiceless" mathematics compare to the mathematics built on Choice?

And so Saharon and I met for a week of a Garden State autumn, and broke some new ground. Before we look at the outcome of our meeting, I need to offer you an excursion to the Land of Choice.

46.2 The Axiom of Choice and its Relatives

> *To choose one sock from each of infinitely many*
> *pairs of socks requires the Axiom of Choice, but*
> *for shoes the Axiom is not needed.*
> — Bertrand Russell

> *At present, set theory has lost its relevance*
> – L. S. Pontryagin[3]

The **AC** was used implicitly throughout the nineteenth century. A careful observation would uncover that it was used for proving even such a classic result as the sequential Bolzano–Weierstrass Theorem (every infinite bounded subset of reals has a sequential limit point). In 1904, while proving the Well-Ordering Principle, Ernst Friedrich Ferdinand Zermelo (1871–1953) formalized and for the first time explicitly used the **AC** [Zer]:

The Axiom of Choice (**AC**): Every family Φ of nonempty sets has a choice function, i.e., there is a function f such that $f(S) \in S$ for every S from Φ.

The newborn axiom prompted a heated debate in the mathematical world. In trying to defend the axiom, in a series of 1908–1909 papers Zermelo developed a system of axioms for set theory. It was improved by Adolf Abraham Halevi Fraenkel (1891–1965) in his 1922 [Fra1], [Fra2] and 1925 [Fra3] papers. Finally, in 1928 John von Neumann named it the *Zermelo–Fraenkel set theory*, or **ZF** [Neu]. The **ZF** with the addition of the **AC** was naturally denoted by **ZFC** and named the *Zermelo–Fraenkel-Choice system of axioms*.

The historian of the **AC**, Gregory H. Moore, opens his remarkable book about it as follows [Moo]:

David Hilbert once wrote that Zermelo's Axiom of Choice was the axiom "most attacked up to the present [1926] in mathematical literature..." To this Abraham Fraenkel later [1958] added that "the axiom of choice is probably the most interesting and, in spite of its late appearance, the most discussed axiom of mathematics, second only to Euclid's axiom of parallels which was introduced more than two thousand years ago."

The Axiom postulated the existence of a choice function, without giving any clue as to how to find it. Therefore it came as no surprise that the Axiom was opposed by constructivists, intuitionists, and other mathematicians, who viewed non-constructive existence results with great suspicion. Moore [Moo] observes that "Despite this initial widespread distrust, today a vast majority of mathematicians accepts the axiom without hesitation and utilize it in algebra, analysis, logic, set theory, and topology." Yes, I agree: vast majority accepts the **AC**, and consequently

---

[3][Pon]. L. S. Pontryagin wrote this *Life of Lev Semenovich Pontryagin...* with the title modeled after the autobiography of the Italian Renaissance sculptor Benevento Cellini. This statement was aimed at A. N. Kolmogorov.

ZFC as the standard foundation of set theory, but is it a good thing for mathematics? A majority—any majority—political, social, mathematical, often loses sensitivity that is so naturally preserved among minorities. We will later look into the consequences of the near universal acceptance of the **AC** as a part of the foundation of mathematics. Here I will introduce other axioms, and first of all, some weaker versions of the **AC**.

Many results in mathematics really need just a countable version of choice:

The Countable Axiom of Choice (**AC$_{\aleph_0}$**) Every countable family of nonempty sets has a choice function.

Much later, in 1942 Paul Isaac Bernays (1888–1977) introduced the following axiom [Bern]:

The Principle of Dependent Choices (**DC**) If E is a binary relation on a nonempty set A, and for every $a \in A$ there exists $b \in A$ with aEb, then there is a sequence $a_1, a_2, \ldots, a_n, \ldots$ such that $a_n E a_{n+1}$ for every $n < \omega$.

The **AC** implies **DC** (see, e.g., Theorem 8.2 in [Jec]), but not conversely. In turn, **DC** implies **AC$_{\aleph_0}$**, but not conversely. **DC** is slightly stronger than **AC$_{\aleph_0}$**, but it is a sufficient addition to **ZF** for creating a foundation for the classical Lebesgue Measure Theory. We observe that, in particular, **DC** is sufficient for Falconer's Theorem (Theorem 9.1).

One—unfortunate in my opinion—consequence of the **AC** is the existence of sets on the line that have no length (I mean, no Lebesgue measure). This "regret" must have given birth to the following axiom:

(**LM**) Every set of real numbers is Lebesgue measurable.

Assuming the existence of an inaccessible cardinal,[4] Robert M. Solovay (nowadays Professor Emeritus at Berkley), using Paul Cohen's forcing, constructed in 1964 (and published in 1970) a model that proved a remarkable theorem [Sol1]. Mitya Karabash and I introduced [KS] the following term in honor of R. M. Solovay.

The Zermelo–Fraenkel–Solovay System of Axioms for set theory, which we denote by **ZFS**, is defined as follows:

$$\mathbf{ZFS} = \mathbf{ZF} + \mathbf{AC}_{\aleph_0} + \mathbf{LM},$$

and **ZFS Plus**, or shorter, **ZFS+** would stand for

$$\mathbf{ZFS+} = \mathbf{ZF} + \mathbf{DC} + \mathbf{LM}.$$

Now the Solovay Theorem formulates very concisely:

---

[4] A cardinal κ is called *inaccessible* if $\kappa > \aleph_0$, κ is regular, and κ is strong limit. An infinite cardinal $\aleph_\alpha$ is *regular*, if cf $\omega_\alpha = \omega_\alpha$. A cardinal κ is a *strong limit* cardinal if for every cardinal λ, $\lambda < \kappa$ implies $2^\lambda < \kappa$.

Solovay's Theorem: 46.1 ZFS+ is consistent.[5]

Solovay reports [Sol1] that "the original problem of showing **ZF+LM** consistent was suggested to the author by Paul Cohen." Here is how Paul Joseph Cohen (1934 – March 23, 2007), the man who completed Kurt Gödel's work and won Fields Medal for it in 1966, described Solovay's Theorem in 1966 [Coh2, p. 142]:

> One of the most interesting results (concerning the relationship of various forms of AC) is due to R. Solovay (as yet unpublished) which says that models N can be constructed in which the countable AC holds and yet every set of real numbers is Lebesgue measurable.

Indeed, this is a profound result, which offers **ZFS+** as a viable alternative to the canonical **ZFC**. In particular, **ZFS+** allows developing the usual Lebesgue Measure Theory. On April 10, 2003 I asked Professor Solovay whether a stronger result is possible, i.e., whether **ZFS** would suffice for building the Lebesgue Measure Theory. The following day Solovay replied [Sol2]:

> I thought about this in the early 60's. The only theorem for which I needed **DC** was the Radon-Nykodim theorem. But I don't know that there isn't a clever way of getting by with just Countable Choice and proving Radon-Nykodim. I just noticed that the usual proof [found in Halmos] uses **DC**.

The Continuum Hypothesis (**CH**) states that there is no cardinal κ such that $\aleph_0 < \kappa < 2^{\aleph_0}$.

The Generalized Continuum Hypothesis (**GCH**) states that for any infinite cardinal λ there is no cardinal κ such that $\lambda < \kappa < 2^\lambda$.

The Axiom of Constructibility (**V = L**) introduced by Gödel in 1940 [Göd2], asserts that every set is constructible, i.e. every set belongs to the constructible universe L.

Kurt Gödel (1906–1978) in 1940 [Göd2] and Paul J. Cohen in 1963–1964 [Coh1] proved independence of **AC** (as well as of the Continuum Hypothesis, **CH**, and the Generalized Continuum Hypothesis, **GCH**) from the rest of the axioms of set theory, **ZF**.

As Saharon Shelah playfully summarized these developments in his 2003 "Logical Dreams" [She3],

> In short: The Continuum Problem asks:
> How many real numbers are there?
>
> G. Cantor proved: There are more reals than rationals. (In a technical sense: "uncountable", "there is no bijection from R into Q").
>
> The Continuum Hypothesis (CH) says: yes, more, but barely so. Every set A $\subseteq$ R is either countable or equinumerous with R.
>
> K. Gödel proved: Perhaps CH holds.
> P. Cohen proved: Perhaps CH does not hold.

---

[5] Assuming the existence of an inaccessible cardinal.

Kurt Gödel also showed that $\mathbf{ZF} + \mathbf{V} = \mathbf{L}$ implies $\mathbf{GCH}$; while the founder of the famous Warsaw school of set theory and topology Wacław Franciszek Sierpiński (1882–1969) proved that $\mathbf{ZF} + \mathbf{GCH}$ implies $\mathbf{AC}$.

Finally, one can remember L. S. Pontryagin not only as a fine mathematician and a fine anti-Semite, but also as a fool, who took his fight against anything Kolmogorov's to the extreme of such a ridiculous statement as this section's tongue-in-cheek epigraph "*At present, set theory has lost its relevance.*" What can be more relevant to mathematics than its very foundation: set theory!

46.3 The First Example

> *The Axiom of Choice differs from other axioms of ZF by postulating the existence of a set ... without defining it ... Thus it is often interesting to know whether a mathematical statement can be proved without using the Axiom of Choice.*
>
> — Thomas Jech [Jec]

> *Theories come and go; examples live forever.*
>
> – I. M. Gelfand

Saharon Shelah's and my week-long joint work (entertainment of the mathematical kind, really) resulted in the first surprising example. (The Conditional Chromatic Number Theorem was obtained too—I will formulate it in the next Chapter.) We dedicated the paper to the memory of our teacher, friend and coauthor Paul Erdős, on the occasion of his 90th birthday. Let us look together at this example.

Our first task is to expand the definition of the chromatic number.[6] How important is it to select a productive definition? Socrates thought highly of this undertaking: "The beginning of wisdom is the definition of terms."[7] And so I took 2 weeks to "sleep" on a definition, and consulted with my coauthors Saharon Shelah and Mitya Karabash before I stopped on the simplest definition, the one that came first to my mind. "Simplest" surely is not a detractor: in fact, such aspects as simple and natural are attributes of definitions that survive the test of time.

Without the $\mathbf{AC}$, the minimum, and thus the chromatic number of a graph, may not exist. In allowing a system of axioms for set theory not to include the $\mathbf{AC}$, we need to come up with a much broader definition of the chromatic number of a graph than the one we have used in Chapter 12—if we want the chromatic number to exist. In fact,

---

[6] It *is* the first task, but we did not think of it then, and so this definition appears here now for the first time in print.

[7] Quoted from [Pet], p. 494.

instead of the chromatic number we ought to talk about the *chromatic set*. There are several meaningful ways to define it. I am choosing the following definition.

Definition 46.1 Let G be a graph and A a system of axioms for set theory. The set of chromatic cardinalities $\chi^A(G)$ of G is the set of all cardinal numbers $\tau \le |G|$ such that there is a proper coloring of the vertices of G in τ colors, and τ is minimal with respect to this property.

As you can easily see, the set of chromatic cardinalities does not have to have just one element as was the case when $A = \textbf{ZFC}$. It can also be empty.

The advantage of this definition is its simplicity. Best of all, we can use inequalities on sets of chromatic cardinalities as follows.

The *inequality* $\chi^A(G) > \beta$, where β is a cardinal number, means that for every $\alpha \in \chi^A(G)$, $\alpha > \beta$. The inequalities $<$, $\le$, *and* $\ge$ are defined analogously. We also agree that the empty set is greater than or equal to any other set.[8] Finally, if β is a cardinal number, $\chi^A(G) = \beta$ means that $\chi^A(G) = \{\beta\}$.

I would like to introduce simple generalizations of the notion of unit distance graph.

A *distance graph* is a graph with the vertex set $V \subseteq R^n$ for some n, and two vertexes v_1, v_2 are adjacent if and only if the distance $|v_1v_2|$ belongs to a fixed set S of distances. In particular, when $S = \{1\}$, we get a unit distance graph.

A *difference graph* is a graph with the vertex set $V \subseteq R^n$ for some n, and two vertexes v_1, v_2 are adjacent if and only if their difference $v_1 - v_2$ belongs to a fixed set $S \subseteq R^n$ of differences. Of course, on the line distance graphs and difference graphs coincide.

As always, Z, Q, and R stand for the sets of integers, rationals, and reals respectively. We are now ready for the first example, which will demonstrate how dramatically the chromatic number of a simple graph we construct depends upon the system of axioms for set theory. It is just 2 in **ZFC** and uncountable in **ZFS**. Let us construct this surprising example and then prove its properties.

Example 46.2 (Shelah–Soifer [SS1]) We define a graph G as follows: the set R of real numbers serves as the vertex set, and the set of edges is $\left\{(s, t) : s - t - \sqrt{2} \in Q\right\}$.

Result 46.3 (Shelah–Soifer [SS1]). For the distance graph G on the line, $\chi^{\textbf{ZFC}}(G) = 2$, while $\chi^{\textbf{ZFS}}(G) > \aleph_0$.

Claim 1 of 46.3: $\chi^{\textbf{ZFC}}(G) = 2$.

Proof Let $S = \left\{q + n\sqrt{2}; q \in Q, n \in Z\right\}$. We define an equivalence relation E on R as follows: $sEt \Leftrightarrow s - t \in S$.

Let Y be a set of representatives for E (in choosing representatives we are using the **AC**). For $t \in R$ let $y(t) \in Y$ be such that $tEy(t)$. We define a 2-coloring $c(t)$

---

[8] I know, this convention seems to be counterintuitive, but it is handy, convenient, and allows to prove meaningful results, as you will soon see.

as follows: $c(t) = l$, $l = 0, 1$ if and only if there is $n \in Z$ such that $t - y(t) - 2n\sqrt{2} - l\sqrt{2} \in Q$. ∎

Without **AC** the chromatic situation changes dramatically:

Claim 2 of 46.3: $\chi^{\text{ZFS}}(G) > \aleph_0$.

We will simplify the proof if we acquire a useful tool first.

Tool 46.4 If $A \subseteq [0, 1)$ and A contains no pair of adjacent vertices of G, then A is null (of Lebesgue measure zero).

Proof Assume to the contrary that A contains no pair of adjacent vertices of G yet A has a positive measure. Then there is an interval I such that[9]

$$\frac{\mu(A \cap I)}{\mu(I)} > \frac{9}{10} \tag{46.1}$$

Choose $q \in Q$ such that $\sqrt{2} < q < \sqrt{2} + \frac{\mu(I)}{10}$.

Let $B = A - \left(q - \sqrt{2}\right) = \left\{x - q + \sqrt{2} : x \in A\right\}$. Then

$$\frac{\mu(B \cap I)}{\mu(I)} > \frac{8}{10}. \tag{46.2}$$

Inequalities (46.1) and (46.2) imply that there is $x \in I \cap A \cap B$. Since $x \in B$, we have $y = x + \left(q - \sqrt{2}\right) \in A$. So, both $x, y \in A$ and $x - y - \sqrt{2} = -q \in Q$. Thus, $\{x, y\}$ is an edge of the graph G with both endpoints in A, which is the desired contradiction. ∎

Proof of claim 2 of 46.3 Assume that the graph G is colored in $\aleph_0$ colors properly (i.e., the adjacent vertices are colored in different colors), and $A_1^1, \ldots, A_n^1, \ldots$ are the corresponding monochromatic sets. Let $A_n = A_n^1 \cap [0, 1)$ for every $n < \omega$.

Since $\mu\left(\bigcup_{n<\omega} A_n\right) = \mu([0, 1)) = 1$ and Lebesgue measure is a countably additive function in **AC**$_{\aleph_0}$, there is a positive integer n such that A_n has a positive measure. By Tool 46.4, A_n contains a pair of adjacent vertices of G, which contradict the assumption that the graph is properly colored. ∎

Remark: This example begs the question: Is **AC** relevant to the problem of chromatic number χ of the plane? The answer depends upon the value of χ which we, of course, do not yet know. However, this example points out the circumstances in which the presence or the absence of **AC** could dictate the value of the chromatic number of the plane.

---

[9] The (Lebesgue) measure $\mu(S)$ of a set S is defined in Chapter 9.

46.4 Examples in the plane

As the main object of our interest has been the good ole Euclidean plane, we aspire to construct a difference graph G_2 on the plane R^2, and thus come much closer to the setting of the chromatic number of the plane problem. The chromatic number of the constructed below graph G is 4 in **ZFC** and uncountable in Zermelo–Fraenkel–Solovay's system of axioms **ZFS**.

Israil' Moiseevich Gelfand once said "theories come and go, while examples live forever." Graphs presented here may prove to be an important illumination in this area of research.

Example 46.5 (Soifer–Shelah [SS2]) We define a graph G as follows: the set R^2 of points in the plane serves as the vertex set, and the set of edges is the union of the four sets $\{(s, t) : s, t \in R^2; s - t - \varepsilon \in Q^2\}$ for $\varepsilon = \left(\sqrt{2}, 0\right)$, $\varepsilon = \left(0, \sqrt{2}\right)$, $\varepsilon = \left(\sqrt{2}, \sqrt{2}\right)$, and $\varepsilon = \left(-\sqrt{2}, \sqrt{2}\right)$ respectively.[10]

Result 46.6 (Soifer–Shelah [SS2]). For the difference graph G in the plane, $\chi^{\textbf{ZFC}}(G) = 4$, while $\chi^{\textbf{ZFS}}(G) > \aleph_0$.

Claim 1 of 46.6: $\chi^{\textbf{ZFC}}(G) = 4$.

Proof Let $S = \left\{\left(q_1 + n_1\sqrt{2}, q_2 + n_2\sqrt{2}\right) : q_i \in Q, n_i \in Z\right\}$. We define an equivalence relation E on R^2 as follows: $s\, E\, t \Leftrightarrow s - t \in S$.

Let Y be a set of representatives for E (we can choose them due to the **AC**). For $t \in R^2$ let $y(t) \in Y$ be such that $t\, E\, y(t)$. We define a 4-coloring $c(t)$ as follows: $c(t) = (l_1, l_2)$, $l_i \in \{0, 1\}$ if and only if there is a pair $(n_1, n_2) \in Z^2$ such that $t - y(t) - 2\sqrt{2}(n_1, n_2) - \sqrt{2}(l_1, l_2) \in Q^2$. ∎

Claim 2 of 46.6: $\chi^{\textbf{ZFS}}(G) > \aleph_0$.

We can create a tool similar to Tool 46.3, and then prove the claim 2 similarly to its counterpart of Result 46.3. ∎

We can define edges of the graph differently.

Example 46.7 (Soifer–Shelah [SS2]) The set R^2 of points on the plane serves as the vertex set for G, and the set of edges is the union of the *two* sets $\{(s, t) : s, t \in R^2; s - t - \varepsilon \in Q^2\}$ for $\varepsilon = \left(\sqrt{2}, 0\right)$ and $\varepsilon = \left(0, \sqrt{2}\right)$ respectively.

Result 46.8 (Soifer–Shelah [SS2]). For the difference graph G in the plane, $\chi^{\textbf{ZFC}}(G) = 2$, while $\chi^{\textbf{ZFS}}(G) > \aleph_0$.

Claim 1 of 46.8: $\chi^{\textbf{ZFC}}(G) = 2$.

Proof Let $S = \left\{\left(q_1 + n_1\sqrt{2}, q_2 + n_2\sqrt{2}\right) : q_i \in Q, n_i \in Z\right\}$. We define an equivalence relation E on R^2 as follows: $s\, E\, t \Leftrightarrow s - t \in S$.

---

[10] Q^2, of course, denotes the rational plane, as in Chapter 11.

Let Y be a set of representatives for E. For $t \in R^2$ let $y(t) \in Y$ be such that $t E y(t)$. We define a 2-coloring $c(t)$ as follows: $c(t) = (\varepsilon_1 + \varepsilon_2)_{\text{mod}2}$ if and only if there is a pair $(\varepsilon_1, \varepsilon_2) \in Z^2$ such that $t - y(t) - \sqrt{2}(\varepsilon_1, \varepsilon_2) \in Q^2$. ∎

Claim 2 of 46.8: $\chi^{\text{ZFS}}(G) > \aleph_0$.

Proof is similar to the one presented in Result 46.3. ∎

You may wonder: what is so special about $\sqrt{2}$ in our constructions? Well, $\sqrt{2}$ is the oldest known irrational number: a proof of its irrationality, apparently, comes from the Pythagoras School. Our reasoning and results would not change if we were to replace $\sqrt{2}$ everywhere with another irrational number.

46.5 Examples in space

> *Space isn't remote at all. It's only an hour's drive*
> *away if your car could go straight upwards.*
> *– Fred Hoyle*

Ideas developed above are extended here to construct difference graphs on the real n-dimensional space R^n, whose chromatic number is a positive integer in **ZFC**, and is not countable in **ZFS**.

Example 46.9 (Soifer [Soi23]) We define a difference graph G_n: the set R^n of points of the n-space serves as the vertex set, and the set of edges is $\bigcup_{i=1}^{n} \left\{ (s, t) : s, t \in R^n; s - t - \sqrt{2}\varepsilon_i \in Q^n \right\}$ where ε_i are the n unit vectors on coordinate axes forming the standard basis of R^n. For example, $\varepsilon_1 = (1, 0, \ldots, 0)$ – we will use this vector in the proof of *Claim 2* below.[11]

Result 46.10 (Soifer [Soi23]). For the difference graph G_n, $\chi^{\text{ZFC}}(G_n) = 2$, while $\chi^{\text{ZFS}}(G_n) > \aleph_0$.

Claim 1 of 46.10: $\chi^{\text{ZFC}}(G) = 2$.

Proof Let $S = \left\{ q + m\sqrt{2} : q \in Q^n, m \in Z^n \right\}$. We define an equivalence relation E on R^n as follows: $s E t \Leftrightarrow s - t \in S$.

Let Y be a set of representatives for E. For $t \in R^n$ let $y(t) \in Y$ be such a representative that $t E y(t)$. We define a 2-coloring $c(\cdot)$ as follows: $c(t) = \|k\|_{\text{mod}2}$ if and only if there is $k \in Z^n$ such that $t - y(t) - \sqrt{2}k \in Q^n$, where $\|k\|$ denotes the sum of all n coordinates of k. ∎

Claim 2 of 46.10: $\chi^{\text{ZFS}}(G) > \aleph_0$.

The proof is similar to the one of result 46.3—we just need an "n-dimensional tool."

---

[11] Z^n is a set of integral n-tuples and Q^n is the "rational n-space."

Tool 46.11 If $A \subseteq [0, 1)^n$ and A contains no pair of adjacent vertices of G, then A is null (of Lebesgue measure zero).

Proof Assume to the contrary that $A \subseteq [0, 1)^n$ contains no pair of adjacent vertices of G_n, yet A has positive measure. Then there is an n-dimensional parallelepiped I, with a side parallel to the first coordinate axis of length, say, a, such that

$$\frac{\mu(A \cap I)}{\mu(I)} > \frac{9}{10} \qquad (46.3)$$

Choose $q \in Q$ such that $\sqrt{2} < q < \sqrt{2} + \frac{a}{10}$. Define a translate B of A as follows:

$$B = A - \left(q - \sqrt{2}\right)\varepsilon_1$$

Then

$$\frac{\mu(B \cap I)}{\mu(I)} > \frac{8}{10} \qquad (46.4)$$

Inequalities (46.1) and (46.2) imply that there is $v \in I \cap A \cap B$. Since $v \in B$, we have $w = v + (q - \sqrt{2})\varepsilon_1 \in A$. So, we have $v, w \in A$ and $v - w - \sqrt{2}\varepsilon_1 = -q\varepsilon_1 \in Q^n$. Thus, $\{v, w\}$ is an edge of the graph G with both endpoints in A, which is the desired contradiction. ∎

We can certainly vary the definition of edges to get new graphs.

Example 46.12 (Soifer [Soi23]) We define a graph G: the set R^n of points of the n-space still serves as the vertex set, but the set of edges is $\bigcup_{0 \leq i \neq j \leq n} \big\{ (s, t) : s, t \in R^n;$
$s - t - \sqrt{2}\left(\varepsilon_i - \varepsilon_j\right) \in Q^n \big\}$ where ε_i are the n unit vectors on coordinate axes forming the standard basis of R^n, and $\varepsilon_0 = 0 \in R^n$.

Result 46.13 (Soifer [Soi23]) For the difference graph G_n, $\chi^{\mathrm{ZFC}}(G) = 2^n$, while $\chi^{\mathrm{ZFS}}(G) > \aleph_0$.

Claim 1 of 46.13: $\chi^{\mathrm{ZFC}}(G) = 2^n$.

Proof Indeed, the 2^n vertices of the n-dimensional unit cube generated by ε_i, $0 \leq i \leq n$, must all be colored in different colors, so the 2^n colors are obviously needed.
    Let Y be a set of representatives for E. For $t \in R^n$ let $y(t) \in Y$ be a representative such that $tEy(t)$. We define a 2^n-coloring $c(t)$ as follows: $c(t) = \left(k^1_{\mathrm{mod}2}, k^2_{\mathrm{mod}2}, \ldots, k^n_{\mathrm{mod}2}\right)$ if and only if there is $k = \left(k^1, k^2, \ldots, k^n\right) \in Z^n$ such that $t - y(t) - \sqrt{2}k \in Q^n$, where $k^i_{\mathrm{mod}2} \in \{0, 1\}$ is the remainder upon division of k^i by 2 for $i = 1, 2, \ldots, n$. ∎

Claim 2 of 46.13: $\chi^{\mathrm{ZFS}}(G) > \aleph_0$.

Proof closely follows the one for Claim 2 of Result 46.10. ∎

Observe: It is certainly possible to construct other examples of difference graphs on R^n whose chromatic number in **ZFC** is any integer between 2 and 2^n, and is uncountable in **ZFS**.

These examples illuminate the influence of the system of axioms for set theory on combinatorial results. They also suggest that the chromatic number of R^n may not exist "in the absolute" (i.e., in **ZF**), but depend upon the system of axioms we choose for set theory. The examples we have seen naturally pose the following open problem:

Open AC Problem 46.14 For which values of n is the chromatic number $\chi(E^n)$ of the n-space E^n defined "in the absolute", i.e., in **ZF** regardless of the addition of the **AC** or its relative?

46.6 AfterMath & Shelah–Soifer Class of Graphs

In the 1910s–1930s the foundations dominated mathematicians' interests. Nowadays, the interest in the foundations in general, and in the **AC** in particular, diminished outside of set theory and set theorists and logicians. Most mathematicians have settled on **ZFC**-based mathematics. Yet, Shelah–Soifer papers seemed to "strike the mathematical heart" [Del]. They received a thought provoking critique [Del] by Jean-Paul Delahaye, a complimentary mention in Ronald L. Graham's articles [Gra5], [Gra6], [Gra7], and [Gra8], entered the column by Joseph O'Rourke [Oro], and were the subject of the column [Szp] by George Szpiro in the newspaper in Zürich, the city where Van der Waerden lived for 45 years. It inspired a series of works by various authors. We will look at one such paper in Section 46.7. Another example was forwarded to me by Professor Branko Grünbaum in February 2005:

From: Janos Pach <pach@CIMS.nyu.edu>
Date: February 27, 2005 8:21:56 PM PST
To: eokoh@gc.cuny.edu, dlazarus@erols.com, sarioz@acm.org, aushakov@mail.ru, herr_strangelove@yahoo.com, mlaufer@gc.cuny.edu, tswaine@gc.cuny.edu, syuan@gc.cuny.edu, msilva@gc.cuny.edu, dmussa@gc.cuny.edu, jharlacher@gc.cuny.edu, Eva@Antonakos.net, mmunn@gc.cuny.edu, raghavan@cs.nyu.edu
Cc: RLandsman@gc.cuny.edu (Robert Landsman)
Subject: Combinatorial Comp. Seminar on Wednesday
SEMINAR ON COMBINATORIAL COMPUTING March 2, Wednesday, 6:30pm
Room 6417, Graduate Center, 365 Fifth Avenue, NY
INDEPENDENCE IN EUCLIDEAN RAMSEY THEORY
Jacob Fox, Massachusetts Institute of Technology

In this talk, I will present several remarkable new developments on independence in Euclidean Ramsey theory. S. Shelah and A. Soifer recently constructed a graph on the real line with chromatic number 2 in the Zermelo–Fraenkel-Choice (ZFC) system of axioms, but with uncountable chromatic number (if it exists) in a consistent system of axioms with limited choice, studied by Solovay in 1970. Motivated by these recent results, Radoicic and I discovered that the statement "every 3-coloring of the non-zero real numbers contains a monochromatic solution to the equation $x\_1 + 2x\_2 - 4x\_3 = 0$" is independent of the Zermelo–Fraenkel axioms for set theory. A system $L : Ax = 0$ of linear homogeneous equations is called a-regular over R if every a-coloring of the real numbers contains a monochromatic solution to L in distinct variables. In 1943, Rado classified those L that are a-regular over R for all finite a. In ZFC, if a is an infinite cardinal, we classify those L that are a-regular. This classification depends on the cardinality of the continuum. In the Solovay model, we classify those L that are aleph_0-regular over R. We also leave several problems concerning the chromatic number of graphs on Euclidean space.

To the best of my knowledge and literature search, the 1970 fundamental work by Robert M. Solovay, has been cited in set theoretic works for decades, but has not been known to or used in combinatorics and Ramsey Theory before [SS1] appearing in 2003. Inspired by our surprising results, Solovay's work and what Mitya Karabash and I named Zermelo–Fraenkel–Solovay System of Axioms **ZFS**, the comparative study of **ZFC** vs. **ZFS** has entered a number of recent combinatorial works, for example by Jacob Fox and Rados Radoicic [FR], Boris Alexeev, Jacob Fox and Ronald L. Graham [AFG], and Boris Bukh [Buk]. Ronald Graham [Gra8] summarizes this group's results as follows:

> An interesting phenomenon has been recently observed by Fox, Radoicic, Alexeev and the author [FR], [AFG] which shows how the axioms of set theory can affect the outcome of some of these questions. For example, consider the linear equation $E : x + y + z - 4w = 0$. This is certainly not partition regular, and in fact, there is a 4-coloring of the integers which prevents E from having any (nontrivial) monochromatic solution. However, suppose we change the question and asked whether E has monochromatic solutions in reals for every 4-coloring of the reals. It can be shown that in ZFC, there exist 4-colorings of the reals for which E has no monochromatic solution. However, if we replace the AC (the "C" in ZFC) by LM..., then in the system ZF + LM (which is consistent if ZFC is), the answer is yes. On other words, in this system every 4-coloring of the reals always contains a nontrivial monochromatic solution to E. On the other hand, this distinction does not occur for the equation $x + y - z = 0$, for example.

Dmytro (Mitya) Karabash, an undergraduate student at Columbia University and a fine mathematician, coined the term for the class of graphs Shelah and I stumbled upon:

Definition 46.15 (Dmytro Karabash) The Shelah–Soifer class S of graphs consists of graphs G, for which $\chi^{\textbf{ZFS}}(G) \cap \chi^{\textbf{ZFC}}(G) = \varnothing$. Let S^c stand for the complement, i.e., class of graphs which are not Shelah–Soifer graphs.

Mitya and I then looked into what causes a graph to belong to this class, and "how many" Shelah–Soifer graphs there are. The following results come from our joint 2007 paper [KS].

Definition 46.16 ([KS]) Let d be the Euclidian metric in R^n. The distance set between $A, B \subseteq R^n$ is defined as follows: $d(A, B) = \{d(x, y) : x \in A, y \in B\}$.

Definition 46.17 ([KS]) Let $D \subseteq R^+ = (0, \infty)$. The symbol G_D^n stands for the graph with the vertex set R^n and the edge set $\{(x, y) : d(x, y) \in D\}$.

Theorem 46.18 ([KS]) If for $D \subseteq R^+$, 0 is a limit point of D in R, then $\chi^{ZFS}(G_D^n) > \aleph_0$.

We can prove Theorem 46.18 using an argument analogous to proof of claim 2 in result 46.3. For the sake of diversifying our tools, we will use the old result of Hugo Steinhaus (1887–1972) instead:

Steinhaus's Lemma 46.19 ([Stei]). If $A \subseteq R$, is a set of positive Lebesgue measure, then the set $A - A = \{x - y : x, y \in A\}$ contains a ball around 0.

Proof of Theorem 46.18 Let us argue by contradiction: suppose $\chi^{ZFS}(G_D^n) \leq \aleph_0$. Then there exists a countable proper coloring $c : R^n \to N$ of G_D^n. Look at the monochromatic sets $A_i = \{x \in R^n | c(x) = i\}$. Since all sets in R^n are measurable in ZFS, we get $\sum_{i=1}^n \mu(A_i) = \infty$. Hence there exists $i \in N$ such that $\mu(A_i) > 0$. Thus, there exists a set of positive measure $A \subseteq R^n$ such that $d(A, A) \cap D = \varnothing$. We reduce to case $n = 1$ by observing that there must exist a line $L \subseteq R^n$ such that $A \cap L$ has positive measure in L by product measure theorem (see, for example, Theorem 2.36 in [Foll]). Now we apply the Steinhaus Lemma 46.19 to see that $d(A, A)$ contains some interval $[0, \varepsilon)$ and since 0 is a limit point of D, we get $d(A, A) \cap D \neq \varnothing$. ∎

Definition 46.20 ([KS]) Set $D \subseteq R$ is called integrally independent mod 2 if for any $n \in N$, $a_1, a_2, \ldots, a_n \in Z$ and $s_1, s_2, \ldots, s_n \in D$, the equality $\sum_{i=1}^n a_i s_i = 0$ implies $2 | \sum_{i=1}^n a_i$.

Theorem 46.21 ([KS]) If $D \subseteq R^+$, $|D| \leq \aleph_0$, then $\chi^{ZFC}(G_D^1) \leq \aleph_0$, i.e., for every $\alpha \in \chi^{ZFC}(G_D^1), \alpha \leq \aleph_0$. If in addition D is integrally independent mod 2, then $\chi^{ZFC}(G_D^1) = \{2\}$.

Proof For any $p \in R$, p lies in the connected component C of G_D^1,

$$C = C(p) = \{x : \exists n > 0, \exists \{x_i\}_{i=1}^n \subseteq R, s.t.$$

$$|x_i - x_{i+1}| \in D, x_0 = x, x_n = p\} = \bigcup_{i=1}^\infty C_i,$$

where $C_i = C_i(p)$ are defined inductively by $C_0(p) = \{p\}$ and $C_i = C_{i-1} + D = \{x + y | x \in C_{i-1}, y \in D\}$. Since $|D| \leq \aleph_0$, for every i, $|C_i| \leq \aleph_0$ and hence $|C| \leq \aleph_0$. Thus we can color the component in $\aleph_0$ colors by coloring each point in different color. The **AC** allows us to similarly color all of the other components. But the chromatic number of the graph is the supremum of the chromatic numbers of its components and hence the first statement of the theorem is proven.

If D is integrally independent mod 2, then $C_i \cap C_j = \varnothing$ if and only if $i - j$ is even. Hence coloring each C_i according to its parity is a well-defined 2-coloring. ∎

Theorem 46.22 ([KS]) For the set of graphs $H = \{G_D^1 : D \subseteq R^n\}$, $|H \cap S| = |H \cap S^c|$, where S is the class of Shelah–Soifer graphs and S^c is its complement.

To prove this theorem, let us first prove two tools:

Tool 46.23 ([KS]) *If $D \subseteq (\varepsilon, \infty)$ for some $\varepsilon > 0$ then $\chi^{ZF}(G_D^n) \leq \aleph_0$.*

Proof We obtain a proper $\aleph_0$-coloring of G_D^n by cutting R^n into n-cubes of side $\frac{\varepsilon}{\sqrt{n}}$ and coloring each cube into a different color. ∎

Tool 46.24 ([KS]) Let G_1, G_2 are graphs with the same vertex set V, and $G = G_1 \cup_{edge} G_2$ be their edge union. Suppose that G_1, G_2 are countably colorable in an axiomatic system A, i.e., $\chi^A(G_i) \leq \aleph_0$ for $i = 1, 2$. Then G is countably colorable in A, i.e., $\chi^A(G) \leq \aleph_0$.

Proof Let c_1, c_2 be proper $-\aleph_0$ colorings of G_1, G_2, respectively. Consider the coloring $c_1 \oplus c_2$ of G in $\aleph_0^2$ colors defined by $c_1 \oplus c_2(v) = (c_1(v), c_2(v))$. This is clearly a proper coloring of G which uses $\aleph_0^2 = \aleph_0$ colors. ∎

Proof of Theorem 46.22 Let $H_D = \{G_{D \cup E}^1 : E \subseteq [1, \infty)\}$.

1. First consider $D \subseteq (0, 1)$ such that $|D| \leq \aleph_0$ and 0 is a limit point of D. Theorem 46.21 implies $\chi^{ZFC}(G_D^1) \leq \aleph_0$ and Tool 46.23 implies $\chi^{ZFC}(G_E^1) \leq \aleph_0$. Hence by Tool 46.24, $\chi(G_{D \cup E}^1) = \chi(G_D^1 \cup_{edge} G_E^1) \leq \aleph_0$. On the other hand, for every $G_{D \cup E}^1 \in H_D$, $\chi^{ZFS}(G_{D \cup E}^1) > \aleph_0$ by Theorem 46.18. Hence $H_D \subseteq S$.
2. Since $|[1, \infty)| = |R^+|$ we get $|H_D| = 2^{|[1, \infty)|} = 2^{|R^+|} = |S|$ and hence $|H| = |H \cap S|$.
3. Now consider $D = (0, 1]$. Since each graph $G_{D \cup E}^1$ in H_D contains a complete subgraph with the vertex set of the cardinality of continuum, we get $\chi^{ZFS}(G_{D \cup E}^1) = \chi^{ZFC}(G_{D \cup E}^1) = |c|$, where c is continuum. Hence $H_D \subseteq S^c$ and similarly to the argument 2 above, $|H| = |H \cap S^c|$. ∎

Theorem 46.22 suggests that the class S is "as big as" the class S^c, whatever "as big" means. Let us make it formal.

Definition 46.25 ([KS]) Let θ be a class of all graphs and α a cardinal number. We define a set θ_α as follows:

$$\theta_\alpha = \{G \in \theta : |V(G)| \leq \alpha\},$$

where $V(G)$ is the vertex set of G, i.e., let θ_α be the set of graphs with the cardinality of their vertex set not exceeding α.

We conjecture:

Conjecture 46.26 ([KS]) For any cardinal $\alpha > \aleph_0$, $|\theta_\alpha \cap S| = |\theta_\alpha \cap S^c|$.

46.7 An Unit Distance Shelah–Soifer Graph

On July 10, 2007, my first day after the 3-day cross-country rally brought me from Princeton to my home in Colorado Springs (the day also remarkable due to personal thunders), I received the following e-mail from Melbourne, Australia:

> Dear Professor Soifer,
>
> I am a student from Monash Uni[versity] in Australia and I have done some work on the chromatic number of the plane problem. I found your various publications on the topic extremely helpful. I particularly liked your recent work with Saharon Shelah and as part of my [Honours] bachelor's thesis I found another example of a graph with 'ambiguous' chromatic number. This graph is a unit distance graph so it may be considered even further evidence that the plane chromatic number may also be ambiguous as you have suggested. It has been submitted for review but you can find a pre-print of it here http://arxiv.org/abs/0707.1177 if you are interested. As you will notice, I am greatly indebted to your work since my proof is essentially analogous to yours.
>
> Kind regards,
> Michael Payne

Indeed, the paper Michael submitted to arXiv the day before his e-mail to me contains a fabulous example. He starts with unit distance graph G_1 whose vertex set is the rational plane Q^2 and, of course, two vertices are adjacent if and only if they are distance 1 apart.

Example 46.27 (Payne [Pay]) The desired unit distance graph G on the vertex set R^2 is obtained by tiling of the plane by translates of the graph G_1, i.e., its edge set is

$$\left\{ (p_1, p_2) : p_1, p_2 \in R^2; p_1 - p_2 \in Q^2; |p_1 - p_2| = 1 \right\}.$$

Claim 1: $\chi^{\mathbf{ZFC}}(G) = 2$.

Proof By Woodall's result 11.2, the chromatic number of the graph G_1 is equal to 2. Since the graph G consists of non-connected components, tiles, each of which isomorphic to G_1, the whole graph G is also 2-colorable (the **AC** is used to select "origin points" for 2-coloring of each tile). ∎

Claim 2: $3 \le \chi^{\mathbf{ZFS}}(G) \le 7$.

Michael Payne shows first that any measurable set S of positive (Lebesgue) measure contains the endpoints of a path of length 3 in G. Of course, this would rule out 2-coloring of S. Payne continues: "We can then proceed in a similar fashion to Shelah and Soifer's proof in [SS1]." Let us look at his proof.

Tool 46.28 ([Pay]) For any point $p \in R^2$ and any $\varepsilon > 0$ there is $q \in Q$ with $|q| < \varepsilon$ such that there is a path of length 3 in G starting at p and ending at $p + (q, 0)$.

Proof We use the fact that the rational points are dense on the unit circle to choose an angle α such that $(\cos \alpha, \, \sin \alpha) \in Q^2$ and

$$\left| \cos \alpha - \frac{1}{2} \right| < \frac{\varepsilon}{3}.$$

The path starting at p and passing through the following three points has the desired property:

$$p_1 = p + (\cos \alpha, \, \sin \alpha),$$
$$p_2 = p + (\cos \alpha - 1, \, \sin \alpha),$$
$$p_3 = p + (2 \cos \alpha - 1, \, 0).$$

From the previous inequality,

$$|2 \cos \alpha - 1| < \frac{2\varepsilon}{3} < \varepsilon,$$

and so we can simply choose $q = 2 \cos \alpha - 1$. ∎

Tool 46.29 ([Pay]) Any measurable set $A \subset R^2$ of positive measure $\mu(A) > 0$ contains a pair of vertices of G that are joined by a path of length 3.

Proof Michael Payne's proof of this central for the example tool is based, according to him, on the use of ideas from Shelah–Soifer examples discussed earlier in this Chapter. Assume that $\mu(A) > 0$. Then there is a unit square S in R^2 with sides parallel to the axes such that

$$\frac{\mu(A \cap S)}{\mu(S)} > \frac{9}{10}.$$

(Since A has positive measure, it must contain points with density equal to 1. Around any such point we can find a square with the desired property.)

By Tool 46.28 we can choose a rational q such that $|q| < 1/10$ and two points $(x, \, y)$ and $(x + q, \, y)$ are joined by a path of length 3. Let A' be a translate of A : $A' = \{(x + q, \, y) : (x, \, y) \in A\}$. We have

$$\frac{\mu\left(A' \cap S\right)}{\mu(S)} > \frac{8}{10}.$$

since the part of A translated out of S has measure at most $\mu(S)/10 = 1/10$. The two inequalities above imply that there exists $u \in A \cap A' \cap S$. Indeed, assume that $A \cap A' \cap S = \emptyset$, then $A \cap S$ and $A' \cap S$ are disjoint, and using additive property of measure, we get:

$$\mu(S) = \mu(A \cap S) + \mu(A' \cap S) + \mu(S \backslash (A \cup A')).$$

Dividing by $\mu(S)$ and using the density bounds, we get

$$1 > \frac{9}{10} + \frac{8}{10} + \frac{\mu\left(S\backslash\left(A \cup A'\right)\right)}{\mu(S)},$$

which is a contradiction. Therefore, there exists a $u \in A \cap A' \cap S$. Since $u \in A'$, it has a pre-image $v \in A$ such that $u = v + (q, 0)$. So $u, v \in A$, and u and v are connected by a path of length 3 in G. ∎

Now *proof of claim 2* of Payne's Example 46.27 is easy: Assume that G is 2-colored, with the corresponding monochromatic sets A_1 and A_2 which together cover the plane. At least one of the sets, say, A_1, has positive measure. Thus by Tool 46.29, A_1 contains a pair of points connected by a path of length 3 in G. However, in a 2-coloring of a graph, points connected by paths of length 3 must have opposite colors. Hence, no 2-coloring of G exists. ∎

47

A Glimpse into the Future: Chromatic Number of the Plane, Theorems and Conjectures

> *The importance of particular axioms being used makes a surprising difference for the question of determining the chromatic number of the plane, as recently shown by Shelah and Soifer*
>
> – Ronald L. Graham [Gra6]

47.1 Conditional Chromatic Number of the Plane Theorem

In the previous chapter, we constructed graphs that highlight the difference between our **ZFC** mathematics and mathematics that could have been, such as the **ZFS** mathematics. But what does it have to do with the main problem of this book, the chromatic number of the plane (CNP)?

Is **AC** relevant to the problem of the chromatic number χ of the plane? The answer depends upon the value of χ which we, of course, do not (yet) know. However, in 2003 Shelah and Soifer published the following conditional result.

Conditional Chromatic Number of the Plane Theorem 47.12 [12] (Shelah–Soifer [SS1]).

Assume that any finite unit distance plane graph has chromatic number not exceeding 4. Then:

$$^{*})\chi^{\textbf{ZFC}}(E^2) = 4;$$
$$^{**})\chi^{\textbf{ZFS+}}(E^2) \geq 5.$$

Proof The claim *) is true due to De Bruijn–Erdős Compactness Theorem 26.1.

---

[12] Due to the use of the Solovay's Theorem, we assume the existence of an inaccessible cardinal.

In the Solovay's system $\mathbf{ZFS+} = \mathbf{ZF} + \mathbf{DC} + \mathbf{LM}$, every subset S of the plane R^2 is Lebesgue measurable. Indeed, S is measurable if and only if there is a Borel set B such that the symmetric difference $S\Delta B$ is null. Thus, every plane set differs from a Borel set by a null set. We can think of a unit segment $I = [0, 1]$ as a set of infinite binary fractions and observe that the bijection $I \rightarrow I^2$ defined as $0.a_1a_2 \ldots a_n \ldots \mapsto (0.a_1a_3 \ldots ; 0.a_2a_4 \ldots)$ preserves null sets. Due to Falconer's Theorem 9.1 of Chapter 9, we can now conclude that the chromatic number of the plane is at least 5 (and, of course, at most 7). ∎

This conditional theorem allows for a certain historical insight. Perhaps, the problem of finding the chromatic number of the plane has withstood for 58 years all assaults in the general case, leaving us with a wide range for χ being 4, 5, 6, or 7, precisely because the answer depends upon the system of axioms we choose for set theory?

It is worth mentioning that I believe that the chromatic number of Euclidean space E^n, $n > 2$, may also depend upon the system of axioms we choose for set theory.

In the end of his 2007 paper, inspired by Shelah–Soifer series of papers, Michael Payne, the author of the important Example 46.27 (see the previous chapter) remarks:

> After demonstrating the existence of graphs whose chromatic number depends on the axiomatisation of set theory, Shelah and Soifer went on to formulate a conditional theorem [which essentially showed that the chromatic number of the plane may be ambiguous in a similar way to the graphs considered here [SS1]]. They showed that the chromatic number of the plane may be 4 with $\mathbf{AC}$ but 5, 6 or 7 with $\mathbf{LM}$. The fact that our new example G [Example 46.27] is a subgraph of P [unit distance plane] makes this possibility seem even more likely.

This also begs a question: was the choice of the mathematical standard $\mathbf{ZFC}$ inevitable? Was this choice the best possible?

In fact, I can formulate an unconditional theorem, which is a consequence of the same ideas as Theorem 47.1:

47.2 Unconditional Chromatic Number of the Plane Theorem

Can we get here anything unconditionally, in the absolute? Yes, we can, but not yet in $\mathbf{ZFC}$.

Unconditional Chromatic Number of the Plane Theorem 47.2 $\chi^{\mathbf{ZFS+}}(E^2) \geq 5$.

47.3 The Conjecture

I trust – all living is related,
The future is my everyday,
As heretic, I end by falling
Into Simplicity, the only way.
                      –Boris Pasternak, *The Waves*, 1931[13]

The great Russian poet provided us with the fitting epigraph about simplicity. Indeed, much of mathematical results are surprisingly simple, as are our conjectures. Just look at Erdős–Szekeres' Happy End Conjecture 29.15!

I have been asked—and asked others—what is the most reasonable expected value of the chromatic number of the plane, and more generally of E^n. We discussed this in the very beginning of this book, in Chapter 3. Now, armed with this book's dozens of Chapters and wisdom of half a century, what should we expect and try to prove? I believe that the chromatic number of the plane is 4 or 7.

Chromatic Number of the Plane Conjecture 47.3

$$\chi\left(R^2\right) = 4 \text{ or } 7.$$

It would be lovely to have four as the chromatic number of the plane: this is when our Conditional Chromatic Number of the Plane Theorem 47.1 would shine. Yet, if you, my reader, were to insist on my choosing just one value, I would choose the latter:

Chromatic Number of the Plane Conjecture 47.4

$$\chi\left(E^2\right) = 7.$$

"OK," you reply, "but then a unit distance 7-chromatic graph must exist in the plane!" This is true, but it could be quite large. In fact, in 1998 Dan Pritikin published the lower bound for the size of such a graph:

---

[13] [Pas]. Translated by Ilya Hoffman and Alexander Soifer. The original Russian text is this (I am adding four lines that follow):

В родстве со всем, что есть, уверясь
  И знаясь с будущим в быту,
  Нельзя не впасть к концу, как в ересь,
  В неслыханную простоту.

  Но мы пощажены не будем,
  Когда ее не утаим.
  Она всего нужнее людям,
  Но сложное понятней им.

Lower Bound for a Unit Distance 7-Chromatic Graph [Pri] Any unit distance 7-chromatic graph G satisfies the following inequality:

$$|G| \geq 6198.$$

In fact, the size of the smallest such graph may have to be even much larger. Now, try to construct it! For the 3-space I conjecture:

Chromatic Number of 3-Space Conjecture 47.5

$$\chi\left(E^3\right) = 15.$$

In general, I believe in the following conjecture:

Chromatic Number of E^n Conjecture 47.6

$$\chi\left(E^n\right) = 2^{n+1} - 1.$$

As Paul Erdős used to say, "This conjecture will likely withstand centuries, but, we will see!"

48
Imagining the Real, Realizing the Imaginary

> *What do you think of the abstract – do you believe
> that one should deduce one's abstraction from the
> forms of nature, or that one should create the form,
> outside of nature? Matisse replied, "There is always
> a measure of reality. The rest, I agree, is
> imagination."*
>
> – Henry Matisse[14]

> *Everything you can imagine is real.*
>
> – Pablo Picasso

> *There is an intimate relationship between the
> order of Nature (which constitutes the basis of life)
> and the order of Art (which constitutes the basis of
> civilization).*
>
> – Herbert Read.[15]

48.1 What Do the Founding Set Theorists Think about the Foundations?

> *In the beginning (if there was such a thing) God
> created Newton's laws of motion together with the
> necessary masses and forces. This is all; everything
> beyond this follows from the development of
> appropriate mathematical methods by means of
> deduction.*
>
> – Albert Einstein, 1946.[16]

---

[14] Interview with R. W. Howe [Fla], p. 186.

[15] [Rea]

[16] [Ein2, p. 19]

A. Soifer, *The Mathematical Coloring Book*,
DOI 10.1007/978-0-387-74642-5_48, © Alexander Soifer 2009

Kurt Gödel and Paul J. Cohen believed that we would eventually identify all of the axioms of set theory and when we have done so, we will no longer be able to choose between **CH** and ¬**CH** (or, similarly, between **AC** and **DC** + **LM**) because the additional axioms would exclude one of the options. Cohen shared his thoughts on the subject in 1966 [Coh2, pp. 150–151]:

> One can feel that our intuition about sets is inexhaustible and that eventually an intu-itively clear axiom will be presented which decides **CH**. One possibility is **V** = **L**, but this is almost universally rejected... A point of view which the author feels may eventually come to be accepted is that CH is *obviously* false.

Saharon Shelah disagreed with this Platonic view in his 2003 *Logical Dreams* [She3]:

> Some believe that compelling, additional axioms for set theory which settle problems of real interest will be found or even have been found. It is hard to argue with hope and problematic to consider arguments which have not yet been suggested. However, I do not agree with the pure Platonic view that the interesting problems in set theory can be decided, we just have to discover the additional axiom. My mental picture is that we have many possible set theories, all conforming to **ZFC**.

Before I queried set theorists about the state of their minds on the foundation of set theory, I read their writings on the subject matter. Robert M. Solovay in his pioneering 1970 paper [Sol1] states:

> Of course, the axiom of choice [**AC**] is true.

Saharon Shelah writes in the introduction to his 1994 classic monograph *Cardinal Arithmetic* [She2]:

> If we interpret "true" [sic] by "is provable in **ZFC**" (the usual axioms of set theory), as I do, then a large part of set theory which is done today does not deal directly with true theorems – it deals, rather, with a huge machinery for building counterexamples (forcing possible universes) or with "thin" universes (inner models). Very often the answer to "can this happen?" is "it depends". Now, I believe that this phenomenon is inevitable, and expresses a deep phase of the development of set theory, which resulted in many fascinating theorems (and also in quite a few proofs of mine). However, there is still some uneasiness about it. A way to express it is to say that if Cantor would have risen from his grave today, he would not just have problems with understanding the proofs of modern theorems – he would not understand what the theorems actually say.

And so I asked some of the leading set theorists, the great contributors to the axiomatics of sets, the following questions:[17]

*) Has **AC** been good for mathematics?
) Ought **AC to be a part of a "standard" system of axioms for set theory?
***) What do you think of the Solovay system of axioms (**ZFS**)?
****) How do you view the standard system of axioms for set theory?

---

[17] June 11–20, 2006 author's e-mal exchanges with Paul J. Cohen, Saharon Shelah, and Robert M. Solovay.

"My opinions shift and I have no obvious candidates [for the standard system of axioms]," Paul J. Cohen replied [Coh3], and added: "Solovay's result on **LM** is very nice, but hardly an axiom." Saharon Shelah [She4] saw a certain value in using the Solovay system **ZFS** and systems weaker than **ZFC**:

> The major question is what is true, i.e., when existence tells you something more if you give an explicit construction. Now, working in **ZF**, **ZF** + **DC** and also **ZF** + **DC** + **LM** and many other systems are ways to explicate the word "construct."

In Shelah's opinion [She4], **AC** has been "definitely" good for mathematics, **AC** is true and "should be in our standard system [of axioms]." Robert Solovay also believes that **AC** is true (and therefore his system, which I admire so much, **ZFS**+ = **ZF** + **DC** + **LM** is false). He writes about **AC** [Sol3]:

a. It's true.

b. It plays an essential role in all sorts of Theorems. e.g., the uniqueness of the algebraic closure; existence of maximal ideals; etc.

This prompted my question: But...what is "true"? Shelah answered it as follows [She5]:

> This is a meta-mathematical question. I will say [it] fits our image of set theory.
> You may say this is circularly, but this is unavoidable.
> You may be Platonist like Cantor then the meaning is clear.
> You may say [it is] what mathematicians who have not been interested in the question will accept.
> You may be a formalist and then this is a definition of **ZFC**.

Shelah is clear on what is not true enough, in his 2003 paper [She3]:

> Generally I do not think that the fact that a statement solves everything really nicely, even deeply, even being the best semi-axioms (if there is such a thing, which I doubt) is a sufficient reason to say it is a "true axiom".

Not surprisingly, there is no rigorous definition of "truth." This elusive notion is subjective, and all we can hope for is to recognize a true axiom when we encounter one. For Shelah, a true axiom is "what I feel/think is self evident."[18] This is a high bar to clear, and only **ZFC** seems to have cleared it for most creators of set theory. Even such serious candidates as **CH**, ¬**CH**, **GCH**, **V** = **L** are termed by Shelah as "semi-axioms", because they are not sufficiently "representative" of all possibilities. Shelah elaborates [She3]:

> Still most mathematicians, even those who have worked with **GCH** [and with other semi-axioms, A. S.] do it because they like to prove theorems and they could not otherwise solve their problems (or get a reasonable picture), i.e., they have no alternative in the short run...

---

[18] [She5].

What are our criterions for semi-axioms? First of all, having many consequences, rich, deep beautiful theory is important. Second, it is preferable that it is reasonable and "has positive measure". Third, it is preferred to be sure it leads to no contradiction...

Jean-Paul Delahaye [Del] believes that Shelah–Soifer results (Chapter 46) may have put a new emphasis on the task of finding which world of sets we think we live in:

It turns out that knowing if the world of sets satisfies the axiom of choice or a competing axiom is a determining factor in the solution of problems that no one had imagined depended on them. The questions raised by the new results are tied to the fundamental nature of the world of sets. Is it reasonable to believe that the mathematical world of sets is real? If it exists, does the true world of sets – the one in which we think we live – allow the coloring of S. Shelah and A. Soifer in two colors or does it require an infinity of colors?...

A series of results concerning the theory of graphs, published in 2003 and 2004 by Alexander Soifer of Princeton University and Saharon Shelah, of the University of Jerusalem, should temper our attitude and invite us to greater curiosity for the alternatives offered by the AC. The observation demonstrated by A. Soifer and S. Shelah should force mathematicians to reflect on the problems of foundations: what axioms must be retained to form the basis of mathematics for physicists and for mathematicians?

48.2 So, What Does It All Mean?

*I know of mathematicians who hold that the axiom of
choice has the same character of intuitive
self-evidence that belongs to the most elementary
laws of logic on which mathematics depends. It has
never seemed so to me.*

– Alonzo Church[19]

Shelah–Soifer's papers and related results we have discussed in this part seem surprising and even strange. How can the presence of the **AC** or its version affect whether we need 2 colors or uncountable infinity of colors for coloring a particular easily understood graph? How can the chromatic number of the plane be, perhaps, 4 in **ZFC** and 5, 6, or 7 in **ZFS+**? What *do* these results mean?

Jean-Paul Delahaye opens his article about these Shelah–Soifer papers in *Pour la Science*, the French edition of *Scientific American*, as follows [Del]:

The axiom of choice, a benign matter for the non-logician, puzzles mathematicians. Today, it manifests itself in a strange way: it takes, depending on the axiom's variants, either two or infinity of colors to resolve a coloring problem.

---

[19] Talk at the International Congress of Mathematicians in Moscow, 1966 [Chu].

Just as the parallels postulate seemed obvious, the axiom of choice has often been considered true and beyond discussion. The inventor of set theory, Georg Cantor (1845–1918), had used it several times without realizing it; Giuseppe Peano (1858–1932) used it in 1890, in working to solve a differential equation problem, consciously; but it was Ernst Zermelo (1871–1953), at the beginning of the 20th century, who identified it clearly and studied it.

When Gödel and Cohen proved independence of **AC** from the rest of the axioms **ZF** of set theory, they created a parallel, so to speak, between **AC** and the parallels postulate. As so, when Shelah–Soifer came out, it showed that various buildings of mathematics can be constructed.

Delahaye observes, "These [Shelah–Soifer's] results mean, as with the parallels postulate that several different universes can be considered," and continues:

> In the case of geometry, the independence of the parallels postulate proved that non-Euclidian geometries deserved to be studied and that they could even be used in physics: Albert Einstein took advantage of these when, between 1907 and 1915, he worked out his general theory of relativity.
>
> Regarding the axiom of choice, a similar logical conclusion was warranted; the universes where the axiom of choice is not satisfied must be explored and could be useful in physics.

Jean Alexandre Eugène Dieudonné (1906–1992), one of the founding members of Nicolas Bourbaki, described the state of the foundations in 1976 as follows:

> Beyond classical analysis (based on the Zermelo–Fraenkel axioms supplemented by the Denumerable Axiom of Choice), there is an infinity of different possible mathematics, and for the time being no definitive reason compels us to choose one of them rather than another.[20]

The Solovay system of axioms **ZFS+** is stronger than the system referred to by Dieudonné. It allows us to build classical analysis, including the complete Lebesgue measure theory; and it eliminates such counter-intuitive objects existing in **ZFC** as non-measurable sets of reals.

Using the **AC** in their 1924 paper [BT], two Polish mathematicians Stefan Banach (1892–1945) and Alfred Tarski (1902–1983) decomposed the three-dimensional closed unit ball into finitely many pieces, and moved those pieces through rotations and translations (pieces were allowed to move through one another) in such a way that the pieces formed two copies of the original ball. Since the measure of the union of two disjoint measurable sets is the sum of their measures, and measure is invariant under translations and rotations, we can conclude that there is a piece in Banach–Tarski decomposition that has no measure (i.e., volume). Having **LM** in the system of axioms for set theory would eliminate this and a good number of other paradoxes.

ZFC allows us to create imaginary objects—or shall I say unimagined objects—such as sets on the line that have no length, sets on the plane that have no area, etc. Are we not paying a high price for the comfort of having a powerful tool, **AC**?

---

[20] Quoted from [Moo, p. 4]

Having lived most of my life in **ZFC** and having enjoyed using transfinite induction, in the course of my work with Shelah I came to a realization that I prefer **ZFS+** = **ZF** + **DC** + **LM** over **ZFC** because it assures that every set of reals is measurable (which is consistent with my intuition: every point set on the line ought to have length), while **DC** gives us as much choice as is consistent with **LM**.

Of course, by downgrading **AC** to **DC**, we would lose such tools as the transfinite induction and the well-ordering of uncountable sets, and would consequently lose some important theorems, such as the existence of basis for a vector space. However, new theorems would be found when mathematicians spend as much time building on the Solovay foundation **ZFS+** as they have on **ZFC**.

48.3 Imagining the Real vs. Realizing the Imaginary

> *As far as the propositions of mathematics refer to reality, they are not certain; and as far as they are certain, they do not refer to reality.*
> – Albert Einstein, 1921 [Ein1]

> *We all know that Art is not truth. Art is a lie that makes us realize the truth.*
> – Pablo Picasso, 1923 [Pic]

> *The mathematician is an inventor, not a discoverer.*
> – Ludwig Wittgenstein, 1937–1944 [Witt, p. 47e]

Einstein, Picasso and Wittgenstein expressed my views in the above epigraphs, and spelled them out with a genius precision and conciseness.

Undoubtedly, vast majority of mathematicians are Platonists.[21] They believe that mathematical objects exist out there independently of the human mind, and mathematicians merely discover them. The Platonists believe that a mathematical statement, such as AC, is objectively either true or false—we simply do not yet know which it is (although in a poll, "AC is true" would win hands down). Likewise, a question, what is the chromatic number of the first Shelah–Soifer graph G, surely, must have a single answer; it cannot be "2 or uncountable infinity." Therefore, for the Platonists either **ZFC** or **ZF** + **DC** + **LM** is true, we just do not know which. Platonists imagine the real.

How does one describe those who hold the view of mathematics that is dual to that of Plato? I propose to call them Artists, especially since Picasso is part of this group. To paraphrase Picasso,

Mathematics is an invention that makes us realize reality.

---

[21] Jim Holt reminisces [Hol]: "Some years ago, while giving a lecture to an international audience of elite mathematicians in Berkeley, I asked how many of them were Platonists. About three-quarters raised their hands."

"The mathematician is an inventor, not a discoverer," agrees Wittgenstein. Mathematics is certain only as an invention, or as Einstein put it," As far as the laws of mathematics refer to reality, they are not certain; and as far as they are certain, they do not refer to reality."

I believe that mathematicians do not only imagine the real, but moreover realize the imaginary. Just like artists, mathematicians create objects that challenge reality in every aspect: beauty, simplicity, intuitiveness, and counter-intuitiveness. The real is but one of inspirations for creating mathematics.

Mathematics of **ZFC** *is* the house that Jack built. Has he built the only possible house? Has he built the best possible house? Must we give up the village for a house, as Richard III offered his kingdom for a horse?

I believe that mathematicians put all their eggs in one **ZFC** basket, and thus have missed out on many alternative mathematical universes that can be built on many alternative foundations, one of which is the Solovay's **ZFS+**. Saharon Shelah also thinks that we ought to build on many foundations, but he puts his main emphasis on building *up* from **ZFC**. It seems we have been too comfortable and too nonchalant about seeing problems with **ZFC** and doing nothing about them. Mathematical results presented in this part may be not important by themselves, but they illuminate the many mathematics that could be built, and serve—I hope—as a wake up call. Delahaye [Del] concludes his analysis of Shelah–Soifer's series of papers with the possibility of the emerging "set-ist revolution":

> In set theory, as in geometry, all axiomatic systems are not equal. Thinking carefully about their meaning and the consequences of each one of them, and asking ourselves (as it is done in geometry) what the particular usefulness of this or that axiom is in expressing and addressing issues of mathematical physics, may be relevant once again and could lead—why not—to a revolution of set theories, similar to the revolution in non-Euclidian geometries.

Starting a revolution? All right! As long as the revolution is imaginary, in mathematics, and not a "real" revolution that causes death and destruction—I am with The Beatles on that:[22]

> You say you want a revolution
> Well, you know
> We all want to change the world.
>
> . . .
>
> You say you got a real solution
> Well, you know
> We'd all love to see the plan.

---

[22] The Beatles, *Revolution*, 1968.

XI
Farewell to the Reader

49
Two Celebrated Problems

If faith can move mountains, disbelief can deny their existence. And faith is impotent against such impotence.

– Arnold Schoenberg, June 1924[1]

Histories of two coloring problems, The Four Color Problem (4CP) and The Chromatic Number of the Plane Problem (CNP), have been strikingly similar on many counts.

Each problem is easy to formulate and hard to solve.

Each problem was created by young students, ages 20 and 18 respectively, born in the year '32:

4CP by Francis Guthrie, born in 1832;
CNP by Edward Nelson, born in 1932.

The 4CP motivated the development of much of the graph theory. The CNP inspired a lot of mathematical work and results in a great variety of fields in which a solution was sought: combinatorics, graph theory, topology, measure theory, abstract algebra, geometry, and combinatorial geometry.

As we have seen on the pages of this book, CNP and 4CP have an essential non-empty intersection: Townsend–Woodall's 5-Color Theorem.

Each of these problems had a chief promoter. For 4CP this was Augustus De Morgan, who kept the problem alive for decades. Paul Erdős's contribution to keeping CNP alive is even greater. First of all, as De Morgan did for 4CP, Erdős kept the flame of the problem lit. He made CNP well-known by posing it in his countless problem talks and many publications, for example, we see it in [E61.22], [E63.21], [E75.24], [E75.25], [E76.49], [E78.50], [E79.04], [ESi], [E80.38], [E80.41], [E81.23], [E81.26], [E85.01], [E91.60], [E92.19], [E92.60] and [E94.60]. Secondly, Paul Erdős created a good number of fabulous related problems, some of which we have discussed in this book.

---

[1] [Scho].

A. Soifer, *The Mathematical Coloring Book*,
DOI 10.1007/978-0-387-74642-5_49, © Alexander Soifer 2009

Both problems require a very long time to be conquered. Victor Klee and Stan Wagon [KW], observing that solving 4CP took 124 years, suggested that CNP might require as long for its solution:

> If a solution of CNP takes as long as 4CC, then we will have a solution by the year 2084.

Will we succeed by 2084? Paul Erdős would have said, "We shall see!" Arnold Schoenberg believed that faith can move mountains. Erdős urged us to believe that the transfinite Book of all the theorems and their best proofs exists. Such a belief led Appel and Haken to success at the breaking point of available computing. Such a belief is needed to conquer my favorite open problem of mathematics: the chromatic number of the plane. We shall overcome!

Thank you for inviting my book into your home and holding it in your hands. Your feedback, problems, conjectures, and solutions will always be welcome in my home. Who knows, maybe they will inspire a new edition in the future and we will meet again!

Bibliography

[Abb1]   Abbott, H. L., *Some Problems in Combinatorial Analysis*, Ph.D. thesis, University of Alberta, Canada, March, 1965.

[Abb2]   Abbott, T. K., *Catalogue of the Manuscripts in the Library of Trinity College, Dublin to which is added a List of Fagel Collection of Maps in the same Library.* Hodges, Figgis, & Co, Dublin, 1900.

[AZ]   Abbott, H. L., and Zhou, B., On small faces in 4-critical graphs. *Ars Combin.* 32 (1991), 203–207.

[Abe1]   Abelin-Schur, H., Talk at *Schur-Gedenkkolloquium at* at the Humboldt University, Berlin, November 15, 1991, manuscript.

[Abe2]   Abelin-Schur, H., Phone conversation of May 18, 1995.

[Aks]   Aksionov, V. A., On continuation of 3-coloring of planar graph (Russian). *Diskret. Analiz.* 26 (1974), 3–19.

[AM]   Aksionov, V. A., and Mel'nikov, L. S., Some Counter examples Associated with the Three-Color Problem, J. Combin. Theory, Series B 28(1980), 1–9.

[AFG]   Alexeev, B., Fox, J., and Graham, R. L., On minimal colorings without monochromatic solutions to a linear equation, *Combinatorial Number Theory: Dedicated to Professor Ron Graham on the Occasion of His Seventieth Birthday.* Walter de Gruyter, Berlin, 2007, 1–22.

[AH0]   Appel, K., and Haken, W., The existence of unavoidable sets of geographically good configurations. *Illinois J. Math.* 20 (1976), 218–297.

[AH1]   Appel, K., and Haken, W., The solution of the four-color map problem. *Sci. Am.* 237(4) (Oct. 1977), 108–121.

[AH2]   Appel, K., and Haken, W., Every planar map is four colorable. Part I. Discharging. *Illinois J. Math.* 21 (1977), 429–490.

[AH3]   Appel, K., Haken, W., and Koch, J., Every planar map is four colorable. Part II. Reducibility. *Illinois J. Math.*, 21 (1977), 491–567.

[AH4]   Appel, K., and Haken, W., *Every Planar Map is Four Colorable.* Amer. Math. Soc., Providence, 1989.

[Arn]   Arnautov, V. I., Nediskretnaya Topologizuemost Schetnykh Kolets, (Russian). *Dokl. Akad. Nauk SSSR* 191 (1970) 747–750. English Translation: Nondiscrete topologizability of countable rings. *Soviet Math. Dokl.* 11 (1970), 423–426.

[Arr]   Arrias, E., In Memoriam Prof. P. J. H. Baudet. *Eigen. Haard.* 48(5) (Jan. 28, 1928), 92–94.

[BaO]   Baker, H., and Oliver, E., *Ericas in Southern Africa.* Purnell, Cape Town.

[BT]   Banach, S., and Tarski, A., Sur la décomposition des ensembles de points en parties respectivement congruentes. *Fundamenta Mathematicae* 6 (1924), 244–277.

[Bau1]   Baudet, P. J. H., *Groepentheoretische onderzoekingen* (doctoral dissertation). Martinus Nijhoff, The Hague, 1918.

[Bau2]   Baudet, P. J. H., *Het Limietbegrip* (inaugural address at Delft). P. Noordhoff, Groningen, 1919.

[Bau3]   Baudet, P. J. H., Een nieuwe theorie van onmeetbare getallen. *Christiaan Huygens* I (1921/1922), 33–47.

[Bau4]   Baudet, P. J. H., Een stelling over rekenkundige reeksen van hoogere orde. *Christiaan Huygens* I (1921/1922), 146–149 (published posthumously).

[Bau5]   Baudet, P. J. H., Gelijktijdige invoering der negatieve en imaginaire getallen. *Christiaan Huygens* I (1921/1922), 226–231 (published posthumously by J. Teixeira de Mattos and F. Schuh on the basis of a lecture and notes by Baudet).

[BII1]   Baudet, letter to A. Soifer of August 24, 1995.

[BII2]   Baudet, H., letter to A. Soifer of October 17, 1995.

[BII3]   Baudet, H., letter to A. Soifer of November 27, 1995.

[BII4]   Baudet, H., letter to A. Soifer of December 18, 1995.

[BII5]   Baudet, H., letter to A. Soifer of January 29, 1996.

[BII6]   Baudet, H., letter to A. Soifer of February 15, 1996.

[BII7]   Baudet, H., letter to A. Soifer of March 6, 1996.

[BII8]   Baudet, H., letter to A. Soifer of March 23, 1996.

[BII9]   Baudet, H., letter to A. Soifer of April 7, 1996.

[BII10]   Baudet, H., letter to A. Soifer of April 18, 1996.

[BII11]   Baudet, H., letter to A. Soifer of May 10, 1996.

[BII12]   Baudet, H., letter to A. Soifer of May 27, 1996.

[BII13]   Baudet, H., letter to A. Soifer of June 12, 1996.

[BW]   Beineke, L. W., and Wilson, R. J., On the edge-chromatic number of a graph, Discrete Math. 5 (1973), 15–20.

[Beh]   Behzad, M., *Graphs and Their Chromatic Numbers*, Ph.D. Thesis, Michigan State University, Michigan, USA, 1965.

[BC1]   Behzad, M., and Chartrand, G., An introduction to total graphs. In P. Rosenstiehl (ed.), *Theory of Graphs. International Symposium, Rome.* July 1966, Gordon and Breach, New York, 1967, 31–33.

[BC2]   Behzad, M., and Chartrand, G., *Introduction to the Theory of Graphs*. Allyn and Bacon, Boston, 1971.

[BCL]   Behzad, M., Chartrand, G., and Lesniak-Foster, L., *Graphs & Digraphs*. Prindle, Weber & Schmidt, Boston, 1979.

[BO]   Belov, A. Ya., and Okhitin, S. B., Ob Odnoj Kombinatornoj Zadaze (On One Combinatorial Problem). *Uspekhi Mat. Nauk* 48(2) (1993), 169–170, (Russian).

[BP]   Benda, M., and Perles, M., Colorings of metric spaces. *Geombinatorics* IX(3) (2000), 113–126.

[BL]   Bergelson, V., and Leibman, A., Polynomial Extension of van der Waerden's and Szemerédi's theorems. *J. Amer. Math. Soc.* 9 (1996), 725–753.

[Ber]   Bergelson, V., Multiplicatively large sets and ergodic ramsey theory. *Israle J. Math.* 148 (2005), 23–40.

[Ber1]   Berhhart, A., Six-rings in minimal five-color maps, *Amer. J. Math.* 69 (1947), 391–412.

[Ber2]   Berhhart, A., Another reducible edge configuration, *Amer. J. Math.* 70 (1948), 144–146.

[Berl]   Berlekamp, E. R., A construction for partitions which avoid long arithmetic progressions. *Canad. Math. Bull.* 11 (1969), 409–414.

[Bern]   Bernays, P., A system of axiomatic set theory III, *J. Symbolic Logic* 7 (1942), 65–89.

[Bert]   Bertsch, S. M., *Nobel Prize Women in Science: Their Lives, Struggles, and Momentous Discoveries*. Carol Pub., Secaucus, N.J., 1993.

[Big]   Biggs, N. L., De Morgan on map colouring and the separation axiom, *Archive Hist. Exact. Sci.* 28 (1983), 165–170.

[BLW]   Biggs, N. L., Lloyd, E. K., and Wilson, R. J., *Graph Theory 1736–1936*, Clarendon Press, Oxford, 1976. Latest version: reprinted with corrections, 1998.

[Bir]   Birkhoff, G. D., The reducibility of maps. *Amer. J. Math.* 35 (1913), 114–128.

[BS]   Boltyanski, V., and Soifer, A., *Geometric Etudes in Combinatorial Mathematics*. Center for Excellence in Mathematical Education, Colorado Springs, 1991.

[BT]   Bóna, M., and Tóth, G., A Ramsey-type problem on right-angled triangles in space. *Discrete Math.* 150 (1996), 61–67.

[Bor]   Borodin, O. V., Structural properties of plane graphs without adjacent triangles and an application to 3-colorings. *J. Graph Theory* 21(2) (1996), 183–186.

[BR]   Borodin, O. V., and Raspaud, A., A sufficient condition for planar graphs to be 3-colorable. *J. Combin. Theory Ser. B* 88 (2003), 17–27.

[BGRS]   Borodin, O. V., Glebov, A. N., Raspaud, A., and Salavatipour, M. R., Planar graphs without cycles of length from 4 to 7 are 3-colorable. *J. Combin. Theory Ser. B* 93 (2005), 303–311.

[BGJR]   Borodin, O. V., Glebov, A. N., Jensen, T. R., and Raspaud, A., Planar graphs without triangles adjacent to cycles of length from 3 to 9 are 3-colorable. *Sib. Electron. Mat. Rep.* 3 (2006), 428–440.

[BMP]   Brass, P., Moser, W., and Pach, J., *Research Problems in Discrete Geomtery.* Springer, New York, 2005.

[BBH]   Brauer, A., Brauer, R., and Hopf, H., Über dies Irreduzibilität einigen spezieller Klassen von Polynomen. *Jahresber. Deutsch. Math.-Verein.* 35 (1926), 99–112.

[Bra1]   Brauer, A., Über Sequenzen von Potenzresten, *Sitzunberichte de Preussischen Akademie der Wissenschaften, Physicalisch-Mathematische Klasse.* Berlin, 1928, 9–16.

[Bra2]   Brauer, A., Combinatorial Methods in the Distribution of kth Power Residues. *Combinatorial Mathematics and its Applications*; Proceedings of the Conference, University if North Carolina, Wilmington, 1969, 14–37.

[Bra3]   Brauer, A., Gedenkrede aufIssai Schur, in Issai Schur. *Gesammelte Abhandlungen*, vol. 1, Springer, Berlin, 1973, v–xiv.

[BraH]   Brauer, H., *Memories*, 114 pages, manuscript.

[BraR]   Brauer, R., *Collected Papers*, vol's I, II, and III, MIT Press, Cambridge, Massachusetts, 1980.

[Bro]   Brooks, R. L., On Colouring the nodes of a network, *Proc. Cambridge Philos. Soc.* 37 (1941), 194–197.

[BR]   Brown, J. I., and Rödl, V., Folkman numbers for graphs of small order. *Ars Combin.* 35(A) (1993), 11–27.

[BDP1]   Brown, N., Dunfield, N., and Perry, G., Colorings of the Plane I. *Geombinatorics* III(2) (1993), 24–31.

[BDP2]   Brown, N., Dunfield, N., and Perry, G., Colorings of the Plane II. *Geombinatorics* III(3) (1993), 64–74.

[BDP3]   Brown, N., Dunfield, N., and Perry, G., Colorings of the Plane III. *Geombinatorics* III(4) (1993), 110–114.

[Bru1]   Bruijn, N. G. de, Letter to P. Erdős, August 18, 1947.

[BE1]   Bruijn, N. G. de, and Erdős, P., On a combinatorial problem. *Indagationes Math.* 10 (1948), 421–423.

[BE2]   Bruijn, N. G. de, and Erdős, P., A colour problem for infinite graphs and a problem in the theory of relations. *Indagationes Math.* 13 (1951), 369–373.

[Bru2]   Bruijn, N. G. de, Letter to B. L. van der Waerden, March 29, 1977.

[Bru3]   Bruijn, N. G. de, Commentary (on B. L. Van der Waerden's article[Wae2]). In E. M. J. Berin et al. (eds), *Two Decades of Mathematics in the Netherlands, 1920–1940*, Mathematical Centre, Amsterdam, 1978, Part I, 116–124.

[Bru4]   Bruijn, N. G. de, e-mail to A. Soifer, May 9, 1995.

[Bru5]   Bruijn, N. G. de, e-mail to A. Soifer, June 21, 1995.

[Bru6]   Bruijn, N. G. de, e-mail to A. Soifer, July 5, 1995.

[Bru7]   Bruijn, N. G. de, e-mail to A. Soifer, January 17, 2004.

[Bru8]   Bruijn, N. G. de, e-mail to A. Soifer, May 26, 2004.

[Bru9]   Bruijn, N. G. de, e-mail to A. Soifer, June 1, 2004.

[Bru10]   Bruijn, N. G. de, e-mail to A. Soifer, June 15, 2004.

[Bru11]   Bruijn, N. G. de, e-mail to A. Soifer, June 23, 2004.

[Bru12]   Bruijn, N. G. de, e-mail to A. Soifer, November 1, 2005.

[Bru13]   Bruijn, N. G. de, e-mail to A. Soifer, November 3, 2005.

[BFS]   Brüning, J., Ferus, D., and Siegmund-Schultze, R., *Terror and Exile: Persecution and Expulsion of Mathematicians from Berlin between 1933 and 1945.* Deutsche Mathematiker-Vereinigung, Berlin, 1998.

[Buk]   Bukh, B., Measurable sets with excluded distances, manuscript, 22pp, January 25, 2007.

[BM]   Burkill, H., and Mirsky, L., Monotonicity. *J. Math. Anal. and Appl.* 41 (1973), 391–410.

[CET]   Calkin, N. J., Erdős, P., and Tovey, C. A., New Ramsey bounds from cyclic graphs of prime order. *SIAM J. Discrete Math.* 10 (1997), 381–387.

[Can1]   Cantwell, K., Finite euclidean ramsey theory. *J. Combin. Theory Ser. A* 73(2), (1996), 273–285.

[Can2]   Cantwell, K., All regular polytopes are Ramsey, *J. Combin. Theory Ser. A* 114 (2007), 555–562.

[Cay1]   Cayley, A., Report of meeting, 13 June 1878, in *Proceedings of the London Mathematical Society*, IX, C. F. Hodgson & Son, Gough Square, London, 1878, 148.

[Cay2]   Cayley, A., Report of meeting, 13 June 1878, in *Nature*, XVIII, MacMillan and CO., London, 1878, 204.

[Cay3]   Cayley, A., On the Colouring of Maps, in *Proceedings of the Royal Geographical Society.* London, 1879, 259–261.

[CL]   Charttrand, G., and Lesniak, L., *Graphs & Digraphs*, 4th ed., Chapman & Hall/CRC, Boca Raton, 2005.

[CRW]   Chen, M., Raspaud, A., and Wang, W., Three-coloring planar graphs without short cycles. *Information Processing Letters* 101 (2007), 134–138.

[Che]   Chetwynd, A. G., Total colourings of graphs – A progress report. In Alavi, Y. et al. (ed.), *Graph Theory, Combinatorics, and Applications*, Proceedings of the Sixth Quadrennial Conference on Graph Theory and Appplications of Graphs, 1988, vol. 1, 233–244, Wiley, New York, 1991.

[Chi1]   Chilakamarri, K. B., Unit-distance graphs in rational n-spaces. *Discrete Math.* 69 (1988), 213–218.

[Chi2]   Chilakamarri, K. B., On the chromatic number of rational five-space. *Aequationes Math.* 39 (1990), 146–148.

[Chi3]   Chilakamarri, K. B., Unit-distance graphs in Minkowski metric spaces. *Geometriae Dedicata* 37 (1991), 345–356.

[Chi4]   Chilakamarri, K. B., The unit-distance graph problem: A brief survey and some new results. *Bulletin of the ICA* 8 (1993), 39–60.

[Chi5]   Chilakamarri, K. B., A characterization in Z^n of finite unit-distance graphs in R^n, (manuscript).

[Chi6]   Chilakamarri, K. B., A 4-Chromatic unit-distance graph with no triangles. *Geombinatorics* IV(3), (1995), 64–76.

[CG]   Chung, F., and Graham, R. L., Forced convex n-gons in thye plane. *Discrete and Computational Geometry* 19 (1998), 3677–371.

[Cho]   Chow, T., Distances forbidden by two-colorings of Q^3 and A_n (manuscript).

[Chow]   Chowla, S., There exists an infinity of 3-combinations of primes in A.P. *Lahore Philos. Soc.* 6(2) (1944), 15–16.

[Chu]   Church, A., and Paul, J., Cohen and the Continuum Problem. In Petrovsky, I. G. (ed), *Proceedings of International Congress of Mathematicians (Moscow – 1966)*, Mir, Moscow, 1968, 15–20.

[Chv]   Chvátal, V., Some unknown van der Waerden numbers, in *Combinatorial Structures and their Applications.* Gordon and Breech, New York, 1969, 31–33.

[CH]   Chvátal, V., and Harary, F., Generalized Ramsey theory for graphs. *Bull. Amer. Math. Soc.* 78 (1972), 423–426.

[Cib]   Cibulka, J., On the Chromatic Number of Real and Rational Spaces, to appear in October 2008 in *Geombiniatorics* XVIII(2).

[Coh1]   Cohen P. J., The independence of the continuum hypothesis, *Proceeding of the National Academy of Science, USA*, part I: 50 (1963), 1143–1148; part II: 51 (1964), 105–110.

[Coh2]   Cohen, P. J., *Set Theory and the Continuum Hypothesis*. Benjamin/Cummings Reading, Massachusetts, 1966.

[Coh3]   Cohen, P. J., e-mail to A. Soifer, June 12, 2006.

[CS]   Conway, J. H., and Sloane, N. J. A., *Sphere packings, lattices and groups*. 3rd ed., with contrib. by E. Bannai, R. E. Borcherds, J. Leech, S. P. Norton, A. M. Odlyzko, R. A. Parker, L. Queen and B. B. Venkov, Springer-Verlag, New York, 1999.

[Cou1]   Coulson, D., An 18-colouring of 3-space omitting distance one. *Discrete Math.* 170(1–3) (1997), 241–247.

[Cou2]   Coulson, D., A 15-colouring of 3-space omitting distance one. *Discrete Math.* 256(1–2) (2002), 83–90.

[Cou3]   Coulson, D., Tilings and Colourings of 3-Space. *Geombinatorics* XII (3) (2003), 102–116.

[Cour]   Courant, R., Reminiscences from Hilbert's Göttingen, *Mathematical Intelligencer* 3 (1981), 154–164.

[CoxJ]   Cox, J., Letter to A. Soifer dated November 21, 1991.

[Cox1]   Coxeter, H. S. M., Map-coloring problems, *Scripta Math.* 23 (1958), 11–25.

[Cox2]   Coxeter, H. S. M., The four-color map problem, 1840–1890, *Math. Teacher* 52 (1959), 283–289.

[Cro]   Croft, H. T., Incidence incidents. *Eureka* (Cambridge) 30 (1967), 22–26.

[CFG]   Croft, H. T., Falconer, K. J., and Guy, R. K., *Unsolved Problems in Geometry*, Springer, New York, 1991.

[CT]   Csizmadia, G., and Tóth, G., Note on a Ramsey-type problem in geometry. J. Combin. Theory, Ser. A 65(2) (1994), 302–306.

[Dal1]   Dalen, D. van, *Mystic, Geometer, and Intuitionist: The Life of L. E. J. Brouwer, Volume I: The Dawning Revolution*, 1999, Oxford University Press, Oxford.

[Dal2]   Dalen, D. van, *Mystic, Geometer, and Intuitionist: The Life of L. E. J. Brouwer, Volume II: Hope and Disillusion*, 2005, Oxford University Press, Oxford.

[DGK]   Danzer, L., Grünbaum, B., and Klee, V., *Helly's Theorem and its Relatives*, Amer. Math. Soc., Providence, 1963.

[Del]   Delahaye, J.-P., Coloriages irréals, *Pour la Science*, Février 2005, 88–93. English translation: Imaginary Coloring. *Geombinatorics* XV(3) (2006), 101–119.

[DR]   Delannoy, H., and Ramsey, A. S., Réponse à question 51, *Interméd. Math.* 1 (1894), 192.

[DeM1]   De Morgan, A., Letter to W. R. Hamilton, October 23, 1852; TCD MS 1493, 668; Trinity College Dublin Library, Manuscripts Department.

[DeM2]   De Morgan, A., Letter to W. Whewell, December 9, 1853, Trinity College Cambridge, Whewell Add. mss., a.202^{125}.

[DeM3]   De Morgan, A., Letter to R. L. Ellis, June 24, 1854, Trinity College Cambridge, Whewell Add. mss., c.67^{111}.

[DeM4]   De Morgan, A., Review of the philosophy of discovery..., *The Athenaeum* 1694, London, 1860, 501–503.

[DeM5]   De Morgan, A., Correspondence to Sir Herschel, in *Memoir of Augustus De Morgan*, by his wife S. E. De Morgan, Longmans, Green, and Co., London, 1882, 333.

[Des1]   Descartes, B., A three-colour problem. *Eureka* 9, April, 21, 1947.

[Des2]   Descartes, B., A three-colour problem, Solutions to Problems in Eureka No. 9, *Eureka* 10, March 24, 1948.

[Des3]   Descartes, B., Advanced Problems and Solutions: Solution of problem 4526, *Amer. Math. Monthly* 61(5) (1954), 352–353.

[Dic1]   Dickson, L. E., On the Last Theorem of Fermat. *Quart. J. Pure Appl. Math.* 40 (1908), 27–45.

[Dic2]   Dickson, L. E., *History of the Theory of Numbers*, vols 1 (1919), 2 (1920), 3 (1923), Carnegie Institution of Washington, Washington, DC.

[Die]   Diestel, R., *Graph Theory*, 2nd ed., Springer-Verlag, New York, 2000.

[Dir]   Dirac, G. A., A property of 4-chromatic graphs and some remarks on critical graphs. *J. London Math. Soc.* 27 (1952), 85–92.

[Dol1]   Dold-Samplonius, Bartel Leendert van der Waerden befragt von Yvonne Dold-Samplonius, *NTM, International Journal of History and Ethics of Natural Sciences, Technology and Medecine*, 2 (1994), 129–147. English translation: Interview with Bartel Leendert van der Waerden, *Notices of the Amer. Math. Soc.* 44(3) (1997), 313–320.

[Dol2]   Dold-Samplonius, In Memoriam: Bartel Leendert van der Waerden (1903–1996). *Historia Mathematica* 24 (1997), 125–130.

[Dre]   Dreyer, C. Th., Thoughts on My Métier (1943), translated by Skoller, D., Gyldendal Boghandel, Nordisk Forlag A/S, Conpenhagen, 1973.

[Dum]   Dumitrescu, A., A Remark on the Erdős-Szekeres Theorem, *Amer. Math. Monthly* 112(10), (2005) 921–924.

[DF]   Dummit, D. S., and Foot, R. M., *Abstract Algebra*, 2nd ed., Wiley & Sons, New York, 1999.

[Dup]   Duparc, H. J. A., Handwritten recollection notes for A. Soifer, Delft, September, 1996.

[Dye]   Dye, D. S., A Grammar of Chinese Lattice, Harvard-Yenching Institute, Harvard Univ. Press, Cambrige, Mass., 1937.

[DU]   Dynkin, E. B., and Uspenskii, V. A., Matematicheskie Besedy, Gosudarstvennoe Izdatel'stvo Tekhniko-Teoreticheskoy Literatury, Moscow, 1952 (Russian). Partial English translation (part 1 of three only): Multicolor Problems, D.C. Heath & Co., Boston, 1963.

[Eck0]   Eckmann, B., e-mail to A. Soifer of December 7, 2004.

[Eck1]   Eckmann, B., e-mail to A. Soifer of December 23, 2004.

[Eck2]   Eckmann, B., e-mail to A. Soifer of December 27, 2004.

[Eck3]   Eckmann, B., e-mail to A. Soifer of December 30, 2004.

[Ein1]   Einstein, A., Geometry and Experience, a lecture before the Prussian Academy of Sciences, January 27, 1921, In Seelig, C. (ed.), *Ideas and Opinions by Albert Einstein*, Crown, New York, 1954.

[Ein2]   Einstein, A. (author), Schillp, P. A. (ed.), *Albert Einstein: Philosopher-Scientist*. Open Court, La Salle, 1970.

[Eis]   Eisenreich, G. B. L., Van der Waerden Wirken von 1931 bis 1945 in Leipzig, in *100 Jahre Mathematisches Seminar der Karl-Marx-Universität Leipzig*. Beckert, H., and Schumann, H. (eds.), VEB Deutscher Verlag der Wissenschaften, Berlin, 1981.

[E42.06]   Erdős, P., Some set-theoretical properties of graphs. *Revista de la Univ. Nac. de Tucumán, Ser A. Mat. y Fis. Teór.*, 3 (1942), 363–367.

[E47/8/4ltr] Erdős, P., Letter to N. G. de Bruijn, August 4, 1947.

[E50.07]   Erdős, P., On integers of the form $2^k + p$ and some related problems, *Summa Brasil. Math.* 2 (1950), 113–124.

[E51.01]   Erdős, P., A colour problem for infinite graphs and a problem in the theory of relations. *Proc. Konink. Nederl. Akad. Wetensch. Amsterdam* 54 (1951), 371–373.

[E52.03]   Erdős, P., Egy kongruenciarendszekröl szóló problémáról (On a problem concerning covering systems). *Mat. Lapok* 3 (1952), 122–128 (Hungarian).

[E57.13]   Erdős, P., Some unsolved problems. *Michigan Math. J.* 4 (1957), 291–300.

[E59.06]   Erdős, P., Graph theory and probability. *Canad. J. Math.* 11 (1959), 34–38.

[E60.09]   Erdős, P., On sets of distances of n points in Euclidean space. *MTA MKI Közl.* 5 (1960), 165–169.

[E61.05]   Erdős, P., Graph theory and probability II. *Canad. J. Math.* 13 (1961), 346–352.

[E61.21]   Erdős, P., Nekotorye Nereshennye Problemy (Some unsolved problems), *Matematika* 7(4) (1963), 109–143. (A Russian translation of [E61.22]).

[E61.22] Erdős, P., Some Unsolved Problems, *Magyar tudomanyos akademia mathematikai kutato intezetenek közlemenyel* 6 ser. A (1–2) (1961), 221–254.

[E71.13] Erdős, P., On the application of combinatorial analysis to number theory, geometry and analysis. *Actes du Congrès International des Mathematiciens* (Nice, 1970), Tome 3, Gauthier-Villars, Paris, 1971, 201–210.

[E73.21] Erdős, P., Problems and results on combinatorial number theory. *A survey of combinatorial theory* (Proc. Internat. Sympos., Colorado State Univ., Fort Collins, Colorado, 1971), 117–138, North-Holland, Amsterdam, 1973.

[E75.24] Erdős, P., On some problems of elementary and combinatorial geometry. *Annali Matematica Pura Applicata*, 103(4) (1975), 99–108.

[E75.33] Erdős, P., Problems and results on finite and infinite graphs. *Recent Advances in Graph Theory* (Proc. Second Czechoslovak Sympos., Prague, 1974), 183–192, Academia, Prague, 1975.

[E76.35] Erdős, P., Problems and results in combinatorial analysis. *Colloquio Internacionale sulle Teorie Combinatorie* (Rome, 1973), Tomo II, Atti dei Convegni Lincei, No. 17, pp. 3–17, Accad. Naz. Lincei, Rome, 1976.

[E75.40] Erdős, P., Some problems on elementary geometry, *Austr. Math. Soc. Gaz.*, 2 (1975), 2–3.

[E76.49] Erdős, P., Problem (p. 681); P. Erdős, Chairman, Unsolved Problems, in *Proceedings of the Fifth British Combinatorial Conference 1975, University of Aberdeen, July 14–18, 1975*, C. St. J. A. Nash-Williams and J. Sheehan, (eds.), Congressus Numerantium XV, Winnipeg, Utilitas Mathematica, 1976.

[E77.26] Erdős, P., Problems and results on combinatorial number theory, II. *J. Indian Math. Soc.* (N.S.) 40(1–4) (1976), 285–298, 1977.

[E77.28] Erdős, P., 1977.28 Problems in number theory and Combinatorics. Proceedings of the Sixth Manitoba Conference on Numerical Mathematics (Univ. Manitoba, Winnipeg, Man., 1976), *Congress. Numer.* XVIII, 35–58, Utilitas Math., Winnipeg, Man., 1977.

[E78.50] Erdős, P., Problem 28 in "Exercises, Problems and Conjectures," Bollobás, B., *Extremal Graph Theory*, London Mathematical Society Monographs, Academic Press, 1978, p. 285.

[E79.04] Erdős, P., Combinatorial problems in geometry and number theory. *Relations between combinatorics and other parts of mathematics, Proc. Sympos. Pure Math.*, Ohio State Univ., Columbus Ohio (1978); *Proc. Sympos. Pure Math.*, XXXIV, *Amer. Math. Soc.*, Providence, RI (1979), 149–162.

[E80.03] Erdős, P., A survey of problems in combinatorial number theory. *Combinatorial Mathematics, Optimal Designs and their Applications* (Proc. Sympos. Combin. Math. and Optimal Design, Colorado State Univ., Fort Collins, Colo., 1978), Ann. Discrete Mathematics 6 (1980), 89–115.

[E80.38] Erdős, P., Some combinatorial problems in geometry. *Geom. & Diff. Geom. Proc.*, Haifa, Israel, (1979); *Lecture Notes in Math.*, 792, Springer (1980), 46–53.

[E80.41] Erdős, P., Some combinational problems in Geometry. *Geometry and Differential Geometry* (Proc. Conf., Univ. Haifa, Haifa, 1979), Lecture Notes in Math., 792, pp. 46–53, Springer, Berlin, 1980.

[E81.16] Erdős, P., On the combinatorial problems which I would most like to see solved. *Combinatorica* 1 (1981), 25–42.

[E81.20] Erdős, P., Problems and results in graph theory. *The Theory and Application of Graphs* (Kalamazoo, Mich., 1980), Wiley, New York, 1981, 331–341.

[E81.23] Erdős, P., Some applications of graph theory and combinatorial methods to number theory and geometry, *Algebraic Methods in Graph Theory (Colloq. held in Szeged, Hungary 1978), vol. I, Coll. Math Soc. J. Bolyai*, 25 (1981), 137–148.

[E81.26] Erdős, P., Some new problems and results in graph theory and other branches of combinatorial mathematics. *Combinatorics and Graph Theory, Proc. Symp. Calcutta 1980, Lecture Notes Math.*, 885, Springer (1981), 9–17.

[E81.32]   Erdős, P., My Scottish Book "problems", *The Scottish Book. Mathematics from the Scotish Café*, Birkhäuser, Boston, 1981, 35–43.

[E83.01]   Erdős, P., Some combinatorial problems in geometry, *Geometry and Differential Geometry, Lecture Notes in Mathematics*, 792, Springer-Verlag, Berlin (1983), 46–53.

[E83.03]   Erdős, P., Combinatorial problems in geometry, *Math. Chronicle* 12 (1983), 35–54.

[E85.01]   Erdős, P., Problems and results in combinatorial geometry. *Discrete Geometry and Convexity, Annals of the New York Academy of Sciences*, 440, The New York Academy of Sciences, New York (1985), 1–11.

[E85.33]   Erdős, P., Some problems and results in number theory. *Number Theory and Combinatorics*, Japan 1984 (Tokyo, Okayama and Kyoto, 1984), World Sci. Publishing, Singapore, 1985, 65–87.

[E85.34]   Erdős, P., Some solved and unsolved problems of mine in number theory. *Topics in Analytic Number Theory* (Austin, Tex., 1982), Univ. Texas Press, Austin, TX, 1985, 59–75.

[E87.12]   Erdős, P., My joint work with Richard Rado. *Surveys in Combinatorics* 1987 (New Cross, 1987), London Math. Soc. Lecture Note Ser., 123, Cambridge Univ. Press, Cambridge-New York, 1987, 53–80.

[E88.28]   Erdős, P., Problems and results in combinatorial analysis and graph theory. *Proceedings of the First Japan Conference on Graph Theory and Applications* (Hakone, 1986), Discrete Math. 72(1–3) (1988), 81–92.

[E89.27]   Erdős, P., Problems and results on extremal problems in number theory, geometry, and combinatorics. *Proceedings of the 7th Fischland Colloquium*, (Wustrow, 1988), Rostock. Math. Kolloq., No. 38 (1989), 6–14.

[E89.35]   Erdős, P., Some problems and results on combinatorial number theory, Graph theory and its applications: East and West (Jinan, 1986), *Ann. New York Acad. Sci.* 576, 132–145, New York Acad. Sci., New York, 1989.

[E89.60]   Erdős, P., Video recording of the talk "Some of My Favorite Problems II". University of Colorado at Colorado Springs, March 16, 1989.

[E89.61]   Erdős, P., Video recording of the talk "Some of My Favorite Problems II". University of Colorado at Colorado Springs, March 17, 1989.

[E89CS]   Erdős, P., An interview in documentary film *Colorado Mathematical Olympiad*, 1989.

[E90.23]   Erdős, P., Problems and results on graphs and hypergraphs: similarities and differences. *Mathematics of Ramsey Theory*, Algorithms Combin., 5, 12–28, Springer, Berlin, 1990.

[E91.31]   Erdős, P., Problems and results in combinatorial analysis and combinatorial number theory, *Graph theory, Combinatorics and Applications*, vol. 1 (Kalamazoo, MI, 1988), Wiley-Intersci. Publ., Wiley, New York, 1991, 397–406.

[E91.60]   Erdős, P., Video recording of the talk "Some of My Favorite Problems I". University of Colorado at Colorado Springs, December 23, 1991.

[E91/7/12ltr] Erdős, P., Letter to A. Soifer of July 12, 1991.

[E91/7/16ltr] Erdős, P., Letter to A. Soifer of July 16, 1991.

[E91/8/10ltr] Erdős, P., Letter to A. Soifer of August 10, 1991.

[E91/8/14ltr] Erdős, P., Letter to A. Soifer of August 14, 1991.

[E91/10/2ltr] Erdős, P., Letter to A. Soifer, received on October 2, 1991 (marked "1977 VII 25").

[E92.14]   Erdős, P., Obituary of My Friend and Coauthor Tibor Gallai. *Geombinatorics* II(1) (1992), 5–6; corrections: II(2) (1992), 37.

[E92.15]   Erdős, P., In memory of Tibor Gallai. *Combinatorica* 12(4) (1992), 373–374.

[E92.19]   Erdős, P., On some unsolved problems in elementary geometry (Hungarian). *Mat. Lapok (N.S.)* 2(2) (1992), 1–10.

[E92.60]   Erdős, P., Video recording of the talk "Some of My Favorite Problems II". University of Colorado at Colorado Springs, January 10, 1992.

[E93.20]   Erdős, P., Some of my favorite solved and unsolved problems in graph theory. *Quaestiones Math.* 16(3) (1993), 333–350.

[E94.21]   Erdős, P., Problems and results in discrete mathematics, Trends in discrete mathematics. *Discrete Math.* 136(1–3) (1994), 53–73.

[E94.22]   Erdős, P., Problems and results on set systems and hypergraphs. *Extremal Problems for Finite Sets* (Visegrád, 1991), Bolyai Soc. Math. Stud., 3, János Bolyai Math. Soc., Budapest, 1994, 217–227.

[E94.26]   Erdős, P., Some problems in number theory, combinatorics and combinatorial geometry. *Math. Pannon.* 5(2) (1994), 261–269.

[E94.60]   Erdős, P., Video recording of the talk "Twenty Five Years of Questions and Answers". *25th South-Eastern International Conference On Combinatorics,Graph Theory and Computing*, Florida Atlantic University, Boca Raton, March 10, 1994.

[E95.20]   Erdős, P., Some of my favourite problems in number theory, combinatorics, and geometry. *Combinatorics Week* (Portuguese), São Paulo, 1994, *Resenhas* 2 (1995), 165–186.

[E95.32]   Erdős, P., On the 120-th Anniversary of the Birth of Schur. *Geombinatorics* V(1) (1995), 4–5.

[E95BR]   Erdős, P., Interview with A. Soifer of March 7, 1995, Boca Raton, Florida.

[E97.18]   Erdős, P., Some of my favorite problems and results. *The Mathematics of Paul Erdős*, vol. I, Graham, R. L., and Nesetril, J., (eds), 47–67, Springer, Berlin, 1997.

[E97.21]   Erdős, P., Some of my favourite unsolved Problems. *Math. Japon.* 46(3) (1997), 527–537.

[EBC]   Erdős, P., Bollobás, B., and Catlin, P. A., Hadwiger's conjecture is true for almost every graph. *European J. Comb.*, 1(3) (1980), 195–199.

[EB]   Erdős, P., and de Bruijn, N. G., A colour problem for infinite graphs and a problem in the theory of relations. *Indag. Math.* 13 (1951), 371–373.

[EFT]   Erdős, P., Harary, F., and W. T. Tutte, On the dimension of a graph. *Mathematika*, London, 12 (1965), 118–122.

[EG0]   Erdős, P., and Graham, R., Old and New Problems and Results in Combinatorial Number Theory: van der Waerden's Theorem and Related Topics. *L'Enseignement Mathématique* 25(3–4), 1979, 325–344.

[EG]   Erdős, P., and Graham, R., *Old and New Problems and Results in Combinatorial Number Theory*. L'Enseignement Mathématique, Université de Genève, 1980.

[EGMRSS]   Erdős, P., Graham, R. L., Montgomery, P., Rothschild, B. L., Spencer, J., and Straus, E. G., Euclidean Ramsey theorems I, II, and III. *J. Combin. Theory Ser. A*, 14 (1973), 341–363; *Coll. Math. Soc. Janos Bolyai*, 10 (1973) North Holland, Amsterdam (1975), 529–557 and 559–583.

[EH1]   Erdős, P., and Hajnal, A., On Chromatic Numbers of Graphs and Set-Systems. *Acta Mathematica Acad. Sci. Hungar.* 17(1–2), 1966, 61–99.

[EH2]   Erdős, P., and Hajnal, A., Research problem 2–5, *J. Combin. Theory* 2 (1967), 104.

[EHT]   Erdős, P., Marary, F. and Tutte, W. T., On the dimension of a graph, *Mathematika* 12 (1965), 118–122.

[EP1]   Erdős, P., and G. Purdy, Some extremal problems in geometry, III. *Proc. Sixth Southeastern Conf. on Combinatorics, Graph Theory and Computing* (Florida Atlantic Univ., Boca Raton, FL, 1975), *Congressus Numerantium*, XIV, *Utilitas Math.*, Winnipeg, Manitoba, (1975), 291–308.

[EP2]   Erdős, P., and Purdy, G., Some extremal problems in combinatorial geometry. In R. Graham, M. Grötschel and L. Lovász, (eds.), *Handbook of Combinatorics*, North Holland, Amsterdam, 1991.

[EP3]   Erdős, P., and Purdy, G., *Some Extremal Problems in Discrete Geometry* (manuscript).

[ESS]   Erdős, P., Sárközy, A., and Sós, V. T., On a conjecture of Roth and some related problems. I. *Irregularities of Partitions (Fertöd, 1986)*, 47–59, Algorithms Combin. Study Res. Texts, 8, Springer, Berlin, 1989.

[ESi]   Erdős, P., and Simonovits, M., On the chromatic number of geometric graphs. *Ars Comb.*, 9 (1980), 229–246.

[ESu1]   Erdős, P., and Surányi, J., *Válogatott Fejezetek a Számelméletbol.* Polygon, Szeged, 1996 (Hungarian).

[ESu2]   Erdős, P., and Surányi, J., *Topics in the Theory of Numbers.* Trans. from the Hungarian 2nd ed. [Esu1], Springer, New York, 2003.

[ES1]   Erdős, P., and Szekeres, G., A combinatorial problem in geometry. *Compositio Math.* 2 (1935), 463–470.

[ES2]   Erdős, P., and Szekeres, G., On some extremum problems in elementary geometry. *Annales Universitatis Scientiarum Budapestinensis de Rolando Eötvös Nominatae, Sectio Math.* 3 (1960/61), 53–62.

[ET]   Erdős, P., and Turán, P., On Some Sequences of Integers, *J. London Math. Soc.* 11 (1936), 261–264.

[EW]   Erdős, P., and Wilson, R. J., On the chromatic index of almost all graphs. *J. Combinatorial Theory Ser.* B 23 (1977), 255–257.

[Eri]   Erickson, Martin, An upper bound for the Folkman number F(3,3;5). *J. Graph Theory* 17(6) (1993), 679–681.

[Err]   Errera, A., *Du Coloriage des Cartes et de Quelques Questions d'Analysis Situs,* Thèse pour le grade de Docteur Special en Sciences Physiques et Mathematiques. Gauthier-Villars & Co, Paris, 1921.

[Ex3]   Exoo, G., Applying Optimization Algorithm to Ramsey Problems. In Y. Alavi (ed.), *Graph Theory, Combinatorics, Algorithms, and Applications.* SIAM Philadelphia, (1989) 175–179.

[Ex4]   Exoo, G., A Lower Bound for $R(5, 5)$, *J. Graph Theory* 13 (1989) 97–98.

[Ex5]   Exoo, G., On Two Classical Ramsey Numbers of the Form $R(3, n)$. *SIAM J. DiscreteMath* 2 (1989), 488–490.

[Ex9]   Exoo, G., Announcement: On the Ramsey Numbers $R(4, 6)$, $R(5, 6)$ and $R(3, 12)$. *Ars Combinatorica* 35 (1993) 85. The construction of a graph proving $R(4, 6) \geq 35$ is presented in detail at http://ginger.indstate.edu/ge/RAMSEY (2001).

[Ex15]   Exoo, G., Some Applications of pq -groups in Graph Theory, *Discussiones Mathematicae Graph Theory* 24 (2004) 109–114. Constructions available at http://ginger.indstate.edu/ge/RAMSEY.

[Ex17]   Exoo, G., Personal communication to S. P. Radziszowski (2005–2006). Constructions at http://ginger.indstate.edu/ge/RAMSEY

[Fal1]   Falconer, K. J., The realization of distances in measurable subsets covering R^n. *J. Combin. Theory (A)*, 31 (1981), 187–189.

[Fal2]   Falconer, K. J., 8-page manuscript written for this book, dated February 17, 2005.

[Fis1]   Fischer, K. G., Additive k-colorable extensions of the rational plane. *Discrete Math.* 82 (1990), 181–195.

[Fis2]   Fischer, K. G., The connected components of the graph $Q\left(\sqrt{N_1}, \ldots, \sqrt{N_d}\right)^2$. *Congressus Numerantium* 72 (1990), 213–221.

[Fla]   Flam, J., *Matisse on Art.* University of California Press, Berkeley, 1995.

[Foll]   Folland, G., *Real Analysis: Modern Techniques and Their Applacations.* 2nd ed, John Wiley & Sons, New York, 1999.

[Fol]   Folkman, J., Graphs with monochromatic complete subgraphs in every edge coloring, *Rand Corporation Memorandum RM-53-58-PR,* October 1967. Also appeared under the same title in *SIAM J. Appl. Math.* 18 (1970), 19–24.

[FR]   Fox, J., and Radoicic, R., The axiom of choice and the degree of regularity of equations over the reals, unpublished.

[Fra1]   Fraenkel, A., Zu den Grundlagen der Cantor-Zermeloschen Mengenlehre. *Math. Ann.* 86(3–4) (1922), 230–237, (German).

[Fra2]   Fraenkel, A., Axiomatische Begründung der transfiniten Kardinalzahlen. I., *Math. Zeitschr.* 13(1) (1922), 153–188, (German).

[Fra3]   Fraenkel, A., Untersuchungen über die Grundlagen der Mengenlehre. *Math. Zeitschr*. 22(1) (1925), 250–273, (German).

[FR1]   Frankl, P., and Rödl, V., All triangles are Ramsey. *Trans. Amer. Math. Soc*. 297 (1986), 777–779.

[FR2]   Frankl, P., and Rödl, V., Large triangle-free subgraphs in graphs without K_4. *Graphs Combin*. 2 (1986), no. 2, 135–144.

[FR3]   Frankl, P., and Rödl, V., A partition property of simplices in Euclidean space, *J. Amer. Math. Soc*., 3 (1990), 1–7.

[FW]   Frankl, P., and Wilson, R. M., Intersection theorems with geometric consequences. *Combinatorica* 1 (1981), 357–368.

[Fre1]   Frei, G., Dedication. Bartel Leendert van der Waerden, Zum 90. Geberstag. *Historia Math*. 20(1), (1993), 5–11.

[Fre2]   Frei, G., Zum Gedenken an Bartel Leendert van der Waerden (2.2.1903–12.1.1996). *Elem. Math*. 53 (1998), 133–138.

[FTW]   Frei, G., Top, J., and Walling, L., A short biography of B. L. van der Waerden, *Nieuw Archief voor Wiskunde*, Vierde serie 12 (3) (1994), 137–144.

[Fri]   Friedman, H., Strong undecidable $\prod_2^0$ statements (in preparation).

[Fur1]   Furstenberg, H., Ergodic behavior of diagonal measures and a theorem of Szemerédi on arithmetic progressions. *J. D'Analyse Math*. 31 (1977), 204–256.

[Fur2]   Furstenberg, H., A polynomial Szemerédi theorem, manuscript dated January 1994.

[Gal]   Galperin, G., A letter to A. Soifer, undated, post-stamped March 1, 2005.

[Gars1]   Garsia, A., Manuscript, Fall 1958, 10pp.

[Gars2]   Garsia, A., e-mail to A. Soifer, February 28, 1995.

[Gar1]   Gardner, M., The celebrated four-color map problem of topology. *Scientific American* 206 (Sept. 1960), 218–226.

[Gar2]   Gardner, M., A new collection of "brain teasers". *Scientific American* 206 (Oct. 1960), 172–180.

[Gar3]   Gardner, M., The coloring of unusual maps leads into Uncharted territory, *Scientific American*, 242, February 1980, 14–21.

[GS]   Gismondi, S. J., and Swart, E. R., A new type of 4-colour reducibility, *Congrs. Numer*. 82 (1991), 33–48.

[Gof]   Goffman, C., And What Is Your Erdős Number? *Amer. Math. Monthly* 76(7), (1969), 761.

[Göd1]   Gödel, K., Über formal unentscheidbare Sätze der Principia Mathematica und verwandter Systeme, I, *Monatshefte für Mathematik und Physik* 38 (1931), 173–198.

[Göd2]   Gödel. K., *The Consistency of the Axiom of Choice and of the Generalized Continuum Hypothesis with the Axioms of Set Theory*. Princeton University Press, 1940.

[Gol1]   Golomb, S. W., Letter to A. Soifer, September 10, 1001.

[Gol2]   Golomb, S. W., Letter to A. Soifer, September 25, 1001.

[Gon]   Gonthier, G., A computer-checked proof of the Four Colour Theorem http://research.microsoft.com/~gonthier/4colproof.pdf

[GO]   Goodman, J. E., and O'Rourke, J., *Handbook of Discrete and Computational Geometry*. CRC Press, Boca Raton, 1997.

[Gou]   Goudsmit, S. A., *Alsos*. Henry Schuman, New York, 1947.

[Gra0]   Graham, R. L., On edgewise 2-colored graphs with monochromatic triangles and containing no complete hexagon, *J. Combin. Theory* 4 (1968), 300.

[Gra1]   Graham, R. L., *Rudiments of Ramsey Theory*, American Mathematical Society, Providence, 1981; revised 1983.

[Gra2]   Graham, R. L., Recent developments in Ramsey theory. *Proceedings of the International Congress of Mathematicians*, vol. 1, 2 (Warsaw, 1983), 1555–1567, *PWN, Warsaw*, 1984.

[Gra3]   Graham, R. L., Recent Trends in Euclidean Ramsey Theory. *Discrete Math*. 136 (1994), 119–127.

[Gra4]   Graham, R. L., A problem talk for the about 200 finalists of the USA Mathematical Olympiad, Massachusetts Institute of Technology, May 4, 2002.

[Gra5]   Graham, R. L., Talk at the Mathematical Sciences Research Institute, Berkeley, August 2003.

[Gra6]   Graham, R. L., Open problems in euclidean Ramsey theory. *Geombinatorics* XIII (4) (2004), 165–177.

[Gra7]   Graham, R. L., Some of my favorite problems in Ramsey Theory. In B. Landman et al. (eds.), *Proceedings of the 'Integers Conference 2005' in Celebration of the 70th Birthday of Ronald Graham, Carrolton, Georgia, USA, October 27–30, 2005*, Walter de Gruyter, Berlin, 2007, 229–236.

[Gra8]   Graham, R. L., Old and new problems and results in Ramsey theory. *Horizon of Combinatorics* (Conference and EMS Summer School), Budapest and Lake Balaton, Hungary, 10–22 July 2006, to appear in 2008–2009.

[GN1]   Graham, R. L., and Nesetril, J., Ramsey Theory in the Work of Paul Erdős. In Graham, R. L., and Nesetril (eds.) The Mathematics of Paul Erdős II, Springer, Berlin, 1997, 193–209.

[GN2]   Graham, R. L., and Nesetril, J., Ramsey Theory and Paul Erdős. In Halász, G., et al. (eds.), *Paul Erdős and his Mathematics II*, Springer, Budapest, 2002, 339–365.

[GR1]   Graham, R. L.; Rothschild, B. L., Ramsey's theorem for *n*-parameter sets. *Trans. Amer. Math. Soc.* 159 (1971), 257–292.

[GR2]   Graham, R. L., and Rothschild, B. L., A survey of finite Ramsey theorems, *Proc. Second Louisiana Conf. on Combinatorics, Graph Theory and Computing*, 1971, 21–40.

[GR3]   Graham, R. L.; Rothschild, B. L., Some recent developments in Ramsey theory. *Combinatorics (Proc. Advanced Study Inst., Breukelen, 1974), Part 2: Graph theory; foundations, partitions and combinatorial geometry*, 61–76. Math. Centre Tracts, No. 56, Math. Centrum, Amsterdam, 1974.

[GRS1]   Graham, R. L., Rothschild, B. L., Spencer, J. H., *Ramsey Theory*, 1st edn., John Wiley & Sons, New York, 1980.

[GRS2]   Graham, R. L., Rothschild, B. L., Spencer, J. H., *Ramsey Theory*. 2nd edn., John Wiley & Sons, New York, 1990.

[GS1]   Graham, R. L., and Spencer, J. H., On small graphs with forced monochromatic triangles. *Recent trends in graph theory (Proc. Conf., New York, 1970)*, 137–141. Lecture Notes in Math., vol. 186. *Springer, Berlin*, 1971.

[GS2]   Graham, R. L., and Spencer, J. H., Ramsey Theory. *Scientific American*, July 1990, 112–117.

[GY0]   Graver, J. E., and Yackel, J., An upper bound for Ramsey numbers. *Bull. Amer. Math. Soc.* 72 (1966), 1076–1079.

[GY]   Graver, J. E., and Yackel, J., Some graph theoretic results associated with Ramsey's theorem. *J.Combin. Theory* 4 (1968), 125–175.

[Grav]   Graves, R. P., Life of Sir William Rowan Hamilton, Knt., LL. D., D. C. L., M. R. I. A., Andrews Professor of Astronomy in the University of Dublin, and Royal Astronomer of Ireland, etc.: including selections from his poems, correspondence, and miscellaneous writings, vol. 1 (1882), 2 (1885) and 3 (1889), Hodges, Figgis, & Co., Dublin.

[GT]   Green, B., and Tao, T., The primes contain arbitrarily long arithmetic progressions, 6th revision, 2007, arXiv:math/0404188v6.

[GG]   Greenwood, R. E., and Gleason, A. M., Combinatorial relations and chromatic graphs. *Can. J. Math.*, 7 (1955), 1–7.

[GR]   Grinstead, C., and Roberts, S., On the Ramsey Numbers *R* (3, 8) and *R* (*3, 9*). *J. Combin. Theory Ser. B* 33 (1982), 27–51.

[Gro]   Groves, L. R., *Now It Can be Told*. Harper, New York, 1962.

[Grö]   Grötzsch, H., Ein Dreifarbensatz ftir dreikreisfreie Netze auf der Kugel, *Wissenschaftliche Zeitschrift der Martin-Luther-Universitat Halle-Wittenberg. Mathematisch-Naturwissenschaftliche Reihe*, 8 (1958–1959), 109–120.

[Grü]   Grünbaum, B., Grötzsch's theorem on 3-colorings. *Michigan Math. J.* 10 (1963), 303–310.

[Gut1]   The Late Professor Guthrie, *Cape Times*, October 23, 1899, Cape Town, South Africa.

[Gut2]   A biography of Francis Guthrie in *Dictionary of South African Biography*. vol. II, Tafelberg-Uitgewers LTD., Editor-in-chief D.W. Krüger, 279–280.

[Gut3]   Ritchie, W., The History of the South African College 1829–1918, T. Maskew Miller, Capetown, 1918.

[GutFr]   Guthrie, Frederick, Note on the Colouring of Maps. *Proceedings of the Royal Society of Edinburgh* 10 (1880), 727–728.

[Guy]   Guy, Richard K., *Unsolved Problems in Number Theory*, 2nd edn., Springer-Verlag, New York, 1994.

[Had1]   Hadwiger, H., Über eine Klassification der Steckenkomplexe. *Vierteljschr. Naturforsch. Ges. Zürich* 88 (1943), 133–142.

[Had2]   Hadwiger, H., Ein Überdeckungssätz für den Euklidischen Raum. *Portugal Math.*, 4 (1944), 140–144.

[Had3]   Hadwiger, H., Uberdeckung des euklidischen Raum durch kongruente Mengen. *Portugaliae Math.*, 4 (1945), 238–242.

[Had4]   Hadwiger, H., Ungelöste Probleme, Nr. 11. *Elemente der Mathematik* 16 (1961), 103–104.

[HD1]   Hadwiger, H., and Debrunner, H., Ausgewählte einzelprobleme der kombinatorishen geometrie in der ebene. *L'Enseignement Mathematique* 1 (1955), 56–89.

[HD2]   Hadwiger, H., and Debrunner, H., *Kombinatorischen Geometrie in der Ebene*. L'Enseignement Mathematique, Geneva, 1959.

[HD3]   Hadwiger, H., and Debrunner, H., *Combinatorial Geometry of the Plane* edited by I. M. Yaglom, Nauka, Moscow, 1965 (extended by the editor Russian Translation version of [HD2]).

[HDK]   Hadwiger, H., Debrunner, H., and Klee, V., *Combinatorial Geometry in the Plane*. Holt, Rinehart and Winston, New York, 1964.

[HN1]   Hadziivanov, N. G., and Nenov, N. D., On the Graham-Spencer number (Russian). *C R Acad Bulg. Sci.* 32 (1979), 155–158.

[HN2]   Hadziivanov, N. G., and Nenov, N. D., Every (3, 3)-Ramsey graph without 5-cliques has more than 11 vertices (Russian). *Serdica* 11 (1985), 341–356.

[HJ]   Hales, A., and Jewett, R., Regularity and positional games, *Trans. Amer. Math. Soc.* 106 (1963), 222–229.

[Hal]   Halmos, P. R., Has Progress in Mathematics Slowed Down? Amer. Math. Monthly 98 (1990), 561–578.

[Ham]   Hamilton, W. R., Letter to A. De Morgan dated October 26, 1852; TCD MS 1493, 669; Trinity College Dublin Library, Manuscripts Department.

[Har0]   Harary, F., *Graph Theory*. Addison Wesley, Reading, Massachusetts, 1969.

[Har1]   Harary, F., The Four Color Conjecture and other graphical diseases. In F. Harary (ed.), *Proof Techniques in Graph Theory*, Academic Press, New York, 1969, 1–9.

[Har2]   Harary, F., A Tribute to Frank P. Ramsey, 1903–1930, *J. Graph Theory* 7 (1983), 1–7.

[Har3]   Harary, F., Letter to A. Soifer, dated February 19, 1996.

[Harb]   Harborth, H., Konvexe Fünfecke in ebenen Punktmengen. *Elem. Math.* 33 (1978), 116–118.

[HaKr]   Harborth, H., and Krause, S., Ramsey Numbers for Circulant Colorings. *Congressus Numerantium* 161 (2003), 139–150.

[HR]   Hatfield, L., and Ringel, G., Pearls in Graph Theory: A Comprehensive Introduc-
            tion, Academic Press, Boston, 1994
[Hav]   Havel, I., On a conjecture of B. Grünbaum, *J. Combin. Theory* 7 (1969), 184–186.
[Hea1]   Heawood, P. J., Map-Colour Theorem. *Quarterly J. of Pure and Applied Math.*
            XXIV (1890), 332–338.
[Hea2]   Heawood, P. J., On the Four-Colour Map Theorem. *Quarterly J. of Pure and Applied*
            Math. XXIX (1998), 270–285.
[Head1]   Headmaster of Clifton College,[Untitled], *J. Educ.* IX, January 1, 1887, 11–12.
[Head2]   Headmaster of Clifton College,[Untitled], *J. Educ.* XI, June 1, 1889, 277.
[Hee1]   Heesch, H., *Untersuchungen zum Vierfarbenproblem* (German), B. I.
            Hochschulscripten, 810/810a/810b, Bibliographisches Institut, Mannheim-
            Vienna-Zürich, 1969.
[Hee2]   Heesch, H., Chromatic reduction of the Triangulations T_e, $e = e_5 + e_7$. *J. Combin.*
            Theory Ser. B 13 (1972), 46–55.
[Het1]   Editorial, Die?? Neen, die niet! *Het Parool* No. 312, January 16, 1946, 3.
[Het2]   Editorial, Prof. Van der Waerden nog niet Benoemd, *Het Parool* No. 313, January
            17, 1946, 1.
[Het3]   Editorial, Prof. dr. B. L. van der Waerden, *Het Parool* No. 320, January 25,
            1946, 1.
[Het4]   Red. (eds.), Untitled commentary on Van der Waerden's letter. *Het Parool* No. 326,
            February 1, 1946, 3.
[Het5]   Editorial, Rondom Van der Waerden, *Het Parool* No. 336, February 13, 1946, 3.
[Hil]   Hilbert, D., Ueber die Irreducibilität ganzer rationaler Functionen mit ganzzahligen
            Coefficienten, *J. Reine Angew. Math.* 110 (1892), 104–129.
[HI]   Hill, R., and Irving, R. W., On group partitions associated with lower bounds for
            symmetric Ramsey numbers. *European J. Combin.* 3(1) (1982), 35–50.
[HH]   Hilton, A. J. W., and Hind, H. R. The total chromatic number of graphs having large
            maximum degree. *Discrete Math.* 17 (1993), no. 1–3, 127–140.
[Hin1]   Hind, H. R. An upper bound for the total chromatic number. *Graphs Combin.* 6
            (1990), no. 2, 153–159.
[Hin2]   Hind, H. R., An upper bound for the total chromatic number of dense graphs.
            J. Graph Theory 16 (1992), no. 3, 197–203.
[HMR]   Hind, H., Molloy, M., and Reed, B., Total coloring with $\Delta + \text{poly}(\log \Delta)$ colors,
            SIAM J. Comput. 28(3) (1999), 816–821.
[Hin]   Hindman, N., Finite sums from sequences within cells of a partition of N, *J. Com-*
            bin. Theory Ser. A 17 (1974), 1–11.
[HO]   Hobbs, A. M., and Oxley, J. G., William T. Tutte, 1917–2002, *Notices Amer. Math.*
            Soc. 51 (2004), 320–330.
[HO]   Hochberg, R., and O'Donnell, P., Some 4-Chromatic unit-distance graphs without
            small cycles. *Geombinatorics* V(4) (1996), 137–14.
[Hod]   Hodge, Sir W., Solomon Lefschetz 1884–1972. *Biographical Memoirs of Fellows*
            of the Royal Society 19 (December 1973), 433–453.
[Hof]   Hofer, H., Letter to A. Soifer of March 3, 1995.
[HS1]   Hoffman, I., and Soifer, A., Almost chromatic number of the plane. *Geombinatorics*
            III(2), 1993, 38–40.
[HS2]   Hoffman, I., and Soifer, A., Another six-coloring of the plane. *Discrete Mathematics*
            150 (1996), 427–429.
[Hol]   Holt, J., Irreligion, a review of the book by Paulos, J. A., *A Mathematician Explains*
            Why the Arguments for God Just Don't Add Up; The New York Times Book Review,
            January 13, 2008, 18.
[Hor]   Horgan, J., The death of proof. *Scientific American*, October 1993, 92–103.
[Hort]   Horton, J. D., Sets with no Empty Convex 7-Gons. *Canad. Math. Bull.* 26(4) (1983),
            482–484.
[HUA1]   Archive of Humboldt University at Berlin, documents UK-Sch 342, Bd. I, B1.1.,
            1R, 2R, 3.

[HUA2]   Archive of Humboldt University at Berlin, document UK-Sch 342, Bd.I, B1.25.

[HZ1]   Huang Y. R., and Zhang K. M., An new upper bound formula for two color classical Ramsey numbers. *J. Combin. Math. and Combin. Computing* 28 (1998), 347–350.

[Isb1]   Isbell, J., Letter to A. Soifer of August 26, 1991.

[Isb2]   Isbell, J., Letter to A. Soifer of September 3, 1991.

[Irv1]   Irving, R. W., On a bound of Graham and Spencer for a graph-coloring constant. *J. Combin. Theory Ser. B* 15 (1973), 200–203.

[Irv2]   Irving, R. W., Generalised Ramsey numbers for small graphs, *Discr. Math.* 9 (1974), 251–264.

[Jec]   Jech, T. J., *The Axiom of Choice.* North-Holland, Amsterdam, 1973.

[Jef]   Jeffrey, R. C., *The Logic of Decision.* McGraw-Hill, New York, 1965.

[JefB]   Jeffreys, B., Letter to the Editor. *Math. Gazette* 63 (1979), 126–127.

[JKSV]   Jelínek, V., Kyncl, J., Stolar, R., and Valla, T. V., Monochromatic triangles in two-colored plane, *Combinatorica* (submitted March 2006); arXiv:math.CO/0701940v1, Jan. 31, 2007.

[JT]   Jensen, T. R., and Toft, B., *Graph Coloring Problems.* Wiley, New York, 1995.

[Jes]   Jessop, J., Some Botanists and Botanical Collectors associated with Southern Africa, in *Lantern*, 1963, 80–85.

[Joh1]   Johnson Jr., P. D., Coloring abelian groups. *Discrete Math.* 40 (1982), 219–223.

[Joh2]   Johnson Jr., P. D., Simple product colorings. *Discrete Math.* 48 (1984), 83–85.

[Joh3]   Johnson Jr., P. D., Two-coloring of real quadratic extensions of Q^2 that forbid many distances. *Congressus Numerantium* 60 (1987), 51–58.

[Joh4]   Johnson Jr., P. D., Two-colorings of a dense subgroup of Q^n that forbid many distances. *Discrete Math.* 79 (1989/90), 191–195.

[Joh5]   Johnson Jr., P. D., Product colorings. *Geombinatorics* I(3) (1991), 11–12.

[Joh6]   Johnson Jr., P. D., Maximal sets of translations forbiddable by two-colorings of abelian groups. *Congressus Numerantium* 91 (1992), 153–158.

[Joh7]   Johnson Jr., P. D., About Two definitions of the fractional chromatic number. *Geombinatorics* V(3) (1996), 11–12.

[Joh8]   Johnson Jr., P. D., Introduction to "Colorings of metric spaces" by Benda and Perles. *Geombinatorics* IX(3) (2000), 110–112.

[Joh9]   Johnson Jr., P. D., Coloring the rational points to forbid the distance one–a tentative history and compendium. *Geombinatorics* XVI(1) (2006), 209–218.

[Juh]   Juhász, R., Ramsey type theorems in the plane. *J. Combin. Theory* (A) 27 (1979), 152–160.

[JRW]   Jungreis, D. S., Reid, M., and Witte, D., Distances forbidden by some 2-coloring of Q^2. *Discrete Math.* 82 (1990), 53–56.

[KK]   Kahn, J., and Kalai, G., A counterxample to Borsuk's conjecture, *Bull. Amer. Math. Soc.* 29(1) (1993), 60–62.

[Kai]   Kainen, P., Is the four color theorem true? *Geombinatorics* III(2) (1993), 41–56.

[Ka1]   Kalbfleisch, J. G., Construction of special edge-chromatic graphs. *Canad. Math. Bull.* 8 (1965), 575–584.

[Ka2]   Kalbfleisch, J. G., *Chromatic Graphs and Ramsey's Theorem*, Ph.D. thesis, University of Waterloo, Canada, January 1966.

[Ka3]   Kalbfleisch, J. G., On an unknown Ramsey number. *Michigan Math. J.* 13 (1966), 385–392.

[KM]   Kakeya, S., and Morimoto, S., On a theorem of M. Bandet[sic] and van der Waerden. *Japan. J. Math.* 7 (1930), 163–165.

[KS]   Karabash, D., and Soifer, A., On Shelah-Soifer Class of Graphs, *Geombinatorics* XVI(4) (2007), 363–366.

[Kei]   Keizer, Madelon de, *Het Parool 1940–1945: Verzetsblad in Oorlogstijd*, 2nd edn., Otto Cramwinckel Uitgever, Amsterdam, 1991.

[KK]   Kelly, J.B., and Kelly, L. M., Paths and circuits in critical graphs. *Amer. J. Math.* 76 (1954), 786–792.

[Kem1]	Kempe, A. B., How to colour a map with four colours. *Nature* 20 (July 17, 1879), 275.
[Kem2]	Kempe, A. B., On the geographical problem of the four colours, *Amer. J. Math.* II (1879), 193–200.
[Kem3]	Kempe, A. B., [Untitled]. *Proc. London Math. Soc.* 10 (1878–1879), 229–231.
[Kem4]	Kempe, A. B., How to colour a map with four colours. *Nature* 21 (February 26, 1880), 399–400.
[Kem5]	Kempe, A. B., Untitled Abstract. *Proc. London Math. Soc.* 22 (1890–1891), 263–264.
[Ker]	Kersten, I., Biography of Ernst Witt (1911–1991), *Contemporary Mathematics* 272 (2000), 155–171.
[Kery]	Kéry, G., On a Theorem of Ramsey. *Matematikai Lapok* 15 (1964) 204–224 (Hungarian).
[KS]	Ketonen, J., and Solovay, R., Rapidly growing Ramsey functions. *Annals of Math.* 113 (1981), 267–314.
[Key]	Keynes, J. M., *Essays in Biography*. W. W. Norton & Co., New York, 1963.
[KSz]	Khalfalah, A., and Szemerédi, E., On the number of monochromatic solutions of $x + y = z^2$. *Combin. Probab. Comput.* 15(1–2) 2006, 213–227.
[Khi1]	Khinchin, A. Y., *Tri Zhemchuzhiny Teorii Chisel*. Gos. Izd-vo tekhn.-teoret, lit-ry, Moskva, 1947.
[Khi2]	Khinchin, A. Y., *Tri Zhemchuzhiny Teorii Chisel*. Gos. Izd-vo tekhn.-teoret, lit-ry, 2nd edn., Moskva, 1948.
[Khi3]	Chintschin, A. J., *Drei Pearlen der Zahlentheorie*. Akademie Verlag, Berlin, 1951.
[Khi4]	Khinchin, A. Y., *Three Pearls of Number Theory*. Graylock, Rochester, New York, 1952.
[Kip]	Kipling, R., *Just So Stories*. 1st edn., Macmillan Publishers Ltd., London, 1902.
[KW]	Klee, V., and Wagon, S., *Old and new Unsolved Problems in Plane Geometry and Number Theory*. Mathematical Association of America, 2nd edn., (with Addendum), 1991.
[Kle]	Kleinberg, E. M., The independence of Ramsey's theory. *J. Symbolic Logic* 34 (1969), 205–206.
[KP]	Kleitman, D. J., and Pachter, L., Finding convex sets among points in the plane. *Discrete and Computational Geometry* 19 (1998), 405–410.
[Kne1]	Knegtmans, P. J., *Socialisme en Democratie: De SDAP tussen klasse en natie (1929–1939)*. Cip-Gegevens Koninklijke Bibliotheek, Den Haag, 1989.
[Kne2]	Knegtmans, P. J., *Een Kweetsbaar Centrum van de Geest: De Universiteit van Amsterdam tussen 1935 en 1950*. Amsterdam University Press, 1998.
[Kne3]	Knegtmans, P. J., e-mail to A. Soifer, May 25, 2004.
[Kne4]	Knegtmans, P. J., e-mail to A. Soifer, May 26, 2004.
[Kne5]	Knegtmans, P. J., e-mail to A. Soifer, May 28, 2004.
[Kne6]	Knegtmans, P. J., e-mail to A. Soifer, June 7, 2004.
[Kne7]	Knegtmans, P. J., e-mail to A. Soifer, June 10, 2004.
[Kne8]	Knegtmans, P. J., e-mail to A. Soifer, October 4, 2004.
[Kri1]	Křiž, I., Permutation groups in Euclidean Ramsey Theory. *Proc. Amer. Math. Soc.* 385 (1991), 899–908.
[Kri2]	Křiž, I., All trapezoids are Ramsey, *Discrete Math.* 108 (1992), 59–62.
[KT]	Kuhn, H. W., and Tucker, A. W., John von Neumann's work in the theory of games and mathematical economics. *Bull. Amer. Math. Soc.* 64 (1958), 100–122.
[Kun]	Kundera, M., *The Curtain: An Essay in Seven Parts*. 1st edn., HarperCollins, New York, 2006.
[LRo]	Landman, B. M., and Robertson, A., *Ramsey Theory on the Integers*. American Mathematical Society, Providence, 2004.
[Lan]	Lang, S., *Algebra*. Addison-Wesley, Reading, Massachisetts, 1965.

[Lar]   Larman, D. G., A note on the realization of distances within sets in euclidean space. *Commentarii Math. Helvetici* 53 (1978), 529–535.

[LR]   Larman, D. G., and Rogers, C. A., The realization of distances within sets in Euclidean space. *Mathematika* 19 (1972), 1–24.

[Led1]   Ledermann, W., Issai Schur and his school in Berlin. *Bull. London Math. Soc.* 15 (1983), 97–106.

[Led2]   Ledermann, W., Ledermann's St Andrews Interview, conducted by O'Connor, J. J., and Robertson, E. F., September, 2000, http://www-gap.dcs.stand.ac.uk/~ history/HistTopics/Ledermann_interview.html#s69

[LN]   Ledermann, W., and Neumann, P., The Life of Issai Schur through Letters and other Documents. In Joseph et al. (eds.) *In Memory of Issai Schur*, Birkhäuser, Boston, 2003, xiv–xci.

[Leh]   Lehto, O., On the life and work of Lars Ahlfors. *Mathematical Intelligencer* 20(4) (1999), 4–8.

[Lin]   Lin, S., On Ramsey numbers and K_r-coloring of graphs. *J. Combin. Theory Ser. B* 12 (1972), 82–92.

[Lit]   Litten, F., Die Carathéodory-Nachfolge in München 1938–1944. *Centaurus* 37 (1994), 154–172.

[Lov1]   Lovász, L., On chromatic number of finite set-systems. *Acta Math.* 19 (1968), 59–67.

[Lov2]   Lovász, L., *Combinatorial Problems and Exercises*. North Holland, Amsterdam, 1979.

[LSS]   Lovász, L., Saks, M., and Schrijver, A., Orthogonal representations and connectivity of graphs. *Linear Algebra Appl.* 114/115 (1989), 439–454.

[LV]   Lovász, L., and Vesztergombi, K., Geometric Representations of Graphs. In Halás, G. et al. (eds.), *Paul Erdős and his Mathematics* II, Springer, 2002, 471–498.

[LS]   Lewin, S. R., *Lise Meitner: A Life in Physics*, Univ. California Press, Berkeley, 1997.

[LX]   Lu, X., and Xu, B., A theorem on 3-colorable plane graphs, *J. Nanjing Norm. Univ. Nat. Sci.* 29 (2006), 5–8.

[LCW]   Luo, X., Chen, M., and Wang, W., On 3-colorable planar graphs without cycles of four lengths. *Information Processing Letters*, to appear in 2007.

[Lov]   Lovász, L., Three short proofs in graph theory. *J. Combin. Theory, Ser. B* 19 (1975), 269–271.

[Mac]   Mackey, J., Combinatorial Remedies, *Ph.D. Thesis*, Department of Mathematics, University of Hawaii, Hawai'i 1994.

[Man1]   Mann, M., A new bound for the chromatic number of the rational five-space. *Geombinatorics* XI(2) (2001), 49–53.

[Man2]   Mann, M., Hunting unit-distance graphs in rational n-spaces. *Geombinatorics* XIII(2) (2003), 86–97.

[May]   May, K., The origin of the four-color conjecture. *ISIS* 56(3), Fall 1965, 346–348.

[MR4]   McKay, B. D., and Radziszowski, S. P., $R(4, 5) = 25$. *J. Graph Theory* 19 (1995), 309–322.

[MR5]   McKay, B. D., and Radziszowski, S. P., Subgraph Counting Identities and Ramsey Numbers. *J. Combin Theory Ser. B* 69 (1997), 193–209.

[MZ]   McKay, B.D., and Zhang, K. M., The value of the Ramsey number $R(3, 8)$. *J. Graph Theory* 16 (1992), 99–105.

[Meh]   Mehrtens, H., Irresponsible purity: the political and moral structure of mathematical sciences in the National Socialist state. In M. Renneberg and M. Walker, (eds.), *Science, Technology and National Socialism*, Cambridge University Press, 1994, 324–338 and 411–413.

[Mel]   Mellor, D. H., Cambridge Philosophers I: F. P. Ramsey. *Philosophy* 70 (1995), 243–262.

[Mil]   Miłosz, C., *The Captive Mind*. Octagon, New York, 1981.

[Min]   Minnaert, M. G. J., *Lilght and Color in the Outdoors*. Springer-Verlag, New
             York, 1993.

[Mir]   Mirsky, L., The combinatorics of arbitrary partitions. *Bull. Inst. Math.* 11 (1975),
             6–9.

[Moo]   Moore, G. H., *Zermelo's Axiom of Choice: Its Origins, Development, and Influence*.
             Springer-Verlag, New York, 1982.

[Mor]   Mordell, L. J., On the rational solutions of the indeterminate equations of the thurd
             and fourth degrees. *Proc. Cambridge Philos. Soc.* 21 (1922), 179–182.

[Morr]   Morrison, P., Alsos: The Story of German Science. *Bull. Atomic Sci.* 3(12) (1947),
             354, 365.

[MM]   Moser, L., and Moser, W., Solution to Problem 10. *Can. Math. Bull.* 4 (1961),
             187–189.

[MP]   Moser, W., and Pach, J., *Research Problems in Discrete Geometry*. McGill Univer-
             sity, 1986.

[Myc]   Mycielski, J., Sur le coloriage des graphes. *Colloquium Mathematicum* 3 (1955),
             161–162.

[Nec]   Nechushtan, O., On the space chromatic number. *Discrete Math.* 256 (2002),
             499–507.

[Nel1]   Nelson, E., Letter to A. Soifer of August 23, 1991.

[Nel2]   Nelson, E., Letter to A. Soifer of October 5, 1991.

[NW]   Nelson, R., and Wilson, R. J.. (eds.), *Graph Colourings*. Longman Scientific and
             Technical, Essex, England, 1990.

[Nen1]   Nenov, N. D., A new lower bound for the Graham-Spencer number (Russian).
             Serdica 6 (1980), 373–383.

[Nen2]   Nenov, N. D., An example of a 15-vertex (3,3)-Ramsey graph with clique number
             4 (Russian). *C R Acad Bulg Sci* 34 (1981), 1487–1489.

[NR]   Nešetřil, J., and Rödl, V., The Ramsey property for graphs with forbidden complete
             subgraphs. *J. Combin. Theory Ser. B* 20 (1976), 243–249.

[Neu]   Neumann, J. von, Über die Definition durch transfinite Induktion und verwandte
             Fragen der allgemeinen Mengenlehre. *Math. Annalen* 99 (1928), 373–391.

[Nil]   Nilli, A., On Borsuk's problem. *Contemporary Math.* 178 (1994), 209–210.

[NM]   Neumann, J. von, and Morgenstern, O., *Theory of Games and Economic Behavior*.
             Princeton University Press, Princeton, 1944.

[Odo1]   O'Donnell, P., A Triangle-Free 4-Chromatic Graph in the Plane. *Geombinatorics*
             IV(1) (1994), 23–29.

[Odo2]   O'Donnell, P., A 40 Vertex 4-Chromatic Triangle-free Unit-Distance graph in the
             plane. *Geombinatorics* V(1) (1995), 30–34.

[Odo3]   O'Donnell, P., High Girth Unit-Distance Graphs, Ph. D. Dissertation, Rutgers Uni-
             versity, May 25, 1999.

[Odo4]   O'Donnell, P., Arbitrary Girth, 4-ChromatiUnit-Distance graphs in the plane part I:
             Graph Description. *Geombinatorics* IX(3) (2000), 145–150.

[Odo5]   O'Donnell, P., Arbitrary Girth, 4-Chromatic Unit-Distance graphs in the plane part
             II: Graph Embedding. *Geombinatorics* IX(4) (2000), 180–193.

[Oro]   O'Rourke, J., Omputational Geometry Column 46. *International J. Comput. Geom.
             & Applic.* 14(6) (2004), 475–478.

[OSe]   Ó Seanór, S., Letter to A. Soifer from March 21, 1997.

[OED]   Oxford English Dictionary, June 2007 edition, http://dictionary.oed.com/

[PH]   Paris, J., and Harrington, L., Mathematical Incompleteness in Peano Arithmetic. In
             Barwise, J. (ed.), *Handbook of mathematical Logic*, North-Holland, Amsterdam,
             1977, 1133–1142.

[Pas]   Pasternak, B. L., *Волны* (*Waves*), 1931, *Complete Works in 11 Volumes*, vol. II,
             50–59, Slovo, Moscow, 2004.

[Pash]   Pash, B. T., *The Alsos Mission*. Award House, New York, 1969.

[Pay]   Payne, M. S., A unit distance graph with ambiguous chromatic number, arXiv: 0707.1177v1[math.CO] 9 Jul 2007.

[Per]   Peremans, W., Van der Waerden Day in Groningen. *Nieuw Archief voor Wiskunde*, Vierde serie 12(3) (1994), 135–136.

[Pet]   Peter, L. J., *Peter's Quotations: Ideas for Our Time*. William Morrow, New York, 1977.

[Pete]   Peters, P., Mijnheer de Red. *Propria Cures*, 56th year, No. 22, February 8, 1946.

[Pic]   Picasso, P., and Ashton, D. (ed.), *Picasso on Art: A Selection of Views*. Da Capo, New York, 1992.

[PRU]   Piwakowski, K., Radziszowski, S. P., and Urbánski, S., Computation of the Folkman number Fe(3, 3; 5). *J. Graph Theory* 32 (1999), 41–49.

[Poi]   Poincaré, H., *The Foundations of Science*. The Science Press, Lancaster, PA, 1946.

[Poisk]   Materialy Konferenzii "Poisk-97", Moscow, 1997, (Russian).

[Pol]   Pólya, G., and Alexanderson, G. L., *The Pólya Picture Album : Encounters of a Mathematician*. Birkha¡??¿user, Boston, 1987.

[Pon]   Pontryagin L. S., wrote *Life of Lev Semenovich Pontryagin, mathematician, compiled by himself. Born 1908, Moscow*, Moscow, 1998.

[Pow]   Powers, T., *Heisenberg's War*: The Secret History of the German Bomb. Little, Brown and Co., Boston, 1993.

[Pre]   Prévert, Jacques, *Paroles*, Le Calligraphe, Paris, 1945[copyright Éditions du Point du Jour, 1946], (French).

[Pri]   Pritikin, D., All Unit-Distance Graphs of Order 6197 Are 6-Colorable. *J. Combin. Theory, Ser. B* 73(2) (1998), 159–163.

[PR]   Prömel, H. J., and Rödl, V., An elementary proof of the canonizing version of Gallai-Witt's Theorem. *J. Combin. Theory Ser. A* 42 (1983), 144–149.

[PR]   Prömel, H. J., and Voigt, B., Aspects of Ramsey-Theory, parts I, II, III, and V, Universität Bonn, unpublished report with part IV missing, 1989.

[Rad1]   Rado, R., Studien zur Kombinatorik. *Math. Zeitschrift* 36 (1933), 424–480.

[Rad2]   Rado, R., Note on Combinatorial Analysis, *Proc. London Math. Soc.* 48(2) (1943), 122–160.

[RT]   Radoicic, R., and Tóth, G., Note on the chromatic number of the space. In B. Aronov, S. Basu, J. Pach, and M. Sharir (eds.), *Discrete and Computational Geometry: the Goodman-Pollack Festschrift*, 695–698, Springer-Verlag, Berlin, 2003.

[Radz1]   Radziszowski, S. P., Small Ramsey Numbers, revision #11, August 1, 2006, *Electronic Journal of Combinatorics*, Dynamic Surveys DS1http://www.cs.rit. edu/~spr/ElJC/abs.pdf

[RX]   Radziszowski, S. P., and Xu, X., On the most wanted Folkman graph. *Geombinatorics* XVI(4) (2007), 367–381.

[RK2]   Radziszowski, S. P., and Kreher, D. L., Upper bounds for some Ramsey numbers R(3, k), *J.Combin. Math. and Combin. Computing* 4 (1988), 207–212.

[Raig1]   Raigorodskii, A. M., On a bound in Borsuk's problem, *Russ. Math. Surv.* 54(2), (1999), 453–454.

[Raig2]   Raigorodskii, A. M., On the chromatic number of a space. *Russ. Math. Surv.* 55(2) (2000), 351–352.

[Raig3]   Raigorodskii, A. M., Borsuk's problem and the chromatic numbers of some metric spaces. *Russ. Math. Surv.* 56(1) (2001), 103–139.

[Raig4]   Raigorodskii, A. M., The Borsuk and Hadwiger problems and systems of vectors with restrictions on scalar products. *Russ. Math. Surv.* 57(3) (2002), 606–607.

[Raig5]   Raigorodskii, A. M., The Borsuk problem for integral polytopes, *Sb. Math.* 193(10) (2002), 1535–1556.

[Raig6]   Raigorodskii, A. M., *Chromatic Numbers* (Russian). Moscow Center of Continuous Mathematical Education, Moscow, 2003.

[Rai]   Raiskii, D. E., Realizing of all distances in a decomposition of the space R^n into $n + 1$ Parts. *Mat. Zametki* 7 (1970), 319–323[Russian]; Engl. transl., *Math. Notes* 7 (1970), 194–196.

[Ram1]   Ramsey, F. P., Mathematical Logic. *Mathematical Gazette* 1926.

[Ram2]   Ramsey, F. P., On a problem of formal logic. *Proc. London Math. Soc.* Ser 2, vol. 30, part 4, 1930, 338–384.

[Ram3]   Ramsey, F. P., and Braithwaite, R. B. (ed. and Introduction), Moore, G. E. (Preface), *The Foundations of Mathematics.* Kegan Paul, Trench, Trubner & Co., London, 1931.

[Ram4]   Ramsey, F. P., Truth and Probability, 1926, in Ramsey. In F. P., Mellor, D. H. (ed.), *Philosophical Papers,* Cambridge University Press, Cambridge, 1990.

[Ram5]   Ramsey, F. P., and Galavotti, M. C., (ed.), *Notes on Philosophy and Mathematics.* Bibliopolis, Napoli, 1991.

[Rea]   Read, H., Gabo, *Naum Gabo: March 2-April 1968.* Albright-Knox Art Gallery, Buffalo, New York, 1968, 5.

[Reid]   Reid, C., *Hilbert.* London, Allen & Unwin, Springer-Verlag, Berlin, 1970.

[Rei]   Reid, M. A., *Undergraduate algebraic geometry.* Cambridge University Press, Cambridge, 1988.

[Rit]   Ritchie, W., *The History of the South African College.* T. Maskew Miller, Capetown, 1, 1918.

[Rob1]   Robertson, N., E-mail to A. Soifer, May 9, 1994.

[RSST]   Robertson, N., Sanders, D. P., Seymour, P. D., and Thomas, R., The four-colour theorem. *J. Combin. Theory Ser. B* 70(1) (1997), 2–44.

[RST]   Robertson, N., Seymour, P. D., and Thomas, R., Hadwiger's conjecture for K_6-free graphs. *Combinatorica* 13(3) (1993), 279–361.

[Rota]   Rota, G.-C., *Indiscrete Thoughts.* Birkhäuser, Boston, 1997.

[Rot]   Roth, K. F., On certain sets of integers. *J. London Math. Soc.* 28 (1953), 104–109.

[Saa1]   Saaty, T. L., Remarks on the Four Color Problem; The Kempe Catastrophe. *Math. Magazine* 40(1) (1967), 31–36.

[Saa2]   Saaty, T. L., Thirteen colorful variations on Guthrie's Four-Color Conjecture. *Amer. Math. Monthly* 79(1) (1972), 2–43.

[SK]   Saaty, T. L., and Kainen, P. C., *The Four-Color Problem: Assaults and Conquest,* McGraw-Hill, New York, 1977.

[San]   Sanders, J. H., *A Generalization of Schur's Theorem,* Ph.D. Dissertation, Yale University, New Haven, 1968.

[SZ]   Sanders, D. P., and Zhao, Y., A note on the three color problem. *Graphs and Combinatorics* 11 (1995), 91–94.

[Sav]   Savage, L. J., *The Foundation of Statistics.* John Wiley & Sons, New York, 1954.

[Schi]   Schiffer, M. M., Issai Schur: Some Personal Reminiscences. In H. Begehr, (ed.), *Mathemaik in Berlin: Geschichte und Documentation,* vol. 2, 1998, 177–181.

[Seg]   Segal, S. L., *Mathematicians under the Nazis.* Princeton University Press, Princeton, 2003.

[Sie]   Siegmund-Schultze, R., Mathematiker auf der Flught vor Hitler, Deutche Mathematiker-Vereinigung, 1998.

[Scho]   Schoenbaum, D., Hitler's Social Revolution: Class and Status in Nazi Germany 1933–1939, W.W. Norton, New York, 1980.

[Scho]   Schoenberg, A., Preface in *Anton Webern, 6 Bagatellen für Streichquartett,* op. 9, Partitur, Universal Edition.

[Schuh]   Schuh, F., In memoriam (Pierre Joseph Henry Baudet). *Christian Huygens* I (1921–1922), 145.

[Sch]   Schur, I., Über die Kongruenz $x^m + y^m \equiv z^m$(mod. p). *Jahresbericht der Deutschen Mathematiker-Vereinigung* 25 (1916), 114–117.

[Seg]   Segal, S., *Mathematicians under the Nazis.* Princeton University Press, 2003.

[Sei]   Seidenberg, A., A simple proof of Erdős and Szekeres, *J. London Math. Soc.*, 34 (1959), 352.

[Sha]   Shader, L. E., All right triangles are Ramsey in E^2! *J. Combin. Theory Ser. A* 20 (1976), 385–389.

[She1]   Shelah, S., Primitive recursive bounds for van der Waerden numbers. *J. Amer. Math. Soc.*1 (1988), 683–697.

[She2]   Shelah, S., *Cardinal Arithmrtic*. Clarendon Press, Oxford, 1993.

[She3]   Shelah, S., Logical Dreams. *Bull. Amer. Math Soc.* 40 (2003), 203–228.

[She4]   Shelah, S., e-mail to A. Soifer, June 12, 2006.

[She5]   Shelah, S., e-mail to A. Soifer, June 21, 2006.

[SS1]   Shelah, S., and Soifer, A., Axiom of choice and chromatic number of the plane. *J. Combin. Theory Ser. A* 103 (2003), 387–391.

[SS2]   Shelah, S., and Soifer, A., Chromatic number of the plane & its relatives part III: Its Future. *Geombinatorics* XIII(1) (2003), 41–46.

[Sie]   Siegmund-Schultze, R., *Rockefeller and the Internationalization of Mathematics Between the Two World Wars*. Birkhäuser, Basel, 2001.

[Sim]   Simmons, G. J., The chromatic number of the sphere. *J. Austral. Math. Soc. Ser. A* 21 (1976), 473–480.

[Sko]   Skolem, T., Ein Kombinatorische Satz mit Anwendung auf ein Logisches Entscheidungsproblem. *Fundamenta Math.* 20 (1933), 254–261.

[Soa]   Soare, R. I., *Recursively Enumerable Sets and Degrees*. Springer-Verlag, 1987.

[SS3]   Soifer, A., and Shelah, S., Axiom of Choice and Chromatic Number: An Example on the Plane. *J. Combin. Theory Ser. A* 105(2) (2004), 359–364.

[Soi0]   Soifer, A., *Mathematics as Problem Solving*. Center for Excellence in Mathematical Education, Colorado Springs, 1987. (A new edition should appear in Springer in 2009.)

[Soi1]   Soifer, A., *How Does One Cut a Triangle?* Center for Excellence in Mathematical Education, Colorado Springs, 1990. (A new edition should appear in Springer in 2009.)

[Soi2]   Soifer, A., Triangles in a two-colored plane. *Geombinatorics* I(1) (1991), 6–7 and I(2) 13–14.

[Soi3]   Soifer, A., Triangles in a three-colored plane. *Geombinatorics* I(2) (1991), 11–12 and I(4) (1992), 21.

[Soi4]   Soifer, A., Chromatic number of the plane: A historical essay. *Geombinatorics* I(3), 1991, 13–15.

[Soi5]   Soifer, A., Relatives of chromatic number of the plane I. *Geombinatorics* I(4) (1992), 13–15.

[Soi6]   Soifer, A., A six-coloring of the plane. *J. Combin. Theory*, Ser A, 61(2) (1992), 292–294.

[Soi7]   Soifer, A., Six-realizable set X_6. *Geombinatorics* III(4), (1994), 140–145.

[Soi8]   Soifer, A., An infinite class of 6-colorings of the plane. *Congressus Numerantium* 101 (1994), 83–86.

[Soi9]   Soifer, A., *Colorado Mathematical Olympiad: The First Ten Years and Further Explorations*. Center for Excellence in Mathematical Education, Colorado Springs, 1994. (A new edition, covering the first 20 years, and including 20 explorations, should appear in Springer in 2009).

[Soi10]   Soifer, A., Issai Schur: Ramsey Theory before Ramsey. *Geombinatorics* V(1) (1995), 6–23.

[Soi11]   Soifer, A., Pierre Joseph Henry Baudet: Ramsey Theory before Ramsey. *Geombinatorics* VI (2) (1996), 60–70.

[Soi12]   Soifer, A., The Baudet-Schur conjecture on monochromatic arithmetic progressions: An Historical Investigation. *Congressus Numerantium* 117 (1996), 207–216.

[Soi13]   Soifer, A., On the border of two years: An interview with Paul Erdős. *Geombinatorics* VI(3) (1997), 99–110.

[Soi14]   Soifer, A., "And you don't even have to believe in G'd, but you have to believe that the book exists," *Geombinatorics* VI(3) (1997), 111–116.

[Soi15]   Soifer, A., Map coloring in victorian age: Problems, history, results, *Mathematics Competitions* 10(1) (1997), 20–31.

[Soi16]   Soifer, A., Competitions, Mathematics, Life. *Mathematics Competitions* 11(2) (1998), 20–41.

[Soi17]   Soifer, A., 50th Anniversary of one problem: Chromatic Number of the Plane. *Mathematics Competitions* 16(1) (2003), 9–41.

[Soi18]   Soifer, A., Chromatic number of the plane & Its relatives part I: The problem & its history. *Geombinatorics* XII(3) (2003), 131–148.

[Soi19]   Soifer, A., Chromatic number of the plane & its relatives part II: Polychromatic Number and 6-coloring. *Geombinatorics* XII(4) (2003), 191–216.

[Soi20]   Soifer, A., In Search of Van der Waerden. Leipzig and Amsterdam, 1931–1951. Part I: Leipzig. *Geombinatorics* XIV(1) (2004), 21–40.

[Soi21]   Soifer, A., In Search of Van der Waerden. Leipzig and Amsterdam, 1931–1951. Part II: Amsterdam, 1945. *Geombinatorics* XIV(2) (2004), 72–102.

[Soi22]   Soifer, A., A Journey from Ramsey theory to mathematical Olympiad to finite projective planes. *Mathematics Competitions* 17(2) (2004), 8–16.

[Soi23]   Soifer, A., Axiom of Choice and Chromatic Number of R^n. *Journal of Combinatorial Theory, Series A*, 110(1) (2005), 169–173.

[Soi24]   Soifer, A., In Search of Van der Waerden. Leipzig and Amsterdam, 1931–1951. Part III: Amsterdam, 1946–1951. *Geombinatorics* XIV(3) (2005), 124–161.

[Soi25]   Soifer, A., On Stephen P. Townshend's 1976 Proof. *Geombinatorics* XVI(4) (2005), 181–183.

[Soi26]   Soifer, A., In Search of Van der Waerden: The Early Years. *Geombinatorics* XIV(3) (2007), 305–342.

[Soi27]   Soifer, A., Chromatic number of the plane theorem in Solovay's set theory. *Geombinatorcs* XVII(2) (2007), 85–87.

[Soi28]   Soifer, A., Editorial dedicated to Paul Erdős's 80th Birthday. *Geombinatorcs* II(3) (1993), 43–45.

[Soi29]   Soifer, A., To Leave, To Die, or To Compromise? A Review of *Constantin Carathéodory: Mathematics and Politics in Turbulent Times*, by Georgiadou, M., *Geombinatorics* XIV(1) (2004), 41–46.

[SL]   Soifer, A., and Lozansky, E., Pigeons in every Pigeonhole. *Quantum*, January 1990, 24–28 and 32.

[Sol1]   Solovay, R. M., A model of set theory in which every set of reals is Lebesgue measurable. *Ann. Math.* 92 *Ser.* 2 (1970), 1–56.

[Sol2]   Solovay, R., M., e-mail to A. Soifer, April 11, 2003.

[Sol3]   Solovay, R., e-mail to A. Soifer, June 12, 2006.

[Sp1]   Spencer, J. H., Introductory Commentary, in Paul Erdös, *The Art of Counting*, MIT Press, Cambridge, Massachusetts, 1973, xxiii.

[Sp2]   Spencer, J. H., Ramsey theory and Ramsey theoreticians, *J. Graph Theory* 7 (1983), 15–23.

[Sp3]   Spencer, J. H., Three hundred million points suffice, *J. Combin. Theory Ser. A* 49 (1988), 210–217.

[Sp4]   Spencer, J. H., Erratum: "Three hundred million points suffice". *J. Combin. Theory Ser. A* 50 (1989), no. 2, 323.

[Spe3]   Spencer, T., University of Nebraska at Omaha, *personal communication with S. P. Radziszowski* (1993), and, Upper Bounds for Ramsey Numbers via Linear Programming, *manuscript*, (1994).

[Ste]   Steinberg, R., The state of the three color problem. In J. Gimbel, J. W. Kennedy, and L. V. Quintas, (eds.), Quo Vadis, Graph Theory? *Annals of Discrete Math.* 55 (1993), 211–248.

[Stei]   Steinhaus, H., Sur les distances des points des ensembles de mesure positive. *Fund. Math.* 1 (1920), 93–104.

[Stee]   Steen, L. A., Solution of the Four Color Problem, Math. Magazine 49(4), 1976, 219–222.

[Ste]   Stern, M., A Review of the book by Olli Lehto. *Korkeat Maailmat. Rolf Nevanlinnan elämä*, Otava, Helsinki, 2001.

[StSh]   Stevens, R. S., and Shanturam, R., Computer generated van der Waerden partitions. *Math. Comp.* 17 (1978), 635–636.

[Sto]   Story, W. E., Note on the Preceding Paper. *American Journal of Mathematics* II (1879), 201–204.

[Str]   Struik, D. J., Letter to A. Soifer, March 3, 1995.

[Sve1]   Sved, M., Paul Erdös – Portrait of Our New Academician. *Austral. Math. Soc. Gaz.* 14(3) (1987), 59–62.

[Sve2]   Sved, M., Old Snapshots of the Young. *Geombinatorics* II(3) (1993), 46–52.

[Sza]   Szabó, Z., An application of Lovász' local lemma – a new lower bound for the van der Waerden number. *Random Structures Algorithms* 1(3) (1990), 343–360.

[Sze1]   Székely, L. A., Remarks on the chromatic number of geometric graphs. In M. Fiedler, (ed.), *Graphs and other Combinatorial Topics, Proceedings of the Third Czechoslovak Symposium on Graph Theory, Prague, August 24–27, 1982*, Teubner Verlaggesellschaft, Leipzig, 1983, 312–315.

[Sze2]   Székely, L. A., Measurable chromatic number of geometric graphs and sets without some distances in Euclidean space. *Combinatorica* 4 (1984), 213–218.

[Sze3]   Székely, L. A., Erdös on unit distances and the Szemerédi-Trotter theorems. In G. Halász, et al. (eds.), *Paul Erdős and his Mathematics II*, János Bolyai Mathematical Society, Budapest and Springer-Verlag, Berlin, 2002, 649–666.

[SW]   Székely, L. A., and Wormald, N. C., Bounds on the measurable chromatic number of R^n. *Discrete Math.* 75 (1989), 343–372.

[Szek]   Szekeres, Gy., A Combinatorial Problem in Geometry: Remeniscences, in Paul Erdös, *The Art of Counting*, MIT Press, Cambridge, Massachusetts, 1973, xix–xxii.

[SP]   Szekeres, G., and Peters, L., Computer Solution To the 17-Point Erdős-Szekeres Problem. *ANZIAM J.* 48 (2006), 151–164.

[Sz1]   Szemerédi, E., On Sets of integers containing no four elements in arithmetic progression. *Acta Math. Acad. Sci. Hungar.* 20 (1969), 89–104.

[Sz2]   Szemerédi, E., On sets of integers containing no k elements in arithmetic progression. *Acta Arithmetica* XXVII (1975), 199–245.

[Szp]   Szpiro, G., Das Auswahlaxiom und seine Konsequenzen: Wie viele Farben braught man zur Kolorierung der Ebene? – Je nachdem. . ., *Neue Zürcher Zeitung*, April 13, 2005, Nr. 85, p. 65.

[Tai1]   Tait, P. G., On the colouring of maps. *Proc. Roy. Soc. Edinb.* 10 (1878–80), 501–503.

[Tai2]   Tait, P. G., Remarks on the previous communication. *Proc. Roy. Soc. Edinb.* 10 (1878–80), 729.

[Tai3]   Tait, P. G., Note on a theorem in the geometry of position. *Trans. Roy. Soc. Edinb.* 29 (1880), 657–660 & plate XVI. Reprinted: *Scientific Papers* 1(54), 1898–1900, 408–411 & plate X.

[Tie1]   Tietze, H., Über das Problem der Nachbargebiete im Raume. *Monatschefte für Mathematik und Physik* 16 (1905), 211–216.

[Tie2]   Tietze, H., *Gelöste und ungelöste mathematische Probleme aus alter und neuer Zeit; vierzehn Vorlesungen für Laien und für Freunde der Mathematik*, München, Biederstein, 1949.
          English translation from the 2nd rev. German ed: *Famous Problems of Mathematics: Solved and Unsolved Mathematical Problems, from Antiquity to Modern Times*. Graylock Press, Baltimore, 1965.

[Tho]   Thomassen, C., On the Nelson unit distance coloring problem. *Amer. Math. Monthly* 106 (1999), 850–853.

[Time]   Whoosh, *Time*, December 26, 1949, 30.

[TV1]   Tóth, G., and Valtr, P., Note on the Erdös-Szekeres theorem, *Discrete and Compu-*
           tational Geometry 19 (1998), 457–459.

[TV2]   Tóth, G., and Valtr, P., The Erdös-Szekeres theorem, upper bounds and general-
           izations. in J. E. Goodman et al. (eds.), *Discrete and Computational Geometry -*
           Papers from the MSRI Special Program, MSRI Publications **52**, Cambridge Univer-
           sity Press, Cambridge (2005), 557–568.

[Tow1]   Townsend, S. P., Every 5-Colouring Map in the Plane Contains a Monochrome Unit.
           J. Combin. Theory, Ser. A, 30 (1981), 114–115.

[Tow2]   Townsend, S. P., Colouring the plane with no monochrome unit. *Geombinatorics*
           XIV(4) (2005), 181–193.

[Tym]   Tymozcko, T., The four-color problem and its philosophical significance. *J. Philos-*
           ophy 76(2) (1979), 57–83.

[Ung]   Ungar, P., Problem 4526, advanced problems and solutions. *Amer. Math. Monthly*
           60 (1953), 123; correction 60 (1953), 336.

[VE]   Vasiliev, N. B., and Egorov, A., *Zadachi Vsesoyuznykh Matematicheskikh*
           Olimpiad (*Problems of Soviet National Mathematical Olympiads*), Nauka, Moscow,
           1988.

[Viz1]   Vizing, V. G., Ob ozenke khromaticheskogo klassa p – grapha (On an estimate of
           the chromatic class of a p-graph). *Diskrenyi Analiz* 3 (1964), 25–30 (Russian).

[Viz2]   Vizing, V. G., Kriticheskie graphy s khromaticheskim klassom (Critical graphs with
           the given chromatic class), *Diskrenyi Analiz* 5 (1965), 9–17 (Russian).

[Viz3]   Vizing, V. G., Nekotorye Nereshennye Zadachi V teorii grafov. *Uspekhi Mat. Nauk*
           XXIII(6) (1968), 117–134 (Russian). English Translation: Some unsolved problems
           in graph theory. *Russian Mathematical Surveys* 23 (1968), 125–141.

[Wae1]   Waerden, B. L. van der, *De algebraiese Grondslagen der meetkunde van het aantal*
           [Ph.D. thesis], Zutphen – W. J. Thieme & Cie, 1926.

[Wae2]   Waerden, B. L. van der, Beweis einer Baudetschen Vermutung. *Nieuw Archief voor*
           Wiskunde 15 (1927), 212–216.

[Wae3]   Waerden, B. L. van der, *Moderne Algebra*, vol. 1 (1930) and 2 (1931), 1st edn.,
           Verlag von Julius Springer, Berlin.

[Wae4]   Waerden, B. L. van der, *Die Gruppentheoretische Methode in der Quanten-*
           mechanik, J. Springer, 1932.

[Wae5]   Waerden, B. L. van der, Nachruf auf Emmy Noether. *Math. Annalen* 111 (1935),
           469–476.

[Wae6]   Waerden, B. L. van der, *Moderne Algebra*, vol. 1 (1937) and 2 (1940), 2nd edn.,
           Verlag von Julius Springer, Berlin.

[Wae7]   Waerden, B. L. van der, Letter to Behnke, May 10, 1940, Erich Hecke's Papers,
           Private German Collection.

[Wae8]   Waerden, B. L. van der, Letter to Hecke, May 16, 1940, Erich Hecke's Papers,
           Private German Collection.

[Wae9]   Waerden, B. L. van der, Prof. Van der Waerden verweert zich. *Het Parool* No. 326,
           February 1, 1946, 3.

[Wae10]   Waerden, B. L. van der, Rechtzetting, *Proria Cures*, February 1, 1946.

[Wae11]   Waerden, B. L. van der, *Moderne Algebra*, vol. 1 and 2, 3rd edn., Springer-Verlag,
           Berlin, 1950.

[Wae12]   Waerden, B. L. van der, *Over de Ruimte*, Rede, 4 December 1950, P. Noordhoff
           N.V., Groningen-Djakarta, 1950.

[Wae13]   Waerden, B. L. van der, Einfall und Überlegung in der Mathematik,1. Mitteilung.
           Elem. Math. VIII(6) (1953), 121–144; 2. Mitteilung, *Elem. Math.* IX(1) (1954),
           1–24; 3. Mitteilung: Der Beweis der Vermutung von Baudet, *Elem. Math.* IX(3)
           (1954), 49–72, (German).

[Wae14]   Waerden, B. L. van der, *Einfall und Überlegung: Drei Kleine Beiträge zur Psycholo-*
           gie des Mathematischen Denkens (*Idea and Reflection: Three Little Contributions*

	to the Psychology of the Mathematical Thinking. In 3 parts; 3rd part: *Der Beweis der Vermutung von Baudet*), Birkhäuser, Basel, 1954 (German).
[Wae15]	Waerden, B. L. van der, *Science Awakening*. P. Noordhoff, Groningen, 1954. Expanded edition, Oxford University Press, New York, 1961.
[Wae16]	Waerden, B. L. van der, Wie der Beweis der Vermutung von Baudet gefunden wurde. *Abhandlungen aus dem Mathematischen Seminar der Hamburgischen Universität* 28 (1965), 6–15.
[Wae17]	Waerden, B. L. van der, letter to Henry Baudet, October 20, 1965.
[Wae18]	Waerden, B. L. van der, How the proof of Baudet's conjecture was found. In L. Mirsky (ed.), *Studies in Pure Mathematics*, Academic Press, London, 1971, 251–260.
[Wae19]	Waerden, B. L. van der, Letter to N. G. de Bruijn, April 5, 1977.
[Wae20]	Waerden, B. L. van der, On the Sources of My Book *Moderne Algebra, Historia Mathematica* 2 (1975), 31–40.
[Wae21]	Waerden, B. L. van der, Beweis einer Baudetschen Vermutung, Bertin, E.M. J. et al. (eds.), *Two Decades of Mathematics in the Netherlands 1920–1940: A Retrospection on the Occasion of the Bicentennial of the Wiskundig Genootschap*, Part I, Mathematical Center, Amsterdam, 1978, 110–115 (with Commentary by N. G. de Bruijn, 116–124).
[Wae22]	Waerden, B. L. van der, *A History of Algebra*. Springer, Berlin, 1985.
[Wae23]	Waerden, B. L. van der, Letter to Soifer, A., March 9, 1995.
[Wae24]	Waerden, B. L. van der, Letter to Soifer, A., April 4, 1995.
[Wae25]	Waerden, B. L. van der, Undated letter to Soifer, A., (Swiss Post Office seal April 24, 1995).
[Wae26]	Waerden, B. L. van der, Wie der Beweis der Vermutung von Baudet gefunden wurde. *Elem. Math.* 53 (1998), 139–148.
[WaD1]	Waerden, Dorith van der, e-mail to A. Soifer, February 22, 2004.
[WaD2]	Waerden, Dorith van der, e-mail to A. Soifer, April 29, 2004.
[WaD3]	Waerden, Dorith van der, e-mail to A. Soifer, June 14, 2004.
[WaD4]	Waerden, Dorith van der, e-mail to A. Soifer, June 14, 2004.
[WaT1]	Waerden, Theo van der, letter to A. Soifer, June 25, 2004.
[WaT2]	Waerden, Theo van der, letter to A. Soifer, November, 2005.
[WaT3]	Waerden, Theo van der, letter to A. Soifer, November, 2005.
[Wag]	Wagner, K., Über eine Eigenschaft der ebenen Komplexe. *Math. Ann.* 114 (1937), 570–590.
[Wal1]	Walker, M., Uranium Machines, Nuclear Explosives, and National Socialism: The German Quest for Nuclear Power, 1939–1949, Ph. D. Dissertation in the History of Science, Princeton University, October 1987.
[Wal2]	Walker, M., Nazi Science: Myth, Truth, and the German Atomic Bomb. Plenum, New York, 1995.
[Wer]	Wernicke, P., Über den kartographischen Vierfarbensatz. *Math. Ann.* 58(3) (1904), 413–426.
[WT]	Whitney, H., and Tutte, W. T., Kempe chains and the four colour problem. *Utilitas Math.* 2 (1972), 241–281.
[Wil1]	Wilson, R. J., New Light on the origin of the Four-color Conjecture. *Historia Mathematica* 3 (1976), 329–330.
[Wil2]	Wilson, R. J., *Four Colours Suffice*. Princeton University Press, 2002.
[WilJ]	Wilson, J., New light on the origin of the four-color conjecture. *Hist. Math.* 3 (1976), 329–330.
[WilJC]	Wilson, J. C., On a supposed solution of the 'four-colour problem'. *Math. Gaz.* 3 (1904–6), 338–340.
[Wit]	Witt, E., Ein kombinatorischer Satz der Elementargeometrie. *Mathematische Nachrichten* 6 (1952), 261–262.

[Witt]	Wittgenstein, L., *Remarks on the Foundation of Mathematics*. Macmillan, New York, 1956.
[Woo1]	Woodall, D. R., Distances realized by sets covering the plane. *J. Combin. Theory Ser. A*, 14 (1973), 187–200.
[Woo2]	Woodall, D. R., The four-colour conjecture is proved. *Manifold* 19 (1977), 14–22.
[Woo3]	Woodall, D. R., The four-colour theorem. *Bull. Inst. Math. & Appl.* 14 (1978), 245–249.
[WW]	Woodall, D. R., and Wilson, R. J., The Appel–Haken Proof of the Four-Color Theorem. In L. Beineke and R. J. Wilson (eds.), *Selected Topics in Graph Theory*, Academic Press, London, 1978, 83–101.
[Wor]	Wormald, N. C., A 4-chromatic graph with a special plane drawing. *J. Austral. Math. Soc. Series A* 28 (1979), 1–8.
[Xu1]	Xu, B., On 3-colorable plane graphs withouit 5- and 7-cycles. *J. Combin. Theory Ser. B* 96 (2006), 958–963.
[Xu2]	Xu, B., A 3-color theorem on plane graphs without 5-circuits. *Acta Mathematica Sinica* (in press).
[Xu3]	Xu, B., A sufficient condition on 3-colorable plane graphs without 5- and 6-circuits, submitted.
[XXER]	Xu, X., Xie, Z., Exoo, G., and Radziszowski, S. P., Constructive lower bounds on classical multicolor Ramsey numbers. *Electronic J. Combinatorics*, #R35, 11 (2004). http://www.combinatorics.org/
[XXR]	Xu, X., Xie, Z, and Radziszowski, S. P., A Constructive Approach for the Lower Bounds on the Ramsey Numbers $R(s, t)$. *J. Graph Theory* 47 (2004) 231–239.
[Zak1]	Zaks, J., On the chromatic number of some rational spaces. *Combinatorics and Optimization*, Research Report Corr 88-26 (1988), 1–3.
[Zak2]	Zaks, J., On four-colourings of the rational four-space. *Aequationes Math.* 27 (1989), 259–266.
[Zak3]	Zaks, J., On the connectedness of some geometric graphs. *J. Combin. Theory (B)*, 49 (1990), 143–150.
[Zak4]	Zaks, J., On the chromatic number of some rational spaces. *Ars Combinatorica* 33 (1992), 253–256.
[Zak5]	Zaks, J., On the connectedness of some geometric graphs, *J. Combin. Theory (B)*, 49 (1990) 143–150.
[Zak6]	Zaks, J., Uniform distances in rational unit-distance graphs. *Discrete Math.* 109(1–3), 1992), 307–311.
[Zak7]	Zaks, J., On odd integral distances rational graphs. *Geombinatorics* IX(2) (1999), 90–95.
[Zer]	Zermelo, E., Beweis dass jede Menge wohlgeordnet werden kann. *Math. Ann.* 59 (1904), 514–516.
[Zyk1]	Zykov, A. A., On some properties of linear complexes, (Russian). *Mat. Sbornik N.S.* 24(66) (1949), 163–188. On some properties of linear complexes. *Amer. Math. Soc. Translation* 1952, (1952). no. 79.
[Zyk2]	Zykov, A. A., Problem 12 (by V. G. Vizing), *Beiträge zur Graphentheorie*, Vorgetragen auf dem Internationalen Kolloquium in Manebach (DDR) vom 9.-12. Mai 1967, p. 228. B. G. Teubner Verlagsgesellschaft, Leipzig, 1968.
[Zyk3]	Zykov, A. A., *Theory of Finite Graphs I*, (Russian), Nauka, Novosibirsk, 1969.

Name Index

Subject Index

W
Wormald's Graph, 87

Z
Zermelo-Fraenkel-Choice System of Axioms, 539, 548, 597

Zermelo-Fraenkel-Solovay System of Axioms, 540, 545, 549, 597

Zermelo-Fraenkel System of Axioms, 60, 515, 597

Zorn's Lemma, 236, 238

Index of Notations

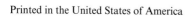

Printed in the United States of America